Telecommunications Network Management

Henry Haojin Wang

McGraw-Hill
New York   San Francisco   Washington, D.C.   Auckland   Bogotá
Caracas   Lisbon   London   Madrid   Mexico City   Milan
Montreal   New Delhi   San Juan   Singapore
Sydney   Tokyo   Toronto

Library of Congress Cataloging-in-Publication Data

Wang, Henry Haojin.
   Telecommunications network management : standards, applications,
and future trends. / by Henry Haojin Wang.
      p.   cm.
   ISBN 0-07-068170-8
   1. Computer networks—Management.
TK5105.5.W355   1999
621.382'1—dc21   99-22024
                                                             CIP

McGraw-Hill

A Division of The McGraw·Hill Companies

1 2 3 4 5 6 7 8 9 0   DOC/DOC   9 0 4 3 2 1 0 9

ISBN 0-07-068170-8

*The sponsoring editor for this book was Stephen S. Chapman, the editing supervisor was
Caroline R. Levine, and the production supervisor was Pamela Pelton. This book was set in
Century Schoolbook per the MHT design by Michele Zito of McGraw-Hill's Professional Book
Group composition unit, Hightstown, N.J.*

Printed and bound by R. R. Donnelley & Sons Company.

This book is printed on recycled, acid-free paper containing
a minimum of 50% recycled, de-inked fiber.

To my wife, my son, my daughter,
and my parents

Contents

Preface

Objectives

In response to the unprecedented technological advances in the telecommunications industry and worldwide telecommunications market liberalization, telecommunications network management is moving away from a predominantly proprietary environment and into an open, standards-based environment. Network management is playing an increasingly important role in the fast-changing industry, and there is a genuine need for a systematic and comprehensive treatment of telecommunications network management-related topics, such as the standards, concepts, applications, and industry trends.

This book aims to achieve the following three goals.

First, it strives to provide a comprehensive treatment of telecommunications network management standards. In addition to a detailed introduction to the established internet-based IP network management protocol SNMP and the OSI management protocol CMIP, an introduction to the telecom network management standards, Telecommunications Management Network, or TMN, is a major focus of this book. The introduction goes beyond the normal realm of network-element-level management by providing detailed discussions of the network level and service level management standards. In addition to standards from formal bodies such as ITU, IETF, and OSI, those from the industry consortium, commonly known as the industry-agreement type of standards are introduced in detail as well.

The second goal of the book is to make the often abstract and complicated standards easy to understand and practical. The management standards are often presented devoid of any concrete application context, and this renders them abstract and difficult to understand. This book presents many application scenarios and examples that demonstrate how a standard is used and where it is used. To further illustrate a standard, a process model is presented to provide a concrete context for the application of the standard presented in the chapter. Each chapter has an outline at the beginning and a summary at the end to facilitate easy grasp and review.

A third objective is to present some of the current trends in the telecommunications network management field. One such trend is the adoption of distributed computing technology from the general computing world into the once proprietary environment of telecommunications networks. A lengthy chapter is devoted to the discussion of how distributed technology such as CORBA is applied to network management. An often-neglected area that has become increasingly important in recent years is service management. A complete life cycle of service management that ranges from service configuration and service ordering, to service provisioning and service performance monitoring is covered.

The convergence of data networks and voice networks is another important trend in the telecommunications industry. From the perspective of network management, one topic covered is the interworking between management systems for data networks and those for voice networks. The numbering plans for both voice networks and data networks are introduced in detail.

Outline of the Book

The book consists of three parts that present different areas of telecommunications network management, as outlined below.

Part 1: Overview of the Internet and OSI management standards

This part consists of five chapters that provide an overview of three areas, i.e., telecommunications networks, the internet network management standards, and the OSI network management standards.

Chapter 1: *Overview of telecommunications networks.* This chapter provides an overview of telecommunications networks in terms of the components, structures, and characteristics of three types of the most common telecommunications networks, i.e., public-switched telephone networks, wireless networks, and Intelligent Networks. An overview of a network *management system* that manages such a telecommunications network introduces the components and functional areas of a telecommunications network management system.

Chapter 2: *Overview of internet management standards—SNMPv1 and MIB-II.* This chapter provides an overview of the most widely used internet network management protocol, Simple Network Management Protocol (SNMP). The overview includes SNMP MIB (Management Information Base) structures, standardized MIB groups, SNMP PDUs (Protocol Data Units), and the procedures for the SNMP message exchanges. The overview focuses on the basic concepts and their applications.

Chapter 3: *SNMP v2 and RMON.* This chapter continues the introduction of the internet network management standards to cover the more recent version of the management protocol SNMP v2 and the widely used Remote Network Monitoring (RMON) which provides the capability to monitor a remote network.

Chapter 4: *OSI Network Management—Management Information Base (MIB) and GDMO.* The OSI network management framework, protocol, and information models have been directly adopted by ITU as part of the telecommunications network management standards. An overview of the OSI MIB structure and Guideline for Definitions of Managed Objects (GDMO), a widely used specification language, lays the groundwork for introduction of the OSI management protocol, CMIP, in the following chapter.

Chapter 5: *OSI network management—Management framework and CMIP.* This chapter continues the review of OSI network management standards with an overview of OSI Common Management Information Protocol (CMIP), with a detailed description of the protocol messages. To facilitate the understanding of this protocol, a simple switch example is used throughout the chapter to illustrate the step-by-step use of the protocol.

Part II: Telecommunications Management Network (TMN) and Five Management Areas

This section has a total of eight chapters. The first two chapters provide a systematic introduction to Telecommunications Management Network (TMN), the ITU network management standards for telecommunications networks. According to TMN, the network management functions are divided into five function areas; fault, configuration, accounting, performance, and security management. The ensuing five chapters are devoted to the five management areas. The last chapter of this section illustrates how to apply the described standard.

Chapter 6: *Telecommunications Management Network (TMN)—Functional, physical, and logical layered architectures.* TMN is intended as the telecommunications network management standard for all network technologies. It adopts the OSI management protocols like CMIP and provides a management architectural framework, including functional, physical, and information architectures. The architectural principles of TMN, and the three architectures, are described in detail.

Chapter 7: *Telecommunications Management Network (TMN)—Information architecture and generic information models.* This chapter continues the introduction to TMN by describing in detail the TMN information architecture and a set of generic information models. The information models introduce five groups of generic managed object classes that represent all management aspects of a telecommunications network. In addition, this chapter provides a high level overview of the published information models for different types of telecom networks such as ATM, SONET, ISDN, and SS7.

Chapter 8: *Configuration management.* This chapter covers a wide range of areas that include network planning and engineering, provisioning, software management, and network element status control. Network planning and engineering describes a model for network capacity planning, the network infrastructure design, and routing design. Software management describes a process of automating the software management tasks such as software deliver, backup, revert, install, and restore. Network provisioning is concerned with the network level and network element level resource provisioning. Network status control includes the network element control, network element resource state management, and status management.

Chapter 9: *Performance management.* The primary goal of performance management is to maintain the desired level of network performance. Performance monitoring collects performance-related information and reports the performance-affecting events. Performance analysis analyzes and understands performance-related data and events and produces estimation on the performance impact of the events. Based on the results of performance analysis, performance management control may take certain preventive measures before performance deterioration reaches a critical level. Each of the areas is described in detail, and applications examples are provided to help explain the basic concepts.

Chapter 10: *Fault management.* Fault management deals with network faults: detect, localize, correct, and recover from network faults. Alarm surveillance is responsible for detecting a fault and reporting it promptly. Fault localization pinpoints the location of the fault by analyzing the information related to the fault. Once the location and causes of a fault have been diagnosed, fault correction takes appropriate actions to correct the fault if possible and recover from the adverse effects caused by the fault.

Chapter 11: *Accounting management.* Accounting management focuses on generating billing records for telecommunications services rendered to customers. The first step is *usage metering*, which uses one of the two metering methods, *multimetering* and *Automatic Message Accounting* (AMA). The next step is *usage metering data processing*, which addresses the issues related to AMA data processing at network level and includes service usage correlation, AMA records formatting and conversion, AMA data aggregation, and usage surveillance. Finally *charging* and *billing* are the processes of charging the network service usage by a customer and generating a final bill.

Chapter 12: *Security management.* Security management deals with fraud to ensure the security of a telecommunications network. The first line of defense against fraud is a fraud prevention process that uses techniques such as customer profiling to prevent fraud. Fraud *detection* identifies a fraud promptly to minimize loss. Once a fraud has occurred, fraud *containment* and *recovery* attempt to limit loss and affected areas and recover from the damage done. Finally, a set of security services (i.e., authentication, integrity, confidentiality, access control, and nonrepudiation) are described in detail.

Chapter 13: *Applications of telecommunications network management standards.* This chapter addresses the practical issues of applying the standards in the areas discussed above. It first presents a step-by-step process of designing a network management system to show the reader how to apply established network management standards. A survey of TeleManagement Forum's Solution Sets provides an overview of a solution set with many ready-to-use solutions. An API architecture shows standardization of implementation technologies for telecommunications network management applications.

Part III: Recent Development and Future Trends

This section comprises three chapters that provide an overall view of the recent developments and trends in telecommunications network management. Three areas that stand out are service management, distributed telecommunications network management, and number portability.

Chapter 14: *Service managment.* As worldwide telecommunications markets are opened up to competition, service management, a once exclusively proprietary area, begins to move into an open environment as well. After an overview of telecommunications services (e.g., Intelligent Network services, wireless services, switch-based services, etc.), service deployment, service ordering process, service maintenance, service provisioning, and service performance monitoring are discussed.

Chapter 15: *Distributed telecommunications network management.* The telecommunications industry is mirroring a computing industry transition that took place over a decade ago: the transition from a centralized, mainframe-based computing environment to a distributed, network-based one. This chapter provides an overview of the ITU's distributed processing architectures, its architectural components, and principles. After an introduction to CORBA, the discussion turns to the applications of CORBA to telecommunications network management, including the issues related to development of CORBA-based TMN, and CORBA working with TMN.

Chapter 16: *Numbering plans and local number portability.* After introducing the ITU numbering plans for public data networks, the chapter describes in detail the ITU numbering plans for telephone networks and the North American numbering plans. Two architectures addressing the issues of local number portability, one IN-based and the other non-IN-based, and their impact on network and service management are discussed, followed by a brief introduction to the concept of number pooling.

In addition to a number of chapter-specific appendices, there are four major appendices at the end of the book:

- *Appendix A: Introduction to ASN.1* provides an introduction to the Abstract Syntax Notation 1 (ASN.1), a specification language widely used for defining abstract syntax types in network management applications.

- *Appendix B: An overview of TCP/IP protocol suite* presents an overview of the internet-based protocol suite as the background information for discussion of the internet-based network management protocol SNMP.
- *Appendix C: OSI network reference model* provides an introduction to the OSI 7 layer network reference model to provide background information for the OSI-based network management protocol.
- *Appendix D: Generic managed object class (MOC) dictionary* provides a central reference repository for all generic managed object classes defined in ITU standards, which are scattered over many individual standards.

Each chapter begins with a succinct outline and ends with a summary and an exercise section that contains problems to reinforce the basic concepts discussed in the chapter.

Intended Audience

The book is intended for a wide range of audiences that includes, but is not limited to, the following groups:

- Telecommunications industry professionals (e.g., engineers and technical managers). For example, a software engineer who is developing network management applications can find applicable standards, steps for applying the standards, and application scenarios and examples. A technical manager or systems engineer may find multiple technology choices to consider in determining the design of a network management system.
- High-level graduate students specializing in the telecommunications fields or any professional who is interested in obtaining a comprehensive grasp of telecommunications network management topics will benefit.
- General computing industry professionals who are interested in having an overview of state-of-the-art telecommunications network management or who wish to own a reference source on the subject will benefit.

It is recommended that all readers, regardless of their level of experience in telecommunications network management, review Chap. 1 because it provides a general context for the rest of the book. If the reader is interested primarily in an overview of current network management standards, Chaps. 2 through 7 cover SNMP, CMIP, and TMN. For all the practitioners interested in a particular area such as fault, configuration, accounting performance, and security management (FCAPS), the FCAP chapters (Chaps. 8 through 12) provide object models for building applications as well as application examples. Chapter 15, *Distributed Telecommunications Network Management*, will be very helpful for those considering distributed computing architectures for the application system. The last three chapters provide the reader with an overview of the new trends and issues in telecommunications network management.

Acknowledgments

Without the generous help of many people, it would have been very difficult for me to have completed this book. I would like to use this opportunity to thank the Department of Computer Science, University of Texas at Dallas, and Prof. Biao Chen for giving me the opportunity to experiment with a new course entitled "Telecom Network Management," which later became the basis of this book. I am deeply indebted to many people at Samsung Telecom America, Network System Lab. To Nancy Korman, for her encouragement, support, and diligent review; to Mike McKinley and Jim Bunch for their encouragement and support; to Mike Bray for his stories of the old telecom days that have made abstract telecom history close and personal; to Craig Miller, Woody Russell, and Maria Sekul for their help in reviewing some of the chapters. Thanks go also to Charles Hannon, Jack Gannon, Rodney Sperull, Kristen Badowski, and Kong Bhat for their valuable comments. I am grateful to McGraw-Hill executive editor Steve Chapman for his tireless efforts in answering numerous questions and concerns I had about the book, and to editing supervisor Caroline Levine and Lucy Mullins, the copy editor, for their diligent efforts. Any errors are solely mine.

Foremost, I would like to thank my family. To my wife Xiaohua, for her support along the way and help in preparing the figures in the book; to my son Andy for his understanding of Dad's numerous absences at his soccer games; and to my daughter Amy for her enthusiastic help at the keyboard that created just the right amount of distraction to keep this process fun.

Henry Haojin Wang

Overview of Telecommunications Networks and Network Management Systems

Outline

- The first section, "Introduction to Telecommunications Networks," includes four parts:
  - "Brief history of telecommunications networks" provides a brief review of the history of telecommunications networks that includes technological innovations and regulatory changes.
  - "Overview of telecommunications networks" covers five different types of networks, including public switched telephone networks, wireless networks, and Intelligent Networks.
  - "Characteristics of telecommunications networks" characterizes the conventional telecommunications network in contrast to the Internet-based data network.
  - "What comes next?" peeks at future trends in the telecommunications industry with respect to network management.
- The second section, "Introduction to Telecommunications Network Management Systems," provides a description of what a telecommunications network management system does and of its components.

Introduction to Telecommunications Networks

The theme of this book is the management of telecommunications networks. It is beneficial to have an overview of telecommunications networks before proceeding to the management of them. Our overview starts with a brief history of telecommunications networks and then covers conventional public switched telephone networks (PSTNs), wireless networks, and Intelligent Networks (INs).

Brief history of telecommunications networks

Telecommunications history starts with the invention of the telephone in 1876 by Alexander Graham Bell. Until recently telecommunications history has been identified exclusively with the public telephone network. The 100-year-plus history is marked by many technological milestones and history-making events, mostly on regulatory and legislative fronts.

Technological innovations. Though the patent for the telephone was granted in 1876, it wasn't until the end of the last century that the first telephone network across a large geographic area was established. The first long-distance phone line between Boston and Delaware was built in 1912. This event signaled the beginning of a rapid infrastructure expansion of telephone networks. During the ensuing half century, the public telephone network reached virtually every corner of the United States. Starting almost half a century ago, and with the advent of modern computing technologies, conventional telephone networks began a transformation that has completely changed what they were meant to be and will continue to challenge the very definitions of telecommunications. Along the way, telecommunications networks have undergone substantial changes in the areas of transmission, transport, switching, and networking technologies.

The dominant transmission medium for telephone networks has been copper wire for almost a century. Recent development of new transmission media like fiber optics and rediscovery of some old media like radio are now challenging the dominance of copper as a medium. Fiber optics, with a traffic-carrying capacity thousands of times that of copper, has become the dominant transmission medium of major backbone networks worldwide. Utilization of fiber-optic transmission has increased at a phenomenal rate since fiber optics' first commercial deployment for carrying telephone traffic in 1977.

The first letter in radio signals was sent across the Atlantic in 1901. During the late 1970s and early 1980s this medium was rediscovered for large-scale commercial development. Since the first commercial cellular network was set up in Japan in 1976, wireless networks have experienced phenomenal growth, particularly in the 1980s and 1990s. There is little sign of this trend letting up as we move into the next century.

Hand in hand with the evolution of transmission media technology is switching technology. Telecommunications switches, also called exchanges, have always been the "brain" of a network. Switches are responsible for setting up the connection between a caller and the called party, maintaining the connection, and routing a call from its originating point to its destination. A switch must route across a variety of network elements, with the goals of minimum delay and maximum utilization of network resources.

Switching technology evolution can be viewed as consisting of three different generations: manual switches, mechanical switches, and digital switches. In 1878 the first commercial telephone exchange was brought into service in New Haven, Connecticut. Calls were manually switched by operators, just as depicted in old films, where operators connected incoming calls by plugging connectors into the switchboard. As the network grew larger and more complicated, manual switching became intolerably tedious and error-prone. A new generation of mechanical switches with fixed relay logic was invented. Network intelligence took the shape of decision-making logic in the relay logic. At that time, the relay logic was built using electrical mechanical devices. State-of-the-art switching technology at the time, a switch with a 50,000-port capacity, could take up several large rooms.

Advancement of computer technologies revolutionized switching technology, as computers did in many other fields, and ushered in the digital switch era. It is interesting to note that the motivation behind the invention of the transistor, the brain cell of the modern computer, was to find an alternative to the bulky, vacuum-tube-based stored program in telephone switches at Bell Laboratories. By many accounts, the first fully functional, commercial digital switch was deployed in the mid-1970s, though numerous vendors have staked the claim of having had the "first digital switch" in various forms. So what are a digital switch and a digital network anyway? In the predigital era, the voice on the network was represented as an electromagnetic wave that had a one-to-one representation of the sound wave created when a person spoke, a form called analog information. Analog signals are difficult to process and inefficient to transport over a network. On the other hand, in a digital network, the voice and all other information are represented, processed, and transported in streams of 0s and 1s. A digital switch essentially has become a special-purpose computer! Digital information can be controlled, processed, and routed by the stored software program in a digital switch much more efficiently. Bulky analog switches are reduced to a fraction of their previous size, and the speed of signal processing is on an ever faster upward spiral. Countless possibilities are open for development of new telecommunications technologies implemented with computing technologies.

Another major milestone of switching technology was the development and standardization of an out-of-band signaling system, called Common Channel Signaling System 7 (CCS7). All the information traveling on a network can be put into two categories: signaling messages and the network payload-like voice of phone calls. Traditionally, the two types of information were bundled together and transported on the same channel. Starting in the late 1970s, efforts were under way to develop a signaling system that separated signaling messages from voice traffic so they could be transported on separate channels through different routes. This signaling system, the CCS7 or Signaling System 7 (SS7) for short, was standardized by the international telecommunications standards body, the Consultative Committee on International Telegraphy and Telephony (CCITT), in 1980. Now most U.S. long-distance carriers and Bell operating companies have migrated to CCS7 networks. It has become the standard implementation around the world. The dedicated signaling channel opens the gate for a whole range of new information to be carried on the network. The new information eventually made possible the development of a wide range of new types of services like caller identification (ID), call forwarding, and free phone. The major events of technological evolution are summarized in Table 1.1.

Of course, telecommunications technology evolution never took place in isolation. Instead, technological advances were interwoven closely with regulatory and legislative changes over the past century.

Regulatory changes. Telephone networks and services were under tight government control in both market-oriented and state-controlled economies alike. This translates into very little competition and domination by a few established players in the industry. By the beginning of this century, AT&T had acquired over 90 percent of the U.S. market. Thus century-long regulatory battles have focused primarily on breaking up the established monopoly and eventually moving toward a global telecommunications market.

The enactment of the 1934 Communications Act set up the basic regulatory structure for the telecommunications industry in the United States. The

TABLE 1.1   A Summary of the Technological Evolution of Telecommunications

1844	Morse sends the first public telegraph message.
1876	Alexander Graham Bell obtains a patent for the telephone.
1881	First long-distance line from Boston to Providence, RI, is built.
1901	First letter is sent across the Atlantic Ocean using radio signals by Guglielmo Marconi.
1915	First transcontinental telephone call in U.S. is made.
1925	Coaxial cable is invented.
1951	Direct long-distance dialing is introduced.
1963	Touch-tone service is introduced.
1960s	First commercial satellite network is deployed.
1971	E-mail is invented by Ray Tomlinson.
1974	Basic ideas of the Internet are presented by Vint Cerf and Bob Kahn, and the Defense Advanced Research Projects Agency (DARPA) Internet project is established with the intention to develop a communications system independent of the Bell System for national security.
1970s	Internet traffic is allowed onto the PSTN, beginning the trend of merging data and voice networks.
1976	First digital switch is installed.
1977	First commercial fiber-optic links are installed by GTE (General Telephone and Electronics) and Bell Systems in the U.S.
1979	First cellular network is operational in Japan.
1980	CCS7 (SS7) standards are published by the CCITT.
Late 1980s	Asynchronous Transfer Mode (ATM) specification is first published.
1990	International Telecommunications Union (ITU) specification of the Intelligent Network is published.

Federal Communications Commission (FCC) was given the primary authority to regulate the telecommunications industry, along with state public utility commissions. This structure remains in place today.

In 1972, an "open-skies" policy was adopted that allowed any entity to own and operate a satellite system. Compared to the land-based communication media, the new frontier air-based communications were more open to competition from the very start.

In 1974, MCI and the Department of Justice both filed a lawsuit against AT&T in an attempt to break the monopoly by the latter over telecommunications. The landmark battle culminated in 1984 with the Modified Final Judgment issued by Judge Harold Green of the U.S. District Court. The settlement dismantled AT&T's monopoly in manufacturing, marketing, and installation of telephone equipment and services. The establishment of seven Baby Bells or Regional Bell Operating Companies (RBOCs) resulted and opened long-distance markets for competition.

The 1996 Telecommunications Act, passed by the U.S. Congress, was the first major overhaul of the 1934 Communications Act and included the following major provisions:

- Removal of barriers to entry into interstate and intrastate telecommunications markets

- Permission given to the RBOCs to offer long-distance telephone services outside their regions

- End of rate regulation for cable television programming services in markets subject to competition on a designated date

In February 1997, nearly 70 countries around the world that together account for over 95 percent of the world telecommunications revenue signed a global telecommunications pact that obligated all signing countries to open their domestic telecommunications markets to foreign competition by a set timetable. The major events of telecommunications regulation in the United States are summarized in Table 1.2.

Overview of telecommunications networks

This section provides an overview of three different telecommunications networks, i.e., public switched telephone networks, wireless networks, and Intelligent Networks. By no means are these the only telecommunications networks there are to discuss. They have been chosen largely because they are commonly seen in the telecommunications industry and the management of these networks is of strong industrywide interest. The emphasis of the overview is on the network structure and on components that are of importance from a network management perspective.

Public switched telephone networks. PSTNs conventionally refer to telephone company networks that provide phone services to the public using switched

TABLE 1.2   A Summary of Regulatory Changes for the Telecommunications Industry

1913	U.S. Government for the first time charges AT&T with violation of the Anti-Trust Act of 1890. As a result, AT&T gives up its controlling interests in Western Union and allows independent telephone companies to connect to the Bell network.
1918	Prompted by World War I, the U.S. Government assumes control of telephone and telegraph networks. The control is returned 1 year later.
1934	The 1934 Communications Act is passed by the U.S. Congress. The FCC is established to regulate interstate and international communications.
1949	United States sues AT&T on the ground of hampering government regulation by price gouging. The lawsuit ends by a consent decree agreement in 1956.
1962	Satellite communications are brought under the jurisdiction of the FCC in the Communication Satellite Act.
1968	The Carterfone decision allows the interconnection of non-Bell equipment to the Bell telephone network with the use of special devices to protect the Bell network from any detrimental effects caused by the non-Bell equipment.
1969	First non-AT&T interstate link between Chicago and St. Louis is installed by MCI.
1972	Open-skies policies are adopted that allow any entity to own and operate a satellite system. Compared to the land-based communication media, the new frontier air-based communications is more open to competition from the very start.
1975	The FCC opens mobile telephone markets by reallocating segments of the radio spectrum for land mobile communications and high-capacity cellular.
1976	MCI and the U.S. Department of Justice file a lawsuit against the AT&T monopoly.
1984	AT&T monopoly over telephone is broken up by the Modified Final Judgment (MFJ). Seven Baby Bells are formed.
1996	Telecommunications Act is passed to allow competition in both long-distance and local service markets in the U.S.
1997	Worldwide telecommunications agreement is signed to obligate signatories to open their telecommunications markets based on a set timetable.

circuits. In the United States, there are basically two types of PSTNs: *interexchange carrier* (IEC) networks provide long-distance services, and *local exchange carrier* (LEC) networks provide local phone services and mainly consist of the Bell operating companies. Delineation between LEC and IEC networks is a direct result of the 1984 AT&T divestiture.

Instead of going through the PSTN component by component, as shown in Figs. 1.1 and 1.2, we introduce the concept of PSTNs by tracing a normal phone call through both the LEC and IEC networks. Of course, this is a high-level overview that leaves out most of the technical details. Assuming that you make a long-distance call to a friend of yours in a different city, let's start the ride on the PSTNs.

LEC networks

Customer premises equipment.   First, you pick up the phone at home. Your phone set (or computer if it is connected to the Internet), phone wire, and any other equipment used in the communication that is located on your premises are called customer premises equipment. A different type of customer premises equipment is that located at an enterprise like the company where you work. Customer premises equipment of this type includes a private branch exchange (PBX), a router, and a wire. A few words about PBXs before we go on with our call. A PBX is a small-scale switch for a business or an institution that allows abridged-number, intraorganization dialing within an enterprise and is directly connected to a central office switch.

Local loop and local loop digital remote terminal.   Once your phone is picked up, an off-hook signal is detected by the network. The next stop of the off-hook signal is a digital loop carrier remote terminal (DLCRT) which converts analog signals coming out of your home into digital ones and multiplexes many users onto one link going to the central office switch. Usually if your phone is not geographically located in the vicinity of the central office switch of the local phone company, it probably is connected to a DLCRT. The DLCRTs, along with the access line, power supply, and other equipment that connects customer premise equipment to an LEC's central office switch, constitute the local access loop, also termed the *last mile* of a telecommunications network.

Central office switch.   The next stop of the off-hook signal of your call is the central office (CO) switch. A CO switch, can be viewed as the brain and nerve center of the network. It processes a call, deciding whether to accept a call, deciding when and how to route a call, and controlling the resources used in a call such as a dial tone, ringing tone, and announcements. Also called local exchanges, or class 5 switches (a legacy of the AT&T Bell System terminology), the CO switch directly interconnects your phone with the outside world by providing access to other CO switches or long-distance carrier switches. A CO switch has two sides: one facing end customers and one facing its network. Connected to the customer side are either DCLRTs or various customer premises equipment including regular phones, computer terminals, and enterprise PBXs. Connected to the network side is either a long-distance carrier switch or a local tandem switch, which will be explained shortly. The channels (the virtual dedicated connections between two points) on the customer side and the network side are distinguished. Thus, the network and customer sides are also called trunk side and line side, respectively, in telecommunications jargon.

Once the CO switch serving your home area receives the off-hook signals from your phone, it first makes sure that the originating line is legitimate by checking its database about the line connecting your phone. It then provides a dial tone to your phone to let you dial your friend's number. From the digits you dialed, a digital message is created that contains the called party's (your friend's) number, your number, your long-distance service carrier, and other information the network added in order to make a routing decision. Assume you live in a large metropolitan area; it is likely that the next stop of your call is not a switch of your long-distance company, but a local tandem switch.

Local tandem switch.   A local tandem switch functions like a hub connecting multiple CO switches. Since the number of CO switches in a large metropolitan area can be potentially very large, it would not be practical to connect every CO switch to every other one. Instead, all CO switches are connected to one or a small number of local tandem switches to reduce the total number of connections. A CO switch wanting to connect to another CO switch or to an outside switch has to go through a tandem switch to do so. Once the local tandem switch receives the digital signal of your call, it routes the signal to a switch of your long-distance company and your call now enters a different network, a network of your long-distance carrier. But before leaving your local phone company's network, let's take a quick look at the transmission equipment that actually carries your call (see Fig. 1.1).

Transmission equipment.   Transmission equipment mainly consists of two types of components, i.e., physical transmission medium and the auxiliary electric and electronic equipment along the path. The transmission medium includes some combination of copper wire, coaxial cable, and fiber optics. Examples of auxiliary equipment include an amplifier to help maintain the signal quality on the medium and a power supply. It is likely that a copper wire connects your home to a DLCRT or CO switch. The transmission medium between a CO switch and local tandem switch or a switch of a long-distance network can be anything from coaxial cable to a fiber-optic link.

IEC networks. The network of your long-distance carrier, also called an interexchange carrier, usually covers a large geographic area. Many switches

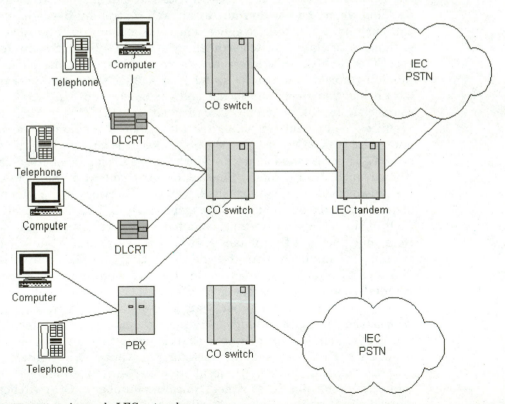

Figure 1.1 A sample LEC network.

used in an IEC network can be placed roughly into three categories, based on their functions: service switch point (SSP), signal transfer point (STP), and gateway switch.

Service switch point. An SSP is a switch that connects to the LEC network. It receives the digital message of your call first. The SSP checks its database about you, the customer; analyzes the message to determine a route within the IEC network; collects billing-related information; and routes the message to the destination SSP that connects the LEC network serving the area in which your friend lives. As you can easily see, the digital message of your call has to go through many intermediate switches between the two SSPs. These intermediate network nodes are called *signal transfer points*, which are explained next. The destination SSP will pass the digital message to the destination LEC serving the area where your friend lives. The destination CO switch, in this case, locates the line connecting your friend's phone. The destination CO switch, after ensuring that the terminating line is valid, sends a message all the way back, again across the IEC network, to the originating CO switch to inform it of the terminating end's readiness to accept the call. The terminating switch then provides the ringing tone to both your phone and your friend's phone. Once the terminating phone is answered, another message is exchanged between the originating and the terminating switches to set up a dedicated channel or circuit for the conversation to begin. This completes the process of setting up your call across two different PSTNs, i.e., LEC PSTN and IEC PSTN. A seemingly simple phone call may have taken many switches and the execution of many millions of lines of software code to set up!

Signal transfer point. STPs are the intermediate nodes that are mainly responsible for passing the digital message of your call to the next switch. Analyzing the message and checking the information about this customer occurs elsewhere. An STP is a packet switch that can only process the signaling messages of your call, not the voice of your call. Recall that only two kinds of information travel on the network: signaling messages and call content. When you start speaking, the voice traffic generated from your conversation travels on a network different from that of the digital message of your call. Voice traffic travels from an SSP to another SSP, not going through any STP.

Gateway switch. A third type of switch is a gateway switch that connects a domestic network to a network in another country. The switch has to perform country code lookup, collect billing-related data, and process other necessary information.

A knowledgeable reader by now must have realized that the IEC network (see Fig. 1.2) that the digital message your call just traveled through is the CCS7 network. Signaling messages are used for setting up, maintaining, and taking down a call, versus the actual voice message of your phone conversation. Obviously the digital message of your call is a signaling message. The network carrying your voice is very mechanical and less interesting. Little network management is involved in carrying the voice information.

An example of network management issues for IEC networks includes having an overall view of the network, coordination of networkwide traffic

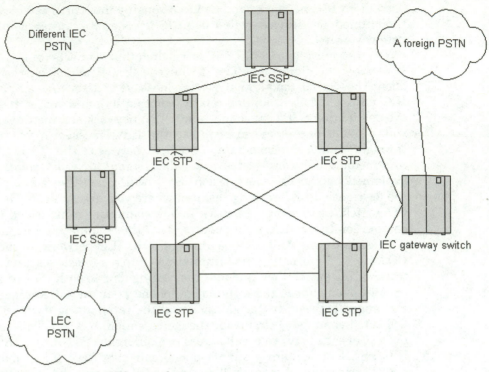

Figure 1.2 A sample IEC network.

management to avoid congestion, and managing individual network nodes such as STPs and SSPs.

A wireless network. Another important telecommunications network requiring management is the wireless network. Again we introduce the components of the network by tracing through a wireless phone call. Note that there are many different wireless technologies such as cellular, Global System for Mobile (GSM), Time Division Multiple Access (TDMA), and Code Division Multiple Access (CDMA), which are not necessarily mutually exclusive, and the wireless network we present here is a generic one void of details of any specific wireless technology.

Again, assume you make a call to one of your friends at a distant city, using a wireless handset. What are the major components of the wireless network your call goes through?

Mobile station. Your call originated from your handset, more formally called a mobile station (MS). When you dial your friend's number and press the send, talk, or other similar button to send the dialed digits, your MS transmits the dialed digits along with other information like a security code and the MS identification number to the service provider's network over a radio channel. Similarly, using radio signals the network sends information back to your MS.

Base-station transceiver.   It is the *base-station transceiver system* (BTS) on the network side that receives your radio signals. A main function of the BTS is to transform the received analog radio signals from the mobile station into digital signals for digital network components. Conversely when it receives digital signals from other network components, it transforms the digital signals into radio signals for the MS. Each BTS covers a specific geographic area called a cell. The maximum number of mobile stations that a BTS can accommodate within a cell is wireless-technology-specific, ranging anywhere from a dozen to a few dozen given current technologies.

Base-station controller.   The next stop of your call is the *base-station controller* (BSC). The digital message containing all the information about your call (e.g., your number, your friend's number, time of call, destination of call) is sent by the BTS to a connected BSC which is responsible for (1) handing you over to another cell if you happen to be driving while you are making the call and have moved out of the original cell (called hand-off) and (2) passing your digit message to a connected mobile switching center to further process your call. A BSC covers a specific geographic area, and the maximum number of cells under a BSC is wireless-technology-specific (see Fig. 1.3).

Mobile switching center.   The switch that receives the digital message of your call from the BSC is called a *mobile switching center* (MSC). An MSC is very similar in function to a CO switch. It sends the digital message to the *authentication center* (AC) connected to this MSC. The AC validates this

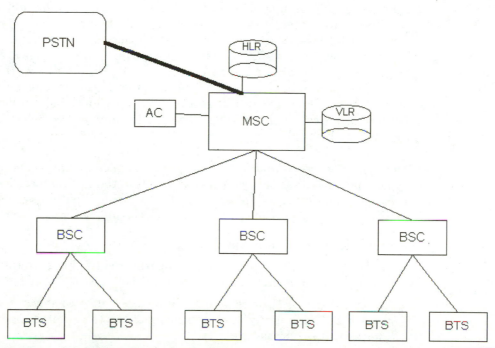

Figure 1.3 A sample wireless network architecture.

subscriber and MS from which the call is originated and checks with a network component called the *home location register* (HLR) that contains a record of what service you have, what types of calls are allowed on this MS, the identification number of your MS, and other relevant information. From this point on, this data record will be placed in a transient database called the *visitor location register* (VLR) within the MSC for access convenience until your call is completed. After all this, the MSC makes a decision as to where to route your call, based on the number you dialed. An MSC covers a specific geographic area and controls a given number of BSCs. If you happen to cross into another BSC area while making the call, the MSC hands you over to the new BSC. If the called party (your friend) is located within the area covered by the same MSC, the MSC will provide the ringing tone to both you and the called party after ensuring the number of the called party is valid. If the called party is located outside the MSC area, say, in another city, then the MSC routes the call to the LEC network it is connected to and, from that point on, the call is no different from a call that originated from a phone at home as described in the PSTN section. It goes through the originating LEC network, then the connected IEC network, and then the terminating LEC network. If the called party happens to be a mobile phone, the terminating LEC switch will route it to the MSC covering the area where the called party is located.

Home location register.   The HLR is essentially a database that stores the information about the subscribers in the area covered by the MSC. All subscriber data necessary to process your call, such as mobile station serial number, phone number, subscribed service features, and allowed service privileges, are stored at the HLR to be used by the MSC.

The VLR stores the records of those subscribers who currently have an active call with this MSC. Their records will be deleted once a subscriber roams out of this MSC area or ends the call. All VLR data are transient.

Authentication center.   The AC is a network component responsible for validating a caller and a mobile station from which a call either has originated or to which a call will terminate. The AC is employed to detect and prevent fraud if at all possible.

Intelligent Network.   *Intelligent Network* is a special term coined to refer to a type of network that is largely an extension to the existing PSTNs in order to provide a specific class of services the existing PSTNs were not capable of providing. The class of services includes the following:

- Free phone that allows the called party to pay
- Calling card, collect, third party, or operator calls that give the caller more control over payment options
- Automatic call distribution that allows the caller to select an option and route a call to a designated internal number (e.g., "press 1 for secretary, 2 for security, 3 for service,...")
- Calling card number verification

Two separate efforts were under way to develop Intelligent Network standards, starting in the mid-1980s. One was by the international telecommunications standards body, ITU, and the other by AT&T and the joint research arm of the RBOCs and Bell Communications Research (Bellcore). The two efforts exerted significant influence on each other but nonetheless produced two separate sets of standards specification. Bellcore's is called *Advanced Intelligent Network* (AIN) which mainly sees its implementation and deployment in the United States by the RBOCs. The other by ITU is termed *Intelligent Network* (IN) which enjoys wide acceptance in Europe and Asia. While AIN efforts have dwindled since the mid-1990s, the ITU IN efforts are on pace to produce the third set of IN standards, called *Capability Set 3* (CS3). For detailed technical differences between the two, see Gaynberg et al. (1997). The IN architecture shown in Fig. 1.4 is a generalization common to both AIN and ITU IN.

In addition to providing capabilities for the IN services listed above, one other often touted driver for IN is to give operating companies more control of what services to create, how services are created, and when and where to deploy them. The relaxed regulatory control enables the operating companies to respond to market demands in a more timely manner.

How does IN shift control from the equipment vendors to the operating companies? A simplified view is that all the functionality and intelligence of a switch can be viewed as performing two types of functions: call processing and service processing, or call logic and service logic. *Call logic* is the part of switch intelligence responsible for setting up, taking down, routing, and processing a call. *Service logic* is the part of the switch intelligence responsible for processing service features such as call forwarding and caller ID. Traditionally the two types of intelligence are indiscriminately interwoven inside the switch software. The basic idea of the Intelligent Network is to separate the two types of intelligence and put them into different network elements. This separation makes it easier and faster to create and deploy new services because the service logic part is separate from the rest of the network intelligence. Being separate it can be changed and extended without the concern for impacting other parts of the network.

Again, we introduce the Intelligent Network by tracing an 800 call through the various components of the network with emphasis on those unique to the Intelligent Network. After you dial an 800 number, the digital message containing the 800 number you dialed along with other information (your phone number, privileges allowed on your phone, etc.) first goes to the CO switch serving your area just as described in the PSTN section. Once the central office detects an 800-number call, it routes the call to the long-distance carrier of your choice, where the tracing of the call through the Intelligent Network starts.

Service switching point. The first stop for the digital message of your call is an SSP of the IEC network. An SSP is where the call processing intelligence resides and functions just like a regular PSTN switch, with one exception: once it detects the call is an IN service type, it transfers control to another network node called the service control point to process the service logic portion of the call.

Service control point (SCP). When the digital message of your 800-number call arrives at the SCP, it translates the 800 number into a regular 10-digit phone number by doing a lookup into a central database which contains an entry for each 800-number to real-phone-number pair while the real phone belongs to the person or enterprise who subscribes to the 800-number services. After performing other needed processing, the SCP sends the real phone number back to the SSP. With a regular phone number, it then becomes just a regular phone call and the SSP routes it just like any other regular call. There are some other IN components as shown in Fig. 1.4 that have not been covered by the 800 call: the intelligent peripheral, service creation environment, and service management system.

Intelligent peripheral (IP). The IP is where shared resources reside. Examples of shared resources include announcements you often hear on your phone and special tones. For a distributed network architecture such as the IN, resources are shared among SSPs and SCPs. The IP presents a uniform representation of resources to all SSPs.

Service creation environment (SCE). The SCE is for developing service logic. Given the increasingly large number of IN services, it requires an impractical amount of effort to write a software module for each new service. Alternatively, a set of independent software building blocks are created using the SCE. To establish a new service, assembly of the building blocks according to the new service requirements is all that is required. For example, for an

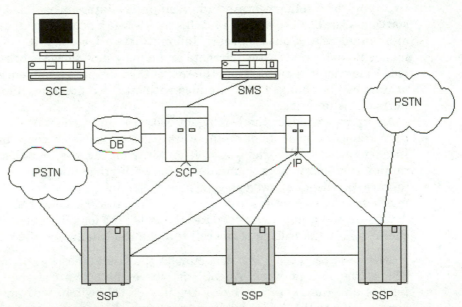

Figure 1.4 A sample Intelligent Network.

800-number call, a database lookup building block would check the database for a given 800 number.

Service management system (SMS).   As shown in Fig. 1.4, the SMS is the functional entity in the Intelligent Network that is responsible for billing information processing, loading the service created at the SCE into SCPs and SSPs, and monitoring service performance and others.

Characteristics of telecommunications networks

The extremely complicated nature of telecommunications networks and the rapid technology changes defy any general characterization. Nonetheless a general characterization of telecommunication networks is presented here to provide a context for network management issues. These characteristics more or less provide a contrast to the Internet.

Circuit-switched network.   Since their inception, telecommunications networks have been exclusively *circuit switched.* A switching method determines how information is routed and transported from the origination to the destination. The circuit-switched method means that whenever two parties communicate a dedicated channel is set up and maintained until the call is completed.

Circuit-based switching owes its origin to the telephone network where communicating parties must have exclusive use of a channel to ensure uninterrupted services. Among the main advantages of the circuit-switched network is its guaranteed quality of service. Among its disadvantages is the inefficient use of network resources. For a telephone user, a fixed channel, standard 64 kilobits per second (kbits/s), for conversation is more than enough.

The counterpart of a circuit-switched network is a packet-switched network where information is transported in packets over shared channels. Data networks, especially the Internet, extensively use a packet-switched scheme for information transport.

Voice-traffic-dominant network.   In the early years when telecommunications networks were identified with telephone networks, voice was the only traffic carried on the network. Only recently have other types of traffic, most notably data traffic, multimedia data in particular, started accounting for higher percentages of total traffic volume. Even in the near future, on U.S. public networks where data traffic surpasses voice traffic in total volume, voice traffic remains dominant in other aspects such as total revenue generated and the number of customers generating the voice traffic.

Separation of signaling messages from voice traffic.   As discussed earlier, the separation of signaling messages from voice traffic (standardized as the CCS7 network) is used to separate channels transporting the two different types of messages. Separation provides a whole range of possibilities for new services, utilizing the potential of the independent signaling channel. Note that separation is also a major departure from the data network, which uses packet headers of a message for control and signaling information.

Fully fault tolerant network.   One thing that distinguishes a telecommunications network from other networks, the data network in particular, is its extremely high degree of fault tolerance and resulting reliability. This is a direct consequence of government regulation over the telecommunications industry. For example, in the United States, the required uptime for a telephone network service is 99.9999 percent or a government-leveled penalty may result.

A switch's fault tolerance is achieved by full duplication. Despite the variation in implementations, the basic method used to achieve the fault tolerance is to have two identical units for each switch component, one as active and another as a standby. When the active unit fails, the standby kicks in immediately. Communication link fault tolerance is achieved by design of alternative routes.

Highly proprietary system.   Although digital switches use general computing technologies, both hardware and software systems are largely proprietary. Traditionally, switch hardware systems are all customer-designed and customer-manufactured. Switching software systems are even more so: each major switching-equipment vendor uses a different home-grown, proprietary programming language, a phenomenon difficult for the general computing world to comprehend.

Part of the reason is historical. Historically, one vendor has tended to dominate the telecommunications equipment market. A typical example is AT&T before divestiture in 1984. The dominant vendor's system was standard for all practical purposes because it was the only game in town. When competition sprang up, the competitive switch vendors developed their own proprietary systems, purposefully being different from the dominant players.

Part of the reason is technical. Highly rigid requirements like performance standards set by government regulations usually could not be met using generic computing equipment available at the market during the early days of digital switches, and the computing platform for switches had to be customer-designed and customer-built. One of the compelling reasons the proprietary programming languages were invented was to meet special performance needs of switching software that general programming languages at that time could not meet. Once a software system is created, it takes on a life of its own and billions of dollars would be required to revamp a large switching software system.

Enormous complexity of switching software.   A switching software system is one of the most complicated systems ever in existence. Any of the systems from major vendors can easily boast more than dozens of millions of lines of code. As a matter of fact, it is even difficult to get a precise count of the size of a switching software system from major vendors, after the decades of evolution and countless patches made on a continuous basis.

The explosion of software system complexity mainly comes from the ever-increasing number of telecommunications services being added to the network. During the early days of telephone networks, all the networks had to offer was basic line-to-line calls plus a few operator services. With digital switches equipped with the capabilities of programmable service logic, new services can be added fairly easily. The number of services such as call forwarding, free phone calls, credit card calls, and caller ID has ballooned to several hundreds, and the number is increasing faster than ever before.

What comes next?

If the above characterization is an indication of the current state of telecommunications networks, then the following describes some emerging or established trends that have important implications for network management.

Data and voice integration. The traditional division between a voice network (telecommunications network) and a data network (the Internet) is fast fading away. Telephone companies are becoming major Internet service providers, and both long-distance and local PSTNs are serving as Internet backbone networks. Although two distinct networks are still used to carry data and voice in major PSTNs, and a majority of switches in use today were designed exclusively for either voice or data, the ATM (Asynchronous Transfer Mode) switches that are capable of switching both voice and data traffic using one switching platform are gaining ground and the new generation of telecommunications networks are being designed with both types of traffic in mind.

A global telecommunications market. As we enter the new millennium, a totally new, global telecommunications market is emerging. The worldwide market is opening up as never before; an increasing number of new competitors are entering both the equipment market and the operating market. Among the implications from the emerging global telecommunications market are

- An increasingly diversified operating environment for service providers as more and more equipment vendors enter the market.
- An increasingly diversified technology base for service providers to deal with that will include a range of technologies such as voice, data, satellite, cable, wireless, and wireline. New technologies are being developed at an increasingly faster pace.
- Increased leverage on the part of customers who have more to say about what services they get and how they get them, as operating companies are competing for customers.
- A faster pace at which new services will be created and deployed to meet new demands of customers.

Network management, a key area. What do these trends mean to the telecommunications network management? Network management has emerged to play a key role in the trends discussed above. Integration of voice and data, circuit-switched networks, and packet-switched networks cannot take place without addressing such key network management issues as integrated billing, service quality assurance, and network security. The emergence of a global telecommunications market demands standards-based network management systems for networks of diverse vendors in vastly different operating environments. Integration of different tele-communications technologies, such as wireless and wireline, local and long-distance services, cable and phone, voice and data, requires an integrated network management system that is standards-based and capable of managing the diverse network technologies. Future trends of the telecommunications industry present enormous challenges as well as opportunities for network management.

Introduction to Telecommunications Network Management Systems

A telecommunications network management system (NMS) can be described by what it does and its components. Although what a network management system does is closely related to the type of network the system manages, we provide an introductory network management system that is general to all network technologies.

Functional areas

First, we present an analogy to help explain what a network management system does. Imagine you are a manager responsible for a group of employees. What then are the managerial tasks? First, you assign each group member a responsibility and make workload adjustments if someone is overworked while some others are idle (configuration management). If an employee calls in sick, or equipment breaks down, you need to find a substitute and make temporary arrangements before the equipment is back up or a replacement arrives (fault management). Also there is a need to monitor and evaluate the performance of each group member for promotion and workload assignment (performance management). Of course, there is the money issue of the salary for each employee (accounting), and there is some confidential information you want to protect (security management).

The overall tasks of telecommunications network management are similar. Network management, traditionally also called *operation, administration, maintenance, and provisioning* (OAM&P) has been divided into five functional areas in the ITU standards. Specifically, fault, configuration, accounting, performance, and security management are abbreviated as FCAPS.

Fault management.   By its very name, fault management deals with fault. A fault is any event that adversely affects the normal functioning of a network and the services it provides. The issues dealt with by fault management follow.

- To manage faults, you have to first detect faults in a timely fashion. Prompt detection is critical because immediate attention is required for some faults and delay in action may cause irreversible loss. Fault detection often requires surveillance of the network, and this is called *alarm surveillance.*

- The next issue of fault management is to locate the faults that have been detected. What caused the fault? Is there another cause behind the apparent cause, a root cause? If multiple faults are detected, are they attributed to the same cause or different ones?

- Finding the root cause of one or more faults often requires test equipment, software, whole system, etc. Thus, test management is a very important part of finding the root cause of a fault.

- Finally, once the root cause is identified, actions are taken to correct the fault and to restore the normal functioning of the network.

Configuration management. Configuration management probably is the most underdefined of the five management areas. If you consult five different documents or books on this subject, including some international standards, it is unlikely that you will find two similar definitions. Part of the reason is that configuration management covers a wide range of issues that may or may not apply to a particular type of network. In essence, configuration management is responsible for planning, engineering, installing a telecommunications network, putting the network into service, and provisioning the service to customers.

- The first step is to plan and design a network. This is called *network planning and engineering* and includes capacity planning, product planning, network design, system design, capacity design to meet the capacity, and product specification. The end product of this process is specification of a network product plan and design of the network.

- The next step is to install the designed network. This includes both hardware and software installation according to the network plan and design.

- Once a network is in place, *resource provisioning* allocates appropriate network resources to realize services.

- The final step is to provision the services to customers and generate revenue.

Accounting management. Accounting management is about generating revenue. To this end, several steps are involved.

- First, the accounting management is responsible for defining methods and granularity for measuring the usage of network resources by a customer. There are various ways of measuring network resource usage by a customer, some crude and some refined, some measuring by time, and some measuring by bandwidth.

- Then, the accounting management is responsible for collecting usage measurement data from switches and other network elements. The initial raw data need filtering, verification, and aggregation before they can be sent to the billing center for further processing.

- Tariffing is the part of accounting management that determines price, rates, discount for a service, and service features.

- At a billing center, customer bills are generated by applying the tax, rate, price policy, and discount to the processed resource usage measurement data.

Performance management.   The goal of performance management is maintaining network performance at a specified level.

- The first order of business for performance management is monitoring network performance on a constant basis to collect network and service performance data and to detect any performance deterioration that requires immediate attention.

- Performance analysis is needed to analyze the performance data for trends and make a decision whether to take preventive measures before performance deterioration crosses a critical threshold. An example is to detect a trend that may lead to traffic congestion and take action to reroute the traffic to avoid the congestion.

- The performance management control is responsible for taking preventive actions to avoid the network and service deterioration.

Security management.   Security management is about preventing and minimizing fraudulent use of network resources.

- Preferably we want to prevent a fraud, instead of recovering from it. Fraud prevention focuses on various security mechanisms for keeping the ill-intentioned user from entering the network in the first place.

- Fraud prevention is not always feasible. The next best thing is to detect a fraud promptly and minimize the damage caused by the fraud. Various tools and facilities can be utilized in fraud detection. For example, analysis of network service usage patterns and network traffic patterns can yield information on deviation from the normal patterns. There are built-in software and hardware checks in a network for the purpose of fraud detection.

- Another part of security management is containing and recovering from damage caused by a fraud, called fraud containment and recovery. *Fraud containment* is a process of isolating the fraud-infected network component and then neutralizing its impact. *Fraud recovery* is restoring the normal functionality interrupted by the fraud.

- An important responsibility of security management is to provide various security mechanisms such as key encryption, authentication, security protocol, and access control.

In short, configuration management is to plan, design, and install a network and have it up and running. Fault management keeps the network running free of trouble. Performance management keeps the network running efficiently. Security management keeps the network running without its security being violated. Accounting management generates network resource usage bills for customers.

Basic components of a network management system

The network management functionality, as expressed in the five functional areas, is realized via a network management system. Although there are many ways of building a network management system and the content of a network management system can vary depending on the network being managed, there is a common, underlying network management system architecture. The five major components in this architecture are shown in Fig. 1.5, i.e., a management station, a manager, one or more management agents, management information bases (MIBs), and a management protocol.

The architecture presented in Fig. 1.5 is an abstracted view of the basic architecture, and there are many variations of this architecture. For example, there can be a system between the agents and the manager that functions as both manager and agent. To the agent, it functions as a manager, and to the manager, it functions as an agent. This hierarchical management system often reflects the hierarchical network architecture. One example of such a hierarchical network is the wireless network shown in Fig. 1.3, with a mobile switching center (MSC) managing a set of BSCs, and each BSC managing a set of BTSs.

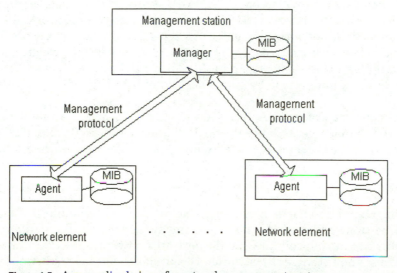

Figure 1.5 A generalized view of a network management system.

Management station. A management station provides a platform where a manager resides, and it provides a manager with

- A hardware platform and an operating system on which manager applications run
- A user interface via which a user can access the management application

Manager. A manager in a network management system functions as a central control that is responsible for making all major management decisions. A manager sends commands to a management agent to carry out the management decisions.

Management agent. A management agent usually is located near the network element being managed and is responsible for

- Carrying out the commands sent from the manager, including setting and checking the value of an attribute of a network element
- Collecting data on the network element and sending the data to the manager when requested
- Reporting an important event that happens at the network element to the manager unsolicited

Note that the agent does not directly act on the network element being managed. Instead, it acts on the attributes of a network element, the logical representation of the elements.

Management information base. An MIB is like a database, containing all the information of interest to the manager about a network element being managed. A network element, such as a switch, is represented as a set of attributes, such as cards, trunks, lines, and services it provides, and all these attributes together make up an MIB. An MIB can be viewed as the common vocabulary understood by both sides communicating with each other.

Management protocol. Another key element of a network management system is the management protocol, which defines the procedure and structure used by a manager and an agent to exchange management information.

Summary

This chapter consists of two introductions: the telecommunications networks to be managed and the management system itself. Four different network types are introduced: LEC public networks, IEC public networks, wireless networks, and Intelligent Networks. Despite differences between those networks, a common thread is that these networks are commonly seen in a typical service provider's domain that needs to be managed as efficiently as possible, hopefully using management systems that are similar in principles and architectures. It is these principles, architectures, standards, and, more importantly, their applications that will be the main thread of this book.

The characteristics of telecommunications networks can be better understood in comparison with data networks, Internet-based data networks in particular. A look at future trends and a review of telecommunications network history all attempt to underline the fact that the telecommunications industry is undergoing tremendous changes and the very definition of telecommunications as well as of the telecommunications networks is changing rapidly as technological innovations and regulatory changes are taking place at an unprecedented pace. Network management is a key area in tomorrow's telecommunications industry.

The overview of network management systems consists of two parts: what a network management system does and what makes up a network management system. The five functional areas of network management have become a well-established standard accepted in the industry. The manager-agent–based network management system is a dominant practice of network management, for both data networks and voice networks.

Exercises

1.1 What are the three types of switches commonly seen in a telecommunications network?

1.2 Describe two major differences between a telecommunications network and a computer network?

1.3 Describe the major common characteristics between telecommunications networks and computer networks.

1.4 What are the five functional areas of telecommunications network management?

1.5 What are the components of a telecommunications network management system?

Internet Network Management: MIB-II and SNMPv1

Outline

- The first section, "Overview of Internet-based Network Management," provides
  - A brief history of the Internet management protocol SNMP
  - An overview of the Internet network management framework that describes the components of an SNMP-based network management system and the basic management operations
  - Network management scenarios that will be used throughout the chapter to illustrate the applications of the SNMP
- The second section, "Internet Management Information Base," provides an introduction to the SNMP MIB and consists of two parts: the basic structure of the management information and the object groups of MIB version 2 (MIB-II), with a detailed description of the managed objects that have been standardized for TCP/IP-based networks.

- The last section, "SNMPv1," provides a discussion of SNMPv1 and consists of two parts: introduction of common terminology and concepts essential for understanding the SNMP and a detailed description of five protocol messages of SNMPv1, their structures, semantics, and applications.

Overview of Internet-based Network Management

First, the network management framework outlines the components of a network management system and the general management operations that can be performed by such a system. Once basic concepts of network management have been introduced, a brief review of the Internet management history will help explain why the Simple Network Management Protocol (SNMP) is what it is today and what the important standards are in this area. Finally, a network management scenario is introduced to help put the network management basics just introduced into a concrete context and to provide an application example that will be used throughout the remainder of the chapter.

Internet management framework

The management framework covers the overall structure of an SNMP-based management system.

Components of a network management system. An Internet network management system in many aspects resembles a management structure we would normally see at our workplace: a manager and a group of employees reporting to the manager. There are certain rules or norms governing the communication between the manager and his or her employees. A network management system consists of a manager, a set of agents, a management protocol for the manager to talk to the agents, and a management information base (MIB) that stores the information about the network elements being managed.

A *manager* resides at a host computer that has a collection of network management applications and plays the role of controlling the network nodes. It is responsible for the management of a network or a subnetwork. Its main responsibilities include maintaining a networkwide view of the network being managed, periodically collecting information from each network element, and analyzing the collected information to make a decision as to whether to take any control actions.

A management *agent* is a management entity that is responsible for a particular network element being managed. Its responsibilities include executing the management instructions sent from the manager and reporting to the manager any abnormal conditions that occurred at the network element. Both the manager and the agents are also referred to as network management entities that can initiate and implement management actions.

An *MIB* essentially is a database for storing the management information. The data stored in an MIB represent the characteristics of the network element being managed that are of interest to the manager from the point of view of management. One point worth emphasizing is that the agent does not directly act on a network element itself, say, a router, but rather on the representation of the router. An MIB is a logical representation of the network element being managed. As shown in Fig. 2.1, MIBs are present at both the manager and agent sides in order for the two sides to communicate with each other, though at the manager side, an MIB may exist only at run time.

A *network management protocol* provides a mechanism for a manager to communicate with agents. The standard Internet-based network management protocol is the SNMP, which resides at the application layer of the Internet protocol hierarchy. The management protocol uses lower-layer protocols for transport of management messages. The main responsibilities of the SNMP include defining the structure of the protocol data units (PDUs), the procedures for exchanging the PDUs, and the management operations permitted by the PDUs. The major components of a network management system and their relationships are shown in Fig. 2.1.

Network management operations. A network management operation can be initiated either by a manager or an agent and can only be completed with participation from both the manager and agent. All the management operations allowed for the SNMP can be put into three generic categories, as described below.

The first category is *query operation*; for example, a manager queries an agent for information on the network element such as status, states, or statistics. The second category is *set operation*; for example, a manager requests that an agent modify the information in the agent's MIB to ultimately achieve the effect of changing the attributes of the network element. How the changed attribute in the MIB eventually is reflected and propagated onto the network element is not covered by the standards and is determined on an element-by-element basis. The third category is *reporting event*; for example, an agent reports an abnormal event to the manager that has occurred at the agent side. The event-reporting messages from the agent to the manager are unsolicited by the manager. The agent's capability to send unsolicited messages to the manager is very limited in the SNMP, as will be explained later. The three types of operations associated with the management protocol are shown in Fig. 2.1.

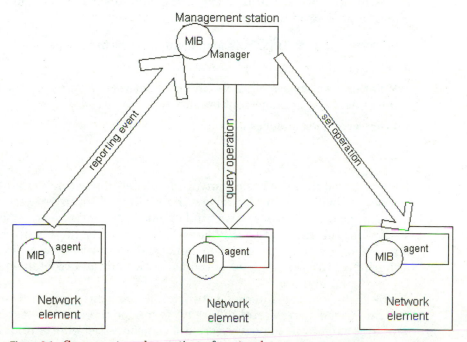

Figure 2.1 Components and operations of a network management system.

A network management scenario

A simple network, shown in Fig. 2.2, and a management scenario are presented to achieve two purposes: (1) help the reader put the Internet-based network management framework in a concrete context, and (2) provide an example that will be used throughout this chapter to illustrate the basic concepts and the application of the SNMP.

As shown in Fig. 2.2, the network consists of two printers, one router, one network server, and four other host computers. The network management system is configured as consisting of a manager residing in a host computer and seven agents residing in other network elements.

The scenario is as follows. Printer 1 stopped printing due to hardware-related failure, and the manager needs to know it immediately. Once having received the notification, the manager checks the job queue of printer 2, and resets the allowed queue size to a larger number to accept more print jobs.

How does the network management system accomplish the described tasks using the SNMP? It suffices at this point to describe network management system operations at a high level to provide a basic understanding of the network management framework just described above. There are three steps of management operations. First, the agent of printer 1 sends a report notifying the manager that the printer went down. Then the manager sends a query to check the other printer's job queue. Finally the manager sends an instruction to the agent of the working printer to set the job queue limit to a larger number.

This scenario will be used throughout this chapter. Some detailed questions that will be answered in the rest of the chapter are listed below to give the reader a preview of the remainder of the chapter.

- What does the MIB for this management system look like?
- What are the steps for building such an MIB?
- What protocol messages are exchanged between a manager and an agent to implement the management operations?
- What are the sequences of protocol message exchanges?

A brief history

Having introduced the basic framework of the Internet network management system, we will now briefly review the history of Internet network management standards. There are two fundamental goals of a network management system for the Internet-based networks: monitoring the health of a network and taking control actions when abnormal conditions have developed that either affect the performance of the network to an unacceptable level or have the potential to do so.

The history of the Internet network management standardization process can be divided into several different stages for the sake of easy description. The first stage was the early period of no concerted efforts to develop a systematic management system. At first it was all a manual process. Since the

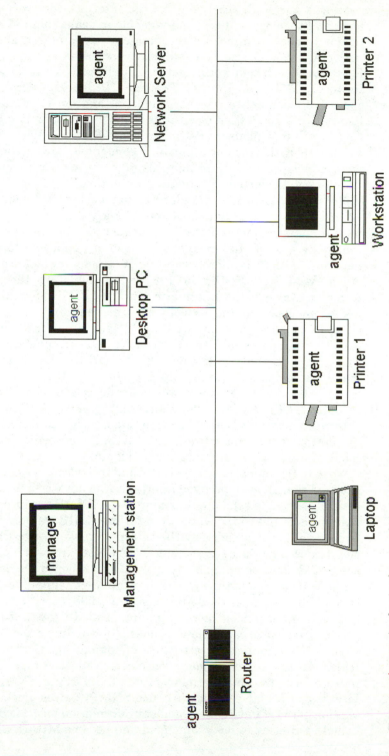

Figure 2.2 A scenario of a simple network and a management system.

networks during the early days of the Internet were very simple in functionality and topology and very manageable in scale, manual processing was sufficient to deal with most of the problems encountered at the time. Then, as networks grew in size and complexity, some ad hoc tools were developed to facilitate the job of detecting network problems. The Internet Control Message Protocol (ICMP), an Internet Protocol (IP) layer protocol, was developed for communication between IP layer entities and for detecting IP layer problems like network congestion or an unreachable port. One tool based on the ICMP that is still in wide use is Packet Internet Groper (PING), which allows a user to query whether or not a network element is down. The protocol and tools were developed mainly to help monitor the network condition. The control actions still depended on human intervention.

Next was the initial standardization period. Starting in the mid-1980s, efforts were under way to develop a standard network management protocol to automate the increasingly complicated network management tasks and to reach the goal of one management system applicable to any Internet-based network, so-called interoperability. A temporary solution was the SNMP, which was first drafted in 1987 while awaiting the long-term solution Common Management Information Protocol (CMIP) to mature. The CMIP was being developed at the time by the international standards organization's (ISO's) Consultative Committee on International Telephony and Telegraphy (CCITT), later renamed the International Telecommunications Union (ITU). The plan was that while the SNMP could fill the short-term need, it was meant to be merged into the CMIP eventually. At the same time, efforts were also under way to develop a protocol called *CMIP over Transmission Control Protocol (TCP)/IP* (CMOT for short) that applies CMIP to the management of Internet networks. The two-pronged approach was formally adopted in 1988 by the Internet Architecture Board (IAB), the governing body for the development of the Internet architecture and standards.

Next is the independent SNMP standardization period. Before long, it was realized that the two-pronged approach was not working well. For example, it severely restricted the development of the SNMP because it had to follow the standards of the CMIP while work on the CMIP was progressing at a slower pace than that demanded by the Internet market. After the IAB made a decision to decouple the development of the SNMP from that of the CMIP in 1988, the SNMP development efforts took off and enjoyed phenomenal success during a short period of time in the market. Meanwhile, the efforts to marry the two all but died after the IAB decision in 1988.

The Internet management standards, called *Request for Comments* (RFCs), are in three basic categories: standards for defining and organizing the management information, or structure of management information (SMI); MIB which defines standard managed objects; and the management information protocol. The first management protocol, SNMP (or SNMPv1 as it was called later because a later version is called SNMPv2), was quickly adopted. The first version of the standard management information base for SNMPv1, sometimes called MIB-I, was quickly amended and became MIB-II. SNMPv1 and MIB-II and the SMI for the MIB-II became the first set of widely used Internet management standards and still have a large installed base. The standards related to SNMPv1, MIB-II, and the accompanying SMI are shown in Table 2.1.

An RFC can have one of the following statuses:

Experimental.   The RFC is in its early infancy, being tried for the various applications. It may never become a standard before it becomes historical.

Enterprise.   The RFC is being used at an enterprise level. The majority of enterprise RFCs don't become standards.

Proposal.   The final stage before an RFC becomes a standard. After the working committee responsible for this RFC recommends to promote this RFC to standard, a 6-month proposal period is used to collect public comments on the RFC before a final decision is made.

Standard.   The Internet Engineering Task Force (IETF) is responsible for making the final decision to promote an RFC to a standard status.

Obsolete.   This RFC is superseded by a new version of the standard on the same topic.

Historical.   The RFC has been commented on but never became a standard.

After this first set of SNMP standards, major additions have been made which resulted in SNMPv2, Remote Network Monitoring (RMON), their corresponding MIBs and SMIs, and now SNMPv3 in the pipe. This chapter focuses only on SNMPv1, MIB-II, and the companion SMI, based on the first set of widely adopted Internet management standards, because the emphasis here is on the basic concept of SNMP rather than on the latest advances.

Internet Management Information Base

An MIB in essence is a database that contains a set of objects for the network management purpose. An MIB

TABLE 2.1   RFCs Related to SNMPv1 and MIB-II

RFC	Title	Year	Status
1155	Structure and Identification of Management Information for TCP/IP Based Internets	1990	Standard
1156	Management Information Base Network Management of TCP Based Internets		Standard
1157	A Simple Network Management Protocol	1990	Standard
1158	Management Information Base Network Management of TCP/IP Based Internets: MIB II		Draft standard
1212	Concise MIB Definitions	1991	Standard
1213	Management Information Base for Network Management of TCP/IP-based Internets: MIB-II	1991	Standard

- Essentially provides a set of common vocabulary for a manager and an agent to communicate with each other and exchange management information
- Allows only a predefined set of operations to be performed
- Is management-protocol-specific in practice
- Is defined using the formal type-specification language Abstract Syntax Notation One (ASN.1)

Introduction to MIBs

An MIB object is a software representation of a managed network element such as a router, a workstation, or even a piece of software that is of interest from the management perspective. A network management system does not directly deal with managed network elements, but rather indirectly through the managed objects.

An MIB provides a vocabulary for a manager and an agent to communicate with each other. This requires that both the manager side and the agent side have an MIB. The relationship between the manager MIB and an agent MIB is a dynamic one. In theory, the manager MIB is a union of all the MIBs of the agents that the manager communicates to, simply because the manager has to know all the agents' vocabulary in order to talk to all of them. However, for practical reasons, this seldom happens and the manager MIB usually contains a subset of the agent MIBs that the manager uses during a particular session.

There are two general categories of management operations that can be performed on an MIB. The first operation, query operation, which is equivalent to the read operation on a database, is for a manager to obtain values of objects stored in an agent MIB. The second type of operation is for a manager to modify the value of an object stored at an agent MIB. The operations on an MIB are initiated by a manager that sends a management protocol message to the agent, and the receiving agent responds by performing the required operation.

For all practical purposes, all MIBs are management-protocol-specific. The major management protocols, such as the SNMP and the open system interconnection (OSI) CMIP, can't interoperate with each other directly without a conversion process. One of the major reasons for this is that the respective MIBs of each management protocol can't interwork with each other.

Overview of ASN.1

ASN.1 is a type-specification language that is used to specify the generic SNMP MIB structure and application-specific MIBs. This provides a high-level overview of ASN.1 with respect to its use for specifying MIB structure. Those readers with little exposure to ASN.1 are encouraged to first read App. A, which provides a more detailed introduction to ASN.1, before proceeding any further. In this overview, we hope to convey to the reader why ASN.1 is needed, what it provides, and what it looks like.

Why is ASN.1 needed? Keep in mind a scenario that a manager sends a command to an agent to carry out some management operation such as modifying an object value in an MIB in the form of a protocol message. The manager's message is transformed into a bit stream before being transferred across the network. When the agent side receives the protocol message and converts from a bit stream back into a platform-specific format, say an ASCCI stream, the agent side will not be able to interpret the message unless there is a structure to put the message in that is agreed upon by all parties involved. This is where ASN.1 comes in: it provides types to structure information being passed across the network to make it possible for the receiving end to interpret the message as meant by the sending end.

ASN.1 is a type-specification language and provides platform- and representation-independent types. There are simple types, constructed types, tagged types, character string types, and others. The examples of simple types include INTEGER, BOOLEAN, BIT STRING, and OCTET STRING. Constructed types are built from the simple types, and detailed definitions of other types are described in App. A. ASN.1 is used to precisely specify the SNMP management information structure.

Structure of management information

The SMI provides (1) a standardized structure for representing managed objects in the MIB, (2) a common object syntax that defines the types allowed in the managed objects, and (3) a naming convention for identifying managed objects in the MIB. The intended effects are to have the same representation for the same network element; for example, a router in every application and a standard MIB can be used in any other application of TCP/IP-based network management. The SMI discussed in this section is based on RFC 1155 (Rose and McCloghrie, 1990).

Note that the term *MIB object,* as used in the context of SNMP management, is different from the objects as defined in object-oriented literature. An object, which will become evident shortly, is merely a variable as we understand in programming language: a place holder for a particular type of values. The choice of the term *object* can be traced back to the early alliance of the SNMP with OSI's management framework, which does employ the concept object in its true object-oriented sense.

There are two related design principles for the SNMP SMI: simplicity and extensibility. The structure of the management information is kept to a bare minimum to simplify the implementation and to enhance the interoperability. A simple MIB structure means a minimum number of constraints for future extensions.

Object naming of SNMP MIB. Each managed object in the MIB is uniquely identified by an object identifier. A naming scheme is used to define how an object identifier is constructed. A key to understanding the object identifiers of the SNMP MIB is the registration hierarchy developed under the auspices of the OSI standards committee.

The registration hierarchy is used to register all ISO-related entities in a hierarchical fashion as shown in Fig. 2.3. Each node represents an entity with two identifiers assigned to it. The first is an integer identifying the relative position of the node at the horizontal level, which we call a *relative identifier*. For example, ITU is assigned 0 and ISO 1 and so on in an ascending order, as shown in Fig. 2.3. The other identifier, which we call an *absolute identifier*, uniquely identifying the position of each node in the entire hierarchy tree, is formed by concatenating all the relative identifiers along the vertical path leading to this node, starting from the very top. The ISO registration hierarchy is adopted for naming the managed objects of the SNMP MIB.

We use Fig. 2.3 to illustrate the naming scheme of the SNMP managed objects. The Internet is always identified as 1.3.6.1, from concatenating node identifiers of ISO, ORG (Organization), DOD (Department of Defense), and the Internet itself. Six nodes are registered under the Internet (Rose and McCloghrie, 1990):

1. *Directory.* Reserved for OSI directory services as defined in the X.500 series standards.

2. *Mgmt* (*management*). Used to identify the Internet management-related objects approved by the IAB.

3. *Experimental.* Used to identify the objects from Internet-related experiments. For example, objects from draft Internet standards might initially be put here and moved to another branch once the drafts are adopted as formal standards.

4. *Private.* For institutes and enterprises to register their organizations.

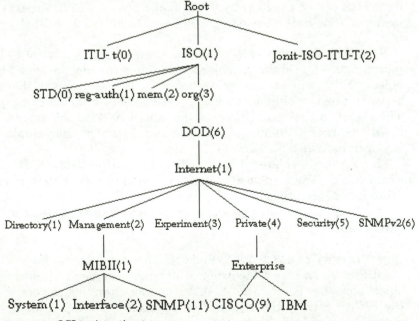

Figure 2.3 OSI registration tree.

These four subtrees were initially specified under the Internet, and two more were added later:

5. *Security.*   For objects identified for the future security feature extension of SNMPv2.

6. *SNMPv2.*   For objects identified for the management operations of SNMPv2, an extension of SNMPv1.

The Internet MIB-II is registered under mgmt with the identifier (1.3.6.1.2.1) and has nine subtrees of its own representing nine object groups of MIB-II which will be discussed in a later section. Some key points about the Internet naming scheme are worth emphasizing. First, the naming scheme, adopted from the OSI registration tree, defines a structure to organize managed objects into a conceptual hierarchy which is very flexible for future extension; second, each managed object has a fixed name and a unique identifier that are standardized and recognizable across all Internet standards-based management applications.

Object syntax.   Object syntax defines the SNMP object structure, types, and syntactic rules for defining managed objects. Object syntax is defined using ASN.1.

Managed object types.   There are two categories of Internet object types, i.e., primitive types and constructor types, as defined in RFC 1155 (Rose and McCloghrie 1990).

Primitive types.   Primitive types are the simplest types that can't be further decomposed into simpler types and that can be used as building blocks for more complicated types. Based on the overriding design principle of the SNMP MIB of keeping it as simple as possible, only a small subset of ASN.1 types are adopted for the SNMP MIB objects.

- *INTEGER.*   The positive and negative whole numbers, including 0
- *OCTET STRING.*   A sequence of 0 or more octets
- *NULL.*   The simple value of null denoting the nonapplicability of choices
- *OBJECT IDENTIFIER.*   A sequence of integers separated by a period, with each integer representing a relative identifier in the registration hierarchy

Constructor types.   These types are used to construct other types.

- *SEQUENCE.*   An ordered list of existing types, with the order fixed; similar to the structure in C++ or C programming language.
- *SEQUENCE OF.*   An ordered list of a single existing type.
- *CHOICE.*   A list of existing, alternative types of which only one will be selected to create a value. It is similar to the union construct of C or C++ programming language.
- *ANY.*   A place holder for any type.

Application types. These types were originally anticipated to be application-wide only, and a very limited few were defined initially in RCF 1155 (Rose and McCloghrie, 1990). Each of these types must map to a primitive type, list, table, or some other application type.

- *NetworkAddress.*   Defined by the CHOICE construct of ASN.1 to represent one of several protocol address families. Currently only the Internet IpAddress is present.

- *IpAddress.*   A 32-bit Internet address of the type OCTET STRING, e.g., *x.x.x.x* where each *x* is an 8-bit nonnegative integer number.

- *Counter.*   A nonnegative integer number that only increases, not decreases. When it reaches the maximum value, it wraps around and starts from 0. The specified maximum value is $2^{32}-1$.

- *Gauge.*   A nonnegative integer. It is like the Counter type with two differences: first, it may increase or decrease; second, it latches at a maximum value when that maximum value is reached. The current specified maximum value is $2^{32}-1$.

- *TimeTick.*   A nonnegative integer that counts the time in hundredths of a second since some epoch. When an object type in the MIB is defined using this type, the object type specifies the epoch reference.

- *Opaque.*   A place holder for an arbitrary ASN.1 data type. The value of this type is encoded using the ASN.1 basic encoding rules into a string of octets.

The ASN.1 definitions of the application types can be found in Fig. 2.4.

Managed object template.   The syntax only defines types, which are components of managed objects, but not the objects themselves. What does a managed object look like? Imagine yourself as a designer of the MIB structure confronted with the following requirements. On the one hand, there is a great variety of network elements and entities that need to be represented by managed objects in the MIB. On the other hand, a managed object type should be defined in such a way that all managed objects should look alike, with the same generic structure, because this would simplify the structure of the MIB and support the interoperability between different network management systems. Remember that the number one design principle of SNMP is keeping it simple. The solution for the problem is a *super* object type defined using a macro that actually consists of a set of more specific types and yet maintains the uniformity of the managed objects in the MIB. This super object type was first proposed in RFC 1155 and enhanced in RFC 1212 (Rose and McCloghrie 1991). Stallings provides a quite thorough discussion on this (Stallings 1992).

The super object type of the SNMP MIB defines two types of notation, i.e., type notation and value notation for specifying detailed component object types and the value a type can take on, respectively, as shown in Fig. 2.4. We examine the type notation first. Each object type consists of seven components, as shown in Fig. 2.5.

```
RFC1155-8MI DEFINITION ::= BEGIN                                          1)
EXPORT — everything                                                       2)
Internet, directory, mgmt, experimental, private, enterprises, OBJECT-    3)
TYPE, ObjectName, OBjectSyntax, SimpleSyntax, ApplicationSyntax,          4)
NetworkAddress, IpAddress, Counter, Gauge, TimeTicks, Opaque;            5)
Internet            OBJECT IDENTIFIER  ::= {iso(1) org(3), dod(6) 1}     6)
directory           OBJECT IDENTIFIER  ::= {Internet 1}                   7)
mgmt                OBJECT IDENTIFIER  ::= {Internet 2}                   8)
Experimental        OBJECT IDENTIFIER  ::= {Internet 3}                   9)
private             OBJECT IDENTIFIER  ::= {Internet 4}                  10)
enterprise          OBJECT IDENTIFIER  ::= {private 1}                   11)
— definition of object types:
OBJECT-TYPE MACRO  ::=                                                   12)
BEGIN                                                                    13)
        TYPE NOTATION ::= "SYNTAX"type (TYPE ObjectSyntax)              14)
                                "ACCESS" Access                          15)
                                "STATUS" Status                          16)
        VALUE NOTATION ::= value (VALUE ObjectName)                     17)
        Access ::= "read-only" | "read-write" | "write-only" |          18)
                "not-accessible"                                         19)
        Status ::= "mandatory" | "optional" | "obsolete"               20)
END — of macro
— Names of objects in the MIB:                                          21)
ObjectName ::= OBJECT IDENTIFIER
— Object syntax definition                                              22)
ObjectSyntax ::= CHOICE {                                               23)
                simple SimpleSyntax,                                     24)
                application-wide ApplicationSyntax }                     25)
SimpleSyntax ::= CHOICE {                                               26)
                number                          INTEGER,                 27)
                string                  OCTET STRING,                    28)
                object                  OBJECT IDENTIFIER,               29)
                empty                           NULL }                   30)
ApplicationSyntax ::= CHOICE {                                          31)
                address                 NetworkAddress,                  32)
                counter                 Counter,                         33)
                ticks                   TimeTicks,                       34)
                arbitrary               Opaque                           35)

}

NetworkAddress ::= [APPLICATION 0] IMPLICIT OCTET STRING(SIZE(4))       36)
Counter ::= [APPLICATION 1] | IMPLICIT INTEGER(0.4294967295)            37)
Gauge ::= [APPLICATION 2] IMPLICIT INTEGER (0.4294967295)               38)
TimeTicks ::= [APPLICATION 3] IMPLICIT INTEGER (0.4294967295)           39)
Opaque ::= [APPLICATION 4] IMPLICIT OCTET STRING                        40)
END — of RFC 1155 SMI module                                            41)
```

Figure 2.4 ASN.1 definitions of SNMP object syntax (RFC 1155).

SYNTAX.   Specifies the abstract syntax for the object that has to resolve to an instance of the ASN.1 type ObjectSyntax defined in RFC 1115. ObjectSyntax can be one of two types: SimpleSyntax or ApplicationSyntax. The SimpleSyntax resolves to one of the four primitive types, and ApplicationSyntax resolves to one of the six application types as discussed above.

```
RFC 1212 MACRO for Managed Objects (1991):
IMPORTS ObjectName, ObjectSyntax FROM RCF-1155-SMI
OBJECT-TYPE MACRO ::=
BEGIN
    TYPE NOTATION ::= "SYNTAX" type(TYPE ObjectSyntax)
                      "ACCESS"      Access
                      "STATUS"      Status
                      DescrPart
                      ReferPart
                      IndexPart
                      DefValPart

    VALUE NOTATION ::= value (VALUE ObjectName)
    Access ::="read-only" | "read-write" | "write-only" | "not-accessible"
    Status ::="mandatory" | "operational" | "obsolete" | "deprecated"
    DescrPart ::="DESCRIPTION" value(description DisplayString)|empty
    ReferPart ::="REFERENCE" value(description DisplayString) | empty
    IndexPart ::="INDEX" `"{" IndexType "}"
    IndexTypes ::=IndexType | IndexTypes "," IndexType
    IndexType ::=value(indexobject ObjectName)
                                        | type(indextype)
    DefValPart ::="DEFVAL" "{" value (defvalue ObjectSyntax)"}" | empty
    DisplayString ::=OCTET STRING (0.255)
END
IndexSyntax ::=CHOICE {number INTEGER (0..MAX),
                                       string OCTET STRING,
                                       object OBJECT IDENTIFIER,
                                       address NetworkAddress,
                                       ipAddress IpAddress
```

Figure 2.5 Enhanced object definition of SNMP MIB (RFC 1212).

ACCESS.   Specifies how an object in the MIB can be accessed by application programs via a protocol message. There are four types of accesses: read-only, write-only, read-write, and not-accessible.

STATUS.   Specifies the implementation requirement for an object. The choices are mandatory, optional, or obsolete. If an SNMP management system vendor intends to claim to be standard-compliant, all the objects with mandatory status will have to be implemented. The obsolete status indicates that the object is no longer considered valid for the SNMP MIB but is left there for historical reasons.

The above three clauses were initially defined in RFC 1115, and four more are added in RFC 1212, among which is an index for the table structure of the MIB.

DescrPart.   Describes the semantics of an object in a textual string. This is an optional clause.

ReferPart.   Specifies a cross reference to another object in another MIB module. This is an optional clause.

IndexPart.   Specifies a row index of a table in an MIB. This is a necessary part to access a table.

DefValPart.   Specifies a default value when an instance of the object type is first created. This is an optional clause.

To help those not familiar with ASN.1 notations, we provide a line-by-line explanation to RFC 1155, shown in Fig. 2.4, and tie together all that has been covered so far on the MIB.

- *Line 1.*   An ASN.1 module identifier with the module name RFC1155 and the module header DEFINITION ::= BEGIN. An ASN.1 module like a module in a programming language serves as an application unit.

- *Lines 2–5.*   Specify the types and object identifiers that can be used by other ASN.1 modules if this module is included in them.

- *Lines 6–11.*   Define the Internet registration hierarchy as shown in Fig. 2.3. The number in parentheses is the relative identifier of each managed object in the registration tree.

- *Lines 12–20.*   Define the supertype OBJECT-TYPE which provides a template for defining MIB objects. TYPE NOTATION specifies the syntax (types) of a managed object and the clauses an object must have in its definition. The clauses are already discussed. VALUE NOTATION indicates the name used to access this object.

- *Line 21.*   Defines the type of ObjectName to be OBJECT IDENTIFIER.

- *Lines 22–24.*   Specify that ObjectSyntax resolves to one of the two syntaxes, SimpleSyntax or ApplicationSyntax.

- *Lines 25–29.*   Specify that SimpleSyntax resolves to one of the four primitive types adopted from ASN.1.

- *Lines 30–35.*   Specify that ApplicationSyntax resolves to one of the five application types.

- *Lines 36–40.*   Define the application types.

Defining MIB tables.   There are two types of managed objects in the SNMP MIB: scalar and columnar objects. A scalar object is like a variable of primitive type, one value per object. The template defined in Figs. 2.4 and 2.5 is for defining scalar objects. A columnar object is a conceptual, two-dimensional table made up of multiple scalar objects indexed by row and column numbers. There are two basic issues related to tables: how to define a table and how to access a table.

Tables are based on a scalar plus the ASN.1 type SEQUENCE or SEQUENCE OF. First, a table is still a managed object, just like any scalar object, and the object syntax is defined using the object template ObjectSyntax. In addition, the conceptual rows and columns are defined using SEQUENCE or SEQUENCE OF. Note SEQUENCE is an ordered set of elements of different types, just like a row in a table, and SEQUENCE OF is an ordered set of elements of the same type, just like a column of a table. The ASN.1 definition of an MIB table will look like the following, and the structure described is shown in Table 2.2.

TABLE 2.2   MIB Table Structure

	Entry 1	Entry 2 · · · Entry N
table row 1		
table row 2		
⋮		
table row N		

```
Table ::= SEQUENCE OF TableRow
TableRow ::= SEQUENCE {
        entry1    ::=  MIB type1,
        entry2    ::=  MIB type2,
        .
        .
        .
        entyN     ::=  MIB typeN
}
```

A table example from RFC 1213 (McCloghrie and Rose 1991) will help illustrate how an MIB table is defined and how the row and column index of a table is determined. Figure 2.6 shows the ASN.1 definitions of an IP address table.

The IP address table consists of a number of rows. Each row has five fields, i.e., an IP address, index of a unique interface associated with the IP address, subnet mask associated with the IP address, the IP broadcast address, and the size of the largest IP datagram that can be reassembled for this IP address.

A line-by-line explanation will help the reader not yet familiar with ASN.1 notations understand the IP address table definitions.

- *Line 1.*   Defines IP address table ipAddrTable as OBJECT-TYPE, the super object type as described in the object template section.

- *Line 2.*   The syntax of the table is defined as SEQUENCE OF IpAddrEntry, i.e., a set of table rows.

- *Line 3.*   The table itself is not accessible by the application.

- *Line 4.*   The table is mandatory for any vendor who claims to be SNMPv1-compliant.

- *Line 5.*   A textual description of the table.

- *Line 6.*   The relative identifier of the table.

- *Lines 7–8.*   The syntax of the table entry is defined by IpAddrEntry, which is defined in the following clause.

- *Lines 9–10.*   The table row itself is not accessible, and the status is mandatory for implementation.

- *Line 11.*   A textual description of the row of the table.

- *Line 12.*   The row index is uniquely determined by the value of the first field of each row, i.e., the object ipAdEntAddr.

- *Line 13.*   The relative identifier of the table entry.

Figure 2.6   ASN.1 definition of an IP address table (RFC 1213).

```
ipAddrTable OBJECT-TYPE                                              1)
      SYNTAX SEQUENCE OF IpAddrEntry                                 2)
      ACCESS    not-accessible                                       3)
      STATUS    mandatory                                            4)
      DESCRIPTION                                                    5)
          "The table of addressing information relevant to this
           entry's IP addresses."
      ::= {ipAddrTable 20}                                           6)
ipAddrEntry          OBJECT-TYPE                                     7)
      SYNTAX         IpAddrEntry                                     8)
      ACCESS         not-accessible                                  9)
      Status         mandatory                                      10)
      DESCRIPTION                                                   11)
      "The addressing information for one of this entity's IP address"
      INDEX          {ipAdEntAddr}                                  12)
      ::= {ipAddrTable 1}                                           13)
IpAddrEntry ::=                                                     14)
      SEQUENCE {                                                    15)
          ipAdEntAddr                     IpAddress,                16)
          ipAdEntIfIndex                  INTEGER,                  17)
          ipAdEntNetMask                  IpAddress,                18)
          ipAdEntBcastAddr                INTEGER,                  19)
          ipAdEntReasmMaxSize       INTEGER(0.65535)               20)
      }                                                             21)
ipAdEntrAddr         OBJECT-TYPE                                    22)
      SYNTAX    IpAddress                                           23)
      ACCESS    read-only                                          24)
      STATUS    mandatory                                          25)
      DESCRIPTION                                                  26)
              "The IP address to which this entry's addressing
               information pertains."
      ::= {ipAddrEntry 1}                                          27)
ipAdEntIfIndex OBJECT-TYPE                                         28)
      SYNTAX    INTEGER                                            29)
      ACCESS    read-only                                          30)
      STATUS    mandatory                                          31)
      DESCRIPTION                                                  32)
          "The index value which uniquely identifies the interface
           to which this entry is applicable. The interface
           identified by a particular value of this index is the same
           interface as identified by the same value of IfIndex."
      ::= { ipAddrEntry 2}                                         33)
ipAdEntEnNetMask OBJECT-TYPE                                       34)
      SYNTAX    IpAddress                                          35)
      ACCESS    read-only                                          36)
      STATUS    mandatory                                          37)
      DESCRIPTION                                                  38)
          "The subnet mask associated with the IP address of this
           entry. The value of the mask is an IP address with all
           the network bits set to 1 and the hosts bits set to 0."
      ::= {ipAdrEntry 3}                                           39)
ipAdEntBcastAddr OBJECT-TYPE                                       40)
      SYNTAX    INTEGER                                            41)
      ACCESS    read-only                                          42)
      STATUS    mandatory                                          43)
      DESCRIPTION                                                  44)
          "The value of the least-significant bit in the IP broadcast
           address used for sending datagrams on the (logical)
           interface associated with the IP address of this entry."
      ::= {ipAddrEntry 4}                                          45)
```

```
ipAdEntReasmMaxSize OBJECT-TYPE                                          46)
     SYNTAX       INTEGER(0.65535)                                       47)
     ACCESS       read-only                                             48)
     STATUS       mandatory                                             49)
     DESCRIPTION                                                        50)
          "The size of the largest IP datagram which this entry can
          re-assemble from incoming IP fragmented datagrams
          received on this interface."
     ::= {ipAddrEntry 5}                                                51)
```

Figure 2.6 (Continued)

- *Lines 14–21.*   The syntax definition of the table row, IpAddrEntry, is defined by a sequence of five table component objects: ipAdEntAddr, ipAdEntIfIndex, ipAdEntNetMask, ipAdEntBcastAddr, and ipAdEntReasmMaxSize. Note that IpAddrEntry defines the row of the table.

- *Lines 22–27.*   The first table component object of the row is defined. The definition syntax specifies that the object ipAdEntAddr resolves to IpAddress, an MIB-II application type. This table component object is the first column of the table, its relative identifier serving as the column index. The table component object is accessible to the application program.

- *Lines 27–51.*   The remaining four fields of the row, i.e., four scalar objects, are defined.

The next issue is how to determine the column and row index of a table in order to access the information in a table entry which is the ultimate goal of using a table to store information. Remember that each component object in a row has an identifier, and this is the column index. For example, in the above IP address table, the first field of each row is ipAdEntAddr, i.e., the IP address entry itself, which has an identifier of 1.3.6.1.2.1.4.20.1.1. The last 1 is the relative identifier of this table component object, and the rest of the identifier is a sequential concatenation of all the identifiers of the node right above ipAdEntAddr in the registration tree. This identifier is the index of the first column of the IP address table.

The next step is to figure out the row index. For MIB-II, the row index is determined by the INDEX clause of the object definition. In Fig. 2.6, line 12 specifies that the row index is determined by the value of the first field, or first component object of each row, i.e., ipAdrEntAddr, the IP address of this row. This means that the row index is table-specific and how it is determined varies from table to table.

A careful reader may have noticed by now that we have not discussed how to specify the number of rows in a table. The simple answer is you don't. The rows are dynamically added and deleted during a session of a management application. Note that no provisioning is present for a table to be nested and that the maximum level of a table is 1.

SNMP MIB-II object groups

The SNMP MIB is intended to provide a standard representation of managed network elements of TCP/IP-based Internet networks for the network management system vendors to follow. The MIB-I, defined in RFC 1156 (McCloghrie and Rose, 1990), has eight object groups, and MIB-II extends it with three more groups: CMOT, transmission, and SNMP, as shown in Fig. 2.7. MIB-II, as defined in RCF 1213 (McCloghrie and Rose, 1991), has an industrywide implementation and is introduced below. For each group, there is a general introduction and sometimes a summary of all the objects of the group in a table. Each object in the table is described from the following aspects:

- *Name.*   A unique name to identify the object.

- *Syntax.*   Abstract syntax of the object, i.e., the type of object.

- *Access.*   The type of access by application programs allowed on this object.

- *ID.*   The relative identifier of this object within the group. The absolute identifier of each object can be obtained by concatenating the relative identifier onto the identifier of the object group which is also provided. The absolute identifier of a table component object can be obtained by concatenating the relative identifier of the object onto the identifier of the table entry, which is the definition of all rows of the table.

To help put the objects to be described into context, the reader should keep in mind a scenario that a manager queries an agent on these managed objects and the agent responds with the information on the objects described below.

System group.   The objects in this group are designed to indicate the overall health of a system (system uptime) and general information about the system (system ID, location, services it provides, etc.). When queried about any of the

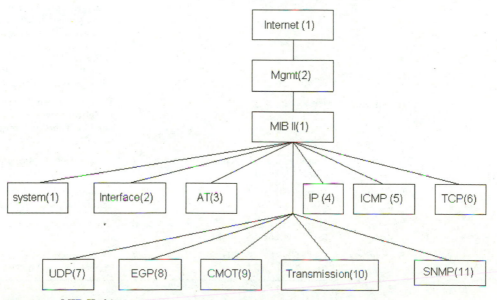

Figure 2.7   MIB-II object groups.

items, an agent can either return the requested object or a string of length 0 to indicate the agent is not configured to provide the requested object. A total of seven objects are in this group; a description of each object can be found in Table 2.3. The absolute identifier of this group is 1.3.6.1.2.1.1.

Interface. This object group provides generic information on the physical interfaces of a network element such as a router or a host computer. The information includes the number of interfaces associated with this network element and the type, capacity, status, and traffic conditions of each interface. See Table 2.4. The implementation of this group is mandatory, and its identifier is 1.3.6.1.2.1.2.

Each row of the interface table describes one physical interface. The identifier of the row template object, ifEntry, is 1.3.6.1.2.1.2.1. The identifier of a table component object is derived by appending its relative identifier to the identifier of the row template object. See Table 2.5.

Address translation. An Internet IP address has to resolve to a physical address of the local network. This object group is for translating a network

TABLE 2.3   MIB-II System Group Objects

Object name	Syntax	Access	ID	Description
SysDescr (SIZE(0..255))	DisplayString	RO (read-only)	1	A textual description of the network node. This value should include the full name and version identification of the P2
SysObjectID	OBJECT IDENTIFIER		2	The vendor's authoritative identification of the network management subsystem contained in the entity. This value is allocated within the SMI enterprise subtree (1.3.6.1.4.1) and provides an easy and unambiguous means for determining what kind of entity is being managed.
sysUpTime	TimeTicks	RO	3	The time in hundredths of a second since the network management portion of the system was last initialized
SysContact	DisplayString (SIZE (0..255))	RW (read-write)	4	Information on the contact person for this network node
SysName	DisplayString (SIZE (0..255))	RW	5	An assigned name for this node, used as a fully qualified domain name
SysLocation	DisplayString (SIZE (0..255))	RW	6	The physical location of this node
SysServices	INTEGER (0..127)	RO	7	An integer value indicating the set of services offered by this node

TABLE 2.4   **MIB-II Interface Group Objects**

Object name	Syntax	Access	ID	Description
ifNumber	INTEGER	RO	1	The number of network interfaces present on this network element
ifTable	SEQUENCE OF IfEntry	NA (not applicable)	2	An interface table containing detailed information on each interface

address, i.e., an IP address, into a subnetwork-specific physical address. This group is deprecated and is kept mainly for compatibility with MIB-I. The function of mapping a network address to the physical address has been assigned to network level protocol-specific object group. One reason for this is that addressing schemes are protocol-specific, and some network element supports multiple protocol and thus multiple addressing schemes. This address translation group is for IP address translation only. The identifier of this group in the registration tree is 1.3.6.1.2.1.3. See Table 2.6.

IP.   This group contains three tables and a number of scalar objects. The scalar objects listed in Table 2.7 are used to monitor the performance and error conditions of the IP layer and include such objects as total number of datagrams received, sent out, and forwarded; total number of datagrams in error; and total number of datagrams reassembled and fragmented at this network element. The three tables are listed here as Tables 2.7 to 2.9. The implementation of this group is mandatory, and its identifier is 1.3.6.1.2.1.4. The IP address table, shown in Table 2.8, has information on the IP addresses related to this network element. A network node may have multiple physical interfaces associated with it, and each interface is assigned an IP address. Each table entry (a row) contains IP address-related information for an interface: a unique interface index, the IP address of this interface, subnetting [a large local area network (LAN) can be divided into subnets for addressing flexibility; see App. B] indication, and broadcasting address from this interface. Note that access allowed on this table is read-only, meaning that it can be monitored but not controlled via the SNMP. This table is indexed by the IP address, i.e., the ipAdEntAddr field, as discussed earlier.

The IP routing table, shown in Table 2.9, contains all the information required to perform Internet routing. It includes all the routes known to this network element, one entry (row) per route. Examples of the information included in each entry are a routing destination, routing protocol, index of the interface through which datagrams are routed, and the age of a route. The table is indexed by the routing destination's IP address.

The IP address translation table of the IP object group, shown in Table 2.10, replaces the address translation table of MIB-I and contains information required to translate an IP address to a local interface-specific physical address.

TABLE 2.5   MIB-II Interface Table

Object name	Syntax	Access	ID	Description
ifIndex	INTEGER	RO	1	A unique value for each interface
ifDescr	DisplayString	RO	2	A textual description of the interface
ifType	INTEGER	RO	3	The type of interface based on the physical/link protocol immediately below the network layer, such as Fiber Distributed Data Interface (FDDI), Ethernet, tokenRing
ifMtu	INTEGER	RO	4	The size of the largest datagram that can be sent or received on the interface
ifSpeed	Gauge	RO	5	An estimate of the interface's current bandwidth in bits per second
ifPhysAddress	PhysAddress	RO	6	The interface's address at the protocol layer below the network layer
IfAdminStatus	INTEGER	RW	7	The desired state of the interface, including up, down, and testing state
IfOperStatus	INTEGER	RO	8	The current operational state of the interface, including up, down, and testing state
IfLastChange	TimeTicks	RO	9	The value of systemUpTime since the last initialization
IfInOctets	Counter	RO	10	The total number of octets received on the interface
ifInUcastPkts	Counter	RO	11	The number of subnetwork-unicast packets delivered to a higher-layer protocol
ifInNUcastPkts	Counter	RO	12	The number of nonunicast packets delivered to a higher-layer protocol
ifInDiscards	Counter	RO	13	The number of inbound packets which were chosen to be discarded for a reason such as freeing up buffer space
ifInErrors	Counter	RO	14	The number of inbound packets which contain errors preventing them from being delivered to a higher-layer protocol
IfInUnknown Protos	Counter	RO	15	The number of packets received via this interface that were discarded because of an unknown protocol
ifOutOctets	Counter	RO	16	The total number of octets transmitted out of the interface, including framing characteristics
ifOutUcastPkts	Counter	RO	17	The total number of packets that higher-level protocols requested to be transmitted to a subnetwork-unicast address
IfOutNUcast Pkts	Counter	RO	18	The total number of packets that higher-level protocols requested to be transmitted to a nonunicast address, including those discarded or not sent

TABLE 2.5 MIB-II Interface Table (Continued)

ifOutDiscards	Counter	RO	19	The number of outbound packets that were discarded for a reason such as freeing up buffer space
ifOutErrors	Counter	RO	20	The number of outbound packets that could not be sent due to errors
ifOutQLen	Gauge	RO	21	The length of the output packet queue
ifSpecific	OBJECT IDENTIFIER	RO	22	A reference to an MIB node specific to the particular medium being used for this interface

TABLE 2.6 MIB-II Address Translation Group Objects

Object name	Syntax	Access	ID	Description
atTable	SEQUENCE OF AtEntry	NA	1	Contains the network address to physical address equivalent
AtIfIndex	INTEGER	RW	1	The interface identified by this index on which the subnetwork is connected. This value is the same as ifIndex of the interface group.
AtPhysAddress	PhysAddress	RW	2	The local physical address of the subnetwork
AtNetAddress	NetworkAddress	RW	3	The IP address corresponding to the physical address

The Internet Control Message Protocol, as defined in RFC 792 (Postel 1981), is an IP layer protocol that provides a number of diagnostic and monitoring functions. The objects in this group, shown in Table 2.11, provide the information related to the performance and fault conditions of ICMP messages. Implementation of this group is mandatory, and its identifier is 1.3.6.1.2.1.5.

TCP. The Transmission Control Protocol is an Internet transport layer, connection-oriented, end-to-end protocol that is responsible for reliable, sequential, and unduplicated delivery of data bytes to the processes at a host. This object group, shown in Table 2.12, includes a number of scalar objects and a TCP connection table. The scalar objects provide the information related to transport layer performance and fault conditions such as the total number of connections supported, the maximum number of retransmissions permitted, and the number of connections in error states. The implementation of this group is mandatory, and its identifier is 1.3.6.1.2.1.6.

The TCP connection table, shown in Table 2.13, contains all the TCP connections this network element currently maintains, one entry per connection. Within each entry are both remote and local IP addresses of this

TABLE 2.7   MIB-II IP Group Objects

Object name	Syntax	Access	ID	Description
ipForwarding	INTEGER	RW	1	Indicates whether this entity is acting as an IP gateway to forward datagrams received by, but not addressed to, this entity. The value 1 for YES, and 0 for NO.
ipDefaultTTL	INTEGER	RW	2	The default value inserted into the Time-to-live field of the IP header of datagrams originated from this entity, if this value is not supplied by the transport layer protocol
ipInReceives	Counter	RO	3	The total number of input datagrams received at this entity
ipInHdrErrors	Counter	RO	4	The number of input datagrams discarded due to error in their IP headers
ipAddrErrors	Counter	RO	5	The number of input datagrams discarded due to an invalid address in the IP header
ipForwDatagrams	Counter	RO	6	The number of input datagrams that are not addressed to this entity as a final destination
ipInUnknown Protos	Counter	RO	7	The number of locally addressed datagrams that are discarded because of unknown protocol
ipInDiscards	Counter	RO	8	The number of input datagrams discarded for reasons other than error, such as lack of buffer space
ipInDelivers	Counter	RO	9	The total number of input datagrams successfully delivered to IP user protocols
ipOutRequests	Counter	RO	10	The total number of IP datagrams which local IP user protocols supplied to IP to be transmitted
ipOutDiscards	Counter	RO	11	The number of output IP datagrams which were discarded for reasons other than error, such as lack of buffer space
ipOutNoRoutes	Counter	RO	12	The number of IP datagrams discarded because no route could be found for transmission to their destination
ipReasmTimeout	INTEGER	RO	13	The maximum number of seconds received fragments are held while they are awaiting reassembly at this entity
ipReasmReqds	Counter	RO	14	The number of IP fragments received at this entity that need reassembly
ipReasmOKs	Counter	RO	15	The number of datagrams successfully reassembled
ipReasmFails	Counter	RO	16	The number of failures detected by the IP reassembly algorithm
ipFragOKs	Counter	RO	17	The number of IP datagrams that have been successfully fragmented at this entity
ipFragFails	Counter	RO	18	The number of IP datagrams that have been discarded because of failure of fragmentation at this entity
ipFragCreates	Counter	RO	19	The total number of IP datagram fragments created at this entity

TABLE 2.7 **MIB-II IP Group Objects (Continued)**

ipAddrTable	SEQUENCE OF IpAddrEntry	NA	20	The table of addressing information for this entity's IP address
ipRouteTable	SEQUENCE OF IpRouteEntry	NA	21	IP routing table with all information needed to route packets to the next hop
ipNetToMedia Table	SEQUENCE OF IpNetToMedia Entry	NA	22	IP address translation table with all the necessary information to translate an IpAddress to a local media-dependent physical address
ipRouting Discards	Counter	RO	23	The number of routing entries chosen to be discarded for reasons other than error, such as lack of buffer space

TABLE 2.8 **MIB-II IP Address Table**

Object name	Syntax	Access	ID	Description
ipAdEntAddr	IpAddress	RO	1	The IP address for this interface
ipAdEntIfIndex	INTEGER	RO	2	The unique index identifying this interface and its value matches that of ifIndex in the interface table.
ipAdEntNetMask	IpAddress	RO	3	The subnet mask associated with the IP address of this entry
ipAdEntBcastAddr	INTEGER	RO	4	The value of the least-significant bit in the IP broadcast address used for sending datagrams on the interface associated with the IP address of this entry
ipAdEntReasm MaxSize	INTEGER	RO	5	The largest IP datagrams this entity can reassemble from incoming datagram fragments

connection, the connection state, and the local and remote port numbers of the connection.

UDP. The User Datagram Protocol is another Internet transport layer protocol that is responsible for delivery of potentially unreliable and unsequenced datagrams to local or remoter users. The objects of this group, shown in Table 2.14, provide generic information on the UDP performance and the fault conditions. This group has a listener table with two columns, shown in Table 2.15, providing information on the local endpoint on which an application is receiving datagrams. Note that the access to this object group is RO, indicating it can only be monitored, not controlled. Implementation of this group is mandatory and its identifier is 1.3.6.1.2.1.7.

EGP. The Exterior Gateway Protocol, as defined in RFC 904 (Mills 1984), is an Internet protocol for exchanging routing information between two autonomous

TABLE 2.9   SNMP MIB-II IP Routing Table

Object name	Syntax	Access	ID	Description
ipRouteDest	IpAddress	RW	1	The destination IP address of this route. An entry with a value of 0.0.0.0 is considered a default route.
ipRouteIfIndex	INTEGER	RW	2	The unique index value for the local interface through which the next hop of this route should be reached
ipRouteMetric1	INTEGER	RW	3	The primary routing metric for this route. The semantics of this metric are determined by the routing protocol specified in the ipRouteProto value.
ipRouteMetric2	INTEGER	RW	4	An alternative routing metric for this route. Again, the semantics of this metric are determined by the routing protocol.
ipRouteMetric3	INTEGER	RW	5	An alternative routing metric for this route
ipRouteMetric4	INTEGER	RW	6	An alternative routing metric for this route
ipRouteNextHop	IpAddress	RW	7	The IP address of the next hop of this route
ipRouteType	INTEGER	RW	8	The type of route: 1 = other, 2 = invalid, 3 = direct, 4 = indirect.
ipRouteProto	INTEGER	RO	9	The routing protocol: 1 = other, 2 = local, 3 = netmgmt, 4 = icmp, 5 = egp, 6 = ggp, 7 = hello, 8 = rip, 9 = is-is, 10 = es-is, 11 = ciscoIgrp, 12 = bbnSpfIgp, 13 = ospf, and 14 = bgp.
ipRouteAge	INTEGER	RW	10	The number of seconds since this route was last updated or checked
ipRouteMask	IpAddress	RW	11	The mask to be logical-ANDed with the destination address before being compared to the value in the ipRouteDes field
ipRouteMetric5	INTEGER	RW	12	An alternate routing metric for this route. The semantics of this metric are determined by the routing protocol. Its value should be set to −1 if it is not used.
ipRouteInfo	OBJECT IDENTIFIER	RO	13	An object identifier in the MIB for the routing protocol that is used for this route

systems. This object group, shown in Table 2.16, provides information on the performance and fault conditions of the EGP entity at this network element. The EGP neighbor table, shown in Table 2.17, furnishes information on the neighboring EGP entities. The implementation of this object group is mandatory, and its identifier is 1.3.6.1.2.1.8.

CMOT. The efforts on the development of CMIP on TCP/IP significantly diminished once SNMP became widely accepted in the industry. This object group has been relegated to historical status and thus is omitted from our discussion.

Transmission.   This group is intended for objects on the transmission media underlying each physical interface on a network element. Initially they will be put under an experimental subtree and then moved into this group as the objects become mature. Currently, no objects of this group are defined for MIB-II.

SNMP.   This group defines objects for support of an SNMP management entity. Depending on the role the management entity takes on, either manager or agent, not all objects are meaningful for any given management entity and some objects are zero-valued. The objects in this group, shown in Table 2.18, are highly SNMP-specific, and the reader is encouraged to revisit this part after reading the discussion of SNMPv1 in the following section. The implementation of this group is mandatory, and its identifier is 1.3.6.1.2.1.11.

An MIB application example

Now let's revisit the management scenario presented in the opening section and depicted in Fig. 2.2. One of the first things in building a management system is building an MIB. What are the major steps involved in developing an MIB? The following list is intended to be an illustration of a generic guideline rather than a step-by-step recipe.

1. First, examine and determine whether the standard MIB, such as the MIB-II we just discussed, is sufficient for the management requirement. Many of the MIB-II object groups, such as system, interface, SNMP, IP, and TCP, can be readily used. However, some management tasks specific to a network cannot be done using the standard MIB alone. For example, no objects are defined for monitoring and controlling printers.

TABLE 2.10   MIB-II IP Address Translation Table

Object name	Syntax	Access	ID	Description
ipNetToMediaIfIndex	INTEGER	RW	1	The interface on which this entry's corresponding physical address is connected
ipNetToMediaPhysAddress	PhysAddress	RW	2	The local media-dependent physical address
ipNetToMediaNetAddress	IpAddress	RW	3	The IpAddress corresponding to the media-dependent physical address
ipNetToMediaType	INTEGER	RW	4	The type of mapping between the physical address and IpAddress: 1 = other, 2 = invalid, 3 = dynamic, and 4 = static.

TABLE 2.11   SNMP MIB-II ICMP Group Objects

Object name	Syntax	Access	ID	Description
icmpInMsgs	Counter	RO	1	The total number of ICMP messages received at this network entity
icmpInErrors	Counter	RO	2	The number of ICMP messages received at this entity that have errors
icmpInDestUnreachs	Counter	RO	3	The number of ICMP Destination Unreachable messages received
icmpInTimeExcds	Counter	RO	4	The number of ICMP Time Exceeded messages received
icmpInParmProbs	Counter	RO	5	The number of ICMP Parameter Problem messages received
icmpInSrcQuenchs	Counter	RO	6	The number of ICMP Source Quench messages received
icmpInRedirects	Counter	RO	7	The number of ICMP Redirect messages received
icmpInEchos	Counter	RO	8	The number of ICMP Echo request messages received
icmpInEchoReps	Counter	RO	9	The number of ICMP Echo Reply messages received
icmpInTimestamps	Counter	RO	10	The number of ICMP Timestamp request messages received
icmpInTimestampReps	Counter	RO	11	The number of ICMP Timestamp Reply messages received
icmpInAddrMasks	Counter	RO	12	The number of ICMP Address Mask Request messages received
icmpInAddrMaskReps	Counter	RO	13	The number of ICMP Address Mask Reply messages received
icmpOutMsgs	Counter	RO	14	The total number of ICMP messages this entity attempted to send
icmpOutErrors	Counter	RO	15	The number of ICMP messages which this entity did not send due to problems within the ICMP
icmpOutDestUnreachs	Counter	RO	16	The number of ICMP Destination Unreachable messages sent
icmpOutTimeExcds	Counter	RO	17	The number of ICMP Time Exceeded messages sent
icmpOutParmProbs	Counter	RO	18	The number of ICMP Parameter Problem messages sent
icmpOutSrcQuenchs	Counter	RO	19	The number of ICMP Source Quench messages sent
icmpOutRedirects	Counter	RO	20	The number of ICMP Redirect messages sent
icmpOutEchos	Counter	RO	21	The number of ICMP Echo request messages sent

TABLE 2.11   SNMP MIB-II ICMP Group Objects (Continued)

icmpOutEchoReps	Counter	RO	22	The number of ICMP Echo Reply messages sent
icmpOutTimestamps	Counter	RO	23	The number of ICMP Timestamp Request messages sent
icmpOutTimestampReps	Counter	RO	24	The number of ICMP Timestamp Reply messages sent
icmpOutAddrMasks	Counter	RO	25	The number of ICMP Address Mask Request messages sent
icmpOutAddrMaskReps	Counter	RO	26	The number of ICMP Address Mask Reply messages sent

2. The next step is to obtain an extension to the standard MIB. It is likely that what you are looking for has already been defined by someone in the network management community, and all it takes is a few Internet searches. However, for the sake of illustration, let's assume that the private portion of the MIB has to be developed. Following is a list of items you might want to consider in developing an MIB in-house.

- Determine the naming scheme for the new objects. The registration hierarchy is a very flexible scheme for accommodating the private MIB extensions. In Fig. 2.3, a *private* subtree is defined under the Internet and under the private subtree; a branch called *enterprise* can be a logical place to derive the object identifiers for your private MIB extension. This is one way to solve the issue of naming the newly created objects.
- Choose a format for your MIB extension. There is more than one format available. For example, you can either choose the original SNMP SMI as defined in RFC 1155, the newer concise MIB definition as defined in RCF 1212, or SMI-II for SNMPv2. The decision may depend on a variety of factors ranging from the vendor of the development tool kit of your choice to the plan for the future extension.
- Determine the syntax of the object you intend to create and the access allowed to each object. The type of access will depend on the type of operation desired, either to monitor or control a particular aspect of the network element under consideration.
- Write ASN.1 MIB definitions.

3. Compile the ASN.1 definitions using an MIB compiler and create an MIB library.

An MIB is like a language; both the agent and manager must know the vocabulary in order to communicate. The private MIB structure must be loaded into the agent system as well as the manager system for them to understand each other. As Stallings pointed out, even if your MIB is standard-compliant, some twiddling may be required before it can be compiled with a vendor's compiler (Stallings 1992).

TABLE 2.12   SNMP MIB-II TCP Group Objects

Object name	Syntax	Access	ID	Description
tcpRtoAlgorithm	INTEGER	RO	1	The algorithm used to determine the time-out value for retransmitting unacknowledged octets: other = 1, constant = 2, rsre = 3, and vanj = 4.
tcpRtoMin	INTEGER	RO	2	The minimum value permitted by a TCP implementation for the retransmission time-out, in milliseconds
tcpRtoMax	INTEGER	RO	3	The maximum value permitted by a TCP implementation for the retransmission time-out, in milliseconds
tcpMaxConn	INTEGER	RO	4	The limit on the total number of TCP connections supported by the entity
tcpActiveOpens	Counter	RO	5	The number of times TCP connections have made a transition to the SYN-SENT state from the CLOSED state
tcpPassiveOpens	Counter	RO	6	The number of times TCP connections have made a transition to the SYN-RCVD state from the LISTEN state
tcpAttemptFails	Counter	RO	7	The number of times TCP connections have made a transition to the CLOSED state from either the SYN-SENT state or the SYN-RCVD state
tcpEstabResets	Counter	RO	8	The number of times TCP connections have made a transition to the CLOSE state from either the ESTABLISHED state or the CLOSE-WAIT state
tcpCurrEstab	Gauge	RO	9	The number of TCP connections that are currently in either the ESTABLISHED or the CLOSE-WAIT state
tcpInSegs	Counter	RO	10	The total number of segments received, including those with errors
tcpOutSegs	Counter	RO	11	The number of segments sent, including those on current connections but excluding those containing retransmitted octets only
tcpRetransSeg	Counter	RO	12	The total number of segments retransmitted, which is the number of TCP segments containing one or more previously transmitted octets
tcpConnTable	SEQUENCE OF TCPConn Entry	NA	13	A table containing TCP connection information
tcpInErrs	Counter	RO	14	The total number of segments received in error
tcpOutRsts	Counter	RO	15	The number of TCP segments sent containing the RST (ReSeT) flag

TABLE 2.13   SNMP MIB-II TCP Connection Table Fields

Object name	Syntax	Access	ID	Description
tcpConnState	INTEGER	RW	1	The state of this TCP connection. The states are closed = 1, listen = 2, synSent = 3, synReceived = 4, established = 5, finWait1 = 6, finWait2 = 7, closeWait = 8, laskAck = 9, closing = 10, timeWait = 11, and deleteTCP = 12.
TcpConnLocalAddress	IpAddress	RO	2	The local IP address for this TCP connection
TcpLocalPort	INTEGER (0..65535)	RO	3	The local port number for this TCP connection
TcpConnRemAddress	IpAddress	RO	4	The remote IP address for this TCP connection
TcpConnRemPort	INTEGER (0..65535)	RO	5	The remote port number for this TCP connection

TABLE 2.14   SNMP MIB-II UDP Group Objects

Object name	Syntax	Access	ID	Description
udpInDatagrams	Counter	RO	1	The total number of the UDP datagrams delivered to the UDP users like an SNMP entity
udpNoPorts	Counter	RO	2	The total number of the UDP datagrams received which don't have an application at the destination port
udpInErrors	Counter	RO	3	The total number of received UDP datagrams that could not be delivered for reasons other than the lack of application at the destination port
udpOutDatagrams	Counter	RO	4	The total number of datagrams sent from this UDP entity
udpTable	SEQUENCE OF UdpEntry	NA	5	Contains information about this entity's UDP endpoint on which an application is currently accepting datagrams

SNMPv1

The Internet network management protocol SNMP, as defined in RFC 1157 (Case et al., 1990), specifies the types of management operations that can be performed on an MIB, and the structure and semantics of five protocol data units (PDUs). The discussion of the SNMP consists of three parts. First, an introduction provides basic background information and gets some terminology out of the way. Second, a protocol specification details the five PDUs in terms

TABLE 2.15   SNMP MIB-II UDP Listener Table

Object name	Syntax	Access	ID	Description
udpLocalAddress	IpAddress	RO	1	The local IP address for this UDP listener
udpLocalPort	INTEGER (0..65535)	RO	2	The local port number for this UDP listener

TABLE 2.16   SNMP MIB-II EGP Group Objects

Object name	Syntax	Access	ID	Description
egpInMsgs	Counter	RO	1	The number of EGP messages received without error
egpInErrors	Counter	RO	2	The number of EGP messages received in error
egpOutMsgs	Counter	RO	3	The number of locally generated EGP messages
egpOutErrors	Counter	RO	4	The number of locally generated EGP messages that are not sent because of resource limitations within an EGP
egpNeighTable	EgpTableEntry	NA	5	Contains information about this entity's EGP neighbors
egpAs	INTEGER	RO	6	The autonomous system number of this EGP entity

of their structure and semantics. Finally an application of the SNMP to the management scenario illustrated in Fig. 2.2 is presented.

Introduction to SNMPv1

First we give a brief description of the SNMP, an application layer protocol, to put it in the context of Internet network architecture. This is followed by an introduction on how object instances are identified, a key part of the SNMP message specification. Then some basic concepts essential for understanding SNMP standards and literature are explained.

SNMP in Internet protocol stack.   The SNMP as an application layer protocol depends on the services provided by the protocols of lower layers to accomplish management information exchange between two management entities; these relationships are shown in Fig. 2.8. The primary responsibilities of the SNMP include generating SNMP PDU messages, encoding them using ASN.1 basic encoding rules (BER), and sending them to the transport layer.

At the transport layer is the UDP, which is responsible for adding header information to correctly identify the communicating application at the transmitting end before sending it down to the network layer. At the receiving end, the UDP is responsible for reassembling the received packets into SNMP packets before passing them on to the receiving application at the application layer. Only minimum error checking is available, and packets may get lost, out of sequence, or duplicated. At the network layer is the IP that is responsible for routing the SNMP packets, masking the difference between underlying networks, and fragmenting the packets if necessary for transmission over the network. Although the SNMP is often used on top of the UDP, it is also possible to combine it with other transport layer protocols such as the TCP.

TABLE 2.17   SNMP MIB-II EGP Neighbor Table

Object name	Syntax	Access	ID	Description
egpNeighState	INTEGER	RO	1	The EGP state of this entity's EGP neighbor with the following definitions: idel=1, acquisition=2, down=3, up=4, and cease=5
egpNeighAddr	IpAddress	RO	2	The IP address of this entity's neighbor
egpNeighAs	INTEGER	RO	3	The autonomous system of this EGP neighbor
egpNeighInMsgs	Counter	RO	4	The number of EGP messages received from this EGP neighbor without error
egpNeighInErrs	Counter	RO	5	The number of EGP messages received from this EGP neighbor in error
egpNeighOutMsgs	Counter	RO	6	The number of EGP messages sent to this EGP neighbor
egpNeighOutErrs	Counter	RO	7	The number of EGP messages generated but not sent to this EGP neighbor due to resource limitations
egpNeighInErrMsgs	Counter	RO	8	The number of EGP-defined error messages received from this EGP neighbor
egpNeighOutErrMsgs	Counter	RO	9	The number of EGP-defined error messages sent to this EGP neighbor
egpNeighStateUps	Counter	RO	10	The number of EGP state transitions to the UP state with this EGP neighbor
egpNeighStateDowns	Counter	RO	11	The number of EGP state transitions from the UP state to any other state with this EGP neighbor
egpNeighIntervalHello	INTEGER	RO	12	The interval between EGP Hello command retransmissions in hundredths of a second
egpNeighIntervalPoll	INTEGER	RO	13	The interval between EGP poll command retransmissions in hundredths of a second
egpNeighMode	INTEGER	RO	14	The polling mode of this EGP entity, either active (1) or passive (2)
egpNeighEventTrigger	INTEGER	RW	15	A control variable used to trigger operator-initiated start and stop events

TABLE 2.18   MIB-II SNMP Group Objects

Object name	Syntax	Access	ID	Description
snmpInPkts	Counter	RO	1	The total number of SNMP messages delivered to the SNMP protocol entity from the transport service
snmpOutPkts	Counter	RO	2	The total number of SNMP messages delivered to the transport service from the SNMP protocol entity
snmpInBadVersion	Counter	RO	3	The total number of SNMP messages delivered to the protocol entity with an unsupported SNMP version
snmpInBad CommunityName	Counter	RO	4	The total number of SNMP messages delivered to the protocol entity with an unknown community name
snmpInBad CommunityUse	Counter	RO	5	The total number of SNMP messages delivered to the protocol entity with an SNMP operation not allowed by the community
snmpInASNParseErrs	Counter	RO	6	The total number of ASN.1 or basic encoding rule errors encountered by the protocol entity when decoding an SNMP message
snmpInTooBig	Counter	RO	8	The total number of SNMP PDUs received by the protocol entity with error-status = tooBig
snmpInNoSuchNames	Counter	RO	9	The total number of SNMP PDUs received by the protocol entity with error-status = noSuchName
snmpInBadValues	Counter	RO	10	The total number of SNMP PDUs received by the protocol entity with error-status = badValue
snmpInReadOnlys	Counter	RO	11	The total number of SNMP PDUs received by the protocol entity with error-status = readOnly
snmpInGenErrs	Counter	RO	12	The total number of SNMP PDUs received by the protocol entity with error-status = genErr
snmpInTotalReqVars	Counter	RO	13	The total number of MIB objects retrieved by an agent in response to GetRequest and GetNextRequest PDUs
snmpInTotalSetVars	Counter	RO	14	The total number of MIB objects set by an agent in response to SetRequest PDUs
snmpInGetRequests	Counter	RO	15	The total number of SNMP GetRequest PDUs accepted and processed by the SNMP agent
snmpInGetNexts	Counter	RO	16	The total number of SNMP GetNext Request PDUs accepted and processed by the SNMP agent

TABLE 2.18   MIB-II SNMP Group Objects (Continued)

snmpInSetRequests	Counter	RO	17	The total number of SNMP GetNext PDUs accepted and processed by the SNMP agent
snmpInGetResponse	Counter	RO	18	The total number of SNMP GetResponse PDUs accepted and processed by the SNMP agent
snmpInTraps	Counter	RO	19	The total number of SNMP Trap PDUs accepted and processed by the SNMP agent
snmpOutTooBigs	Counter	RO	20	The total number of SNMP PDUs generated by the protocol entity with error-status = tooBig
snmpOutNoSuch Names	Counter	RO	21	The total number of SNMP PDUs generated by the protocol entity with error-status = noSuchName
snmpOutBadValues	Counter	RO	22	The total number of SNMP PDUs generated by the protocol entity with error-status = badValue
snmpOutGetErrs	Counter	RO	24	The total number of SNMP PDUs generated by the protocol entity with error-status = genErr
snmpOutGetRequests	Counter	RO	25	The total number of SNMP GetRequest PDUs generated by the protocol entity
snmpOutGetNexts	Counter	RO	26	The total number of SNMP GetNext PDUs generated by the protocol entity
snmpOutSetRequests	Counter	RO	27	The total number of SNMP SetRequest PDUs generated by the protocol entity
snmpOutGetResponse	Counter	RO	28	The total number of SNMP GetResponse PDUs generated by the protocol entity
snmpOutTraps	Counter	RO	29	The total number of SNMP Trap PDUs generated by the protocol entity
snmpEnableAuten Traps	INTEGER	RO	30	Indicates whether the SNMP agent process is permitted to generate authentication-failure traps with enabled = 1 and disabled = 2. The value of this object overrides any other setting.

Object instance identification. The MIB-II SMI defines the structure and syntax of the MIB, and the standard MIB-II object groups specify what is in the MIB. It is the instances of those objects in the MIB that the SNMP PDUs use to specify management operations. The following subsection focuses on how the instances of both scalar and columnar objects are identified.

The ultimate purpose of defining an MIB is to let applications access the objects in the MIB. To do so, a way to uniquely identify each instance in the MIB

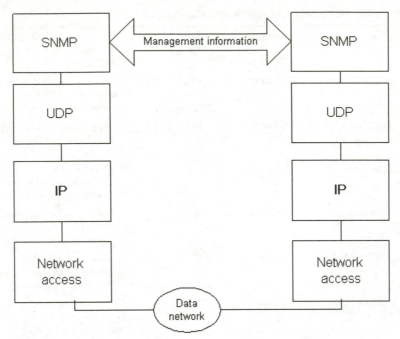

Figure 2.8 SNMP and the Internet protocol stack.

at run time is needed. The naming scheme using the registration tree discussed a while ago provides a basis for identifying each instance. Note that the object syntax described above provides a means to define an object template, not an instance. A template has to be instantiated to create an instance, and both the template and its instance need to be uniquely identified. The detailed syntax for accessing the object instances is part of the SNMP message.

Scalar object instance identification. The instance identification of a scalar object is a straightforward extension of its template identification: a 0 is appended to the template identifier derived from the registration. This is possible because there is only one instance per scalar object. Table 2.19 shows the instance identifier of the UDP group objects.

When a manager sends a protocol message asking for information on an object in the MIB, the object template itself is not accessible by the application. Only the instance of the object is accessible. The object instances can be viewed as the leaf nodes of the registration tree.

Columnar object instance identification. Columnar or tabular object instance identification is a little more complicated. As in any two-dimensional table, an element in an MIB table is identified by the combination of a row index and a column index. The column index is simply the columnar object identifier, formed by appending its relative identifier to the absolute identifier of its parent node. Refer to Fig. 2.9 for an example IP address table tree. The identifier of the second-column object, ipAdEntIfIndex, is 1.2.6.1.2.1.4.20.1.2; it is made up of the identifier of its parent node (ipAddrEntry) table entry, which is (1.2.6.1.2.1.4.20.1), appended with the relative identifier of this columnar object, which is a 2.

Next we need to figure out the row index in order to uniquely identify an entry in a table. For MIB-II, the row index is determined by the INDEX clause of the object template, as described in the SMI section. For the IP address table, it is the value of the first-column object, ipAdEntrAddr, the IP address of a physical interface that uniquely distinguishes one row from another.

We use examples from Table 2.20 to illustrate MIB-II table indexing. For example, 1.3.6.1.2.1.4.20.1.2.10.11.12.14 refers to the entry in the second column and the second row. The column index is 1.3.6.1.2.1.4.20.1.2, which is the second-column object identifier, and the row index is the value of ipAdEntAddr at the second row, 10.11.12.14.

Basic concepts.   Just like in any other field, there are a few specialized terms and concepts used throughout the SNMP standards and literature whose clarification will greatly aid the understanding of the SNMP.

Management, application, and protocol entity.   A *management entity* refers to that portion of a management system that can independently initiate a management operation and respond to a management operation request. Physically a management entity may or may not be an independent entity. A concrete example of such an entity is a manager or agent. The term *management application entity,* or simply *application entity,* is also used more or less to refer to the same thing.

TABLE 2.19   Scalar Object Instance Identification Example

Object name	Object identifier	Object instance identifier
udp	1.3.6.1.2.1.7	Table itself can't be instantiated
udpInDatagram	1.3.6.1.2.1.7.1	1.3.6.1.2.1.7.1.0
udpNoPorts	1.3.6.1.2.1.7.2	1.3.6.1.2.1.7.2.0
udpInError	1.3.6.1.2.1.7.3	1.3.6.1.2.1.7.3.0
udpOutDatagrams	1.3.6.1.2.1.7.4	1.3.6.1.2.1.7.4.0

TABLE 2.20   An Example IP Address Table

ipAdEntAddr	ipAdEntIfIndex	IpAdEntNetmask	ipAdEntBcastAddr	ipAdEntReasm MaxSize
10.11.12.13	4	0.0.0.0	1	1024
10.11.12.14	5	255.255.255.255	1	512
10.11.99.57	3	1.1.1.1	1	512
10.12.95.102	2	33.34.35.36	1	1024

A *protocol entity* is the part of a management entity that is responsible for handling protocol-related tasks such as receiving and sending a PDU, interpreting the semantics of a PDU, and maintaining peer-to-peer relationships between protocol entities.

Lexicographical order. The lexicographical order is a very important concept for operations on MIB tables. If we view the MIB as a tree and require that the child nodes of a parent node follow the ascending numerical order, then the lexicographical ordering is what you get when you do a depth-first search on the tree. You visit the root first and then traverse the subtrees from left to right recursively. We use the IP address table in Table 2.20 to build the partial MIB tree shown in Fig. 2.9; the lexicographical ordering of the tree is shown in Table 2.21.

SNMP community. The community concept provides a basic security mechanism for SNMPv1 management operations. The basic idea is this: every management entity wishing to communicate with an agent in order to gain access to the agent's MIB has to be in the agent's community. Otherwise, the management protocol messages from this unauthorized management entity will be rejected. A community is a pairing of an SNMP agent with a set of SNMP application entities (e.g., a set of managers). A unique community name is defined from an agent's perspective, and those managers who know the community name are said to be in this community and therefore authorized to communicate with the agent. As shown in Fig. 2.10, an agent communicates with only those in the community, and a community is associated with an MIB view, which is explained next.

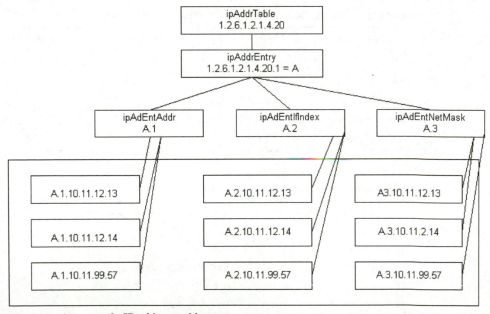

Figure 2.9 An example IP address table tree.

TABLE 2.21   Lexicographical Ordering of an Abbreviated IP Address Table

Object	Object identifier	Object instance of lexicographical successor
ipAddrTable	1.3.6.1.2.1.4.20	1.3.6.1.2.1.4.20.1.1.10.11.12.13
ipAddrEntry	1.3.6.1.2.1.4.20.1	1.3.6.1.2.1.4.20.1.1.10.11.12.13
ipAdEntAddr	1.3.6.1.2.1.4.20.1.1	1.3.6.1.2.1.4.20.1.1.10.11.12.13
ipAdEntAddr.10.11.12.13	1.3.6.1.2.1.4.20.1.1.10.11.12.13	1.3.6.1.2.1. 4.20.1.1.10.11.12.14
ipAdEntAddr.10.11.12.14	1.3.6.1.2.1.4.20.1.1.10.11.12.14	1.3.6.1.2.1. 4.20.1.1.10.11.99.57
ipAdEntAddr.10.11.99.57	1.3.6.1.2.1.4.20.1.1. 10.11.99.57	1.3.6.1.2.1.4.20.1.2.10.11.12.13
ipAdEntIfIndex	1.3.6.1.2.1.4.20.1.2	1.3.6.1.2.1.4.20.1.2.10.11.12.13
ipAdEntIfIndex.10.11.12.13	1.3.6.1.2.1.4.20.1.2. 10.11.12.13	1.3.6.1.2.1.4.20.1.2.10.11.12.14
ipAdEntIfIndex.10.11.12.14	1.3.6.1.2.1.4.20.1.2. 10.11.12.14	1.3.6.1.2.1.4.20.1.2.10.11.99.57
ipAdEntIfIndex.10.11.99.57	1.3.6.1.2.1.4.20.1.2. 10.11.99.57	1.3.6.1.2.1.4.20.1.3.10.11.12.13
ipAdEntNetMask	1.3.6.1.2.1.4.20.1.3	1.3.6.1.2.1.4.20.1.3.10.11.12.13
ipAdEntNetMask.10.11.12.13	1.3.6.1.2.1.4.20.1.3. 10.11.12.13	1.3.6.1.2.1.4.20.1.3.10.11.12.14
ipAdEntNetMask.10.11.12.14	1.3.6.1.2.1.4.20.1.3. 10.11.12.14	1.3.6.1.2.1.4.20.1.3.10.11.99.57
ipAdEntNetMask. 10.11.99.57	1.3.6.1.2.1.4.20.1.3. 10.11.99.57	NULL

MIB view.   A subset of objects in the MIB that is visible to a management entity is called an SNMP MIB view. The MIB view reflects the administrative domain of a management entity. For example, an agent for a host computer is only concerned about the objects representing the attributes of the computer, and these objects constitute the MIB view of this agent. As shown in Fig. 2.10, the MIB view of the agent represents a single managed network element. The objects in an MIB view need not belong to a single subtree of the object name space.

An SNMP access mode can be read-only or read-write. A pairing of an SNMP access mode with an SNMP MIB view is called an SNMP community profile, which represents specified access privileges to objects in a specified MIB view. A pairing of an SNMP community with an SNMP community profile is called an SNMP access policy. Each SNMP community has an access policy that specifies access privileges allowed to the MIB view by all community members wishing to access the MIB view. If a management entity violates the access policy by requesting an unauthorized type of access, the access will be rejected by the agent.

The concept of proxy and proxy agent.   A different type of MIB access issue arises when the access policy defined on an MIB view is different from the access policy allowed on the network element that the agent is supposed to manage. This is where the concept of proxy and proxy agent comes in. As shown in Fig. 2.11, the MIB view is defined based on the standard SNMP MIB structure and the management system (a community member) knows the access policy

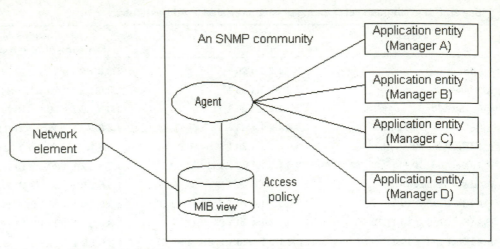

Figure 2.10   Illustration of community and MIB view concept.

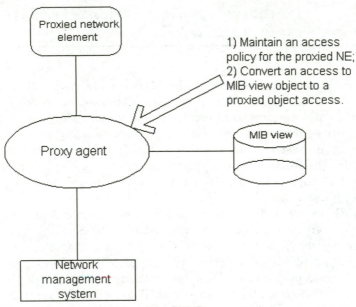

Figure 2.11   An illustration of the SNMP proxy service.

on the MIB view. But the managed network element does not support the SNMP MIB, and the set of accesses it will accept may be different from what the MIB view access policy stipulates. Therefore, an access request by the management system has to be mediated by the agent or by the proxy agent that translates the requested access to a type of access that is understood and allowed by the network element.

SNMP specification

This section discusses the details of the SNMP, i.e., the management operations provided by the SNMP, the procedure of exchanging PDUs between two peer management entities, and the structure and semantics of the five SNMPv1 PDUs. SNMPv1 is primarily defined in RFC 1157 (Case et al. 1990) on which this section is based.

Protocol message exchange procedure.   We start out by looking at the process of generating SNMP PDUs. The actions of the protocol entity are as follows:

1. The protocol entity constructs the appropriate PDU, e.g., the GetRequest PDU as an ASN.1 object.
2. It passes the constructed ASN.1 object along with a community name and the source and destination transport addresses to the part of the management entity that implements the desired authentication scheme. The entity returns another ASN.1 object.
3. The protocol entity then constructs an ASN.1 message object, using the community name and the ASN.1 object.
4. It then serializes the new ASN.1 object, using the basic encoding rules of ASN.1, and sends it to the transport layer to be transmitted over to the peer protocol entity at the receiving end.

The actions of the protocol entity at the receiving end are as follows:

1. The protocol entity performs a preliminary parse on the incoming datagrams to build an ASN.1 message object. If the parse fails, it discards the datagram and performs no further actions.
2. Otherwise, it verifies the version number of the SNMP message. If a mismatch is detected, it discards the datagram and stops further processing.
3. Otherwise, the protocol entity passes the community name and the user data found in the ASN.1 message object, along with the datagram's source and destination transport addresses, to the part of the management entity responsible for authentication.
4. It then parses the authenticated ASN.1 objects and builds an ASN.1 PDU object. The application entity picks out the community name, obtains the associated access policy, and carries out the management operation stated in the received SNMP PDU.

The process is depicted in Fig. 2.12.

Generic PDU constructs.   Each SNMP message has the same generic structure as shown in Table 2.22a with three fields: a version number, a community name, and one of the five SNMP PDUs. The version number indicates the version of SNMP protocol, mainly for compatibility reasons. For example, one

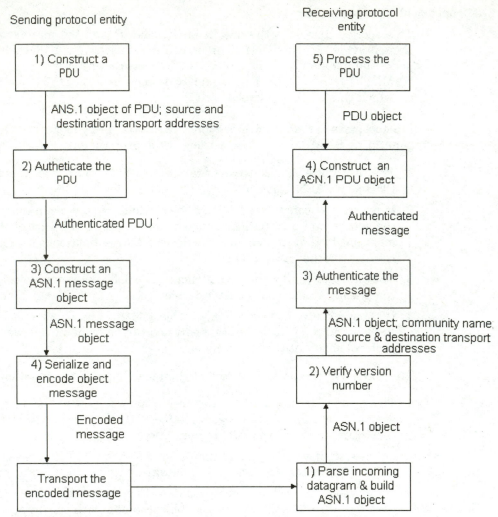

Sending protocol entity

Receiving protocol entity

1) Construct a PDU

ANS.1 object of PDU; source and destination transport addresses

2) Autheticate the PDU

Authenticated PDU

3) Construct an ASN.1 message object

ASN.1 message object

4) Serialize and encode object message

Encoded message

Transport the encoded message

5) Process the PDU

PDU object

4) Construct an ASN.1 PDU object

Authenticated message

3) Authenticate the message

ASN.1 object; community name; source & destination transport addresses

2) Verify version number

ASN.1 object

1) Parse incoming datagram & build ASN.1 object

Figure 2.12   The procedure of implementing SNMP PDUs.

of the widely used versions is SNMPv2. A community name is an identifier defining all legitimate management entities that are allowed to communicate with a particular agent. The first two fields of an SNMP message, the version and community name, are also called the wrapper or header of an SNMP message.

There are five SNMP PDUs: GetRequest, GetNextRequest, SetRequest, GetResponse, and Trap. The first three are request PDUs, which are used by a manager to request an agent to perform certain management operations. Out of the first three, the first two PDUs, GetRequest and GetNextRequest, are used for requesting the values of a set of named object instances in the agent's MIB view. The third PDU, SetRequest, is used to request that a set operation be performed on the values of a set of named object instances in the agent's MIB view. The fourth PDU, GetResponse, is for an agent to respond to any of the above three request PDUs. The last one, Trap PDU, is for an agent to report an event unsolicited by the manager.

The five PDUs can be put in two different categories with respect to the PDU format. The first four share the same general format, as shown in Table 2.22b and c, and the corresponding ASN.1 definitions are provided in Fig. 2.13. The second category of SNMP PDUs, the Trap PDU, is discussed later.

Each field in the structure is explained briefly.

- *PDU-type.* An integer identifying the type of an SNMP PDU, i.e., GetRequest, GetNextRequest, SetRequest, GetResponse, or Trap.

- *request-id.* An integer to uniquely identify a request (i.e., GetRequest, GetNextRequest, or SetRequest) sent from a manager to an agent.

- *variable-bindings.* A set of object-name–object-value pairs. An object name is an object instance identifier, and the value is the object instance value.

- *error-status.* An integer to indicate an error condition. Its value is assigned by a management entity, normally an agent, in response to a request to indicate one of the following conditions: noError (0)—no error is present; tooBig (1)—the protocol message generated locally exceeds the maximum size allowed on the local host computer; noSuchName (2)—at least one of the object names requested by the manager does not exist in the agent's MIB view; badValue (3)—the type of value found in the agent's MIB view does not match that requested by the manager; readOnly (4)—a write request is made by a manager on an object whose access is read-only; genErr (5)—any other error not listed above.

- *error-index.* An integer to identify which object in a set caused the error condition identified in the error-status field.

SNMPv1 PDU specification. The description of each of the five SNMPv1 PDUs includes an ASN.1 definition, semantics of each field, rules for carrying out the

TABLE 2.22   Generic Format of SNMP Protocol Messages and PDUs

Table 2.22a   Generic Format of an SNMP Protocol Message

Version	Community name	SNMP PDU

Table 2.22b   The Structure of the GetRequest, GetNextRequest, SetRequest, and GetResponse PDUs

PDU-type	request-id	error-status	error-index	variable-bindings

Table 2.22c   Variable-bindings Fields

name 1	value1	name2	value2	· · ·	nameN	valueN

SOURCE: Stallings (1992).

```
Message ::= SEQUENCE {
    version              INTEGER { version-1(0)},
    community            OCTET STRING,
    data                 ANY     —e.g., PDUs if trivial authentication is
                                         being used.
}


PDUs ::= CHOICE
{
    get-request          GetRequest-PDU,
    get-next-request     GetNextRequest-PDU,
    get-response         GetResponse-PDU,
    set-request          SetRequest-PDU,
    trap                 Trap-PDU
}
GetRequest-PDU           ::= [0] IMPLICIT PDU
GetNextRequest-PDU       ::= [1] IMPLICIT PDU
GetResponse-PDU          ::= [2] IMPLICIT PDU
SetRequest-PDU           ::= [3] IMPLICIT PDU
PDU                      ::= SEQUENCE {
                                request-id       INTEGER,
                                error-status     INTEGER
                                {
                                         noError(0),
                                         tooBig(1),
                                         noSuchName(2),
                                         badValue(3),
                                         readOnly (4),
                                         genErr (5)
                                },
                                error-index       INTEGER,
                                variable-bindings VarBindList
}
varBind         ::= SEQUENCE {
    name        ObjectName,
    value       ObjectSyntax
}
VarBindList     ::= SEQUENCE OF VarBind
```

Figure 2.13 ASN.1 definitions of a generic PDU structure.

management operation by the receiving protocol entity, and some examples for illustration.

GetRequest PDU. The ASN.1 definitions of this PDU are

```
GetRequest-PDU ::= [0] IMPLICIT SEQUENCE
    {
                request-id         RequestID,
                error-status       ErrorStatus,
                error-index        ErrorIndex,
                variable-bindings VarBindList
}
```

The 0 in brackets, i.e., [0], is an application tag used to uniquely identify an ASN.1 definition within an application. The key word IMPLICIT is an ASN.1

type to indicate that the fields of the SEQUENCE type will have implicit tags, as opposed to explicit ones. See App. A for more details.

- *Request-ID.* An integer assigned by the management entity, normally a manager, to keep track of each outstanding request and to detect whether a request is lost.
- *ErrorStatus.* An integer to indicate an error condition, always set to noError (0) for this PDU.
- *ErrorIndex.* An integer value to indicate which object in a list of objects caused the error indicated in error-status; it is always set to 0 when the error-status field is set to noError (0).
- *VarBindList.* A set of object-name–object-value pairs. Each object name identifies an object instance whose value is to be retrieved. The object value fields are left empty to be filled by the responding agent.

The manager sends this PDU to an agent to request information on a set of named objects in the agent's MIB view. Upon receipt of this PDU, the receiving protocol entity of an agent responds according to the following rules:

- If any of the requested objects named in the variable-bindings field does not match exactly any of the names in an agent's MIB view, or the requested object type is of an aggregate type (e.g., a table), the agent returns a GetResponse PDU that is identical to the received GetRequest PDU except that the error-status field is set to noSuchName and the error-index field to the index of the offending object name.
- If the size of the GetResponse PDU generated by the agent exceeds a local limit, the agent returns the GetResponse PDU that is identical in content to the received GetRequest PDU except that the error-status field is set to tooBig and the error-index field to 0.
- If the agent cannot retrieve any of the objects named in the variable-bindings field for any reason other than those listed above, the agent returns a GetResponse PDU that is identical in content to the received GetRequest PDU except that the error-status field is set to genErr and the error-index field to the index of the offending object name.
- If none of the above applies, the agent retrieves from its MIB view the value of each object named in the variable-bindings field of the received GetNextRequest PDU and returns a GetResponse PDU that is the same as the received GetNextRequest PDU except for the variable-bindings field that now has the retrieved object values.

The following examples illustrate the use of the GetRequest PDU to retrieve both scalar and columnar object instance values. A manager sends a GetRequest PDU to request the total number of received UDP datagrams (see Table 2.19):

GetRequest (GetRequest-PDU, 1, 0, 0, (1.3.6.1.2.1.6.7.1.0, NULL))

The agent returns a GetResponse-PDU, assuming there is no error of any kind detected:

GetResponse (GetRequest-PDU, 1, 0, 0, (1.3.6.1.2.1.6.7.1.0, 235))

Note that the PDU type, request-id, error-status, and error-index fields of the GetResponse PDU are identical to those of the GetRequest PDU except that in the variable-bindings field, the retrieved total number of received UDP datagrams (i.e., 235) is returned.

The following GetRequest PDU is used to request the second IP address' interface index in Table 2.20 which is the element in the second row and second column of the table.

The manager sends the following request PDU:

GetRequest (GetRequest-PDU, 2, 0, 0, (1.3.6.1.2.1.4.20.1.2.11.12.13.14, NULL))

The agent responds by sending the following in return, assuming there is no error:

GetResponse (GetRequest-PDU, 2, 0, 0, (1.3.6.1.2.1.4.20.1.2.11.12.13.14, 5))

Note that the PDU type, request-id, error-status, and error-index fields of the GetResponse PDU are identical to those of the GetRequest PDU except that in the variable bindings field, the interface index value (i.e., 5) is returned.

GetNextRequest.   The ASN.1 definition of the GetNextRequest PDU is as follows:

```
GetNextRequest-PDU ::= [1] IMPLICIT SEQUENCE {
        request-id        RequestID,
        error-status      ErrorStatus,
        error-index       ErrorIndex,
        variable-bindings VarBindList
}
```

A protocol entity generates the GetNextRequest PDU at the request of a management application entity. This PDU is usually sent from a manager to an agent to query for the values of a set of named object instances that each is next in the lexicographical order to the one named in the PDU.

- *request-id.*   An integer assigned by the management entity that created this PDU, normally a manager, to uniquely identify each outstanding request.
- *error-status.*   An integer to indicate any error condition. This field is always set to noError (0) for this PDU.
- *error-index.*   An integer to identify which object in a set caused the error condition identified in the error-status field. This field is always set to noError (0) for the GetNextRequest PDU.

- *variable-bindings.* A set of object-name–object-value pairs. Each object name identifies an object instance whose lexicographical successor's value is to be retrieved. All object value fields are left empty to be filled by the responding agent.

The receiving protocol entity responds to a GetNextRequest PDU according to the following rules:

- If the object named in the variable-bindings field does not precede some object in lexicographical order in the MIB view, meaning that the next object does not exist for retrieval operation, then the receiving entity returns a GetResponse PDU with error-status=noSuchName.

- If the size of the GetResponse PDU generated by the receiving protocol entity exceeds the local limitation, then the receiving entity returns a GetResponse PDU with error-status = tooBig and error-index = 0.

- If the value of the lexicographical successor of any object named in the variable-bindings field of the GetNextRequest PDU cannot be retrieved for any reason other than those listed above, the receiving entity returns a GetResponse PDU with error-status = genErr and error-index set to the index of the offending object name.

- If none of the above applies, the agent retrieves from its MIB view the value of the lexicographical successor of each object named in the variable-bindings field of the received GetNextRequest PDU and returns a GetResponse PDU that is the same as the received GetNextRequest PDU except for the variable-bindings field that now has the retrieved object values and their corresponding object names, the lexicographical successor of the one named in the received GetNextRequest PDU.

An important usage of the GetNextRequest PDU is for manipulating the conceptual tables. It allows a manager to discover the structure of an agent's MIB view by traversing the whole agent MIB tree; it allows for searching for a table whose entry names are unknown to the manager. We use the abbreviated IP address table, shown in Table 2.23, to illustrate how the GetNextRequest PDU is used to traverse the table. For reading convenience, the following information on the table is repeated here:

The index of the IP address table is defined as {INDEX ipAdrEntAdr }, as in Fig. 2.6.

The identifier of the IP address table, ipAddrTable = 1.3.6.1.2.1.4.20.

The identifier of the table entry template, ipAddrEntry = 1.3.6.1.2.1.4.20.1.

The identifier of the first column object, ipAdrEntAdr = 1.3.6.1.2.1.4.20.1.1.

The identifier of the second column object, ipAdrEntIfIndex = 1.3.6.1.2.1.4.20.1.2.

TABLE 2.23   An Abbreviated IP Address Table

ipAdEntAddr	ipAdEntIfIndex	ipAdEntNetMask
10.11.12.13	4	0.0.0.0
10.11.12.14	5	255.255.255.255
10.11.99.57	3	1.1.1.1

The identifier of the third column object, ipAdrEntNetMask = 1.3.6.1.2.1.4.20.1.3.

All but the variable-bindings field are omitted for brevity. At first, the manager does not know what is in the table. It requests that the first row be examined by specifying the three columnar object identifiers which themselves are not accessible:

GetRequest ((1.3.6.1.2.1.4.20.1.1, NULL), (1.3.6.1.2.1.4.20.1.2, NULL) (1.3.6.1.2.1.4.20.1.3, NULL))

The agent returns the instance identifier of the lexicographical successor of each requested object and its value, using Tables 2.21 and 2.23.

GetResponse ((1.3.6.1.2.1.4.20.1.1.10.11.12.13, 10.11.12.13)
(1.3.6.1.2.1.4.20.1.2.10.11.12.13, 5)
(1.3.6.1.2.1.4.20.1.3.10.11.12.13, 0.0.0.0))

Note that the object names in the GetResponse PDU are the lexicographical successors of the ones in the received GetNextRequest PDU. Now that the manager knows the names and values of the object instances of the first row, it uses them to request for the next row:

GetResponse ((1.3.6.1.2.1.4.20.1.1.10.11.12.13, NULL)
(1.3.6.1.2.1.4.20.1.2.10.11.12.13, NULL)
(1.3.6.1.2.1.4.20.1.3.10.11.12.13, NULL))

The agent returns the three object names and their values in the next row:

GetResponse ((1.3.6.1.2.1.4.20.1.1.10.11.12.14, 10.11.12.14)
(1.3.6.1.2.1.4.20.1.2.10.11.12.14, 5)
(1.3.6.1.2.1.4.20.1.3.10.11.12.14, 255.255.255.255))

In this fashion, a manager can traverse a whole table without prior knowledge of the number of rows and row index.

SetRequest. This SNMPv1 PDU is used for a manager to request that an agent modify the values of a set of named objects in the agent's MIB view. The ASN.1 definitions of this PDU are as follows

```
SetRequest-PDU ::= [3] IMPLICIT SEQUENCE
{
    request-id        RequestID,
    error-status      ErrorStatus,
    error-index       ErrorIndex,
    variable-bindings VarBindList
}
```

The meanings of the PDU fields are similar to that of the GetRequest PDU:

- *request-id.*   An integer to uniquely identify each outstanding request and to detect whether a request or its response is lost.

- *error-status.*   An integer to indicate any error condition, always set to noError (0) for this SetRequest PDU.

- *error-index.*   An integer to indicate which object out of a set caused the error condition identified in the error-status field. This field is always set to 0.

- *variable-bindings.*   A set of object-name–object-value pairs. Each object name identifies an object instance whose value is to be set. Each object value field specifies a desired value to be set by the responding agent for the name object.

At the receiving end, the agent responds to the SetRequest PDU according to the following rules:

- If the set operation cannot be performed on any of the objects named in the variable-bindings field, the agent returns a GetResponse PDU with error-status = noSuchName and the error-index field set to the index of the offending object name.

- If the type, length, or value of any object named in the variable-bindings field is not consistent with that found in the agent's MIB view, the agent returns a GetResponse PDU with error-status = badValue and the error-index field set to the index of the offending object name.

- If the size of the GetResponse PDU generated by the agent exceeds the local limit, the agent returns a GetResponse PDU with error-status = tooBig and error-index = 0.

- If the agent cannot set the value of any object named in the variable-bindings field for any reason other than the ones listed above, the agent returns a GetResponse PDU with error-status = genErr and the error-index field set to the index of the offending object name.

- If none of the above applies, the agent assigns the desired value to each object specified in the variable-bindings field in its MIB view and returns a GetResponse PDU that is the same as the received SetRequest PDU.

We still use the IP address table shown in Table 2.23 to illustrate the use of the SetRequest PDU. The manager sends the following to update the table entry at the first row and second column of the table, an IP address interface index. Again, all but the variable-bindings fields are omitted for brevity.

SetRequest (1.3.6.1.2.1.4.20.1.2.10.11.12.13, 9)

The agent changes the value from 4 to 9 and returns the following, assuming that no error occurred.

GetResponse (1.3.6.1.2.1.4.20.1.2.10.11.12.13, 9)

Note that it indicates the success of a set operation when a GetResponse PDU is identical to the received GetRequest PDU in content.

The next example is using a SetRequest PDU to add a new row into the IP address table as shown in Table 2.22. The manager sends the following SetRequest PDU in an attempt to create a new row in the table:

SetRequest ((1.3.6.1.2.1.4.20.1.1.12.13.14.15, 12.13.14.15)
    (1.3.6.1.2.1.4.20.1.2.12.13.14.15, 7)
    (1.3.6.1.2.1.4.20.1.3.12.13.14.15, 128.128.128.128))

When the agent realizes that the specified object names are not defined in the MIB view, it can create a new row and returns the following GetResponse PDU; the resulting table is shown as Table 2.24:

GetResponse ((1.3.6.1.2.1.4.20.1.1.12.13.14.15, 12.13.14.15)
    (1.3.6.1.2.1.4.20.1.2.12.13.14.15, 7)
    (1.3.6.1.2.1.4.20.1.3.12.13.14.15, 128.128.128.128))

RFC 1157 does not specify the action to take when an undefined object name is received in the SetRequest PDU. RFC 1212 suggests three options, one of which is adding a new row. A vendor may or may not choose to implement this option.

GetResponse PDU. The GetResponse PDU is used to respond to a request PDU. All three request PDUs require a GetResponse PDU in return, and thus the GetResponse PDU serves as an acknowledgement to a request PDU. The exchange of protocol messages between two protocol entities, e.g., between a manager and an agent, is shown in Fig. 2.14. The ASN.1 definitions of the PDU are the same as those of the request PDUs:

TABLE 2.24   IP Address Table After Addition of a New Row

ipAdEntAddr	ipAdEntIfIndex	ipAdEntNetMask
10.11.12.13	9	0.0.0.0
10.11.12.14	5	255.255.255.255
10.11.99.57	3	1.1.1.1
12.13.14.15	7	128.128.128.128

```
GetResponse-PDU ::= [2] IMPLICIT SEQUENCE
{
    request-id        RequestID,
    error-status      ErrorStatus,
    error-index       ErrorIndex,
    variable-bindings VarBindList
}
```

A GetResponse PDU is always generated by the receiving protocol entity, normally an agent, in response to the receipt of a request PDU. The meanings of most fields should be clear by now.

- *request-id.* The same request-id of the request PDU for which this GetResponse PDU is generated

- *error-status.* Determined by the rules of each request PDU as described above

- *error-index.* Determined by the rules of each request PDU as described above

- *variable-bindings.* Determined by the rules of each request PDU as described above

Trap PDU. The Trap PDU is used for an agent to report an exception event to the manager. The ASN.1 definitions of the Trap PDU are

```
Trap-PDU ::= [4] IMPLICIT SEQUENCE {
    enterprise        OBJECT IDENTIFIER,
    agent-addr        NetworkAddress,
    generic-trap      INTEGER {
                      coldStart (0),
                      warmStart (1),
                      linkDown (2),
                      linkUp(3),
                      authenticationFailure (4),
                      egpNeighborLoss (5),
                      enterpriseSpecific (6)
    }
    specific-trap     INTEGER,
    time-stamp        TimeTicks,
    variable-bindings VarBindList
}
```

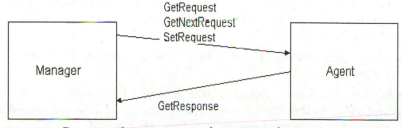

Figure 2.14 Request and response protocol message exchanges.

TABLE 2.25   **Structure of the Trap PDU**

PDU-type	enterprise	agent-address	generic-trap	specific-trap	time-stamp	variable-bindings

The structure of this PDU is illustrated in Table 2.25, and the details of the fields are as follows.

- *PDU type.*   The Trap PDU.

- *enterprise.*   An object identifier identifying the enterprise that owns the network element whose network management application entity emitted this Trap PDU.

- *agent-addr.*   The IP address of the enterprise.

- *generic-trap.*   Standardized trap types as explained below.

- *specific-trap.*   Specific code that provides further information on the trap.

- *time-stamp.*   Time elapsed between the last (re)initialization of this network element and the generation of the trap.

- *variable-bindings.*   A set of object-name–object-value pairs. Implementation of this field is left to vendors.

The following generic trap types have been defined:

- *coldStart (0).*   Signifies that the sending protocol entity is reinitializing itself such that the agent's configuration or the protocol entity implementation may be changed.

- *warmStart (1).*   Signifies that the sending protocol entity is reinitializing itself in a way that neither the agent configuration nor the protocol entity implementation is changed.

- *linkDown.*   Signifies to the manager that a failure in one of the communication links represented in the agent's MIB view has been detected. The name and value of the ifIndex instance of the interface object group for the affected interface is put in the variable-bindings field of this Trap PDU as the first element.

- *linkUp.*   Notifies a manager that one of the communication links represented in the agent's MIB view has come up. Similarly, the name and value of the ifIndex instance for the affected interface is put in the variable-bindings field of this Trap PDU as the first element.

- *authenticationFailure.*   Signifies that a received protocol message is not properly authenticated.

- *egpNeighborLoss.*   Signifies that a peer EGP neighbor is down. The name and value of the egpNeighAddr instance of the affected neighbor is the first element of the variable-bindings field of this PDU.

- *enterpriseSpecific.* Indicates that some enterprise-specific event has occurred at this network element.

The trap protocol message is not acknowledged in the sense that the receiving protocol entity does not reply upon the receipt of a trap message, as shown in Fig. 2.15. All the traps are mandatory for implementation. However, the mechanism to suppress any of the traps should be provided by all implementations.

An end-to-end application example

Now we revisit the network management scenario depicted in Fig. 2.2 and illustrate how the SNMP can be used to accomplish the required management task. First, we give a brief review of the scenario. One printer stopped printing, and the agent at the printer notified the management station of the event. The manager responded by querying the current job queue status of printer 2, and upon receiving the status information, it requests that printer 2 set its threshold to a higher value to take more print jobs. We use the exchange of SNMP PUDs between the manager and the agents to illustrate the use of the SNMP (see Fig. 2.16).

The SNMP PDU exchanges include the following:

1. The agent at printer 1 sends a Trap PDU to the manager at the management station to report the printer failure.

2. Upon receiving the Trap PDU, the manager sends a GetRequest PDU to the agent of printer 2 to request the printer job queue size.

3. The agent at printer 2 responds by sending a GetResponse PDU with the requested information.

4. The manager sends a SetRequest PDU to request that the agent at printer 2 change the job queue limit to a higher value.

5. The agent sets the job queue to the specified value and returns a GetResponse PDU to the manager to report the success of the operation.

The above management scenario assumes that no abnormality occurs during the SNMP PDU exchanges and leaves out many of the details of the PDU messages to focus on the protocol message exchange.

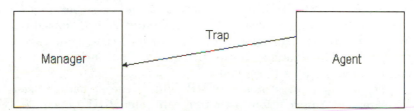

Figure 2.15 Trap protocol message.

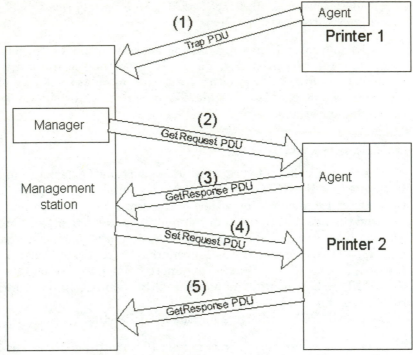

Figure 2.16 An example of a PDU exchange.

Summary

There are three basic components of the SNMP network management frame-work: SMI, MIB, and management protocol.

The SNMP SMI defines three basic elements required for building an MIB. The first is an overall structure to organize management information (conceptual tree and tables) represented as managed objects. Next is a syntax notation to specify how to define managed objects and what types of operations are permitted on each object. Third is a naming convention to uniquely identify each object and object instance so that it is possible to put managed objects in the structure as described and retrieve them as needed.

The SNMP MIB contains domain knowledge about the network elements being managed and defines the management operations that are allowed on each specific managed object. The SNMP MIB-II standardizes the representation of the TCP/IP-based network and makes it possible for one SNMP network management system to manage any TCP/IP-based network. There are only two types of operation allowed on MIB-II objects: read-only and read-write. This translates into two types of operations on the network elements: monitoring and control.

The management protocol SNMP defines the structure used to exchange management information with five different PDU types, i.e., GetRequest, GetNextRequest, SetRequest, GetResponse, and Trap. This includes the structure of the PDUs and semantics of each PDU.

The underlying design principle of the SNMP is simplicity. This is reflected in its simple MIB structure, very limited number of PDUs, and extremely simple PDU structure. Its wide acceptance and the short period it took to establish its dominant presence are largely due to its simplicity.

As the Internet is changing and evolving at a breathtaking pace, it begins to be used to provide a variety of services and applied to applications it was not originally meant for. This presents enormous opportunities for growth as well as challenges. The SNMP was meant to just monitor and control networks. This functionality will soon prove to be far from adequate for today's Internet and the services it provides. It remains to be seen how the SNMP will evolve in response to the new challenges and the opportunities.

Exercises

2.1 Internet network management standards have three basic components. What are they?

2.2 Use a network example as illustrated in Fig. 2.2 to explain the following terms: network element, network management entity, and protocol entity.

2.3 Describe the Internet protocols the SNMP depends on. Explain what services these protocols provide to the SNMP.

2.4 Refer to Fig. 2.2 for this question. Describe the SNMP message exchanges between the manager and the agent at the network server for the following scenario: during one of its regular polls the manager finds out that the network server is overloaded. It takes action by setting the message window to a lower threshold.

2.5 Explain the main difference between the concept of object in the Internet MIB and the object concept in object-oriented design and analysis.

2.6 What are the major steps involved in building an SNMP PDU by a protocol entity?

2.7 Given the following UDP table and related information, illustrate the steps to retrieve the third row of the table using the GetNextRequest PDU.

The identifier of the UDP table, updTable = 1.3.6.1.2.1.7.5.

The identifier of the row template, udpEntry = 1.3.6.1.2.1.7.5.1.

The identifier of the first-column object, udpLocalAddress = 1.3.6.1.2.1.7.5.1.1.

The identifier of the second-column object, udpLocalPort = 1.3.6.1.2.1.7.5.1.2.

The index clause is INDEX {udpLocalAddress, udpLocalPort}.

udpLocalAddress	UdpLocalPort
22.23.24.24	3
120.12.13.14	3
125.10.76.11	4

2.8 Show the SetRequest PDU and GetResponse PDU for changing the udpLocalPort value from 3 to 5 at the second column and second row in the table of Exercise 2.7.

2.9 Refer to Exercise 2.7, and list the lexicographical successors of all objects involved in the udp table, including udpTable and udpEntry.

2.10 Use a network example such as the one shown in Fig. 2.1 to explain the concept of community and the MIB view.

2.11 Define a printer object group using ASN.1 for the management scenario illustrated in Fig. 2.2. The example objects include maxJobQueueSize, currentJobQueueSize, and tonerStatus.

2.12 Explain why the *error-status* and *error-index* fields are always set to noError for GetRequest, GetNextRequest, and SetRequest.

2.13 Assuming an agent successfully performed a set operation requested by a manager in a SetRequest PDU, explain how the agent can let the manager know that the requested operation is successful.

Introduction to
SNMPv2 and RMON

Outline

- The first section, "SNMPv2 SMI and MIB," provides an overview of the SNMPv2 structure of management information with an emphasis on the extensions to the SNMPv1 SMI and an introduction of the SNMPv2 MIB, i.e., its three object groups. This section concludes with a comparison between the SNMPv2 SMI and SNMPv1 SMI, and between SNMPv2 MIB and SNMPv1 MIB, and the interoperability-related issues.

- The second section, "SNMPv2 Protocol Operations," introduces SNMPv2 protocol operations with emphasis on the new additions of SNMPv2 PDUs, namely, GetBulkRequest and GetInformationRequest, and on the differences between SNMPv2 PDUs and SNMPv1 PDUs.

- The last section, "Remote Network Monitoring," first introduces the RMON principles and RMON1 MIB, and then provides a high-level view of RMON2.

SNMPv2 SMI and MIB

The SNMP was originally designed as a short-term management solution to provide bare-bones functionality. As the Internet-based networks experienced a phenomenal growth, the newer networks had outgrown what the SNMPv1 was intended to manage both in functionality and complexity. The second generation of SNMP standards, known as SNMPv2, was initially completed in early 1993 and was formally adopted in 1996 after several years of trials. SNMPv2 expands the functionality of SNMPv1 to include OSI-based as well as TCP/IP-based networks. Introduced in this section are two categories of enhancements:

■ Enhancement in the SMI, especially the definition of the conceptual table

■ Enhancement in the MIB that includes objects for SNMPv2 entities

The SNMPv2-related standards are listed in Table 3.1.

SNMPv2 SMI

The SMI for SNMPv2, as it's counterpart for SNMPv1, specifies the syntax and semantics for defining managed objects. The SNMPv2 SMI, specified in RFC 1901, consists of four parts: object definitions, conceptual tables, notification definition, and information modules.

Object definitions. Recall that in the SNMPv1 SMI, the ASN.1 macro OBJECT-TYPE MACRO is used to specify a managed object. A quick comparison of the following OBJECT-TYPE macro against the one listed in Chap. 2 and defined in RFC 1212 reveals that the structures of the two macros are identical except that there is an additional optional field UNITS in the SNMPv2 SMI. An SMI defines the syntax of managed objects and types of fields in an object. The SNMPv2 SMI is based on the SNMPv1 SMI, and it suffices to explain only the differences.

For both SNMP and SNMPv2, the type of an object may either be simple or application-based. The simple types of the two SMIs are quite similar with one exception. The BIT STRING simple type is added to the SNMPv2 SMI. For the application types, counter64, NsapAddress, and Uinteger62 are new additions of the SNMPv2 SMI. In the SNMPv2 specification, both counter32 and counter64 are read-only and don't have any initial values. The value of a single counter is not meaningful, and only the difference between two readings of a counter is significant. The new application type NaspAddress (network service access point) accommodates the OSI type of address scheme. A comparison between SNMPj93
v1 and SNMPv2 data types is shown in Table 3.2.

```
ASN.1 OBJECT-TYPE macro for SNMPv2 SMI (RFC 1901)
OBJECT-TYPE MACRO ::= BEGIN
TYPE NOTATION ::= "SYNTAX" type (ObjectSyntax)
      UnitsPart
      "MAX-ACCESS" Access
      "STATUS" Status
      ReferPart
      IndexPart
      DefvalPart
VALUE NOTATION ::= value (VALUE ObjectName)
UnitsPart ::= "INITS" Text ¦ empty
Access ::= "not accessible" ¦"read-only" ¦"read-write"¦ "read-create"
Status ::= "current" ¦ "deprecated" ¦ "obsolete"
ReferPart::= "REFERENCE" Text ¦ empty
IndexPart ::= "INDEX" "{" IndexTypes "}" ¦"AUGMENTS" "{" Entry "}" ¦
empty
IndexTypes ::= IndexType ¦ IndexTypes "," IndexType
IndexType ::= "IMPLIED" Index¦Index
Index ::= value (indexobject ObjectName)
Entry ::= value (entryobject ObjectName)
DefValPart ::= "DEFVAL" "{" value (Defval ObjectSyntax) "}"¦ empty
Tex ::= """"string """"
END
```

TABLE 3.1   SNMPv2-related Standards*

RFC	Title
1902	Structure of Management Information for Version 2 of the Simple Network Management Protocol (SNMPv2)
1903	Textual Conventions for Version 2 of the Simple Network Management Protocol
1904	Conformance Statements for Version 2 of the Simple Network Management Protocol (SNMPv2)
1905	Protocol Operations for Version 2 of the Simple Network Management Protocol (SNMPv2)
1907	Management Information Base for Version 2 of the Simple Network Management Protocol (SNMPv2)
1908	Coexistence between Version 1 and Version 2 of the Internet-standard Network Management Framework

*All dated January 1996.

TABLE 3.2   Comparison of SNMP and SNMPv2 Simple Data Types

Type	Present in SNMP SMI	Present in SNMPv2 SMI
Simple types		
INTEGER	X	X
OCTET STRING	X	X
OBJECT IDENTIFIER	X	X
BIT STRING		X
Application types		
Counter32	X	X
Counter64		X
Gauge32	X	X
IpAddress	X	X
NsapAddress		X
Opaque	X	X
TimeTicks	X	X
UInteger32		X

The syntax of the SNMPv2 SMI is very similar to that of the SNMPv1 except for the following differences. An SNMPv2 OBJECT-TYPE macro contains an optional Units Part clause, which provides textual definition of the units of measurement. This is useful for those objects that have associated measurement. For example, the units for a time interval can be seconds, minutes, or hours. A MAX-ACCESS clause is used in place of the SNMPv1 ACCESS clause. The prefix MAX emphasizes that this is the maximal level of access allowed, independent of what any administrative policy specifies. The SNMPv2 SMI does not have the write-only access category. A new access category, read-create, is added; it is used in operations on conceptual tables and will be explained later.

SNMPv2 table definition.   As with SNMPv1 SMI, there are only two types of data structures for SNMPv2 SMI: scalar and conceptual tables. Conceptual tables can represent more complicated information and relationships between information elements. The definition of scalar objects has already been explained in Chap. 2; this discussion focuses on the extensions to the table definition.

SNMPv1 table definition.   A table has zero or more conceptual rows, and each row contains one or more scalar objects. Typically a row of a table represents a unit of information with multiple attributes. For example, a TCP connection has a state, a local address, a remote address, a local port, and a remote port. A single TCP connection can be represented as a row of a TCP connection table with the five attributes represented as scalar objects in a row.

The structure of a table for SNMPv2 is the same as that for SNMPv1, which can be summarized as follows.

- A conceptual table is made up of a set of rows represented as a sequence of table entries with the following syntax:

  SEQUENCE OF <entry>

  where each <entry> is a conceptual row.

- A conceptual row consists of a set of scalar objects represented as a sequence of types with the following syntax:

  SEQUENCE { <type1>, <type2>, ..., <typeN>}

  where there is a <type> for each columnar object and a <type> has the form

  <descriptor> <syntax>

  where <descriptor> is the application-specific scalar object name and <syntax> specifies the type of the scalar object such as INTEGER, OBJECT IDENTIFIER, or IPAddress.

SNMPv2 extensions.   The conceptual table of SNMPv2 has the following major extensions:

- Optional use of the IMPLIED modifier to facilitate the construction of an object identifier in a table

- An AUGMENT mechanism to extend a table to have additional columns

- A new textual convention called RowStatus for creation and deletion of table entries

IMPLIED key word.   The key word IMPLIED may precede a variable-length object name to indicate that the length of the variable is implicit, instead of being the first component of a constructed object identifier, in reference to a scalar object in a table. The following example shows an object with a variable-length identifier that is of type OCTET STRING. Assume that the octet string has the value 67.78.89.

Identifier with IMPLICIT key word: 67.78.89.

Identifier without IMPLICIT key word: 3.67.78.89.

Thus the IMPLIED key word enables a small savings in space in the instance identifier when one of the index variable is of variable length.

AUGMENT clause.   As an alternative to the INDEX clause, a conceptual row definition may include the AUGMENT clause, referred to as a conceptual row extension. Simply put, the AUGMENT mechanism is used to increase the number of columns in a table without changing the table definition. The resulting two tables are treated in the same way since they have been defined as a single table, although the two tables may reside in two different identifier hierarchies. The AUGMENT feature is useful in cases where a basic set of objects is stored in a table and some possible extensions depend on the instantiation of a table row. For example, a management application may have some attributes in addition to those specified in the standard and can conveniently represent them as table extensions.

RowStatus textual convention.   Textual conventions are used to represent new types for SNMPv2 MIBs. Each of the new types has a different name, a syntax similar to that of SMI, but with more precise semantics. It is an alternative for defining MIB types, only easier to understand. For example, the DisplayString is an SNMP type that can be expressed in textual convention as follows:

```
DisplayString ::= TEXTUAL-CONVENTION
DISPLAY-HINT "255a"
Status     current
DESCRIPTION "Represents textual information taken from the ASCII
character set."
```

The RowStatus, a new type to be used in the SYNTAX clause, is defined as follows:

```
RowStatus ::= TEXTUAL-CONVENTION
STATUS     current
DESCRIPTION
   "The RowStatus textual convention is used to manage the creation and
deletion of conceptual rows and is used as the value of the SYNTAX
clause for the status column of a conceptual row."
```

In brief, every SNMPv2 table must have a column with SYNTAX set to RowStatus. The value of this column controls the creation, deletion, and suspension of each corresponding row. RowStatus can have one of six possible values:

- *active.*   The conceptual row is available for use.
- *notInService.*   The conceptual row exists in the agent but is unavailable for use.
- *notReady.*   The conceptual row exists in the agent but is missing information necessary for it to be ready for use.
- *createAndGo.*   One of the two methods for creating a conceptual row that is indicated by a management station. When used, this method indicates that the management station wishes to create a conceptual row and set its status to active automatically, making it available for use.
- *createAndWait.*   The other method for creating a conceptual row.
- *destroy.*   If set by a management station, the agent should immediately delete all the object instances in the conceptual row.

Table operations.   The following discussion provides a high-level view of the steps to be taken for creating, deleting, and suspending a conceptual row of an SNMPv2 table.

Row creation.   A manager uses a combination of SetRequest and GetRequest PDUs to create new rows for tables based on the SNMPv2 SMI. To support row creation, there must be one columnar object in the table with a SYNTAX clause *RowStatus* and a MAX_ACCESS clause *read-create*.

The normal steps of creating a new row using the CreateAndWait method are as follows.

1. A manager sends a SetRequest PDU to the agent to construct a new row with a given index value.

2. The agent creates the requested row and assigns values to those objects that have default values. If all read-create objects have default values, the state of the row is set to notInService. If any of the read-create objects do not have default values, the row is set to the notReady state.

3. The agent sends a response PDU that includes a noSuchInstance error code for each read-create object that does not have a default value, and a default value is assigned to each read-create object.

4. The manager sends a GetRequest PDU to the agent to determine the state of each read-create object. It assigns values to all those objects that don't have default values and may also change the default values if it desires to do so.

5. Finally, the manager, after making sure all read-create objects have been created using GetRequest as many times as needed, sends a SetRequest PDU to set the state of the row to active.

The second method of creating a new row, the createAndGo method, is similar to the createAndWait method with exceptions in two aspects. First, step 3 of the createAndWait method is omitted in the createAndGo method, meaning that the manager does not automatically know which read-create object does not have a default value and has to query with GetRequest. Second, this method can only be applied to tables whose objects can fit into a single SetRequest or Response PDU.

Row deletion.   The steps for deleting a row are relatively simple, as follows.

1. A manager sends a SetRequest to the agent to set the value of the status object of the table row to destroy.

2. The agent immediately removes the entire conceptual row from the table. It is not mandatory for the agent to send a Response PDU to the manager reporting the success of the operation.

Row suspension.   Suspension of a row results from negotiation between a manager and an agent.

1. First, the manager sends a SetRequest PDU to the agent to set the RowStatus value to inActive.

2. If the agent is unwilling to comply, a response is sent back to the manager stations with an error of wrongValue; otherwise, the agent sets the RowStatus value as requested and returns a response with the value noError.

SNMPv2 MIB

The SNMPv2 MIB, defined in RFC 1907, specifies objects that represent the behavior of SNMPv2 entities. An SNMP entity can act in either an agent role or a manager role. The SNMPv2 MIB is an extension to the SNMP MIB-II, as defined in RFC 1212, and consists of the following object groups.

- *System group.*   An extension of the original MIB-II system group to include a set of objects to allow an SNMPv2 agent to keep track of their dynamically configurable object resources.

- *SNMP group.*   A narrowed-down version of the original SNMPv1 group with focuses on those objects essential to SNMPv2 operations.

- *MIB objects group.*   A collection of objects related to SNMPv2-Trap PDUs that allows the cooperating SNMPv2 managers to coordinate their use of SNMPv2 set operations.

System group.   As shown in Table 3.3, this is an extension to the system group of SNMPv1 MIB-II with the addition of a scalar object sysORLastChange and an object resource table sysORTable. The scalar object is used to record the time of the last change to any of the object instances contained in the object resource table. The added table contains several objects related to system object resources that are used by the SNMPv2 entity acting in an agent role to describe the object resources that are subject to dynamic reconfiguration by a manager.

TABLE 3.3   SNMPv2 System Group

Object name	Syntax	Access	ID	Description
SysDescr	DisplayString (SIZE(0..255))	RO	1	A textual description of the entity. This value should include the full name and version identification of the system's hardware type, software operating system, and networking software.
SysObjectID	OBJECT IDENTIFIER	NA	2	The vendor's authoritative identification of the network management subsystem contained in the entity. This value is allocated within the SMI enterprise subtree (1.3.6.1.4.1) and provides an easy and unambiguous means for determining what kind of box is being managed.
SysUpTime	TimeTicks	RO	3	The time in hundredths of a second since the network management portion of the system was last initialized
SysContact	DisplayString (SIZE (0..255))	RW	4	Information on the contact person for this network node
SysName	DisplayString (SIZE (0..255))	RW	5	An assigned name for this node, used as a fully qualified domain name
SysLocation	DisplayString (SIZE (0..255))	RW	6	The physical location of this node
SysServices	INTEGER (0..127)	RO	7	An integer value indicating the set of services offered by this node
SysORLast Change	TimeStamp	RO	8	The value of sysUpTime at the time of the most recent change in state or value of any instance of sysORID
SysORTable	Sequence of Sys OREntry	NA	9	A table listing the capabilities of the local SNMPv2 entities in an agent role with respect to various MIB module-supporting dynamically configurable objects

Each row in the object resource table, as shown in Table 3.4, represents a system instance and contains objects describing the system characteristics, identifier, and the time this entry is created for the system instance.

SNMP group.   This is the same group defined in MIB-II (RCF 1212) but with the additional objects as shown in Table 3.5. The SNMP group provides the capabilities to control and maintain an SNMPv2 entity. It enables a manager to keep count of the parameters such as total number of SNMPv2 messages received at an SNMPv2 entity, the number of bad messages out of the total, and total number of parse errors. In addition, the object snmpEnableAutheTraps controls the generation of traps by an agent for authentication failure.

Compared to the MIB-II SNMP group, the SNMP group has far fewer objects. The main reason for this is that the detailed statistics of the MIB-II SNMP group add a considerable burden to the agent system and are not essential for debugging the majority of problems in an SNMP entity.

MIB objects group.   The MIB objects group contains two sets of objects related to SNMP traps generated by agents and to the coordination of uses of the set operation by multiple managers. There are two subgroups. The first one is snmpTrap, which contains two objects that are included in every SNMPv2-Trap PDU.

- *snmpTrapOID.*   A unique identifier of the notification that is currently being sent. This variable is included in every SNMPv2-Trap PDU and InformationRequest PDU.

- *snmpTrapEnterprise.*   A unique identifier of the enterprise associated with the trap current being sent. This variable is included in every SNMPv2-Trap PDU and InformationRequest PDU.

The second subgroup contains only one object:

- *snmpSetSerialNo.*   An advisory lock object that allows several cooperating SNMPv2 managers to coordinate their use of the SNMPv2 set operations. This object is created on a per-MIB basis.

SNMPv2 Protocol Operations

SNMPv2 is an extension of the SNMPv1. In addition to the five SNMPv1 PDUs, SNMPv2 defines two more: GetBulkRequest and InformationRequest PDUs.

TABLE 3.4   System Object Resource Table

Object name	Syntax	Access	ID	Description
SysORIndex	INTEGER	RO	1	An auxiliary variable used for identifying an instance of the columnar object in this table
SysORID	OBJECT IDENTIFIER	RO	2	An authoritative identification of capabilities supported by the local SNMPv2 entity acting in an agent role in various MIB modules
SysORDescr	DisplayString	RO	3	A textual description of the capabilities identified by the sysORID
SysORUptime	TimeStamp	RO	4	The value of sysUpTime at the time this conceptual row is last instantiated

TABLE 3.5 SNMPv2's SNMP Object Group

Object name	Syntax	Access	ID	Description
snmpInPkts	Counter32	RO	1	The total number of SNMP messages delivered to the SNMP protocol entity from the transport service
snmpInBadVersion	Counter32	RO	3	The total number of SNMP messages delivered to the protocol entity with an unsupported SNMP version
snmpInBad CommunityName	Counter32	RO	4	The total number of SNMP messages delivered to the protocol entity with an unknown community name
snmpInBad CommunityUse	Counter32	RO	5	The total number of SNMP messages delivered to the protocol entity with an SNMP operation not allowed by the community
snmpInASN ParseErrs	Counter32	RO	6	The total number of ASN.1 or BER errors encountered by the protocol entity when decoding an SNMP message
snmpEnable AutenTraps	INTEGER	RO	30	Indicates whether the SNMP agent process is permitted to generate authentication-failure traps with enabled = 1 and disabled = 2. The value of this object overrides any other setting.
snmpSilentDrops	Counter32	RO	31	The total number of GetRequest PDUs, GetNextRequest PDUs, GetBulkRequest PDUs, SetRequest PDUs, and InformRequest PDUs that were delivered to the SNMPv2 entity but were silently dropped because the size of the Response PDU exceeds either a local limitation or maximum message size of the originator of the request
snmpProxyDrops	Counter32	RO	32	The total number of GetRequest PDUs, GetBulkRequest PDUs, SetRequest PDUs, and InformRequest PDUs that were delivered to the SNMPv2 entity but were silently dropped because the transmission of the message to a proxy target failed in a manner such that no Response PDU could be returned

SNMPv2 PDU formats

The SNMPv2 PDU message has the same generic structure as SNMPv1 PDU messages, shown in Table 3.6, with three fields: a version number, a community name, and one of the seven SNMPv2 PDUs. The version number indicates the version of SNMP protocol, mainly for compatibility reasons. A community name is an identifier defining all legitimate management entities that are allowed to communicate with a particular agent. The first two fields of an SNMP message, the version and the community name, are also called the wrapper or header of an SNMP message.

Six SNMPv2 PDUs, i.e., GetRequest, GetNextRequest, SetRequest, GetResponse, SNMPv2-Trap, and InforRequest, share the same structure as shown in Table 3.7. This is the same structure of the SNMPv1 PDUs GetRequest, GetNextRequest, SetRequest, and GetResponse. A new SNMPv2 PDU, GetBulkRequest PDU, has a different format as shown in Table 3.8. The variable bindings field remains the same as shown in Table 3.9. An overall comparison between SNMPv1 and SNMPv2 PDUs is summarized in Table 3.10.

Five of the SNMPv2 PDUs are identical to their SNMPv1 counterparts in syntax and semantics with some subtle differences. The focuses of the discussion will be first on these differences and then on the two new SNMPv2 PDUs, InformRequest and GetBulkRequest.

GetRequest PDU

The SNMPv2 GetRequest PDU is identical in both syntax and semantics to the SNMPv2 counterpart. The only major difference is in the way the responses to this PDU are defined. There are two possible responses to the SNMPv1 GetRequest: either all requested values are retrieved or none is retrieved. In contrast, the SNMPv2 GetRequest is designed in such a way that even though not all values of the requested variables can be retrieved, those retrievable values are returned in a variable-bindings list, in pairs of a variable name and a variable value. If the value of a requested variable can't be retrieved due to

TABLE 3.6   Generic Format of an SNMP Message

Version	Community name	SNMP PDU

TABLE 3.7   The Structure of GetRequest, GetNextRequest, SetRequest, GetResponse, SNMPv2-Trap, and InforRequest PDUs

PDU-type	request-id	error-status	error-index	variable-bindings

TABLE 3.8   The Structure of GetBulkRequest PDU

PDU-type	request-id	non-repeater	max-repetitions	variable-bindings

TABLE 3.9   Variable-bindings Fields

name1	value1	name2	value2	· · ·	nameN	valueN

TABLE 3.10   Comparison between SNMP and SNMPv2 PDUs

SNMPv2 PDU	SNMP PDU	Comments
GetRequest	GetRequest	SNMPv2 GetRequest and SNMPv1 GetRequest share the same semantics and structure, as shown in Table 3.7.
GetNextRequest	GetNextRequest	SNMPv2 GetNextRequest and SNMPv1 GetNextRequest share the same semantics and structure, as shown in Table 3.7.
SetRequest	SetRequest	SNMPv2 SetRequest and SNMPv1 SetRequest share the same structure and semantics.
Response	GetResponse	SNMPv2 Response and SNMPv1 GetResponse share the same structure but have different PDU names. Also different are the ways SNMPv2 Response is generated. See the discussion in this section.
SNMPv2-Trap	Trap	SNMPv2-Trap and SNMPv1-Trap have different names and structures, though the semantics are similar. SNMP Trap has a unique structure of its own that is different from that of the request PDUs. The SNMPv2-Trap PDU shares the same structure with other SNMPv2 PDUs such as GetRequest, Response, GetNextRequest, SetRequest.
InformRequest	No equivalent	This new SNMPv2 PDU has the same structure as all other SNMPv2 PDUs except for GetBulkRequest.
GetBulkRequest	No equivalent	This new SNMPv2 PDU has a unique structure of its own, as shown in Table 3.8. See the discussion in this section for details.

an error condition, the error condition (e.g., noSuchObject, noSuchInstance, endOfMibView) accompanied by the variable is returned along with those retrievable values. Specifically, a response PDU in response to an SNMPv2 GetRequest PDU is constructed according to the following rules, as specified in RCF 1905:

1. If the name of a requested variable in the variable-bindings name field of the SNMPv2 Request PDU exactly matches the name of the variable that is accessible by the request in the MIB, then the variable-bindings value field is set to the value of the named variable.

2. Otherwise, if the variable-bindings name field does not have an OBJECT IDENTIFIER prefix that exactly matches the OBJECT IDENTIFIER prefix of any variable in the MIB that is accessible by this request, the value field is set to noSuchObject.

3. Otherwise, the variable-bindings value field is set to noSuchInstance.

Another possible error scenario is that the size of the protocol message that contains the generated response PDU exceeds a local limitation, the response PDU is discarded, and a new response PDU is generated. The PDU has an error-status field of tooBig, an error-index field of zero, and an empty variable-bindings field.

If retrieval of the value of any variable in the variable-bindings field fails for any reason other than the ones listed above, a response PDU is reconstructed. The new response PDU will have the same values in its request-id and variable-bindings fields as the received GetRequest PDU, with an error-status field of genError and a value in the error-index field that is the index of the problem object in the variable-bindings field.

GetNextRequest PDU

The SNMPv2 GetNextRequest PDU is identical to the SNMPv1 counterpart in both syntax and semantics. As with the GetRequest PDU just discussed above, the only difference is that the SNMPv1's GetNextRequest PDU is atomic, i.e., either retrieves everything or nothing, and SNMPv2's is best-effort, i.e., retrieves as many values as possible.

Specifically, an SNMPv2 response PDU for the GetNextRequest is generated by processing each variable in the incoming variable-bindings list, according to the following rules, as specified in RFC 1905:

- For each named variable in the request list, the object instance that is next in lexicographical order to the named variable is located and a pair of object name and object value is put in the resulting variable-bindings list.

- If no lexicographical successor can be found for a named object in the incoming requested variable-bindings list, then the corresponding variable-bindings field in the response PDU has its value field set to endOfMibView and its name field set to the variable-bindings name field in the request.

SetRequest PDU

Again, the SNMPv2 SetRequest PDU is identical to the SNMPv1 counterpart in syntax and semantics except for the way in which the responses are processed. First, upon receiving an SNMPv2 SetRequest PDU, the agent side determines the size of the message encapsulating a response PDU. If the size is greater than a local limitation, or than the maximum message size allowed by the SNMPv2 entity that originated the SetRequest PDU, a new response PDU is constructed. The response PDU has the same values in its request-id field as the received SetRequest PDU, error-status field of tooBig, and error-index field of 0. The new response PDU is encapsulated in a message. If the size of the resulting message is smaller than or equal to both the local limitation or the maximum message size of the originator, it is sent to the originator of the SetRequest PDU. Otherwise, the snmpSilentDrops counter is incremented and the new response message is discarded.

If none of the above occurs, the receiving SNMPv2 entity processes each variable-bindings field in the incoming variable-bindings list in two logical phases, in order to produce a response PDU. In the first phase, all variables are validated; if the validations are successful, each variable is updated in the second phase, as described below.

During the validation phase, the SNMPv2 entity, acting in an agent role, performs the following validation steps until either all steps are successful or one step fails.

- If the specified variable is not accessible, the value of the error-status field is set to noAccess.

- If the specified variable is not found, or a value can't be assigned, the value of the error-status field is set to notWritable.

- If the type of the value to be assigned is inconsistent with the type of the specified variable, the value of the error-status field is set to wrongType.

- If the length of the value to be assigned is inconsistent with the length of the specified variable, the value of the error-status field is set to wrongLength.

- If the ASN.1 encoding of the value to be assigned is inconsistent with the field's ASN.1 tag, the value of the error-status field is set to wrongEncoding.

- If the value can't be assigned to the specified variable under any circumstances, the value of the error-status field is set to wrongValue.

- If the specified variable does not exist and cannot be created, the value of the error-status field is set to noCreation.

- If the specified variable does not exist and cannot be created under the current condition, the value of the error-status field is set to inconsistentName.

- If the value can be assigned to the specified variable under another circumstance but is currently inconsistent, the value of the error-status field is set to inconsistentValue.

- If the assignment of the value to the specified variable requires a resource that is currently unavailable, the value of the error-status field is set to resourceUnavailable.

- If the processing of the variable-bindings list fails for a reason other than those listed above, the value of the error-status field is set to genErr.

If any of the above validation steps fails, the agent sends a response PDU with the corresponding error-status value to the management station. Otherwise, the processing proceeds to the variable updating phase.

During the updating phase, the SNMPv2 entity, acting in an agent role, performs the following steps.

- If necessary, first create the named variable; otherwise, go to the next step.

- Assign the specified value to the named variable.

- If any assignment fails, undo all completed assignments and generate a response PDU with the value of the error-status field set to commitFailed and the value of the error-index field set to the index of the failed variable. If it is not possible to undo all the assignments, then set the value of the error-status field to undoFailed and the value of the error-index field to zero.

Response PDU

An SNMPv2 Response PDU is generated by an SNMPv2 entity in an agent role in response to the receipt of a GetRequest, GetNextRequest, GetBulkRequest, or SetRequest PDU. As discussed above, the SNMPv2 Response PDU is different from the SNMPv1 counterpart mainly in the ways a response is generated. Detailed differences are described in each request PDU.

Some conventions regarding the error fields of the Response PDU are as follows. If the error-status field of the Response PDU is nonzero, the value fields of the associated variable-bindings field are ignored. If both the error-status field and error-index field of the Response PDU are nonzero, the value of the error-index field is the index of the variable-bindings field of the corresponding request PDU that has failed.

GetBulkRequest PDU

This is one of the major enhancements of SNMPv2. The purpose of the GetBulkRequest PDU is to minimize the potentially large number of protocol message exchanges in the retrieval of a large amount of data, especially large tables.

As shown in Table 3.8, the GetBulkRequest PDU has two fields not found in the other PDUs: non-repeaters and max-repetitions. The non-repeaters field specifies the number of variables in the variable-bindings list for which a single lexicographical successor is to be returned. The max-repetitions field specifies the number of lexicographical successors to be returned for the rest of the variables in the variable-bindings list.

In general terms, the GetBulkRequest PDU operation works as follows. The GetBulkRequest PDU includes a list of $(N + (M * R))$ variable names in the variable-bindings list. Each of the first N variables, as specified in the non-repeaters field, is treated like a variable in GetNextRequest PDU: the next object instance in lexicographical order is retrieved. If no lexicographical successor is found, the name variable and a value of endOfMibView are paired and put in the corresponding variable-value slot of the response PDU's variable-bindings list. For the remaining R named variables, i.e., $(N + 1)th$ through the end of the incoming variable-bindings list, multiple lexicographical successors are retrieved, according to the number specified in the max-repetitions field.

The following relationships hold:

$$N = \text{Max} \, (\text{MIN} \, (\text{non} - \text{repeaters}, L), 0)$$

$$M = \text{Max} \, (\text{max} - \text{repeaters}, 0)$$

$$R = L - N$$

where L = total number of variable names in variable-bindings field of GetBulkRequest PDU

N = number of variables, starting with first variable in variable-bindings list, for which a single lexicographical successor is requested and ending at Nth variable, as specified in non-repeaters field

R = number of variables following first N variables for which multiple lexicographical successors are to be retrieved

M = number of lexicographical successors requested for each of the last R variables, as specified in max-repetitions field of GetBulkRequest PDU

According to the above rules, the total number of object instances to be put in the Response PDU is $N + (M * R)$. But fewer than $N + (M * R)$ pairs of variable name and variable value may be generated for one of the following three reasons:

- The size of the message that encapsulates the Response PDU exceeds the local limitation or the maximum size allowed by the SNMPv2 entity that sent the GetBulkRequest PDU.

- For some value i such that i is greater than 0 and less than or equal to M, all the generated variable-bindings fields have the value of endOfMibView. In this case, the variable-bindings fields may be truncated after the $(N + (I * R))$th variable-bindings field.

- If retrieving all $(N + (R*M))$ object instances requires a significantly greater amount of processing time and power than a normal request, an agent may terminate the request with less than the full number of requested objects, given at least one repetition as the specified R is completed.

SNMPv2-Trap PDU

An SNMPv2-Trap PDU is generated and transmitted by an SNMPv2 entity acting in an agent role when an error condition arises. This function is the same as with the SNMPv1 Trap PDU, but the SNMPv2-Trap PDU has a format that is different from that of its SNMPv1 counterpart. The format of the SNMPv2-Trap PDU is the same as that for the request PDUs, thus reducing the number of PDU formats by one. The first two variable-bindings fields in the variable-bindings list of an SNMPv2-Trap PDU are sysUpTime and snmpTrapOID, respectively.

As with an SNMPv1 Trap PDU, no response is issued to an SNMPv2-Trap PDU by an SNMPv2 entity acting in a manager role.

InformRequest PDU

This PDU is a new addition to the SNMPv2 PDU set. An InformRequest PDU is generated and sent by an SNMPv2 entity acting in a manager role, on behalf of an application, to another SNMP entity acting in an agent role, to provide management information. The format of the InformRequest PDU is shown in Table 3.7. As with the SNMPv2-Trap PDU, the first two variable-bindings fields in the variable-bindings list are sysUpTime.0 and SNMPv2EventId.i, respectively, as specified in the manager-to-manager MIB.

The receiving SNMPv2 entity, upon receipt of an InformRequest PDU, checks for any violation in the size of the received message that encapsulated the InformRequest PDU: the size should not exceed either the local limitation or the maximum size allowed by the originator of the InformRequest PDU. In case of a violation, the receiving SNMPv2 entity constructs a response PDU that contains the same request-id field as the received InformRequest PDU, an error-status field of tooBig, an error-index field of 0, and an empty variable-bindings list.

In case of no error, the receiving SNMPv2 entity

- Presents the contents of the received InformRequest PDU to appropriate application

- Generates a response PDU that contains the same values in the fields of the request-id and variable-bindings list as the received InformRequest PDU, and an error-status field of noError and an error-index field of 0

- Transmits the response PDU to the originator of the InformRequest PDU

Remote Network Monitoring (RMON)

As local networks are connected into an ever increasingly larger network, the need is strongly felt for the capabilities not only to know about the local area networks, but to look beyond a few local nodes and to monitor internetworks.

A remote network monitoring (RMON) MIB provides such capabilities and is one of the most important additions to the basic set of SNMP standards. What is remarkable about RMON is that it provides for an important extension to the SNMP functionality without any change to the existing SNMP. In a nutshell, RMON is a specification of a special-purpose MIB, and it uses the underlying SNMP. Actually, there are two RMONs, referred to as RMON1 and RMON2. The RMON-related standards are listed in Table 3.11.

TABLE 3.11   RMON-Related Standards

RFC	Date	Title
1513	Sept. 1993	Token Ring Extension to the Remote Network Monitoring MIB
1757	Feb. 1995	Remote Network Monitoring Management Information Base (RMON1)
2021	Jan. 1997	Remote Network Monitoring Management Information Base, Version 2, using SMIv2 (RMON 2)
2074	Jan. 1997	Remote Network Monitoring MIB Protocol Identifiers

An overview of RMON

Basic principles of RMON. In essence, RMON is a special-purpose MIB plus a dedicated management entity, acting in an agent role, to collect and manage an RMON MIB. This dedicated management entity along with a network device in which the management entity resides is called a *remote monitor* or *probe*. The term *probe* will be used throughout this section to refer to the management entity. The device can be dedicated or shared with other management functions. A probe monitors the health and behavior of a segment of a network, reducing the burden on the management station and other agents. The following design goals of RMON are stated in RFC 1757.

- *Off-line operation.* The RMON MIB allows a probe to be configured to continuously collect performance, fault, and configuration data independent of whether or not a manager is in constant contact with the probe. Sometimes the manager is not in constant contact with the probe by design when an attempt is made to lower communications costs, or the situation can be created by accident as network failures affect the communications between a management station and the probe. From the perspective of a manager, this could be advantageous in certain situations.

- *Proactive monitoring.* If the probe has sufficient resources, it can be configured to continuously run diagnostic and log network performance. This makes it possible for the probe to notify a management station of any failure as the failure occurs and for management to diagnose the root cause of a problem using the historical data collected by the probe.

- *Problem detection and reporting.* Even in the passive observing mode, the probe can be configured to recognize certain conditions, most notably the error conditions, such as traffic congestion. When one of the conditions occurs, the probe can log the event and report it to the management station.

- *Value-added data.* The probe can be configured to analyze and categorize the data it collects and extract a networkwide view from the data. For example, a probe can determine the cause of traffic congestion on a subnetwork by analyzing the data on the subnetwork as well as by looking beyond the subnetwork itself. This information is of greater value to a management station than just the collection of raw data.

- *Multiple managers.* A probe can be configured to accommodate more than one management station concurrently. In an internetworking environment, multiple management stations are often desired for reasons such as improving reliability or performing different functions (e.g., operation and engineering).

RMON is completely based on the existing SNMP architecture, without any additional protocol messages or operations. The only addition is a remote monitor or probe that is either a dedicated device or a shared device that contains RMON capabilities. An RMON probe consists of an agent that is no different from any other SNMP agent and the local RMON MIB. In summary, an RMON system consists of

- A management station, that should be the same as those seen in an SNMP-based management system
- A set of RMON probes
- A management protocol that is identical to the one used in an existing SNMP management system.

In turn, an RMON probe consists of

- An RMON MIB that is specific for the RMON system
- An agent that is the same in structure and functions as an agent in an existing SNMP-based management system

An example configuration of RMON is shown in Fig. 3.1 where two RMON probes are responsible for monitoring two subnetworks. The management station exchanges SNMP messages (e.g., GetRequest and GetNextRequest PDUs) with the RMON probes to retrieve the statistics collected by the probes. This scenario will be used to illustrate the RMON1 and RMON2 in the rest of the section.

Structure of an RMON MIB. An RMON MIB consists of a set of object groups. In general, two types of data are contained in an object group: control data and statistical data. The control data specify how and when to collect the statistical data, and the latter provide a statistical view of various aspects of the subnetwork being monitored by the RMON probe. The two types of data can be in two separate tables or combined into one table.

When a table has a *status* column, it indicates that read, write, and create operations are allowed on the table. The status column is of the type RowStatus of SNMPv2 SMI, as discussed earlier, but has a different set of possible values:

- *valid (1).* Indicates the associated entry is a valid one.
- *createRequest (2).* Indicates this entryStatus object is being created. Once the creation is completed, the value of this object shall be changed to either underCreation or valid.
- *underCreation (3).* Indicates the associated table entry is in the process of being created and may be in an inconsistent and incomplete state.
- *invalid (4).* Marks the associated table entry as invalid.

Setting this object to the value invalid has the effect of invalidating the associated entry. This is the way of deleting an entry. It is a local matter whether to delete the invalidated entry or leave it in the table. A management station must be equipped to tell an invalidated entry from the valid ones in the retrieved data.

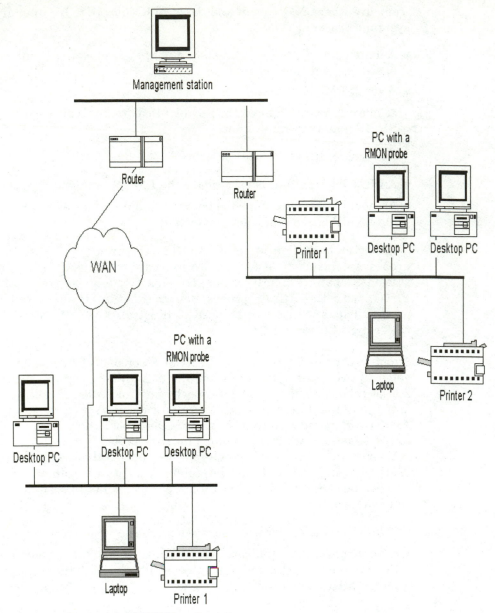

Figure 3.1 An example of RMON configuration.

If two tables are used for control and data, respectively, the general structure of the tables and one frequently seen relationship between the two tables
are shown in Fig. 3.2. It is often the case that one entry of a control table maps
to multiple entries in the data table. For example, if an entry in the control
table represents a network interface and an entry in the data table represents
the count of data packets transmitted between a pair of hosts on the interface,

one interface has multiple pairs of hosts attached to it. The control status column in the control table is the status column just discussed, and it represents the status of the corresponding control table entry. A data table may have a status column of its own, which is not shown in the figure. The data index of the data table is an index number uniquely identifying an entry within the data table.

The remainder of this section will present an introduction of RMON1 and RMON2, i.e., two RMON MIBs that provide two distinct sets of monitoring capabilities.

Introduction to RMON1

The initial version of RMON1 was published in 1991, and the formal standards version became available in 1995. As discussed earlier, since the bulk of RMON1 standards is devoted to the definitions of RMON MIBs, so is our introduction. An RMON1 MIB consists of the following object groups.

- *Statistics.* Maintains statistics collected by an RMON1 probe from the monitored Ethernet interface. This group consists of an Ethernet statistics table.

- *History.* Records periodic statistical sampling of data collected in the statistics group.

Example index table

Control index	Control parameter 1	Control parameter n	Control owner	Control status
1			Monitor A	valid
2			Manager A	valid
3			Manager B	underCreation

Example data table

Data control index	Data index	Data value
1	1	43
1	2	45
1	3	2034
3	4	78
3	5	49
3	6	45

Figure 3.2 Relationship between control and data table of RMON MIB.

- *Alarm.* Allows the alarm thresholds and intervals for collecting performance data to be set by an interested party such as a management station.

- *Host.* Contains the traffic data to and from the hosts in the subnetwork monitored by the probe.

- *HostTopN.* Contains sorted statistical data for the top *N* hosts based on the parameters that can be specified by a management station.

- *Matrix.* Presents traffic data collected for each pair of addresses (e.g., hosts) of the monitored subnetwork.

- *Filter.* Allows the probe to selectively observe traffic data based on filtering criteria that can be specified by a management station.

- *Packet capture.* Captures the data that meet the criteria specified in the filter object.

- *Event.* Allows all the events generated by the probe to be logged.

Statistics group. This is a table that contains one entry per Ethernet interface. The control information and data are combined into one table. Other than etherStatsIndex, etherStatsDataSource, etherStatsOwner, and etherStatsStatus, all the objects in this group are of Counter type. The data objects in this group collectively shall provide an overall view of the traffic condition on each of the Ethernet interfaces associated with this RMON probe. The following parameters can be found in the table:.

- *etherStatsIndex.* An integer index for a row. One row per monitored Ethernet interface on the device is created in the statistics table.

- *etherStatsDataSource.* Identifies the interface associated with this row.

- *etherStatsDropEvents.* The total number of events that have caused the RMON probe to drop packets because of lack of resources. This is not necessarily the number of packets dropped, but the number of times the event has occurred.

- *etherStatsOctets.* The total number of received octets of data including those in bad packets.

- *etherStatsPkts.* The total number of received packets of data including bad packets, broadcast packets, and multicast packets.

- *etherStatsBroadcastPkts.* The total number of good broadcast packets received.

- *etherStatsMulticastPkts.* The total number of good multicast packets received.

- *etherStatsCRCAlignErrors.* The total number of packets received that were between 64 and 1518 octets in length and had either a CRC (cyclic redundancy checking) error or an alignment error.

- *etherStatsUndersizePkts.* The total number of packets of data received with a length less than 64 octets.

- *etherStatsOversizePkts.*   The total number of packets of data received with a length greater than 1518 octets.

- *etherStatsFragments.*   The total number of packets of data received that were less than 64 octets long and had a bad frame check sequence or an alignment error (not an integral number of octets).

- *etherStatsJabbers.*   The total number of packets received that were longer than 1518 octets and had either a bad frame check sequence or an alignment error.

- *etherStatsCollisions.*   The estimated total number of collisions.

- *etherStatsPkts64Octets.*   The total number of packets received (including bad packets) that were 64 octets long.

- *etherStatsPkts65to127Octets.*   The total number of packets received (including bad packets) that were between 65 and 127 octets in length.

- *etherStatsPkts128to255Octets.*   The total number of packets received (including bad packets) that were between 128 and 255 octets in length.

- *etherStatsPkts256to511Octets.*   The total number of packets received (including bad packets) that were between 256 and 511 octets in length.

- *etherStatsPkts512to1023Octets.*   The total number of packets received (including bad packets) that were between 512 and 1023 octets in length.

- *etherStatsPkts1024to1518Octets.*   The total number of packets received (including bad packets) that were between 1024 and 1518 octets in length.

- *etherStatsOwner.*   A string that identifies the entity that owns the interface associated with this entry.

- *etherStatsStatus.*   The status of the interface associated with this table entry.

History group.   This group consists of two tables: the history control table and the history data table. Each entry (i.e., a row) in the history control table represents a network interface monitored by the probe with the following parameters:

- *historyControlIndex.*   An integer index that uniquely identifies an entry in the history control table

- *historyControlDataSource.*   An identifier of the type OBJECT IDENTIFIER that specifies the source of the historical data that were collected

- *historyControlBucketsRequested.*   An integer number representing requested discrete time intervals over which data are saved

- *historyControlBucketsGranted.*   An integer number representing actual discrete time intervals over which data are saved

- *historyControlInterval.*   The data sampling interval in seconds for each bucket

- *historyControlOwner.*   A string identifying the entity that configured this entry for the associated Ethernet interface

- *historyControlStatus.*   The status of this control entry

The history data table contains sampling of a wide range of traffic data on an Ethernet-based subnetwork for a given sampling period, which is specified in the history control table.

- *etherHistoryIndex.*   An integer index identifying a table entry of history samples.

- *etherHistorySampleIndex.*   An index identifying the particular sample this entry represents among all samples associated with the same history control table entry.

- *eitherHistoryIntervalStart.*   The value of sysUpTime at the start of the interval over which the sample of this table entry was taken.

- *etherHistoryDropEvents.*   The total number of events that have caused the RMON probe to drop packets because of lack of resources. This is not necessarily the number of packets dropped, but the number of times the event has occurred.

- *etherHistoryOctets.*   The total number of received octets of data including those in bad packets.

- *etherHistoryPkts.*   The total number of received packets of data including bad packets received during the sampling period.

- *etherHistoryBroadcastPkts.*   The total number of good broadcast packets received during the sampling period.

- *etherHistoryMulticastPkts.*   The total number of good multicast packets received during the sampling period.

- *etherHistoryCRCAlignErrors.*   The total number of packets received during the sampling period that had a length of between 64 and 1518 octets and with either a CRC error or an alignment error.

- *etherHistoryUndersizePkts.*   The total number of packets of data received during the sampling period that had a length less than 64 octets.

- *etherHistoryOversizePkts.*   The total number of packets of data received during the sampling period that had a length greater than 1518 octets.

- *etherHistoryFragments.*   The total number of packets of data received during the sampling period that were less than 64 octets long and had a bad frame check sequence or an alignment error (not an integral number of octets).

- *etherHistoryJabbers.*   The total number of packets received during the sampling period that were longer than 1518 octets and had either a bad frame check sequence or an alignment error.

- *etherHistoryCollisions.*   The estimated total number of collisions at this segment of Ethernet during the sampling period.

- *etherHistoryUtilization.*   The best estimate of the mean physical layer network utilization on this interface during the sampling period, in hundredths of a percent.

Alarm group.   The alarm group defines a set of thresholds for generating alarms on network traffic. In general terms, it works as follows. The RMON probe periodically takes statistical samples and compares them against the thresholds that have been configured by the management personnel. If a sample is determined to have crossed the threshold values, an alarm event is generated. The RMON probe only monitors for threshold crossing variables that are of ASN.1 primitive types such as INTEGER, Counter, Gauge, or TimeTicks.

Two threshold values are defined: rising and falling. Any sampled object value that falls between the two, as shown in Fig. 3.3, is deemed to have crossed the threshold values and will trigger an alarm event to be generated.

The alarm group consists of a single alarm table, alarmTable. As indicated by the presence of the status column, this table is configurable by a management station. Each entry in the table represents a variable to be monitored and has the following parameters:

- *alarmIndex.*   A number that uniquely identifies an entry in the alarm table.

- *alarmInterval.*   The interval in seconds over which data are sampled and compared with rising and falling thresholds.

- *alarmVariable.*   The object identifier of the variable to be sampled.

- *alarmSampleType.*   The method of sampling: either absolute value or delta value. While an absolute value of the selected variable is directly compared with the threshold values, the difference between the values of the variable at the last sample and its current value is compared with the threshold values.

- *alarmValue.*   The value of the statistics during the last sampling period.

- *alarmStartupAlarm.*   The alarm that may be sent when this entry is set to the entry status valid. If the first sample after this table entry becomes valid is greater or equal to the risingThreshold value and the alarmStartupAlarm value is equal to that of risingAlarm or risingOrFallingAlarm, then a single

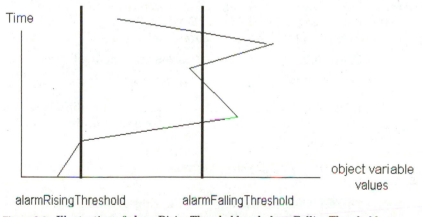

Figure 3.3   Illustration of alarmRisingThreshold and alarmFallingThreshold.

rising alarm will be generated. If the first sample after this entry becomes valid is less than or equal to the fallingThreshold value and the alarmStartupAlarm value is equal to that of fallingAlarm or risingOrFallingAlarm, then a single falling alarm is generated.

- *alarmRisingThreshold.*   A threshold for sampling statistics.
- *alarmFallingThreshold.*   A threshold for sampling statistics.
- *alarmRisingEventIndex.*   The index of an event in the event table that is used when a rising threshold is crossed.
- *alarmFallingEventIndex.*   The index of an event in the event table that is used when a falling threshold is crossed.
- *alarmOwner.*   The entity responsible for configuring this table.
- *alarmStatus.*   The status of each entry in the table.

Host group.   This object group is for discovering new hosts on the subnetwork monitored by this RMON probe. This is accomplished by keeping a list of source and destination addresses discovered in good packets. For each of these addresses, the host group keeps a set of statistics.

There are three tables in this group: host control table, host entry table, and host time table. The host control table keeps track of the interfaces through which the new hosts are discovered. Each entry in this table represents a network interface monitored by the probe with the following parameters:

- *hostControlIndex.*   An index number to uniquely identify an entry in this table.
- *hostControlDataSource.*   The object identifies the source of the data for the host table. The source can be any interface associated with this RMON probe.
- *hostControlTableSize.*   The number of host entries in the host control table and host time table associated with this host control table.
- *hostControlLastDeleteTime.*   The system time in the sysUpTime variable in the system group of SNMP MIB when the last entry was deleted from the host control table. A zero indicates that no entry has been deleted.
- *hostControlOwner.*   The entity that is responsible for configuring this table.
- *hostControlStatus.*   The table entry status.

The host table contains an entry for each address discovered on an interface. This table is indexed by the host address and contains various traffic data to and from a host.

- *hostAddress.*   The physical address of the host for which this table entry is created.
- *hostCreationOrder.*   The relative ordering of creation time for this host. This value should be between 1 and N where N is the value of hostControlTableSize in the host control table.

- *hostIndex.*   The set of collected host statistics of which this entry is a part. The value of this object matches the value of hostControlIndex for one of the rows of the host control table. Therefore, all entries in the host table with the same hostIndex value contain statistics for hosts on the same subnetwork.

- *hostInPkts.*   The number of good packets transmitted to this host address since it was added to the host table.

- *hostOutPkts.*   The number of packets, including the bad packets, transmitted by this host address since it was added to the host table.

- *hostInOctets.*   The number of octets transmitted to this host address since it was added to this host table.

- *hostOutOctets.*   The number of octets transmitted by this host address since it was added to this host table.

- *hostOutErrors.*   The number of bad packets transmitted by this host address since this host was added to the host table.

- *hostOutBroadCastPkts.*   The number of good packets transmitted by this host address that were directed to the broadcast address since this host was added to the host table.

- *hostMultiCastPkts.*   The number of good packets transmitted by this host address that were directed to a multicast address since this host was added to the host table.

A third table in this group, the host time table, contains the same set of hosts and organizes the data on the hosts in the same format as the host table. But this table is indexed by hostTimeCreationOrder, the order in which hosts are discovered and inserted into the table, rather than by the host address. This table has two important uses. First, it allows fast downloading of this potentially large table. This is because the index of this table runs from 1 to the size of the table, and this allows efficient packing of variables into SNMP PDUs and allows table transfer to have multiple packets outstanding. The other use is the efficient discovery by the management station of the new hosts discovered since the last time the RMON probe was queried.

- *hostTimeAddress.*   The physical address of the host for which this table entry is created.

- *hostTimeCreationOrder.*   The relative ordering of creation time for this host. This value should be between 1 and N where N is the value of hostControlTableSize in the host control table.

- *hostTimeIndex.*   The set of collected host statistics of which this entry is a part. The value of this object matches the value of hostControlIndex for one of the rows of the host control table. Therefore, all entries in the host table with the same hostIndex value contain statistics for hosts on the same subnetwork.

- *hostTimeInPkts.*   The number of good packets transmitted to this host address since it was added to the host table.

- *hostTimeOutPkts.* The number of packets, including the bad packets, transmitted by this host address since it was added to the host table.

- *hostTimeInOctets.* The number of octets transmitted to this host address since it was added to this host table.

- *hostTimeOutOctets.* The number of octets transmitted by this host address since it was added to this host table.

- *hostTimeOutErrors.* The number of bad packets transmitted by this host address since this host was added to the host table.

- *hostTimeOutBroadCastPkts.* The number of good packets transmitted by this host address that were directed to the broadcast address since this host was added to the host table.

- *hostTimeMultiCastPkts.* The number of good packets transmitted by this host address that were directed to a multicast address since this host was added to the host table.

HostTopN group. The hostTopN group is used to prepare reports that describe the top *N* hosts that are ordered by one of the statistics specified by the management station. A report is a set of statistics for one host object group on one interface or subnetwork, collected during one sampling interval. For example, a management station may select the top five hosts that have the most bad packets transmitted or that have received the most number of packets. This group depends on the host group to derive the data required in a report.

The hostTopN group consists of one control table and one data table. In the control table, each entry represents one top-*N* report prepared for one network interface with the following parameters:

- *hostTopNControlIndex.* The index uniquely identifying an entry in the control table

- *hostTopNHostIndex.* The host table for which a top-*N* report will be prepared on behalf of this entry

- *hostTopNRateBase.* Specifies which one of the seven variables from the host table will be used as the statistical parameter for the hostTopNRate report

- *hostTopNTimeRemaining.* The number of seconds left in the sampling interval for the report currently being collected

- *hostTopNDuration.* The sampling interval, in seconds, for this hostTopN report

- *hostTopNRequestedSize.* The maximum number of hosts requested for the top-*N* table for this report

- *hostTopNGrantedSize.* The maximum number of hosts in the top-*N* table for this report

- *hostTopNStartTime.* The value of sysUpTime when this top-*N* report was last started

- *hostTopNOwner.*   The entity that is responsible for configuring this table
- *hostTopNStatus.*   The entry status

The data table, hostTopNTable, contains one entry for each of the top-N hosts with the following parameters:

- *hostTopNReport.*   This integer variable identifies the top-N report of which this entry is a part.
- *hostTopNIndex.*   An index that uniquely identifies an entry in the Top-N table among those in the same report.
- *hostTopNAddress.*   The physical address of the host for which the host table entry is created.
- *hostTopNRate.*   The amount of change in the selected variable during this sampling period. The variable is specified by the hostTopNRateBase parameter in the hostTopN control table.

Matrix group.   The matrix group is used to record information about traffic between a pair of hosts on a subnetwork monitored by this RMON probe. This group of objects is useful when a management station desires to find out which pairs of hosts are making the most use of a network server or have most of the traffic between them.

The matrix group consists of three tables: a matrix control table and two data tables, one recording traffic from sources to destinations (matrixSDTable), and one recording traffic from destinations to sources (matrixDSTable). For the matrix control table, one entry is created per interface and hence per subnetwork with the following parameters:

- *matrixControlIndex.*   An index that uniquely identifies an entry in the matrix control table.
- *matrixControlDataSource.*   This object identifies the interface and hence the subnetwork that is the source of the data in this row.
- *matrixControlTableSize.*   The number of entries in matrixSDTable for this interface. This must also be the number of entries in matrixDSTable.
- *matrixControlLastDeleteTime.*   The value of systUpTime when the last entry was deleted from either matrixSDTable or matrixDSTable.
- *matrixControlOwner.*   The entity that is responsible for configuring this table.
- *matrixControlStatus.*   The entry status of this table.

The first data table, matrixSDTable, records the traffic data from a particular source host, one entry per host on a particular interface, with the following parameters:

- *matrixSDSourceAddress.*   The physical address of a source host.
- *matrixSDDestAddress.*   The physical address of a destination host.

- *matrixSDSDIndex.* The index value for a matrix control table entry. All pairs of hosts on the same interface will have the same index value.

- *matrixSDPkts.* The number of packets transmitted from the source address to the destination address.

- *matrixSDOctets.* The number of octets transmitted from the source address to the destination address.

- *matrixSDErrors.* The number of bad packets transmitted form the source address to the destination address.

The second data table, matrixDSTable, is the mirror image of the first table, recording the traffic in the other direction between a pair of hosts on a network interface.

Filter group. The filter group provides a mechanism for a management station to instruct an RMON probe to selectively observe traffic on an interface according to a filtering expression. There are two kinds of filters: data and status. A data filter allows an RMON probe to screen collected data by comparing the bit pattern in observed data against a specified bit pattern. The status filter allows a probe to screen collected packets on the basis of their status such as valid, invalid, certain error condition, or others. The two kinds of filters can be combined using logical AND and OR to form more complex logical expressions.

A stream of packets that are passed through the filtering expressions is referred to as a channel. Filtering expressions are defined on a per-channel basis. A channel can be configured to generate events, as defined in the event group to be discussed later, when packets pass through a channel and the channel is in the enabled state. A management station can also use the mechanism defined in the packet capture group to capture packets passing through a channel. A logical expression using the two kinds of filters and logical operators like AND and OR can become a very powerful tool for screening data. A detailed discussion on filter logic can be found in Stallings (1995).

This group of objects consists of two tables: a filter table and a channel table. There is one entry per filter in the filter table and one channel per entry in the channel table. Each entry in the filter table contains both data and status filters, as described below.

- *filterIndex.* An index that uniquely identifies an entry in the table

- *filterChannelIndex.* The index of the channel of which this filter is a part

- *filterPktDataOffset.* The offset that indicates the starting point to apply a data filter for a bit pattern match

- *filterPktData.* The data of a data filter that is to be matched with the input packets

- *filterPktDataMask.* The mask to be applied to the match process

- *filterPktDataNotMask.*   The inversion mask that is applied to the match process

- *filterPktDataStatus.*   The status of a status filter that is to be matched with the input packet

- *filterPktDataStatusMask.*   The mask to be applied to the status match process

- *filterPktDataStatusNotMask.*   The inversion that is applied to the status match process

- *filterOwner.*   The entity that is responsible for configuring this table

- *filterStatus.*   The entry status of the table

Each channel table entry defines one channel, a logical data and event stream, and can have multiple filters associated with it.

- *channelIndex.*   An index uniquely identifying an entry in the channel table.

- *channelIfIndex.*   An index identifying an interface, and hence the subnetwork, to which the associated filters are applied to allow data into this channel.

- *channelAcceptType.*   This object controls the action of the filters associated with this channel. If this object has the value acceptMatched (1), packets will be accepted to this channel only if they pass both data and status matches of at least one of the associated filters. If this has the value acceptFailed (2), packets will be accepted to this channel only if they fail either the packet match or status match of every associated filter.

- *channelDataControl.*   This object controls the flow of data through the associated channel: the value on (1) will allow data, status, and events to flow through this channel and the value off (2) will prevent data, status, and events from flowing through the channel.

- *channelTurnOnEventIndex.*   This object specifies the event that is configured to turn the associated channelDataControl object from off to on when the event is generated.

- *channelTurnOffEventIndex.*   This object specifies the event that is configured to turn the associated channelDataControl object from on to off when the event is generated.

- *channelEventIndex.*   The object identifies an event that will be generated when the associated channelDataControl object is on and a packet is matched. This is the same event identified by eventIndex of the event group.

- *channelEventStatus.*   The event status of this channel, which can be one of the following values:
    - *eventReady (1).*   A single event may be generated.
    - *eventFired (2).*   The probe sets the channelEventStatus object to this value after an event is generated; no new event can be generated while in this state.
    - *eventAlwaysReady (3).*   Allows events to be generated at will without any restriction.

- *channelMatches.*   The number of times this channel has matched a packet. This object is updated even when the channelDataControl object is set to off.

- *channelDescription.*   A textual string describing the characteristics of this channel.

- *channelOwner.*   The entity that is responsible for configuring this channel.

- *channelStatus.*   The entry status of this entry.

Packet capture buffer group.   This object group is used to capture packets upon a filter match. It consists of two tables: a buffer control table and a capture buffer table. One entry of the buffer control table is created for each capture buffer. Specified in the control table are the maximum size of the octets in a capture buffer, the action to take when the buffer reaches full status, the channel that serves as the data source for the capture buffer, and others.

- *bufferControlIndex.*   An index uniquely identifying an entry in this control table.

- *bufferControlChannelIndex.*   An index identifying a channel that is the source of packets for this buffer control table. This is the same index value of the channelIndex object in the channel table.

- *bufferControlFullStatus.*   This object indicates whether the buffer has room to accept new packets with one of the two possible values:
  - *spaceAvailable (1).*   The buffer is accepting new packets.
  - *full (2).*   The buffer will accept new packets if the bufferControlFullAction object has the value wrapWhenFull by deleting the old packets in first-in–first-out fashion. The buffer stops receiving packets if the bufferControlFullAction object has the value lockWhenFull.

- *bufferControlFullAction.*   Controls the buffer action when it becomes full. As described above, there are two values: lockWhenFull (1) and wrapWhenFull (2).

- *bufferControlCaptureSliceSize.*   The maximum number of octets of each packet that will be saved in this capture buffer.

- *bufferControlDownloadSliceSize.*   The maximum number of octets of each packet in this capture buffer that will be returned in an SNMP message.

- *bufferControlDownloadOffset.*   The offset of the first octet of each packet in this capture buffer that will be returned in an SNMP retrieval.

- *bufferControlMaxOctetsRequested.*   The requested maximum number of octets to be saved in this capture buffer, including the implementation-specific overhead.

- *bufferControlMaxOctetsGranted.*   The actual maximum number of octets that can be saved in the capture buffer, including the implementation-specific overhead. If this object has the value −1, the capture buffer shall save as many octets as possible.

- *bufferControlCapturedPackets.*   The number of packets currently in this capture buffer.

- *bufferControlTurnOnTime.*   The value of the system group object systUpTime when this capture buffer was first turned on.

- *bufferControlOwner.*   The entity responsible for configuring this control table.

- *bufferControlStatus.*   The entry status of this table.

The second table, capture buffer table, contains the captured packets. The captured packets are organized as one packet per entry in the table, and each entry has an index back into the buffer control table indicating the channel from which the packet is captured.

- *captureBufferIndex.*   The index of the bufferControlEntry object with which this packet is associated.

- *captureBufferIndex.*   The index uniquely identifying an entry in the capture buffer table.

- *captureBufferPacketID.*   An index that indicates the order in which packets are received on the associated interface.

- *captureBufferPacketData.*   The data inside a packet, starting at the beginning of the packet plus any offset specified in the associated bufferControlDownloadOffset object.

- *captureBufferPacketLength.*   The actual length of the packet stored in this entry.

- *captureBufferPacketTime.*   The time elapsed in milliseconds since this capture buffer was first turned on when this packet was captured.

- *captureBufferPacketStatus.*   A value indicating the error status of the captured packet with the following definitions:
  - *0.*—Packet is longer than 1518 octets.
  - *1.*—Packet is shorter than 64 octets.
  - *2.*—Packet has a CRC or alignment error.
  - *3.*—The first of a set of packets that may not be processed properly.
  - *4.*—Packet's order in buffer is only approximate.

Event group.   The event group supports the definitions of event and control of the generation and notification of events from this device. An event can be triggered by a condition defined somewhere else. For example, the condition that a capture buffer is full and the bufferControlFullAction object has the value of wrapWhenFull will cause the buffer to receive additional packets. This event may also cause the information about the event to be logged in this group and an SNMP Trap message to be sent.

This group consists of an event control table and a log table. The event control table, eventTable, contains the parameters of the event that can be triggered. Each entry of a table represents one event to be generated and contains the following objects:

- *eventIndex.*   An index number that uniquely identifies an entry in the event table.

- *eventDescription.*   A textual string describing this event entry.

- *eventType.*   The type of notification the RMON probe will make on this event that takes on one of the following values:
  - *none (1).*   No notification is generated.
  - *log (2).*   An entry is made in the log table for each event.
  - *snmp-trap (3).*   An SNMP Trap message is sent to one or more management stations.
  - *log-and-trap (4).*   Both log and SNMP trap are generated.
- *eventCommunity.*   It specifies the SNMP community to which SNMP Trap messages are to be sent.

- *eventLastTimeSent.*   The value of sysUpTime at the time an event was last generated based on the parameters specified in this event entry.

- *eventOwner.*   The entity that is responsible for configuring this entry.

- *eventStatus.*   The entry status of this event table. If this object is a value other than valid (1), all associated log entries shall be deleted by the agent.

The second table, logTable, contains one entry for each logged event and has the following objects:

- *logEventIndex.*   The entry in the event table that generated this log entry.

- *logIndex.*   An index that uniquely identifies an entry in the log table among those generated for the same event type.

- *logTime.*   The value of sysUpTime when this log entry was created.

- *logDescription.*   A textual description of the event that activated this log entry.

Overview of RMON2

Started in 1994, an extension to the RMON MIB was published in RFC 2021, January 1997, and became known as RMON2 while the original RMON is referred to as RMON1. Put in general terms, RMON2 extends the scope of monitoring from the physical medium access layer to protocol and application layers. The significance of the extensions can be seen from the following two aspects, as pointed out in Stallings (1995).

- The capabilities to monitor network level traffic enable a probe to look beyond the connected LAN segment and see traffic coming into the LAN via routers. This opens up a whole range of network management possibilities for an RMON probe, as illustrated in the following scenario:

1. When there is excessive traffic load on the LAN, the probe is able to determine what portion of traffic is coming from an incoming router and what portion is coming from hosts of the LAN.
2. When a router is overloaded due to a high volume of outgoing traffic, the probe is able to determine what local hosts account for the bulk of the outgoing traffic and to what destination network and hosts the traffic is directed.
3. When a high volume of pass-through traffic (coming via one router and going out via another router connected to the LAN of the probe) occurs, the probe can decide what networks or hosts are responsible for the bulk of the traffic.

- The capability to monitor traffic above the network (IP) layer enables a probe to describe network traffic on a per-protocol and per-application basis and thus manage network traffic at a much finer level of granularity.

The RMON2 MIB is quite lengthy and comprehensive, consisting of nine object groups as shown in Fig. 3.4. The remainder of this subsection will

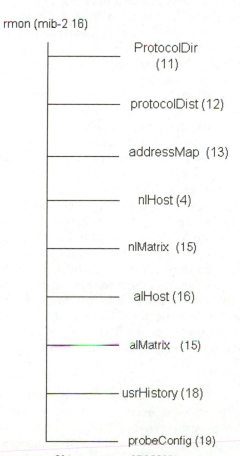

rmon (mib-2 16)

ProtocolDir (11)

protocolDist (12)

addressMap (13)

nlHost (4)

nlMatrix (15)

alHost (16)

alMatrix (15)

usrHistory (18)

probeConfig (19)

Figure 3.4 Object groups of RMON2.

provide a high-level description of each object group without describing individual objects within an object group.

Protocol directory group. The protocol directory group allows a probe to monitor protocols and to manage protocol entries by adding, deleting, and configuring the entries in the table. For example, all protocols that this probe is capable of monitoring are contained in the protocolDirTable table. The monitoring capability of the probe for each protocol, such as tracking sessions or counting fragments, is also specified in the table.

Protocol distribution group. The protocol distribution group enables a probe to monitor the traffic statistics on a subnetwork based on a protocol supported on the subnetwork. For example, it keeps a record of the number of packets received for a protocol. In addition, it also counts the number of octets transmitted to a host on the subnetwork connected to a particular interface.

Address map group. This object group maintains a set of mapping between media access control (i.e., physical) layer address and network layer address. This is helpful for a probe to discover new hosts on an interface and to pinpoint the specific paths of network traffic. For example, included in the address data table is the object addressMapSource that identifies the interface or port on which a network address was last seen. The probe continuously checks the physical layer address in the data packets it observes and the corresponding network layer address to update the mapping between the addresses of the two layers.

Network layer host group. The network layer host object group enables the user to decode packets based on their network layer address. This allows the network manager to look beyond a router and into the connected hosts. This group collects statistics similar to those collected in the RMON1 host group. The main difference is that this group collects traffic data based on network layer addresses, rather than on the physical layer addresses. For example, included in the host table are network layer host addresses and data packets and octets that are transmitted to and sent out from a network layer address.

Application layer host group. The application layer host group keeps track of application-level protocols discovered at each known network layer address. Note that in RMON2, the term *application level* refers to all protocols above the network layer. It counts the amount of traffic by protocol, sent from and to each network address. For example, it records the number of error-free packets and the number of octets of a particular protocol that were transmitted to and from an address.

Network layer matrix groups. This object group enables the probe to keep track of the amount of traffic between each pair of network layer addresses monitored by the probe. The parameters monitored at the network layer fall into two categories: a source data table and a destination data table.

Included in the tables are source and destination addresses and data packets and octets that are sent out from a source and transmitted to a destination address.

The probe also maintains the traffic count for the top N pairs of network layer addresses that received or transmitted most data packets and octets for a given sampling period, while the number N is settable by a manager. The probe also generates reports on the top N addresses for management stations, and the report variables are specified by the management station. This is very similar to the RMON1 table hostTopNTable, with the exception that the monitored addresses are of the network layer instead of the physical access layer.

Application layer matrix group. This group enables the probe to track the amount of traffic by protocol that was sent between each pair of network addresses discovered by the probe. The data fall into two tables: one is associated with source addresses and one is associated with destination addresses. Among the parameters monitored by the probe for the source table are source addresses of each communicating pair, data packet, and octet sent out from a source address. The parameters are very similar for the destination addresses: destination addresses of each communicating pair, data packet, and octet transmitted to a destination address.

The probe also maintains the application layer top-N statistics, which are very similar to those of the network layer with the exception that the application layer protocols are taken into account.

User history collection group. This group periodically polls specified MIB object instances and variables and logs the data based on user-defined parameters. This group takes over the function traditionally performed by network management applications. With this powerful feature, a network manager can configure any counter in the system such as a history data on a specified host, router, or file server. For example, the userHistoryObjectVariable object allows the user to specify an arbitrary object instance in the MIB to be sampled. The great flexibility allowed to the user also results in the most complex data and control table structures among all RMON2 object groups.

Probe configuration group. This object group controls the configuration of a range of operating parameters of the probe so that a standardized set of parameters is used by all probe vendors and they can interoperate with each other. For example, the configurable operation environment variables can include any of the following:

- *probCapabilities.*   Indicates which RMON groups are supported
- *probSoftwareRev.*   The software revision of this device
- *probeDataTime.*   Probe's current data and time

An application example

The purpose of this application example is to illustrate how an RMON probe can be used to monitor traffic conditions of a remote LAN. This application example refers back to Fig. 3.1, where one management station is connected to two RMON probes that monitor two LANs. The application scenario is described as follows. The manager is interested in knowing the top two hosts in the LAN depicted at the bottom of the figure that transmitted the most bad packets for a specified period of time. The steps required of an RMON system to accomplish this are described at a very high level with many of the details left out.

1. First, the management station sends a SetRequest PDU to the agent to create a new row in the host control table with the parameter values that include
   - hostControlIndex = 1.
   - hostControlDataSource = index of the network interface attached to the probe.
   - hostControlTableSize = 5, the number of hosts on the LAN.
   - hostControlOwner = ID of the management station.

   Other protocol message exchanges between the management station and the agent used to accomplish the new row creation are left out for brevity.

2. The management station sends a SetRequest PDU to the agent to create a new row in the hostTopN control table with the parameter values that include
   - hostTopNControlIndex = 10 (a unique index number for this entry).
   - hostTopNHostIndex = 1 (same as the value of hostControlIndex).
   - hostTopNRateBase = hostOutErrors (the variable to create a report on).
   - hostTopNDuration = a specified sampling period.
   - hostTopNRequestedSize = 2 (the maximum number of hosts requested in the top-N table for this report).
   - hostTopNGrantedSize = 2 (the maximum number of hosts granted by the agent in the top-N table for this report).

   After this row is successfully created with more protocol exchanges between the management station and the agent, the management station leaves it to the agent and will not come back until it is time to retrieve the report from the agent.

3. The agent continuously collects traffic data on each host and creates new rows for the interface in the host table in a round-robin fashion, meaning the oldest row is deleted as a new row is added. The data parameters in the host table include
   - *hostInPkts.*   The number of packets received at the host
   - *hostOutPkts.*   The number of packets transmitted from a host
   - *hostOutErrors.*   The number of error packets transmitted from a host

4. At the end of the sampling period as specified in the top-N control table, the agent creates a report based on the specification in the top-N control table: the variable to be monitored (i.e., hostOutErrors), number of hosts for this report (i.e., 2), and other specifications.

5. The management station sends a GetBulkRequest PDU to retrieve the whole table.

This application example is illustrative rather than prescriptive, with the intention to demonstrate some important goals of RMON. First, it shows how the probe off-loads the burden of having to constantly poll the network devices from the management station. Once the two new rows in the control tables are set up, the manager does not return until the desired time for retrieving the report. Second, the probe extracts the information of interest to the management station from the potentially large amount of data. The management station receives the report on the top 2 hosts that transmitted the most bad packets, instead of raw data from which it has to extract the desired information.

Summary

This chapter provides a high-level view of two areas of Internet-based network management: SNMPv2 and remote network monitoring (RMON). SNMPv2 covers the SMI, MIB, and protocol operations. The remote network monitoring portion introduces both RMON1 and RMON2.

SNMPv2 SMI, which defines the structure of management information, is a straight extension to its predecessor, SNMPv1 SMI, with two major additions. One is the AUGMENT clause that enables additional columns to be added to an existing table without rewriting the table definitions. This feature is very useful when some vendor-specific or application-specific parameters need to be added onto a standard table. The second significant addition is the definition of a new SNMPv2 MIB type RowStatus that greatly increases the flexibility for creating and deleting conceptual table rows.

The SNMPv2 MIB consists of three object groups, i.e., a system group, an SNMP group, and an MIB object group. Together, they instrument both SNMPv2 and SNMPv1 by defining objects that describe the behavior of an SNMPv2 entity.

Although the SNMPv2 PDUs are based on the SNMPv1 PDUs, there are two major aspects that distinguish the former from the latter: the ways in which a response PDU is generated in response to SNMPv2 GetRequest, SetRequest, GetNextRequest, and GetBulkRequest, and the addition of two new SNMPv2 PDUs. In the previous version of SNMP, an all-or-nothing approach is used to generate a response PDU for a GetRequest, SetRequest, or GetNextRequest PDU: either all requested variables are retrieved or none is retrieved. In SNMPv2, the best-effort approach is used: as many variables are retrieved as possible. The PDU additions are GetBulkRequest and GetInformRequest. The GetBulkRequest PDU enables a manager to retrieve a large amount of data while minimizing the number of protocol exchanges. The GetInformRequest PDU allows a manager to exchange management information with another manager.

RMON is a major extension to the original SNMP standards. RMON significantly extends the functionality of SNMP without requiring any changes to the SNMP and SMI. In essence, RMON is a special-purpose MIB that allows an SNMP entity to monitor network traffic in a proactive or off-line mode and then analyze and report the traffic statistics to a management station. RMON2, an extension to the original RMON, expands the RMON capabilities by allowing a probe to look beyond the physical medium access layer and into the network and application layers.

Exercises

3.1 Explain the main differences between the SNMPv2 SMI and the SNMPv1 SMI.

3.2 Describe the effect of having a conceptual table column with SYNTAX set to RowStatus. Compare the way in which a table row is created or deleted in SNMPv2 with the way a table row is created or deleted in SNMPv1 as described in Chap. 2.

3.3 What is the main purpose of the clause AUGMENT of SNMPv2? Discuss the alternatives if no such mechanism is available for extending a standard MIB table.

3.4 Are the formats of SNMPv2-Trap PDU and SNMP Trap PDU the same? If not, describe the differences.

3.5 Assume that in a GetBulkRequest PDU, we have the following parameter values: $L = 5$, $N = 3$, $R = 2$, and $M = 2$; what is the maximum number of object instances in a response PDU? Explain the reasons that may make the actual number of object instances smaller than the maximum number.

3.6 Describe what makes up an RMON system and an RMON probe.

3.7 Explain how the new MIB type RowStatus is used in RMON MIB.

3.8 In general terms, summarize the functions of an RMON probe from the RMON1 object groups.

3.9 Describe the major aspects in which RMON2 extends the functionality of RMON1.

4

Introduction to
OSI Management Information Models

Outline

- The first section, "Introduction to the OSI Network Management Framework," provides an overview of the open system interconnection (OSI) network management framework, its basic components, its function areas, and the context in which the network management information exchanges take place.

- The second section, "Structure of the OSI MIB," defines the concepts of managed objects, how each object instance is uniquely identified in the MIB, and how the objects are organized in the MIB.

- The last section, "An Introduction to Guidelines for Definition of Management Objects," discusses the GDMO, which is a specification language used to define the OSI MIB, and includes detailed syntax, explanations, and application examples.

Introduction to the OSI Network Management Framework

This section introduces the basic concepts and requirements of open system interconnection network management, with a focus on the OSI management model that can be characterized from three aspects, i.e., informational, functional, and organizational. At the end of this section, a scenario for managing a telecommunications switch is introduced to provide an application example.

Introduction

This subsection provides background information on the OSI network management framework by introducing the concepts of OSI and the goals of the OSI network management framework and by providing an overview of OSI network management standards.

The concept of OSI. Initially, computer systems were stand-alone functioning units, a closed world. As systems evolved, computers were networked. As networking evolved, networking became a major function of computer systems. Since the mid-1980s, efforts have been undertaken by standards organizations such as the International Standards Organization (ISO) to provide standardized interfaces for computer systems. This provides the capability for computer systems from different vendors to communicate with each other though an open and standardized interface. The term *open systems* was coined to refer to those systems that are open to other systems by virtue of their adoption of common standards of communication. Thus, standards are primarily concerned with the interconnection between open systems, that is, on the exchange of information between open systems rather than on internal functions and implementation issues.

The large number of standards that have been developed and continue to be developed for OSI are collectively referred to as OSI standards. OSI standards are also cross-listed under the CCITT and the International Electrotechnical Committee due to the fact that the ISO is the umbrella standards organization of the United Nations and joint efforts from the CCITT, the International Electrotechnical Organization, and the ISO have been devoted to activities related to the OSI standards. For more information on the ISO and other related standards organizations, see App. 4B.

Goals of the OSI network management. The OSI network management standards, as discussed in this chapter and in Chap. 5, are part of the overall OSI standards. The OSI network management resides at the top layer of the OSI seven-layer network reference model and uses the services provided by the lower layers. The basic requirements of OSI management are to

- Provide capabilities that enable managers to plan, organize, supervise, control, and account for the use of interconnection services
- Be flexible in responding to changing network management requirements
- Provide the facilities to ensure reliable communications of management information between two open management systems
- Provide the facilities to ensure that management information exchanged between two management systems is secure and protected from unauthorized access

A *managed system* in OSI terminology refers to an open system that can recognize and accept OSI-based management commands from a managing system.

A *managing system* refers to a management system that performs management operations on a managed system. From the very beginning, the OSI network management standards were meant to be generic to cover all types of computer networks, including telecommunications networks. This requirement creates an inherent paradox. On the one hand, the OSI management standards are applicable to a wide range of networks, because they are not tailored to any particular network. On the other hand, their comprehensiveness and generality can result in a high degree of complexity resulting in unwieldy implementation that may hinder their otherwise wide range of implementations.

Overview of network management standards. The large number of network management-related standards defined on OSI can be classified into four categories as shown in Table 4.1.

The first category, OSI management framework, describes the general framework of OSI management architecture, its components, and the context in which OSI management applications are run. The second category, the OSI information model, focuses on the structure of the OSI management information base, its naming convention, organizational scheme, and notation for defining managed objects. The third category, the management communication services and protocol, describes the application layer management protocol termed the *Common Management Information Protocol* (CMIP) used to transport management information from one management entity to another. The last category, system management functions, is a set of generic management functions that can be utilized by user applications. The first two categories are covered in this chapter, while the third category is covered in Chap. 5. System management functions will be introduced in later chapters on the five Telecommunication Management Network (TMN) management areas, where system functions are introduced in the context of a particular management area.

OSI model for network management

This subsection provides an overall view of the OSI network management model by presenting the following four aspects:

- Model components
- Management information
- Communication and management functions
- Functional groups

Components of the OSI network management model outline the required component systems that constitute an OSI management system and describe their relation to each other. Management information presents a high-level overview of the types of management information and the structure of the information that is exchanged between system components in order to accomplish management tasks. Communication and management functions

TABLE 4.1   Four Categories of OSI Management Standards

1. OSI Management Framework

Open System Interconnection—Model and Notation	X.200
Management Framework for Open Systems Interconnection for CCITT Application	X.700
System Management Overview	X.701
Application Context for Systems Management with Transaction Processing	X.702

2. OSI Management Information Model

Structure of Management Information: Management Information Model	X.720
Structure of Management Information: Definition of Management Information	X.721
Structure of Management Information: Guideline for the Definition of Managed Objects	X.722

3. OSI Management Protocols

Common Management Information Service Definition for CCITT Applications	X.710
Common Management Information Protocol Specification for CCITT Applications	X.711

4. System Management Functions

Object Management Function	X.730
State Management Function	X.732
Attributes for Representing Relationships	X.733
Alarm Reporting Function	X.734
Event Reporting Management Function	X.735
Log Control Function	X.736
Security Alarm Reporting Function	X.737
Workload Monitoring Function	X.739
Security Audit Trail Function	X.740
Objects and Attributes for Access Control	X.741
Accounting Meter Function	X.742
Software Management	X.744

introduce the mechanism for system components to communicate with each other in order to exchange management information. Functional groups describe and catalog the management functions of an OSI management system.

OSI management system components.   An OSI management system, as shown in Fig. 4.1, consists of the following basic components:.

- A manager application that resides at a management station. The manager system is responsible for managing a set of network resources by issuing management operation commands and controlling and coordinating the overall management operations.

- An agent application that resides at a managed network resource such as a switch, transmission equipment, or software module. An agent system receives management instructions from a manager, carries out the requested management operations, and replies back to the requesting manager with the operation results and status. The agent is also responsible for emitting notifications of the events at the network resource that are of interest to the manager.

- An MIB made up of a set of managed objects that present a logical view of the network resources being managed.

- A management protocol that provides a communication mechanism for the manager to send management operation instructions to the agent and for the agent to send replies and notifications to the manager.

The relationship between a management application system and its management role is not fixed: an open system can serve as a manager, an agent, or both. A management application scenario will help illustrate the interactions between the various components just described. As shown in Fig. 4.1, the manager sends an instruction to the agent requesting that an attribute value be retrieved from a named object. The agent identifies the named object in its MIB and performs the requested operation. If the operation is successful, the agent reports the result of the operation back to the manager via the appropriate protocol message. As you may recall, the fundamental structure of an OSI management system is very similar to that of the SNMP-based management system. As a matter of fact, the SNMP management structure is based on the OSI model. The major differences between the SNMP- and OSI-based management systems are the MIB structure, the underlying management protocol, the management environment, and, as a result, the types of management applications that can be implemented.

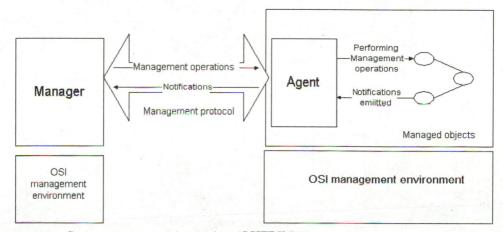

Figure 4.1   System management interactions (CCITT X.701).

For a developer, the main task is to design and develop the manager side application, agent(s), and the MIB. The management protocol is normally purchased off the shelf and the management environment comes with the management protocol stack.

Management information. Management information provides a high-level introduction of the concept of managed objects and the structure of the MIB that allows an application system to organize, store, and retrieve the managed objects. A detailed description of both managed objects and the MIB is provided in the following section.

Managed objects. A key concept in the OSI management framework is that of a managed object. A *managed object* is an abstraction of a network resource to be managed such as a connection, a protocol layer entity, or a physical network equipment. A managed object is characterized by a set of attributes and management operations. The attributes represent properties of the network resource that are of interest to the OSI management system. The management operations represent the management behavior of the objects. Therefore, the network resource is visible to the management system only through the managed object, and only those parts of the resources that are explicitly represented in a managed object can be managed through an OSI management system. Note that the management operations are performed on the managed object instead of on the network resource. How a management operation upon a managed object representing a network resource is implemented is a local matter and not subject to standardization. One managed object can represent a whole system, a small portion of a system, a network layer, or more than one layer. How a managed object represents a network resource is not subject to standardization either. The relationships between managed objects and network resources are shown in Fig. 4.2.

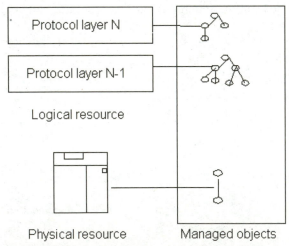

Figure 4.2 Relationships between managed objects and network resources.

Management information base. An MIB is a set of managed objects within an open system that may be transferred or affected through the use of OSI management protocols. An OSI MIB has the following characteristics:

- A standard logical structure that defines how managed objects are structured and organized.

- A standard naming convention that enables an individual managed object to be referenced unambiguously.

- A standard specification language used to specify the syntax and semantics of the managed object. However, this does not imply any form of physical or logical storage for the information, and its implementation remains a matter of local concern. Details of the MIB will be provided shortly.

Communications aspect. The communications services available to the user management application and the OSI management environment are introduced in this subsection.

OSI communications services. The OSI system management is based on the agent-manager paradigm. The interactions between an agent and a manager are realized through the exchange of management information. The OSI communications service that provides the support of information exchanges between an agent and a manager is called *Common Management Information Services* (CMIS). Specifically it provides the following three types of support, as shown in Fig. 4.3. First, the OSI communication services support the transfer of requests for management operations and notifications. A set of communications primitives is defined specifically for transferring the requests, which will be explained in Chap. 5. Second, the OSI communication services provide an access control mechanism that can either grant or deny management operation requests from a manager on selected managed objects. Third, the OSI communication services provide the mechanism to identify the destination and matching criteria for sending the notifications. The notifications originate

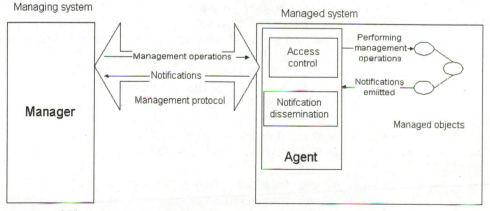

Figure 4.3 OSI communications services.

from the managed object, and it is via the agent that they are sent to the appropriate manager.

Management protocols. The OSI management protocol, CMIP, is the underlying mechanism for providing the communication services. It is the CMIP PDUs that carry the management operation requests from a manager to an agent and the notifications from an agent to a manager. The capabilities of the PDUs determine the type of management operation requests and the parameters of the requests that can be exchanged between a manager and an agent. This is in part why the CMIP is sometimes equated to the OSI management environment to emphasize its prominent position.

Shared management knowledge. In order for manager and agent systems to communicate with each other, they must share management knowledge. Just like two people talking to each other, they must share a language and a common vocabulary set. The shared management knowledge for management communications includes but is not limited to

- *Protocol knowledge.* Without a management protocol, communication never takes place between an agent and a manager.

- *Function knowledge.* For example, a manager has to know what operations an agent is capable of performing and what are illegal operations for the agent.

- *Managed object knowledge.* For example, a manager must know the identifier of an object in an agent's MIB in order to request that a management operation be performed on the object.

- *Constraints of the management functions and relationships between those functions and managed objects.* For example, if only read access is allowed on a managed object, the manager cannot request a write or set operation on the object.

Managed objects must be present to support management functions. The shared knowledge of an open system can be organized into domains based on the manager-agent pairs. As shown in Fig. 4.4, an open system assumes both agent and manager roles; thus its shared knowledge can be conceptually partitioned into two parts based on the role it plays. There are two issues related to the maintenance of the shared management knowledge: when the shared knowledge is established and how to modify the knowledge. Part of the shared knowledge is built into an open system, which we call *a priori knowledge,* and part of the knowledge can be established at run time after the association between a manager and an agent has been established, which is termed *dynamic knowledge.* Examples of a priori knowledge include protocol knowledge and function knowledge, while an example of dynamic knowledge includes object identifiers that are created after the association establishment.

OSI management functional aspect. Functionally, the OSI management environment provides the following sets of capabilities to user applications:

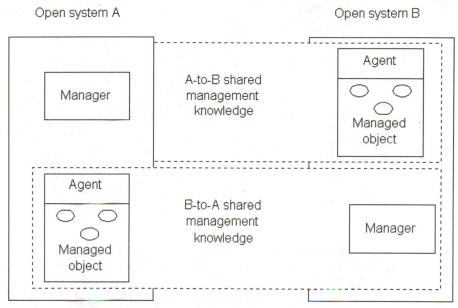

Figure 4.4   Shared management knowledge.

- The capability that allows an agent to notify a manager of an event
- The capability that allows a manager to request an agent to perform a set of management operations such as retrieving or setting a managed object's attribute values in the agent's MIB

A management scenario

A scenario for managing a simple, hypothetical switch is presented here to illustrate the basic concepts of OSI management that have been discussed so far. It also serves as an example of the application to be used throughout this chapter and Chap. 5. As shown in Fig. 4.5, the simplified switch consists of a central processing card containing a software computing module. The card performs all the switching functions, and the software computing module is the switching application. The switch consists of two identical units, one as an active unit and one as a standby unit for reliability. An OSI-based management system consists of a manager station, an agent residing in the switch, a set of managed objects that represent the network resources to be managed, and an underlying OSI environment that provides communication capabilities for the manager and agent. The manager has the responsibility of monitoring the health of the switch and taking the necessary actions when needed, such as switching to the standby unit in case the active unit fails. An agent system resides in the switch and communicates with both the manager and the managed objects that represent the software and hardware switch resources to accomplish the management tasks.

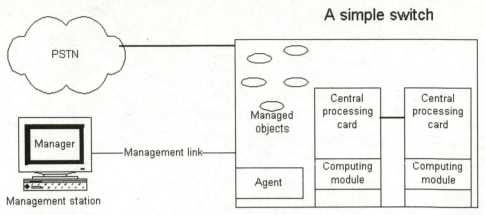

Figure 4.5 A simple switch and a management system.

Although this is a simplified scenario of telecommunications network management, it is sufficient to illustrate what an OSI-based management system looks like in a real network and how network management frameworks, information models, and management protocols can be applied to a real-world management problem.

Structure of the OSI MIB

This section introduces the OSI management information base and related concepts. Fundamental to the OSI MIB is the concept of managed objects, a basic concept of object-oriented information modeling. A brief overview of object-oriented concepts is presented here to provide some necessary background information, followed by a description of the components of a managed object as specified in the *Guideline for Definition of Managed Objects*. (GDMO). The GDMO is an OSI-specified object-oriented management information modeling language. Then the discussion shifts to the OSI MIB structure, the object naming scheme, and related issues.

The building blocks of OSI MIB—managed objects

The central concept of the OSI MIB is the managed object that is an application of the generic object-oriented concepts to the network management information modeling. The generic object-oriented concepts are reviewed first.

Overview of object-oriented concepts

Object. When the term *object* is used in a generic sense, it refers to a software unit that represents an entity in the domain to be modeled. Any entity, physical or conceptual, concrete or abstract, can be an object. An object is characterized by a set of attributes that represents properties of the entity and a set of operations that can be performed on the attributes. Each object has a state and an identity. For example, switch, router, and software modules can all be objects. When the term object is used in a narrow sense, it is interchangeable with object instance.

Class. An *object class* is an abstraction of a set of objects sharing the same attributes and common behavior. A class is like a type of object that can be instantiated, and sometimes the terms *type* and *class* are used interchangeably. An instantiated object class is called an *object instance* or *object*. An example of a class is a category of switches such as a central office switch. An object instance of this class is a particular central office switch.

Encapsulation. *Encapsulation* refers to the fact that the attributes (data) and the operations on the data are encapsulated in the object and the only way to affect the object is to perform an operation on it. A related concept is *information hiding*; that is, the internal structure is hidden from the outside world. In order to use an object, we do not need to know how the object's behavior or attributes are represented or implemented internally.

Inheritance. An important relationship between classes is inheritance. *Inheritance* occurs when a child class shares the structure or behavior defined in one (single inheritance) or more (multiple inheritance) parent classes. The child class is also termed a *subclass,* and the parent class or classes are termed *superclasses.* A subclass usually specializes its superclass by augmenting or redefining existing structure or behavior. Inheritance is a way to reuse the structure and behavior without redefining them again and again if many classes share the same attributes and operations.

For the purpose of illustration, the simple switch described above can be a class with attributes such as operational states (e.g., in_service, out_of_services, idle), the number of ports supported, and operations the switch is capable of, such as start-up and shutdown. A local switch can be a subclass of the generic switch class and defines an additional attribute such as total number of line cards (remember that a local switch has two sides: the line side faces the customer and the trunk side faces other switches).

Aggregation. This is another important relationship between classes, where one class consists of one or more other classes. The component classes are also termed *contained classes* when one class is contained by another class as an attribute of the containing class. This is also called "has_a" relationship: the containing class has a contained class as a component. The containment relationship is to model the real-world hierarchies of is_part_of relationship (e.g., engine is part of a car) or real-world organizational hierarchies (e.g., a directory contains files, a file contains records, and a record contains fields). For example, a switch can have a central computing card, the card itself is a class, and both can have attributes and behaviors.

Polymorphism. *Polymorphism* is a concept in type theory, according to which a name may denote objects of many different classes that are related by some common superclass. Any object denoted by this name is able to respond to some common set of operations in different ways. For example, an object name can be created for the class animal. There are subclasses for a variety of specific types of animals such as dog, cat, or cow, and each has its own behavior. It is only at the run time that the specific type of animal can be determined. The object name can be bound to a specific animal type and takes on its behavior at run time. See Meyer (1988), Booch (1994), Jacobson (1995), and Wirfs-Brock (1990) for more details on object-oriented concepts and object-oriented analysis and design.

Structure of managed objects. This subsection introduces the structure of managed objects as defined in the GDMO, the OSI-specified object-oriented management information modeling language. A managed object class consists of a set of packages, and a package contains attributes, attribute groups, actions, and notifications, as shown in Fig. 4.6. All these can be either mandatory or optional. Each instance of a managed object class must have all the mandatory portions of the class. A brief description of each component provides an overall view of the managed object, while detailed syntax and semantics are in the GDMO section.

Package. A package is a collection of attributes, attribute groups, notifications, and behavior which form an integral part of an object class definition. A package is either mandatory or optional in a managed object class definition. A mandatory package must be present in all instances of the given managed object class. A conditional package is included in a managed object instance only if the condition associated with the package is true. The condition for a conditional package is usually related either to the capability of the underlying resource being modeled by the managed object or the

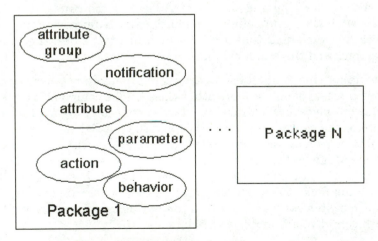

OSI managed object class
structure

Figure 4.6   Structure of a managed object class.

presence or absence of certain management functions supported by the managed system. A package is a logical grouping of attributes, attribute groups, actions, notifications, and behavior; once encapsulated in a managed object, it becomes an integral part of the managed object and becomes accessible only through the interface of the managed object. The following rules apply to a package:

- Only one instance of a given package can be present in a managed object.

- A package can never be instantiated without the object class that encapsulates it, and it has to be instantiated as part of the object class; it has to be deleted at the same time as the managed object.

Attribute. An *attribute* represents a property of the network resource being modeled. Any access to an attribute has to go through the interface of the managed object that contains the attribute. The value of an attribute can be either retrieved or modified, again via the containing object interface. The permitted operations on an attribute are defined in the containing managed object class. An attribute can be single-valued or set-valued. If set-valued, each of the elements of the set is of the same data type and the set can be empty.

Attribute group. An *attribute group* is a logical grouping of a set of attributes for organizational and operational convenience. Two types of attribute groups may be defined: fixed and extensible. The set of attributes in a fixed attribute group are determined in the initial definition and cannot be changed later, while new attributes can be added into the extensible attribute group with the behavior specific to the group.

Action. An *action* is an operation that can be performed on one or more managed objects by an agent as requested by a manager. An action can be in either a confirmed or an unconfirmed mode. If in a confirmed mode, the agent is required to send a message to the manager to confirm the execution of the action along with any result. The parameters required for action are specified in a separate parameter construct.

Notification. A *notification* is a report emitted by a managed object and sent to a manager via the associated agent. A notification is an unsolicited operation initiated by the managed object. Like an action, it can be in either a confirmed or an unconfirmed mode. If in a confirmed mode, the receiving manager sends an acknowledgment to the agent upon receipt of the notification.

Behavior. *Behavior* is the part of a package that provides a textual description of the semantics of a package. A notification or an action can also have a behavior part describing its behavior. Two types of relationships (inheritance and aggregation) exist between managed object classes in the OSI MIB. The inheritance relationship is defined in the definition of a managed object class, and the containment relationship is defined by the name-binding construct.

Name binding. The *name-binding construct* defines the containment relationship between managed object classes by specifying a containing class or classes.

OSI MIB structure

Managed objects must be organized in the MIB in such a way that each object instance can be uniquely identified and objects can be easily stored into and retrieved from the MIB when needed. There are four hierarchical structures that are important for understanding the OSI MIB structure. They are the OSI registration hierarchy, inheritance hierarchy, containment hierarchy, and naming tree.

OSI registration hierarchy.   For the OSI MIB, the registration hierarchy is as its name implies: registration of managed object classes and their components so that each managed object class and its components can be uniquely referenced and used by management applications. Note that unlike the Internet MIB, the OSI registration tree does not represent the actual structure of an OSI MIB. As shown in Fig. 4.7, each node registers an entity with two identifiers assigned to it. The first is an integer identifying the relative position of the node at the horizontal level, which we call a relative identifier. For example, ITU is assigned a 0 and ISO a 1 and so on in an ascending order, from left to right. The other identifier, which we call an absolute identifier, uniquely identifies the position of each node in the entire tree and is formed by concatenating all the relative identifiers along the vertical path leading to this node, starting from the very top. For example, the managed object class AttributeValueChange (AVC) is assigned the identifier

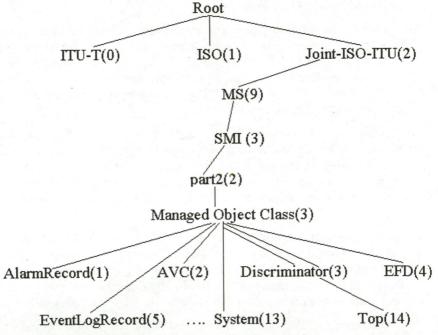

Figure 4.7 A partial OSI registration tree.

2.9.3.2.3.2 and EventForwardDiscriminator (EFD) the identifier 2.9.3.2.3.4. They are formed by concatenating all node identifiers from the top of the tree to the node in question, respectively.

Inheritance hierarchy.   The *inheritance hierarchy* represents the inheritance relationship between managed object classes. A child's managed object class inherits all the attributes, actions, and notifications of its parent, and in addition, it can have attributes of its own.

Figure 4.8 shows an inheritance hierarchy of the managed object class defined in ITU X.721. The managed object class Top is at the top of the hierarchy and has no parent class. The rest of the managed object classes are derived from it either directly or indirectly. The detailed syntax of inheritance relationships is presented in the GDMO section.

Containment hierarchy.   This hierarchy represents the containment relationship between managed object classes. As shown in Fig. 4.9, the System managed class contains the Log managed object class, which in turn contains the LogRecord object class. The containing class is called the *superior class,* while the contained class is the *subordinate class..*

Naming tree.   The final hierarchy structure is perhaps the most important structure of the OSI MIB. The *naming tree,* or the instance tree of the containment hierarchy, represents the containment relationships between object instances rather than classes. The naming tree is important because it is the actual structure of the OSI MIB. As shown in Fig. 4.10, an object

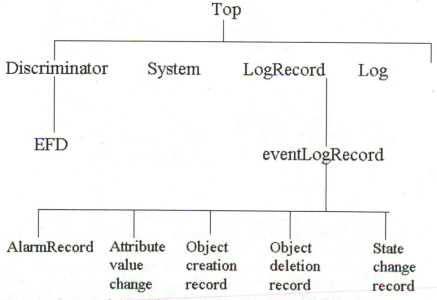

Figure 4.8 A managed object inheritance tree.

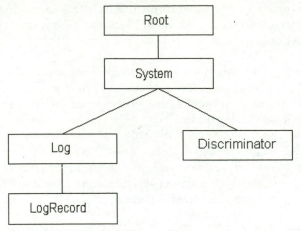

Figure 4.9 A containment hierarchy.

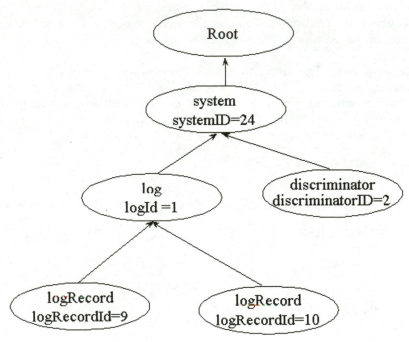

Figure 4.10 A naming tree.

instance of a containing managed object class and an object instance of a contained managed object class are represented as nodes in a directed graph with an arrow going from a subordinate object instance to a superior object instance. The top level of the tree is referred to as the root, which is a null object that always exists. Each object instance in the naming tree, or in an OSI MIB, is uniquely identified by its distinguished name, which is made up of a set of relative distinguished names.

Relative distinguished name (RDN). When an attribute takes on a value, the whole statement of "attribute identifier = attribute value" is termed an *attribute value assertion* (AVA). The RDN of an object is the AVA that must unambiguously identify a single managed object within the scope of its superior object. In addition, the attribute used as the RDN must

- Be part of a mandatory package
- Be testable for equality
- Permit its value to remain fixed for the lifetime of the managed object that uses it for naming
- Not have one of the following ASN.1 types as the attribute syntax:
  - The real type
  - The set type
  - The set-of type
  - The any type
- Have a choice type that includes any of the above types as a choice
- Have a type derived from any of the above types by tagging, by subtyping, or by use of the selection type

For example, in the naming tree shown in Fig. 4.10, the relative identifier of a log object instance is "logId = 1".

Distinguished name (DN). The OSI MIB is structured according to the naming tree. Each object is uniquely identified in the entire tree by its DN. The DN of an object is formed by concatenating the RDNs of all the nodes starting from the top of the tree all the way down to the node of the object. For example, the DN of the object that has the RDN "LogRecordId = 10" is "systemId = 14, logId = 1, LogRecordId = 10". Table 4.2 lists all DNs and RDNs of the tree in Fig. 4.10.

Comparison between the OSI MIB and the SNMP MIB. A comparison between the OSI and the Internet SNMP MIB will further shed light on the OSI structure. Table 4.3 provides a summary of Internet and OSI MIB characteristics.

Operations on managed objects. The operations that can be performed on managed objects in the OSI MIB fall into two categories: attribute and object level. Both types of operations are directly supported by the OSI management

TABLE 4.2   Examples of RDNs and DNs

Relative distinguished name	Distinguished name
systemId=24	{}
logId=1	SystemId=24, logId=1
eLogRecordId=9	systemId=24, logId=1, eLogRecord=9
aLogRecordId=10	systemId=24, logId=1, aLogRecord=10

TABLE 4.3   Comparison between OSI MIB and SNMP MIB

	OSI MIB	Internet MIB
Object-oriented	Yes	No
Inheritance	Yes	No
Containment	Yes	No
Polymorphism	Yes	No
MIB structure is based on	Naming tree, an instantiation of the managed object class containment hierarchy	The registration tree
The registration tree is used for	Registration of managed object class and its components, not the instance of objects	Generating a distinct object identifier, with no distinction between managed object class and management object instance
Static vs. dynamic MIB structure	May be different	Same

protocol CMIP. Only a high-level introduction of the operations is presented here, and the detailed description is left to Chap. 5.

Attribute-level operations. Bear in mind that a manager sends the requests to a managed object to perform the following management operations on its attributes:

- *Get attribute value.* A manager supplies a list of attribute identifiers. The managed object determines whether the class definition allows the encapsulated attribute values to be read. If yes, a list of requested values is returned. Otherwise, an indication of an error is returned.

- *Replace attribute value.* A manager supplies a list of attribute identifiers and corresponding values. The managed object determines whether the replacement of the attribute values is allowed. If yes, the attribute values of the named object in the MIB are replaced. If the attribute values are non-writable or if incorrect types of attribute values are supplied, an indication of an error is returned.

- *Replace-with-default value.* A manager sends a request to the managed object with a list of attribute identifiers. The values of the named attributes are replaced with the default values. The default values may be defined as part of the managed object class specification or may be left as a choice of the implementation.

- *Add a member to a set-valued attribute.* A manager sends a request to a managed object with a list of attribute identifiers and the associated values containing the members to be added. For each named, set-valued attribute, this operation replaces the attribute value with the set union of the set of existing values with the set of values supplied by the manager.

- *Delete members from a set-valued attribute.* A manager sends a list of attribute identifiers and the associated values containing the members to be deleted. The operation replaces the attribute's value with the set difference of the set of existing members and the set of members supplied by the manager. In other words, the resulting attribute value set is the previous set minus the ones supplied by the manager.

Object-level operations.   The operations that can be applied to an object as a whole:

- *Create.*   A manager sends an instruction to an agent to create a specified object instance, and that instance is added to the agent's MIB.

- *Delete.*   A manager sends an instruction to an agent to delete one or more specified object instances. The MIBs at both the agent and manager sides are updated to reflect the change.

- *Action.*   A manager sends an instruction for an agent to execute a specified action. This indicates that the manager has knowledge of the capabilities of all the associated agents to perform given actions.

Both the attribute- and object-level operations are illustrated in Fig. 4.11.

An example of an OSI MIB

Figure 4.5 shows a simple scenario of a network manager managing a simple switch. The switch consists of a central processing card and a computing module running on the card. The general steps for building an OSI MIB for this management scenario are outlined below.

The very first step of building an MIB is to model the network resource, the switch in this case. Decide how to represent the switch; determine what should

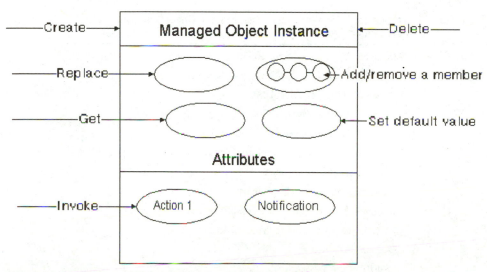

Figure 4.11 Attribute- and object-level operations on the OSI MIB.

be represented as an object and what as attributes. How to model a network resource largely depends on the management requirements and consequently the view of the network resource in which the management system is interested.

For our simple switch, we identified three managed object classes and their associated attributes, actions, and notifications, as shown in Table 4.4.

The second step is to determine the relationships between the managed object classes that have been identified in the first step. There are two basic types of relationships that need to be explicitly modeled: inheritance and containment. The inheritance and containment hierarchies for the simple switch example are shown in Figs. 4.12 and 4.13, respectively. An important decision in designing the containment hierarchy is to designate an attribute of each managed object class that will be used to generate an RDN for each object of the class. The AVA generated from the attribute must satisfy the conditions such as being unique among all objects subordinate to a given object and its value capable of remaining constant for the lifetime of the object, as discussed earlier. The attribute name of a class that will be used to generate RDNs for the objects of the class is included in parentheses of the containment hierarchy shown in Fig. 4.13.

The third step is to formally represent the classes and the relationships between classes identified above using a formal specification language and then to compile the representation into a run-time MIB that can be used by management applications. The next section describes the formal specification language that is widely used for specifying OSI-based MIBs.

Introduction to Guideline for Definition of Managed Objects

Because of GDMO's wide acceptance and deployment in telecommunications network management applications, this section presents a detailed discussion

---

TABLE 4.4   Managed Object Classes and Class Components for the Simple Switch

Managed object class: simpleSwitch

   Attribute: switchId, hwState, swState, errorState

   Action: switchToBackupUnit

   Behavior: simpleSwitchBehavior

Managed object class: computingModule

   Attribute: computingModuleId, lastBackupTime, lastRestoreTime

   Action: executeProgram

   Notification: autobackupReport

Managed object class: centralComputingCard

   Attributes: cardID, cardType, cardState

   Action: testCard

   Notification: cardDown

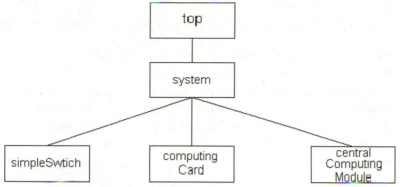

Figure 4.12 Inheritance hierarchy for the simple switch system.

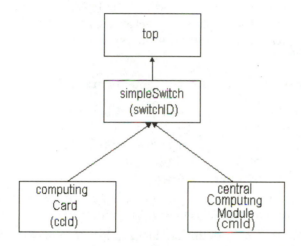

Figure 4.13 Managed object class containment hierarchy.

of GDMO, its syntax, and its constructs and concludes with an application example. GDMO was first published in 1992 as a joint ISO/International Electrotechnical Committee and CCITT standard specification language to be used for defining managed object classes for OSI MIB (see ITU X.722). GDMO is a specification language with well-defined syntax and semantics. The syntax defines a set of rules specifying what constitutes a legal or illegal statement, like a statement in a regular programming language. Like a programming language, the definitions of a managed class in GDMO can be compiled. The difference is that the output of the GDMO compiler is a predefined representation of data that can be used by a management application, while the output of a programming language compiler is the management application itself.

A GDMO consists of a set of templates that defines a standard format and structure of the managed object class and a component of a managed object class such as a package, an attribute, or a notification. A brief introduction of the syntax rules of GDMO is followed by a detailed description of each template.

GDMO syntax

The following syntactic rules are observed while writing a GDMO template.

- All terminal symbols and key words that are part of the template definition are case-sensitive.

- A template consists of a set of template elements, which are separated by a blank space, the end of a line, a blank line, or a comment.

- A comment starts with the character pair - - and terminates with the same character pair or the end of a line, whichever comes first.

- The semicolon is used to mark the end of a template construct within a template except for the REGISTERED AS and DEFINED AS template constructs and to mark the end of the template itself.

- The legal character set consists of uppercase or lowercase alphabetic characters, the digits 0 to 9, the hyphen character -, and the right slash character / in any order, with the exception that no two consecutive hyphens can be present.

The conventions for defining the GDMO elements are as follows:

- Strings surrounded by square braces [ ] delimit parts of a template that may be present or absent. If the closing brace is followed by an asterisk (i.e., [ ]*), the content in the braces may appear zero or more times. The circumstances under which these parts of the definition may be omitted are dependent upon the definition of the template type.

- A string surrounded by < > is similar to a parameter name in a programming language in which each instance will be replaced in the template.

- The character | is used as a delimiter for an alternative string in the syntactic definition of a supporting production.

- A template label shall be unique at least within the GDMO module.

Description of the construct template

Figure 4.14 shows the three types of GDMO constructs: managed object classes, components of a managed object class, and name binding. Name binding defines the containment relationships between managed object classes.

This section describes the template of each construct. The description consists of four parts: an overview of the construct, a *template structure* section providing the exact syntax of the construct, a *template elements* section explaining each element of the construct, and a *template example* illustrating the construct just described.

Managed object class template.   The managed object class template is the basis of the GDMO representing a type of underlying managed network resource. Besides defining the components of a managed object class, it also specifies the

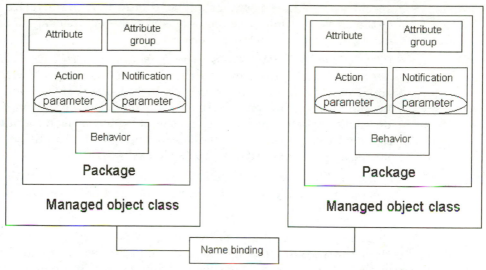

Figure 4.14 GDMO constructs and their relationships.

inheritance relationships between this class and others so it can be placed at the appropriate node of the inheritance tree.

Template structure

```
<class-label> MANAGED OBJECT CLASS
    [DERIVED FROM          <class-label> [,<class-label>]*;]
    [CHARACTERIZED BY      <package-label> [,<package-label>]*;]
    [CONDITIONAL PACKAGE   <package-label> PRESENT IF condition-
            definition [<,package-label> PRESENT IF condition-
            definition]*;]
REGISTERED AS object-identifier;
```

Supporting productions:

```
condition-definition -> delimited string
```

Template elements

- *<class-label> MANAGED OBJECT CLASS*.   A managed object class name is in place of <class-label>. The managed object class name has to be unique in this GDMO module and will be used in the management protocol message to identify this class.

- *[DERIVED FROM <class-label> [,<class-label>]*;]*.   This element specifies the superclass or superclasses from which the class is derived. Multiple inheritances are allowed, and this element must be present in all managed object class definitions.

- *[CHARACTERIZED BY <package-label> [,<package-label>]*;]*.   The mandatory package or packages are used to specify the managed object's behaviors, attributes, notifications, and operations that provide a complete specification of the operations and attributes that characterize all instances of the class. Multiple package names in place of <package-label> [,<package-label>*;] can be present here.

- *[CONDITIONAL PACKAGE <package-label> PRESENT IF condition-definition [<,package-label> PRESENT IF condition-definition]*;].* The conditional package or packages are included only if the condition defined in condition-definition is true. The *condition definition* is a textual description of a condition such as the presence or absence of certain management capabilities or elements. An example of a condition is "the corresponding protocol entity supports Class 4 operation."

- *REGISTERED AS object-identifier.* The object class identifier has to be a globally unique identifier that is used to register this class in the registration hierarchy.

A package, an attribute, an action, or a notification in a package can be included more than once due to either inheritance or multiple inheritances, and the resulting set of package, attributes, action, or notification is a set union of all packages, attributes, actions, and notifications included in the class definition.

Template example

```
system MANAGED OBJECT CLASS
    DERIVED FROM top;
    CHARACTERIZED BY systemPackage;
    CONDITIONAL PACKAGES
       administrativeStatePackage PRESENT IF "an instance supports it.",
       supportedFeaturePackage PRESENT IF "an instance supports it.";
    REGISTERED AS {smi2ObjectClass 13};
```

The managed object class named *system* is inherited from the class *Top* and has one mandatory package and two conditional packages. Besides the object name *system,* it has a registration identifier that is made up of another identifier smi2ObjectClass and the number 13.

Name-binding template.   The name binding as described earlier defines the containment relationships between a subordinate managed object class and a superior managed class that contains the subordinate class. The name binding identifies the naming attribute whose AVA is the RDN of all the instances of the subordinate class, which is part of the DN that uniquely identifies each instance of the object class in the entire MIB.

The name binding is not part of the definition of either the subordinate nor the superior managed object class. Instead, it is the only construct in the GDMO that defines the relationship between the classes.

Template structure

```
<name-binding-label> NAME BINDING
    SUBORDINATE OBJECT CLASS <class-label> [AND SUBCLASSES];
    NAMED BY SUPERIOR CLASS <class-label> [AND SUBCLASSES];
```

```
WITH ATTRIBUTE <attribute-label>;
[BEHAVIOR <behavior-definition-label> [,<behavior-definition-
        label]*;]
[CREATE <create-modifier [, create-modifier]] [<parameter-label>]*;]
[DELETE [delete-modifier] [<parameter-label>]*; ]
```

Template elements

- *SUBORDINATE OBJECT CLASS <class-label> [AND SUBCLASSES];*
  The subordinate class is named here by class-label whose instances may be named by instances of the object class defined by the NAMED BY SUPERI-OR OBJECT CLASS element. As you recall, the name of an instance of this subordinate class is constructed by concatenating the RDN of the object instances of its superior class with the RDN of the subordinate class instance. If AND SUBCLASS is present, the name binding also applies to all subclasses of this subordinate class.

- *NAMED BY SUPERIOR CLASS <class-label> [AND SUBCLASSES];*
  This identifies the superior object class whose instance may name the instances of the subordinate class defined by the SUBORDINATE OBJECT CLASS template element. If AND SUBCLASS is present, the name binding also applies to all subclasses of the specified superior object class.

- *WITH ATTRIBUTE <attribute-label>;* This is the attribute that will be used to construct the relative distinguished name of each instance of this subordinate object class. The values of this attribute shall be restricted to a single-valued data type such as INTEGER. If no suitable attribute is available for use as a naming attribute, the designers shall provide a naming attribute of the type GraphicString.

- *[BEHAVIOR <behavior-definition-label> [,<behavior-definition-label]*;].*
  If present, this element allows the designer to name one or more behavior definitions identified by <behavior-definition-label> [,<behavior-definition-label]*;] that each provide a textual description of the impact created by the name binding.

- *[CREATE <create-modifier [, create-modifier]] [<parameter-label>]*;]* If present, this element specifies that the creation of new instances of the object class identified by SUBORDINATE OBJECT CLASS is permitted in a system management operation such as a manager sending a request to an agent. The create-modifier specifies the options available on creation of the object instances. The permitted values the create-modifier is allowed to have are

  - *WITH-REFERENCE-OBJECT.* If present, a reference object class instance can be specified on the creation of an instance of the subordinate class as a source of default values and as a means to specify the choices of conditional packages.

  - *WITH-AUTOMATIC-INSTANCE-NAMING.* If present, the create request sent from a manager to an agent may omit the name of the object instance to be created because the system will automatically generate a name.

- *[DELETE [delete-modifier] [<parameter-label>]*;].* A scenario of the delete operation is that a manager sends a protocol message requesting an agent to perform a delete operation on certain objects in the agent's MIB. If this element (a template element) is present, it is permitted to delete the instances of the object class identified by the SUBORDINATE OBJECT CLASS element. The delete-modifiers clause, if present, further defines how the delete operation should be performed and can have the following values:

  - *ONLY-IF-NO-CONTAINED-OBJECT.* The delete operation is permitted only if the object to be deleted does not contain any other object. Otherwise an error will result.

  - *DELETE-CONTAINED-OBJECT.* If this delete modifier is present, the delete operation will delete this and all contained objects. But if any contained object has the ONLY-IF-NO-CONTAINED-OBJECT delete modifier, the delete operation will fail. The parameter-labels clause identifies name-binding-specific error parameters associated with the delete operation. The syntax of the error parameters is defined in the parameter template.

Template example

```
log-system NAME BINDING
     SUBORDINATE OBJECT CLASS log AND SUBCLASSES;
     NAMED BY SUPERIOR OBJECT CLASS system AND SUBCLASSES;
     WITH ATTRIBUTE logId;
     CREATE WITH-REFERENCE-OBJECT,  WITH-AUTOMATIC-INSTANCE-NAMING;
     DELETE ONLY-IF-NO-CONTAINED-OBJECTS;
REGISTERED AS {smi2NameBinding 2};
```

In this example, the managed object class *log* and its subclasses are named (contained) by the object class *system*. The RDN of each instance of the log object class is determined by the attribute logId. When a log object instance is created, a reference object instance can be specified to provide the default values for the new object instance and the name of the new object can be omitted because the system will automatically assign one. A log object instance can be deleted successfully only if the object does not contain any other object or the delete operation will fail.

Package template. A package is a logical grouping of a set of attributes, attribute groups, notifications, behaviors, and actions, and it can't be instantiated itself without the instantiation of the class containing this package.

Template structure

```
<package-label> PACKAGE
  [BEHAVIOR              <behavior-definition-label> [,behavior-
                        definition-label]*;]
  [ATTRIBUTES           <attribute-label>propertylist [<parameter-label>]*
                        [,<attribute-label>propertylist [<parameter-
                        label>]*]*;]
  [ATTRIBUTE GROUPS     <group-label> [<attribute-label>]* [,<group-
                        label> [attribute-label]*];]
```

```
        [ACTIONS                <action-label> [<parameter-label>]*
                                [<action-label> [<parameter-label>]*]*;]
        [NOTIFICATION           <notification-label> [<parameter-label>]*
                                [,<notification-label> [<parameter-label>]*]*;]
REGISTERED AS object-identifier;
```

Supporting productions:

```
propertylist       →      [REPLACE-WITH-DEFAULT]
                          [DEFAULT VALUE value-specifier]
                          [INITIAL VALUE value-specifier]
                          [PERMITTED VALUE type-reference]
                          [REQUIRED VALUE type-reference]
                          [get-replace]
                          [add-remove]
value-specifier    →      value-reference | DERIVATION RULE <behavior-
                          definition-label>
get-replace        →      GET | REPLACE | GET-REPLACE
add-remove         →      ADD | REMOVE | ADD-REMOVE
```

Template element

- *BEHAVIOR <behavior-definition-label> [,behavior-definition-label]*;].* A unique behavior name should be in place of <behavior-definition-label>. The BEHAVIOR construct allows the semantics of the package to be described in detail in textual format. An instance of the behavior template is intended to help the reader to understand the package as well as to describe the required behavior for the system implementation.

- *[ATTRIBUTES   <attribute-label>propertylist   [<parameter-label>]* [,<attribute-label>propertylist [<parameter-label>]*]*;].* This element allows attributes to be included in the package. Each attribute is uniquely named by an attribute name in place of attribute-label. The propertylist clause defines the operations permitted on this attribute and defines any default, initial, permitted, or required value.

- *[ATTRIBUTE GROUPS <group-label> [<attribute-label>]* [,<group-label>. [attribute-label>]*];]].* This template element defines a set of attribute groups, each of which is uniquely identified by group-label.

- *[ACTIONS <action-label> [<parameter-label>]* <action-label> [<para-meter-label>]*]*;].* The template element defines a set of action definitions identified by <action-label>. An action is an operation one management entity (e.g., manager) requests another entity (e.g., agent) to perform and the agent replies back to the manager on the operation performed. The parameter-labels clause specifies either the parameter of the requested operation or the reply or any error parameters associated with the action.

- *[NOTIFICATION <notification-label> [<parameter-label>]* [,<notifica-tion-label>[<parameter-label>]*]*;].* This template element identifies the notifications to be included in the package. When a notification is sent from an agent to a manager, the manager sends back an acknowledgment. Like ACTION, <parameter-label> specifies either the parameter of the notifica-tion, the reply, or any error parameters associated with the NOTIFICATION.

Supporting productions:

- *propertylist* → *[REPLACE-WITH-DEFAULT]*.   As described above, propertylist defines the operations permitted on an attribute included in the package. The REPLACE-WITH-DEFAULT propertylist specifies that the attribute has a default value that can be set by the replace-with-default-value operation.

- *[DEFAULT VALUE value-specifier]*.   The DEFAULT VALUE propertylist specifies a default value for the attribute at the time the object class is instantiated. If a default value is not present but the REPLACE-WITH-DEFAULT propertylist is, the default value is determined either by value-reference or a DERIVATION RULE that specifies how the default value should be determined.

- *[INITIAL VALUE value-specifier]*.   The INTIAL VALUE propertylist specifies a mandatory initial value for the attribute at the time of the object class instantiation. The value is determined either by value-reference or a DERIVATION RULE that specifies how the default value should be determined.

- *[PERMITTED VALUE type-reference]*.   The PERMITTED VALUE propertylist restricts the possible value the attribute can take to the type specified by type-reference.

- *[REQUIRED VALUE type-reference]*.   The REQUIRED VALUE propertylist specifies the type of values that the attribute is capable of taking.

- *[get-replace]*.   The operations permitted on this attribute are either GET (read), REPLACE (write), or both, GET-REPLACE.

- *[add-remove]*.   The operations permitted on the attribute are ADD (create), REPLACE (delete), or both, ADD-REPLACE.

Template example

```
weekScheduling PACKAGE
   ATTRIBUTES
     weekMask REPLACE-WITH-DEFAULT
             DEFAULT VALUE Attribute-ASN1Module.defaultWeekMask
             GET-REPLACE ADD-REMOVE
REGISTERED AS {smi2Package 29}
```

The package weeklyScheduler has only one attribute, weekMask, and the attribute instance at the time of instantiation of the containing managed object class shall take on the default value of the type defaultWeekMask defined in the ASN.1 module Attribute-ASN1Module. The permitted operations on this attribute are GET-REPLACE and ADD-REMOVE.

Attribute template.   An attribute represents a property of a managed network resource and can take on as its value a set, a singleton, or another attribute. It is possible that the value of one attribute depends on the values of one or more other attributes in the same managed object, and when the value of the attribute that provides value to other attributes changes, a synchronization is required. A synchronization requirement such as this is specified in the behavior of the attribute.

Template structure

```
<attribute-label> ATTRIBUTE
     derived-or-with-syntax-choice;
     [MATCHES FOR qualifier [,qualifier]*;]
     [BEHAVIOR <behavior-definition-label>
               [,<behavior-definition-label>]*;]
     [PARAMETERS <parameter-label> [,<parameter-label>]*;]
REGISTERED AS object-identifier;
```

Supporting productions:

```
qualifier                          →    EQUALITY | ORDERING | SUBSTRING |
                                        SET-COMPARISON | SET-INTERSECTION
derived-or-with-syntax-choice    →    DERIVED FROM <attribute-label> |
                                        WITH ATTRIBUTE SYNTAX type-reference
```

Template elements

- *<attribute-label> ATTRIBUTE*. The attribute-label provides a globally unique attribute identifier.

- *derived-or-with-syntax-choice* where *qualifier → DERIVED FROM <attribute-label> | WITH ATTRIBUTE SYNTAX type-reference*. This element allows the definition of this attribute to be derived from an existing attribute identified by attribute-label, and the additional constraints can also be added to the existing attribute definitions. The WITH ATTRIBUTE SYNTAX type-reference, present only if DERIVED FROM <attribute-label> is absent, defines the ASN.1 syntax that defines the data type of the attribute. If the base type of the syntax is the set-of type, the attribute is a set-valued attribute. All other ASN.1 data types, including the set type, sequence type, and sequence-of type define a single-valued attribute.

- *[MATCHES FOR qualifier [,qualifier]*;]* where *qualifier → EQUALITY | ORDERING | SUBSTRING | SET-COMPARISON | SET-INTERSECTION*. This element defines the types of tests that may be applied to the value of the attribute as part of a filtering operation. The filtering operation provides for the selection of objects based on certain specified criteria. Assuming the attribute value being tested is X and a given value is Y, the following types of tests can be applied to X and Y:
  - *EQUALITY*. The attribute value may be tested for equality against a given value (i.e., is $X = Y$?).
  - *ORDERING*. The attribute value may be tested against a given value to determine which value is greater (i.e., is $X > Y$?).
  - *SUBSTRING*. The attribute value may be tested against a given substring value to determine whether the substring is present or absent in the attribute value (i.e., is Y a substring of X?).
  - *SET-COMPARISON*. The attribute value may be tested against a given value to determine whether one is a subset of other (i.e., is X a subset of Y or vice versa?).

- *SET-INTERSECTION.*   The attribute value may be tested against a given value to determine whether the two values intersect (i.e., does *X* intersect *Y?*).

- *[BEHAVIOR <behavior-definition-label> [, <behavior-definition-label>]*;].* Each behavior element, identified by a behavior-definition-label, is a textual specification of the behavior generic to this attribute type.

- *[PARAMETERS <parameter-label> [, <parameter-label>]*;].*   This element provides the parameters that are used to define various failures associated with the behaviors of this attribute.

Template example

```
logId ATTRIBUTE
    WITH ATTRIBUTE SYNTAX Attribute-ASN1Module.SimpleNameType;
    MATCHES FOR EQUALITY, SUBSTRING;
    BEHAVIOR rDNIdBehavior;
REGISTERED AS {smi2AttributeID 3};
```

The attribute labeled logId has the syntax of SimpleNameType which is defined in an ASN.1 definition module named Attribute-ASN1Module. Whenever a filtering operation is performed on this attribute, the value of this attribute is checked to see whether it is either equal to or a substring of a given value. A textual description of the generic behavior of this attribute is provided in a behavior construct named rDNIdBehavior.

Attribute group template.   The attribute group construct is a logical grouping of attributes that are often addressed as a group.

Template structure

```
<group-label> ATTRIBUTE GROUP
    [GROUP ELEMENTS    <attribute-label> [,<attribute-label>]*;]
    [FIXED; ]
    [DESCRIPTION delimited-string; ]
REGISTERED AS object-identifier;
```

Template elements

- *<group-label> ATTRIBUTE GROUP.*   The group-label provides a globally unique identifier for this group of attributes.

- *[GROUP ELEMENTS <attribute-label> [, <attribute-label>]*;].*   All the attributes that constitute the elements of the attribute group are identified here by attribute-labels, and they must be present in all instances of the attribute group. Each attribute identified in the group shall be a member of the managed object class that contains this group. For example, it should be defined in an ATTRIBUTE construct in one of the packages of the class.

- *[FIXED;].*   This indicates that the attribute group is of fixed membership. That is, all elements must be present as a group.

- *[DESCRIPTION delimited-string;].*   This is a textual description of the semantics of the attribute group.

Template example

```
scheduleGroup ATTRIBUTE GROUP
    GROUP ELEMENTS schedulerName, startTime, stopTime;
    FIXED;
    DESCRIPTION "This group is for scheduling of alarm reporting."
REGISTERED AS {smi2AttributeGroup 1}
```

The attribute group labeled scheduleGroup has three attributes, schedulerName, startTime, and stopTime, and they must all be present if the group is used.

Notification template. The notification is a GDMO construct that defines the structure of event information to be sent from an agent to a manager, event reply information to be sent from a manager back to the agent, or specific error parameters associated with the event or event reply information.

Template structure

```
<notification-label> NOTIFICATION
    [BEHAVIOR <behavior-definition-label> [,behavior-definition-label]*;]
    [PARAMETERS <parameter-label>[,<parameter-label>]*;]
    [WITH INFORMATION SYNTAX type-reference
    [AND ATTRIBUTE IDS <field-name><attribute-label>
                      [,<field-name><attribute-label>]*];]
    [WITH REPLY SYNTAX <syntax-label>;]
REGISTERED AS <object-identifier>;
```

Template elements

- *<notification-label> NOTIFICATION.* <Notification-label> is a globally unique identifier for this notification.

- *[<BEHAVIOR <behavior-definition-label> [,behavior-definition-label]*;].* If present, this element defines a set of behavior descriptions of the notification each referenced by a behavior-definition-label.

- *[<PARAMETERS <parameter-label>[,<parameter-label>]*;].* This element identifies the parameters associated with either event information that an agent sends to a manager or an event reply a manager sends back to the agent or the processing error of this notification.

- *[WITH INFORMATION SYNTAX type-reference [AND ATTRIBUTE IDS <field-name><attribute-label> [,<field-name><attribute-label>]*];].* The type-reference provides an ASN.1 data type for the event notification information that an agent sends to a manager. If the AND ATTRIBUTE IDS is present, the field-name shall be a label identified in the ASN.1 data type referenced by the type-reference. The data type labeled by the field-name is used to carry values of the attribute referenced by the attribute-label, and, of course, the data type of the attribute should be the same as that identified by the field-name. No label within a SET OF or SEQUENCE OF type can be used as a field name because a label is repeated within such a data type and cannot always refer unambiguously to a single instance of a data type.

- [*WITH REPLY SYNTAX <syntax-label>;*].   If present, this element identifies an ASN.1 data type used in a notification reply a manager sends to an
agent after receiving an instance of this notification from the agent system.

Template example

```
objectCreation NOTIFICATION
    BEHAVIOR objectCreationBehavior;
    WITHIN INFORMATION SYNTAX Notification-ASN1Module.ObjectInfo
        AND ATTRIBUTE IDS
        sourceIndicator            sourceIndicator,
        attributeList              attributeList,
        notificationIdentifier     notificationIdentifier,
        correlatedNotifications    correlatedNotifications,
        additionalText             additionalText,
        additionalInformation      additionalInformation;
REGISTERED AS {smi2Notification 6}
```

This is a notification template for object creation. An agent system sends an
instance of this notification to a manager system to report the event of creating an object. The notification information syntax is provided by the ASN.1
SEQUENCE type ObjectInfo defined in the ASN.1 module Notification-
ASN1Module. In each field-name–attribute-label pair, for example, the
"sourceIndicator sourceIndicator" pair, the first element (i.e., the first
sourceIndicator) is a field name within the ASN.1 SEQUENCE type
ObjectInfo and the second element of the pair (i.e., the second sourceIndicator)
is an attribute name defined in a separate ASN.1 attribute module. Note that
no reply syntax is defined in this notification.

Action template.   An action is an operation requested by a manager and to be
performed by an agent system. The action template specifies the parameters
required in the operation and the action reply, if the action confirmation is
required. The semantics of the operation are described in text for
implementation by the system developer.

Template structure

```
<action-label> ACTION
    [BEHAVIOR <behavior-definition-label> [,<behavior-definition-
    label>]*;]
    [MODE CONFIRMED;]
    [PARAMETERS <parameter-label>[,<parameter-label>]*;]
    [WITH INFORMATION SYNTAX type-reference;]
    [WITH REPLY SYNTAX type-reference;]
REGISTERED AS object-identifier;
```

Template elements

- *<action-label> ACTION*.   A globally unique identifier for this ACTION
template.

- *[<BEHAVIOR   <behavior-definition-label>   [,<behavior-definition-
label>]*;]*.   This element identifies a set of behavior definitions describing
the semantics of an ACTION template.

- *[MODE CONFIRMED;].* If present, the action shall be in a confirmed mode; that is, the agent shall confirm to the manager upon completing the requested operation.

- *[PARAMETERS <parameter-label>[,<parameter-label>]*;].*   This element identifies the parameters of the action, the action reply, or processing error.

- *[WITH INFORMATION SYNTAX type-reference;].*   If present, this element identifies the ASN.1 data type required in the action.

- *[WITH REPLY SYNTAX type-reference;].*   If present, this element identifies the ASN.1 data type required in the reply.

Template example. There are few ACTIONS that have been standardized because actions are mostly application-specific. The following example of ACTION is for the management scenario illustrated in Fig. 4.5.

```
switchToBackupUnit ACTION
    BEHAVIOR
        switchToBackupUnitBehavior BEHAVIOR
            DEFINED AS " if the active unit fails, switch to the backup
            unit of the switch";
        MODE CONFIRMED;
REGISTERED AS { switchAction 1};
```

A manager sends this action request to ask an agent to switch the normal call processing to the backup unit. Note that no parameter is specified for this action.

Parameter template. This template defines parameters that are used in ACTION, NOTIFICATION, or other GDMO templates.

Template structure

```
<parameter-label> PARAMETER
    CONTEXT context-type;
    syntax-or-attribute-choice;
    [BEHAVIOR <behavior-definition-label> [,<behavior-definition-
    label>]*;]
REGISTERED AS object-identifier;
```

Supporting productions:

```
context-type → context-keyword | ACTION-INFO | ACTION-REPLY | EVENT-
INFO | EVENT-REPLY | SPECIFIC-ERROR
context-keyword → type-reference<identifier>
syntax-or-attribute-choice → WITH SYNTAX type-reference | ATTRIBUTE
<attribute-label>
```

Template elements

- *<parameter-label> PARAMETER.* This element provides a globally unique parameter identifier.

- *CONTEXT context-type* where *context-type* → *context-keyword | ACTION-INFO |ACTION-REPLY | EVENT-INFO | EVENT-REPLY | SPECIFIC-ERROR* and *context-keyword* → *type-reference<identifier>*. This element defines the context in which the parameter is applicable. The options are

  - *context-keyword.* This provides a reference to a context defined externally to the template. The type reference identifies a CMIS primitive, and the identifier is a field identifier in the PDU.
  - *ACTION-INFO.* This option means that the parameter may be used in the CMIS Action.
  - *ACTION-REPLY.* This option means that the parameter may be used in the CMIS Action reply.
  - *EVENT-INFO.* This option means that the parameter may be used in the CMIS Event Notification.
  - *EVENT-REPLY.* This option means that the parameter may be used in the CMIS Event reply.
  - *SPECIFIC-ERROR.* This option means that the parameter may be used in the representation or generation of a CMIS processing failure error.

- *syntax-or-attribute-choice*; where *syntax-or-attribute-choice* → *WITH SYNTAX type-reference | ATTRIBUTE <attribute-label>*. If present, type-reference identifies the ASN.1 type of the parameter that is carried in a management protocol message. Alternatively, the name and syntax of an attribute template identified by attribute-label can be used as the type of the parameter.

- *[BEHAVIOR <behavior-definition-label> [,<behavior-definition-label]*;].* This construct identifies the textual behavior descriptions of the semantics of the parameter.

Template example

```
miscellaneousError PARAMETER
    CONTEXT SPECIFIC-ERROR;
    WITH SYNTAX Parameter-ASN1Module.MiscellaneousError
    BEHAVIOR
    miscellaneousErrorBehavior BEHAVIOR
    DEFINED AS "When a processing error failure has occurred and the
    error condition encountered does not match any of the object's
    defined specific error types, this value is used.";;
REGISTERED AS {smi2Parameter 1}
```

The context of this parameter is SPECIFIC-ERROR, meaning that a miscellaneous protocol message processing error was encountered. The syntax of the parameter is the MiscellaneousError type defined in the ASN.1 module Parameter-ASN1Module.

GDMO application example

In the management scenario shown in Fig. 4.5, a network manager manages a simple switch and the switch consists of a set of hardware and software modules. The hardware module is a central processing card, and the software module is the switching application software running on the processing card. At the end of the last section, the managed object classes, inheritance, and containment relationships between the classes, and a number of class components such as attributes, notifications, actions, and parameters have been identified. The next step of building an OSI MIB is to formally represent the classes, relationships, and class components in GDMO. A complete listing of the GDMO definitions for this application can be found in App. 4A.

Discussion on GDMO

GDMO is perhaps the most widely used specification language not only for the development of OSI MIBs but also for the specification of international standards on computer networking technologies. GDMO is based on the same object-oriented concepts that are currently widely used in the software industry. These all underline the importance of GDMO and help explain the dedication of the pages to this topic in this chapter.

On the other hand, GDMO is protocol-dependent in the sense that the constructs of GDMO are specifically for CMIP, the OSI management protocol. This will become evident after reading the following chapter on CMIP. As alternative management protocols, SNMP in particular, are being adopted in the telecommunications industry as the result of the continuing trend of telecommunications and data communication networks converging onto one network, the limitations of GDMO become apparent. Efforts are under way to adapt generic object-oriented notations and specification languages such as the Unified Modeling Language (UML) or Object Modeling Techniques (OMT) as a complement or an alternative to GDMO.

Table 4.5 provides a simple mapping between the GDMO constructs and the object-oriented concepts in the general computing world. Among others, there are two major differences between the GDMO and generic object-oriented concepts. One difference is that the package construct of GDMO (which in essence is a placeholder for attributes, action, and notification) has no counterpart in the general object-oriented programming languages. Second, in GDMO there is

TABLE 4.5   Mapping between Generic Object-oriented Concepts and GDMO Constructs

Generic object-oriented concepts	GDMO constructs and comments
Class	MANAGED OBJECT CLASS Construct
Object	An instance of a managed object class
Object attribute	ATTRIBUTE
Method parameters	PARAMETER
Textual comments	BEHAVIOR
Inheritance relationship	Supported by DERIVED FROM in the class template
Containment relationship	Name binding
Object method	ACTION and NOTIFICATION

no distinction between private and public attributes within a managed object class. All attributes are assumed public and can be inherited by a subclass.

Summary

There are three focuses in this chapter: overviews of the OSI management framework, OSI MIB structure, and GDMO. Out of various topics discussed in the overview of the OSI management framework, an understanding of what the OSI is intended to achieve is fundamental to comprehending the complicated OSI management framework. The concept of OSI is centered on the interconnections and communications between different computer systems. Vastly different computers can communicate with each other by adopting standard interfaces. For the network management, the standard interface includes two parts: a management protocol and management information model. The latter, management information model, is the focus of this chapter. The OSI management model consists of a manager, a set of agents, an MIB, and a management protocol that carries the management information between an agent and a manager by using the underlying OSI seven-layer protocol stack. This basic management model is also adopted for SNMP-based management systems. The key to understanding the OSI MIB structure is understanding the concept of managed objects and a naming tree. A naming tree is the run-time representation of the containment hierarchy. This implies that each and every managed object class with the exception of the class Top is contained in another class. The way to uniquely identify a managed object instance in the MIB is based on the RDN and the DN. An RDN is an instance of a naming attribute [i.e., an attribute identifier with the assigned value, also termed attribute value assertion (AVA)]. A DN of an object in the MIB is the concatenation of all RDNs of the nodes along the vertical path from Top to the node associated with the object in the MIB tree. GDMO is one of the most widely used specification languages for developing management MIBs. This is why many pages are devoted to the detailed explanation of the GDMO constructs.

Exercises

4.1 Compare the OSI MIB structure with the SNMP MIB structure, and list some major differences.

4.2 Compare the object in an SNMP MIB with the object in an OSI MIB, and discuss some major differences.

4.3 Explain what RDN is and how to generate an RDN.

4.4 Explain why a DN can uniquely identify each object in an OSI MIB.

4.5 List two major components of an OSI management model, and briefly describe the functionality of each.

4.6   Compare the attribute construct of GDMO with the object attribute in a generic object-oriented programming language.

4.7   Compare the package construct of GDMO with those in a generic object-oriented programming language. What construct does the package best map to?

4.8   Provide some examples of shared management knowledge that is dynamically determined.

4.9   Discuss how the inheritance and containment relationships are represented in GDMO.

4.10   With which part of the OSI management environment does a user management application directly interface?

4.11   Write a GDMO model to represent the managed object classes shown in Fig. 4.15.

4.12   Make up a management scenario similar to the one shown in Fig. 4.5 and some attributes, notifications, and actions.

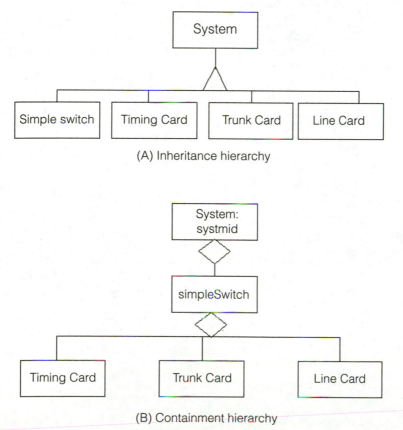

(A) Inheritance hierarchy

(B) Containment hierarchy

Figure 4.15 The inheritance and containment hierarchies.

Appendix 4A

Managed object classes

```
simpleSwitchClass MANAGED OBJECT CLASS
    DERIVED FROM system
    CHARACTERIZED BY simpleSwitchPackage
REGISTERED AS {switchClass 1} ;
processingCardClass MANAGED OBJECT CLASS
    DERIVED FROM system
    CHARACTERIZED BY processingCardPackage
REGISTERED AS {switchClass 2} ;
centralComputingModuleClass MANAGED OBJECT CLASS
    DERIVED FROM system
    CHARACTERIZED BY centralComputingModulePackage
REGISTERED AS {switchClass 3} ;
```

Name-binding definition

```
simpleSwitchNameBinding NAME BINDING
    SUBORDINATE OBJECT CLASS simpleSwitchClass ;
    NAMED BY
    SUPERIOR OBJECT CLASS "CCITT Rec. X.721 (1992) ¦ ISO/IEC 10165-2:
        1992":top ;
    WITH ATTRIBUTE switchID;
    DELETE DELETES-CONTAINED-OBJECTS ;
REGISTERED AS {switchNameBinding 10} ;
centralComputingModuleNameBinding NAME BINDING
    SUBORDINATE OBJECT CLASS centralComputingModuleClass ;
    NAMED BY
        SUPERIOR OBJECT CLASS simpleSwitchClass
    WITH ATTRIBUTE computingModuleId ;
    DELETE DELETES-CONTAINED-OBJECTS ;
REGISTERED AS {switchNameBinding 11} ;
processingCardNameBinding NAME BINDING
    SUBORDINATE OBJECT CLASS processingCardClass ;
    NAMED BY
        SUPERIOR OBJECT CLASS simpleSwitchClass
    WITH ATTRIBUTE cardId;
    DELETE DELETES-CONTAINED-OBJECTS ;
REGISTERED AS {switchNameBinding 12} ;
```

Package definition

```
simpleSwitchPackage PACKAGE
    ATTRIBUTES              switchID GET ,
                           errorType GET;
    ACTIONS                switchToBackupUnit;
    BEHAVIOR               simpleSwitchBehavior;
REGISTERED AS {switchPackage 30};
centralComputingModulePackage PACKAGE
    ATTRIBUTES              computingModuleId GET,
                           lastBackupTime GET-REPLACE,
                           lastRestoreTime GET-REPLACE;
    ACTIONS                executeProgram;
    NOTIFICATIONS          autobackupReport;
    BEHAVIOR               computingModuleBehavior;
REGISTERED AS {switchPackage 31};
processingCardPackage· PACKAGE
```

```
        ATTRIBUTES              cardID      GET,
                                cardType    GET,
                                cardState   GET,
        ACTIONS                 testCard;
        NOTIFICATIONS           cardDown;
        BEHAVIOR                processingCardBehavior;
    REGISTERED AS {switchPackage 32};
```

Attribute definitions

```
        switchID ATTRIBUTE
            WITH ATTRIBUTE SYNTAX MyASN1module.Idtype;
            MATCHES FOR EQUALITY ;
        REGISTERED AS {switchAttribute 40} ;
        errorType ATTRIBUTE
            WITH ATTRIBUTE SYNTAX myASN1module.ErrorType;
            MATCHES FOR EQUALITY ;
        REGISTERED AS {switchAttribute 41};
        computingModuleId ATTRIBUTE
            WITH ATTRIBUTE SYNTAX MyASN1module.IdType;
            MATCHES FOR EQUALITY ;
        REGISTERED AS {switchAttribute 42};
        lastBackupTime ATTRIBUTE
            WITH ATTRIBUTE SYNTAX MyASN1module.TimeType;
            MATCHES FOR EQUALITY ;
        REGISTERED AS {switchAttribute 42};
        lastRestoreTime ATTRIBUTE
            WITH ATTRIBUTE SYNTAX MyASN1module.TimeType;
            MATCHES FOR EQUALITY ;
        REGISTERED AS {switchAttribute 43};
        cardState ATTRIBUTE
            WITH ATTRIBUTE SYNTAX MyASN1module.CardState;
            MATCHES FOR EQUALITY ;
        REGISTERED AS {switchAttribute 44};
```

Action definitions

```
        switchToBackupUnit ACTION
            BEHAVIOR
                switchToBackupUnitBehavior BEHAVIOR
                    DEFINED AS " if the current unit fails, switch to the backup
                    unit of the switch";
                MODE CONFIRMED ;
        REGISTERED AS { switchAction 50};
        executeProgram ACTION
            BEHAVIOR
            executeProgramBehavior BEHAVIOR
                    DEFINED AS "Execute the software module once receiving an
                    instruction from the manager.";;
            MODE CONFIRMED;
        REGISTERED AS {switchAction 51 };
        testCard ACTION
            BEHAVIOR
                testCardBehavior BEHAVIOR
                    DEFINED AS " Test this card once receiving an instruction
                    from manager";
                MODE CONFIRMED ;
        REGISTERED AS { switchAction 52};
```

Notification definitions

```
autobackupReport NOTIFICATION
    BEHAVIOR communicationErrorBehavior ;
    WITH REPLY SYNTAX MyASN1module.BackupStatus;
REGISTERED AS { switchNotification 60} ;
cardDown NOTIFICATION
    BEHAVIOR
    cardDownBehavior BEHAVIOR
        DEFINED AS "sends a notification to the manager once the card
        is down.";;
    WITH REPLY SYNTAX MyASN1module.CardType;
REGISTERED AS {switchNotification 61};
```

Behavior definitions

```
simpleSwitchBehavior BEHAVIOR
    DEFINED AS "The QOS Error Cause attribute indicates the reason for a
    failure in quality of service associated with the managed object.";
computingModuleBehavior BEHAVIOR
    DEFINED AS "The QOS Error Cause attribute indicates the reason for a
    failure in quality of service associated with the managed object.";
processingCardBehavior BEHAVIOR
    DEFINED AS "The QOS Error Cause attribute indicates the reason for a
    failure in quality of service associated with the managed object.";
```

ASN.1 modules

```
MyASN1module DEFINITIONS  ::=  BEGIN
    TimeType              ::=  INTEGER(0..4,294,967,295)
    BackupStatusType      ::=  EMUMERATED {successful(0), failed(1),
                               notPerformed(2)}
    IdType                ::=  INTEGER
    ErrorType             ::=  ENUMERATED { hwError(0), swError(1),
                               powerError(2)}
    CardType              ::=  ENUMERATED {timingCard(0),
                               signalingCard(1), computingCard(2)}
    CardState             ::=  ENUMERATED {In-service (0), out-of-
                               service(1), suspended(2)}
    END
```

Appendix 4B

The international standards organizations related to network management—ISO/International Electrotechnical Committee

Founded in 1949 and based in Geneva, the International Organization for Standards (ISO) is a voluntary, nontreaty organization chartered by the United Nations and made up of regional and member country's standards bodies. For example, the U.S. representative to the ISO is the American National Standards Institute (ANSI). The primary mission of the ISO is to promote the development of international standards to facilitate trade in goods and services, and its focus has been on the standards for national and international data communications. The ISO is best known for the seven-layer OSI Reference Model. All standards published by the ISO become international standards.

The International Electrotechnical Committee is an international standards body dedicated to the electrotechnical areas. The International Electrotechnical Committee and ISO established a Joint Technical Committee 1 (JTC 1) in 1987 to develop international standards in the area of information technology.

International Telecommunication Union (ITU)

The ITU owes its origin to the Union Telegraphique founded in 1865. In 1947, it became a specialized agency of the United Nations charged with the responsibilities of developing worldwide standards for telecommunications, under UN Charter Articles 57 to 63, and later it was renamed the ITU. The ITU consists of three sectors: ITU-R for radio technology, ITU-D for development, and ITU-T for telecommunications.

Consultative Committee on International Telephone and Telegraph (CCITT) and ITU

CCITT, renamed ITU-T in 1992, is the primary international standards organization for the telecommunications industry. The ITU-T consists of more than a dozen study groups defining telecommunications-related standards which are cataloged in a lettered series from A to Z. The study groups responsible for network management standards are study groups 4, 7, 11, and 13. ITU-T has adopted a majority of the ISO's OSI network management standards as its X-series standards.

OSI Network Management

CMIS and CMIP

Outline

- The first section, "Common Management Information Services," introduces three types of services: association services, notification services, and management operation services. The association service is provided by the association control service element and is passed through the CMIS to the user application. The latter two services are provided directly by the CMIS.

- The second section, "Common Management Information Protocol," presents the OSI management protocol from three aspects: CMIP PDUs, CMIP procedures, and the transfer of CMIP PDUs through the remote operation service element.

Introduction

The previous chapter provides an overview of the OSI management model and a detailed introduction to the OSI MIB and the specification language (GDMO) for defining an OSI MIB. Recall that an OSI network management system consists of four basic elements: an agent system, a manager system, an MIB, and a communication protocol for exchanging management information. It is the communication protocol that carries the management information from one management entity (a manager) to another (an agent). It provides a basic set of services that allow a management application in a manager role to instruct another management application in an agent role to carry out specified management operations or allow an agent to report an event to a manager.

These are also called *Common Management Information Service element* (CMISE) services. The CMISE is the cornerstone of the OSI network management framework. It specifies an interface between a user management application and the OSI management environment by providing a set of management services for the applications to exchange management information through the CMIP.

As shown in Fig. 5.1, the CMISE consists of two parts: the CMIS and the CMIP. The CMIS provides an interface for the application program to use the OSI environment by providing a set of management services and passes the association service from the association control service element (ACSE) to the application. Note that the management application systems (e.g., agent and manager systems) directly interface with the CMIS, not the CMIP. CMIS services can be classified into three categories: association services, management event notification services, and management operation services, as shown in Fig. 5.1.

The CMIP is the management protocol that implements the management services provided by the CMIS by carrying the CMIS service primitives in CMIP PDUs. This supplies the mechanism that allows two management entities to exchange management information. The protocol has two portions: the PDU and predefined procedures for passing the management information.

Common Management Information Service

The management operation services of the CMIS are provided by a set of CMIS service primitives: M-GET, M-SET, M-DELETE, M-CREATE, M-ACTION, and

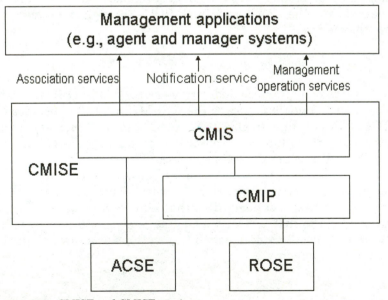

Figure 5.1 CMISE and CMISE services.

M-CANCEL-GET. The notification service is provided by a single CMIS primitive, M-EVENT-REPORT. The association service that provides the capability for two communicating entities to establish channels to exchange management information is provided by the ACSE and is passed to the application through the CMIS.

Selection of managed objects

CMIS services allow management applications to perform operations on managed objects in an MIB. However, the first issue to be addressed is a systematic way to select managed objects for an operation. As discussed in the previous chapter, the OSI MIB structure is based on the naming tree, which is an instantiation of the class containment hierarchy. Figure 5.2 shows an example of an MIB tree with 18 managed objects.

The selection process consists of two steps: scoping and filtering. *Scoping* applies a criterion or boundary condition to a set of objects. *Filtering* applies a criterion or filtering condition to the selected subset of objects using specified tests. A simple analogy is catching fish. Scoping is like throwing a net to catch a group of fish. Filtering is to select only those desired ones from the net using certain test criteria (size, types, etc.) and to discard the remaining ones.

Scoping.   There are four different ways to specify scoping levels:

- The base object alone where only one object is selected
- The nth-level subordinates of the base object
- The base object and all its subordinates down to and including the nth level
- The base object and all its subordinates (whole subtree)

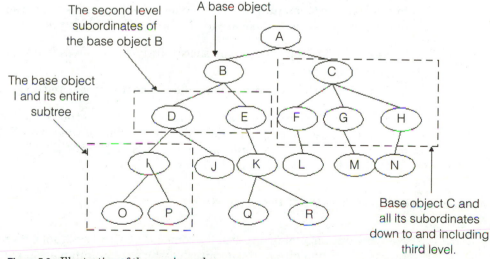

Figure 5.2 Illustration of the scoping rules.

Figure 5.2 shows the application of the different ways of scoping.

Filtering.   Filtering uses a set of assertions that define the filter test to be applied to the scoped managed object(s). Multiple assertions may be grouped using the logical operators AND, OR, and NOT. Each assertion may be a test for equality, ordering, presence, or a set of comparisons of attributes. The Boolean expression of assertions is part of the CMIP PDU parameter. The following matching rules can be applied to object attributes.

- *Equality*.   Attribute value is equal to that in the CMIS parameter.
- *Greater or equal*.   Attribute value in the CMIS parameter is greater than or equal to that of the tested object.
- *Less or equal*.   Attribute value in the parameter is less than or equal to that of the tested object.
- *Present*.   Whether the attribute value is present in the tested object.
- *Substring*.   Attribute value of the tested object includes the substring specified in the parameter.
- *Subset of*.   All members of a set specified in the parameter are present in the test object.
- *Superset of*.   All attribute values of the tested object are present in the set specified in the parameter.
- *Non-null-set intersection*.   At least one of the attribute values specified in the parameter is present in the tested object.

Synchronization.   After scoping and filtering, multiple objects may be selected for application of a CMIS operation. Then there is the issue of in what order the operation will be applied and the action to be taken in case the operation fails for one or more objects. The order in which the operation is applied is left to the local implementation. Two modes with regard to the action to be taken in case of operation failure are available to CMISE service users.

1. *Atomic*.   The operation can successfully be applied to either all selected objects or to none.
2. *Best effort*.   The operation is applied to as many selected objects as possible.

Filtering and synchronization are applicable only to those management operations that can be applied to multiple objects. Out of the seven CMIS services, four are applicable to multiple objects: M-GET, M-SET, M-ACTION, and M-DELETE, as shown in Table 5.1.

Association service—ACSE (CCITT Recommendation X.710)

Network management eventually comes down to two applications out of two processes on two different computer platforms communicating with each other to accomplish network management tasks. The association between the two applications is the precondition of any management operation involving more than one application entity. The application level association is achieved via the ACSE, which provides three different services: application association, application context negotiation, and authentication. Only the first two are discussed in the context of CMISE services.

An application association is a communication channel established between two application entities. The CMIS (or, for that matter, CMISE) does not establish the association itself and instead uses the ACSE to establish the application association. For an application association, the ACSE of the connection-oriented mode provides the four services as shown in Table 5.2. There are four ACSE service primitives used to carry out an ACSE service: request, indication, respond, and confirm.

There are two parts involved in establishing an application association between two CMISE service users using ACSE: association establishment itself and negotiation of service parameters. The first part is to establish a communication channel between the two CMISE users, using the four ACSE service primitives. The initiator first sends an A-Associate.request primitive to request an association. The request arrives at the responding application

TABLE 5.1   Scope, Filter, and Synchronization with CMIS Services

	Scope	Filtering	Synchronization
M-GET	X	X	X
M-SET	X	X	X
M-ACTION	X	X	X
M-EVENT-REPORT	X		
M-CREATE	X		
M-DELETE	X	X	X
M-CANCEL-GET	X		

TABLE 5.2   ACSE Services

Service name	Type	Description
A-Associate	Confirmed	Causes the start of an association between two applications
A-Release	Confirmed	Causes the release of an established association without losing information in transit
A-Abort	Nonconfirmed	Causes the abnormal release of an established association unconditionally and may lose information in transit
A-P-Abort	Undefined	Causes the abnormal release of an established association unconditionally as a result of an action by the presentation layer operation with possible loss of information in transit

as an A-Associate.indication. The responding application returns an A-Associate.response to indicate the agreement on association. Finally, the response arrives at the requesting application as an A-Associate.confirmation. The process of exchanging ACSE service primitives to establish a management connection is shown in Fig. 5.3.

The second part is negotiation of services to be carried on this association. The initiating party can send an A-Associate.indication to propose the additional CMIS services to be used on this association, besides those default six CMIS services (M-GET, M-SET, M-CREATE, M-ACTION, M-DELETE, M-CANCEL-GET). Additional services include scoping, filtering, and multiple reply. The responding application returns an A-Associate.confirm to confirm or reject the request for additional services.

An association can be ended by either party by sending an A-Release (orderly disconnection) or an A-Abort (abrupt disconnection as circumstances warrant).

Management notification service

This CMIS service is used by a management application to report an event to a peer management application. Mostly, it is an agent system that uses this service to report an event to a manager. There is only one CMIS service in this service category: M-REPORT-EVENT. The GDMO construct Notification is meant to specify the information required for this service.

Parameters for the M-EVENT-REPORT operation. The parameters required for this operation are listed in Table 5.3. The generic definitions of the parameters can be found in Table 5.4. Conventions used in this and the following CMIS services are as follows:

- *M.*   Mandatory.
- *(=).*   Value of the parameter is equal to the value of the parameter in the column to the left.
- *C.*   Conditional.

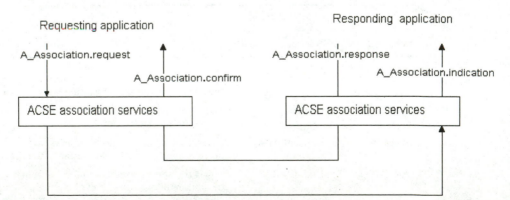

Figure 5.3   The process of establishing an application association using ACSE service primitives.

TABLE 5.3   M-EVENT-REPORT

Parameter name	Request/ indication	Response/ confirm
Invoke identifier	M	M(=)
Mode	M	—
Managed object class	M	U
Managed object instance	M	U
Event type	M	C(=)
Event time	U	—
Event information	U	—
Current time	—	U
Event reply	—	C
Errors	—	C

SOURCE: CCITT Recommendation X.710, 1991.

- *U.* Use of the parameter is the service user's option.
- —. This parameter is not present.

The possible errors for this notification are listed as follows, and their definitions can be found in Table 5.5.

- Duplicate invocation
- Invalid argument value
- Mistyped argument
- No such argument
- No such event type
- No such object instance
- Processing failure
- Resource limitation
- Unrecognized operation

M-EVENT-REPORT procedure. The procedure of event reporting in terms of message exchanges between two CMISE service users is shown in Fig. 5.4. The service can be requested in either the confirmed or nonconfirmed mode. The procedure of the service in the nonconfirmed mode is shown in Fig. 5.5.

Management operation services

The third category of CMIS services, management operation services, has six services: M-GET, M-SET, M-CREATE, M-DELETE, M-ACTION, and

TABLE 5.4   Parameter Definitions

Parameter name	Definitions
Access control	Information of an unspecified form to be used as the input to access control functions
Attribute identifier list	A set of attribute identifiers whose values are to be returned
Attribute list	A set of attribute identifiers and the corresponding values that are returned by the CMISE user that performed the requested operation
Base object class	The class of a base object that is used as the starting point of object selection in scoping
Base object instance	The base object that is used as the starting point of object selection in scoping
Current time	The time at which a request or a response to a request is generated
Errors	The error notification included in the failure confirmation. See Table 5.5 for the error types.
Event information	The information about the event provided by the invoking CMISE service user that sends the event
Event reply	The reply to an event notification
Event time	The time at which the event is generated
Event type	The event type
Filter	A set of assertions defining the filtering test to be applied to the scoped objects
Invoke identifier	An identifier assigned to a CMIS operation that has to be unique to distinguish this from all other outstanding operations
Linked identifier	An identifier provided by the performing CMISE service user when multiple replies are to be sent to the requesting CMISE service user to indicate that the multiple replies are for the same request. Its value shall be the same as that of the invoke identifier provided in the operation request.
Managed object class	The managed object class whose instances are the target of the management operation specified in a CMIS primitive
Managed object instance	The object that is the target of the management operation specified in a CMIS primitive
Mode	Specification of the mode requested for an operation. Two possible values are confirmed and nonconfirmed. If in a confirmed mode, the recipient of the operation request must send a reply back to the sender to confirm the receipt of the request.
Modification list	The list of objects to be modified
Scope	Specification of one of the four scoping methods discussed earlier
Synchronization	Specification of one of the two synchronization methods, i.e., atomic or best-effort

M-CANCEL-GET. Normally these services are used by a manager to request a management operation be performed by an agent, while the services can also be requested from a CMIS user to a peer CMIS user such as from manager to manager.

M-GET.   This service is invoked by a CMISE service user to request the retrieval of management information from a peer CMISE service user. The request can be for the information of a single object or multiple objects. The service can only be requested in a confirmed mode, and a reply is required.

TABLE 5.5   Definitions of Error Types

Error name	Definitions
Access denied	Access is denied because of security violation.
Class instance conflict	The specified object instance is not a member of the specified class.
Complexity limitation	The request operation could not be performed because the parameters are too complex.
Duplicate invocation	The invoke identifier specified is already allocated to another notification or operation.
Get list error	One or more attribute values could not be retrieved because access is either denied or there is no such attribute.
Invalid argument value	The argument value of event information is invalid (e.g., out of range, type mismatch).
Invalid filter	The filter parameter has an invalid assertion or logical operator.
Invalid scope	The value of the scope parameter is invalid.
Mistyped argument	One of the parameters supplied has not been agreed upon for use on the association by the two CMISE service users.
No such argument	The argument specified in a parameter is unrecognized.
No such attribute	A specified attribute is not recognized.
No such event type	The event type specified is not recognized.
No such object instance	The object instance specified is not recognized.
No such reference object	The specified reference object instance is not recognized.
Processing failure	A general failure in processing an operation request or notification has occurred.
Resource limitation	The operation or notification is not processed due to the local resource limitation.
Set list error	The requested operation, such as modification, cannot be performed on one or more attribute values for a variety of reasons depending on the operation in question.
Synchronization not supported	The requested synchronization method is not supported.
Unrecognized operation	The operation requested is not one of those agreed upon by the two CMISE service users.

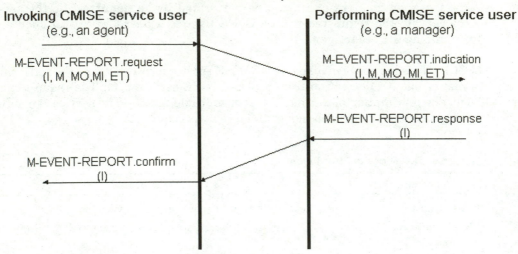

CMISE service provider

Invoking CMISE service user
(e.g., an agent)

Performing CMISE service user
(e.g., a manager)

M-EVENT-REPORT.request
(I, M, MO,MI, ET)

M-EVENT-REPORT.indication
(I, M, MO, MI, ET)

M-EVENT-REPORT.response
(I)

M-EVENT-REPORT.confirm
(I)

I--Invoke Identifier,   M -- Mode, MC -- Managed Object Class
MI - Managed Object Instance,   ET -- Event Type

Figure 5.4   M-EVENT-REPORT procedure in confirmed mode.

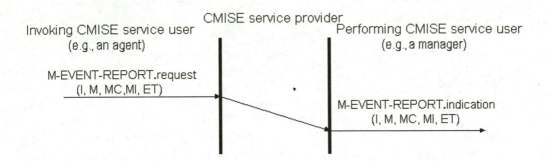

CMISE service provider

Invoking CMISE service user
(e.g., an agent)

Performing CMISE service user
(e.g., a manager)

M-EVENT-REPORT.request
(I, M, MC,MI, ET)

M-EVENT-REPORT.indication
(I, M, MC, MI, ET)

I-- Invoke Indentifier,   M -- Mode, MC -- Managed Object Class
MI - Managed Object Instance,   ET -- Event Type

Figure 5.5   M-EVENT-REPORT in the nonconfirmed mode.

Parameters for the M-GET operation.   The parameters of the operation for this service are shown in Table 5.6, and the generic definitions of the parameters can be found in Table 5.4. The following specific provisions are defined for the parameter of the M-GET operation:

■ If no synchronization method is specified, the best-effort synchronization is used.

■ The default scoping method is the base object only.

- If an attribute value assertion is present in the filter but the attribute is not present in the scoped objects, the result of the test for the attribute value is FALSE. All scoped objects must be evaluated to TRUE in the filter test in order to be selected. If no filter is specified, all scoped objects are selected.

- The errors parameter may contain one of the following errors:
  1. Access denied
  2. Class instance conflict
  3. Complexity limitation
  4. Duplicate invocation
  5. Get list error

- The attribute values that could be read are returned despite the presence of one of the following errors: invalid filter, invalid scope, mistyped argument, no such object class, no such object instance, processing failure, resource limitation, synchronization not supported, or unrecognized operation.

Procedure of the M-GET operation.   Figure 5.6 shows the procedure of M-GET in the confirmed mode with multiple replies. The link identifier parameter of each reply has the value of the invoke identifier provided in the M-GET.indication primitive to indicate it is one of the multiple replies. The end of the reply is signaled by an M-GET.response primitive with no link identifier in it.

The case where the M-GET request is rejected because of error is shown in Fig. 5-7. Note the presence of the error parameter in M-GET.confirm.

TABLE 5.6   M-GET Parameters

Parameter name	Request/ indication	Response/ confirm
Invoke identifier	M	M
Linked identifier	—	C
Base object class	M	—
Base object instance	M	—
Scope	U	—
Filter	U	—
Access control	U	—
Synchronization	U	—
Attribute identifier list	U	—
Managed object class	—	C
Managed object instance	—	C
Current time	—	U
Attribute list	—	C
Errors	—	C

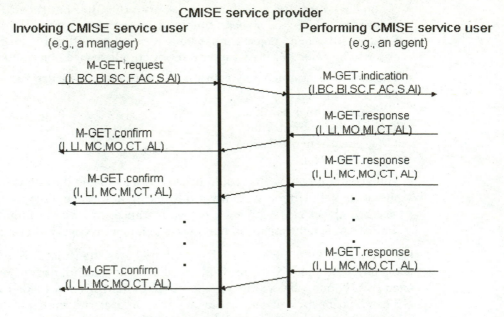

I-- Invoke Identifier, BC-- Base Object Class, BI--Base Object Instance,
SC--Scope, F-- Filter, AC--Access Control, S--Synchronization
AI- Attribute Identifier List,   LI--Linked Identifier, MC -- Managed Object Class
MI - Managed Object Instance, AL -- Attribute List.

Figure 5.6   The M-GET procedure in a confirmed mode with multiple replies.

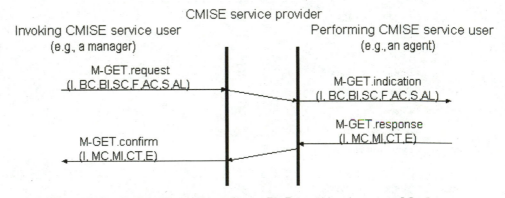

I-- Invoke Identifier,   BC-- Base Object Class, BI--Base Object Instance, SC--Scope,
F--Filter, AC--Access Control, S--Synchronization,
AI- Attribute Identifier List, MC -- Managed Object Class,
MI - Managed Object instance, CT--Current Time, E --Errors.

Figure 5.7   Procedure of M-GET request rejected.

M-SET. The M-SET service is invoked by a CMISE service user to request the modification of management information by a peer CMISE service user. The operation of this service can be in either the confirmed or nonconfirmed mode. The modification can be performed on either a single object or multiple objects in an MIB.

The parameters for the M-GET operation. The parameters of this service are listed in Table 5.7.

In addition to the generic definitions, the following special provisions are specified for the parameters.

- The default scope is the base object alone.
- The default synchronization method is the best-effort if no synchronization is specified.
- If an attribute value assertion is present in the filter but the attribute is not present in the scoped objects, the result of the test for the attribute value is FALSE. All scoped objects must be evaluated to TRUE in the filter test in order to be selected. If no filter is specified, all scoped objects are selected.
- The modification list parameter contains a set of modification specifications each of which has the following three fields.
  1. An attribute identifier of the attribute or attribute group whose value(s) are to be modified. An attribute group identifier is specified only when the *set to default* operation is specified.

TABLE 5.7   M-SET Parameters

Parameter name	Request/ indication	Response/ confirm
Invoke identifier	M	M
Linked identifier	—	C
Mode	M	—
Base object class	M	—
Base object instance	M	—
Scope	U	—
Filter	U	—
Access control	U	—
Synchronization	U	—
Managed object class	—	C
Managed object instance	—	C
Modification list	M	—
Attribute list	—	U
Current time	—	U
Errors	—	C

2. A set of attribute value(s), which are the values to be used in modification of the attribute.

3. A modify operator which can be one of the following:

- *Replace.* The attribute value(s) specified in the parameter shall replace the current value(s) in the MIB.

- *Add values.* The attribute value(s) specified in the parameter shall be added to the current value(s) of the attribute in the MIB. This operator can only be applied to a set-valued attribute, and a set union shall be performed between the attribute value(s) specified in the parameter and the attribute value(s) in the MIB.

- *Remove value.* The attribute value(s) specified in the parameter shall replace the current value(s) of the attribute in the MIB. This operator can only be applied to a set-valued attribute, and a set difference shall be performed between the specified attribute values and those in the MIB.

- *Set to default.* When applied to a single-valued attribute, the value of the attribute is set to its default; when applied to a set-valued attribute, only as many values as defined by the default shall be assigned to the attribute; and when applied to an attribute group, each member of the group shall be set to its default value(s). If there is no default value defined, the *invalid operation* error shall be returned. The modify operator is optional, and if it is absent in the parameter, the replace operator shall be assumed.

- If any of the following errors occurs, an error notification shall be included in the failure confirmation: access denied, class instance conflict, complexity limitation, duplicate invocation, invalid filter, invalid scope, mistype argument, no such object class, no such object instance, processing failure, resource limitation, or set list error.

The last item on the possible error list, set list error, means that one or more attribute values were not modified for one of the following reasons: access denied, invalid attribute value, invalid operator, invalid operation, or no such attribute.

M-SET procedure. The procedure of the M-SET operation for a nonconfirmed mode is shown in Fig. 5.8. The procedure of the M-SET operation with multiple replies is shown in Fig. 5.9. Note that the end of the reply is signaled by an M-SET.response primitive with no link identifier in it. The procedure of the M-SET operation rejected because of an error is shown in Fig. 5.10. Once an error occurs, the performing CMISE service user rejects the operation request and returns an M-SET response with an error notification parameter. The procedure for the M-SET operation with a single reply is identical to this except that the error notification parameter is absent.

M-ACTION. The M-ACTION service is for a CMISE service user (e.g., a manager) to request a peer CMISE service user (e.g., an agent) to perform an action on one or more managed objects. It can be in either the confirmed or nonconfirmed mode.

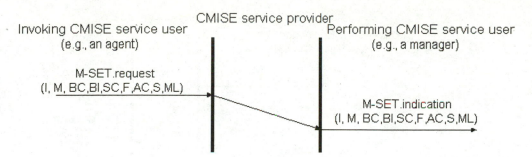

I-- Invoke Identifier,   M -- Mode, BC-- Base Object Class,
BI - Base Object Instance,   SC -- Scope, F--Filter, AC--Access Control,
S--Synchronization, ML--Modification list.

Figure 5.8   Procedure of the M-SET operation in nonconfirmed mode.

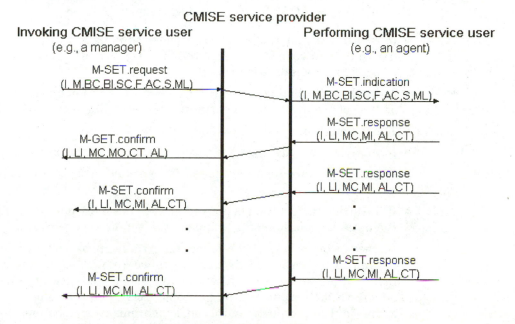

I-- Invoke Identifier, M--Mode, BC-- Base Object Class,
BI--Base Object Instance, SC--Scope, F-- Filter, AC--Access Control,
S--Synchronization ML--Modification List, LI--Linked Identifier,
MC -- Managed Object Class, MI - Managed Object Instance, AL -- Attribute
List, CT -- Current Time.

Figure 5.9   Procedure of the M-SET operation in confirmed mode with multiple replies.

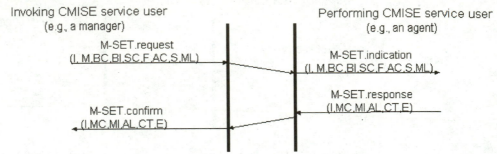

I-- Invoke Identifier, M--Mode, BC-- Base Object Class, BI--Base Object
Instance, SC--Scope, F-- Filter, AC--Access Control, S--Synchronization,
ML- Modification List, MC -- Managed Object Class, MI - Managed Object Instance,
AL- Attribute List, CT--Current Time, E -- Errors.

Figure 5.10 Procedure for the M-SET operation with error.

M-ACTION parameters. The parameters for the M-ACTION operation are listed in Table 5.8.

In addition to the generic definitions of the parameters as listed in Table 5.4, the following M-ACTION specific provisions are defined.

- In case of multiple replies, each response shall have a unique invoke identifier, and the invoke identifier of the final reply shall be the same as the one provided in the indication primitive.

- The value of each linked identifier except for the final one shall be the same as that of the invoke identifier provided in the indication primitive.

- The default scope is the base object only.

- If no filter is specified, all the scoped managed objects are selected.

- If no synchronization parameter is specified, the best-effort method is assumed.

- If the action information is present, it specifies more detailed information on the nature, variation, or operands of the action to be performed.

- In case of error, one of the following errors can be included in the errors parameter: access denied, class instance conflict, complexity limitation, invalid argument, invalid filter, invalid scope, mistyped argument, no such action, no such argument, no such object class, no such object instance, processing failure, resource limitation, synchronization not supported, or unrecognized operation.

M-ACTION procedure. The sequence of protocol message exchanges for the M-ACTION operation with multiple replies and no error is shown in Fig. 5.11. The procedures for the cases of single reply without error and operation request rejected because of error are the same as those for the M-GET services except for the parameters. The correct parameters for those cases can be found in the M-ACTION parameter table shown in Table 5.8.

M-CREATE. The M-CREATE service is for one CMISE service user (e.g., a manager) to request a peer CMISE service user (e.g., an agent) to create an instance of a managed object class, along with the value of its identification and the initial attribute values. This service is always in the confirmed mode.

Parameters for the M-CREATE operation.   The parameters for this service are defined in Table 5.9. In addition to the generic definitions of the parameters, the following special provisions are defined for this service.

- The superior object instance parameter specifies an existing object instance, which will be the superior of the object instance to be created. Note that the superior object in the OSI naming tree is the one that contains the object to be created.

- The reference object instance parameter specifies an existing object instance whose attribute values will be used as initial values for the object instance to be created.

- When the attribute list parameter is provided in the M-CREATE request by the invoking CMISE service user, it contains a set of attribute names

TABLE 5.8   M-ACTION Parameters

Parameter name	Request/ indication	Response/ confirm
Invoke identifier	M	M
Linked identifier	—	C
Mode	M	—
Base object class	M	—
Base object instance	M	—
Scope	U	—
Filter	U	—
Managed object class	—	C
Managed object instance	—	C
Access control	U	—
Synchronization	U	—
Action type	M	C(=)
Action information	U	—
Current time	—	U
Action reply	—	C
Errors	—	C

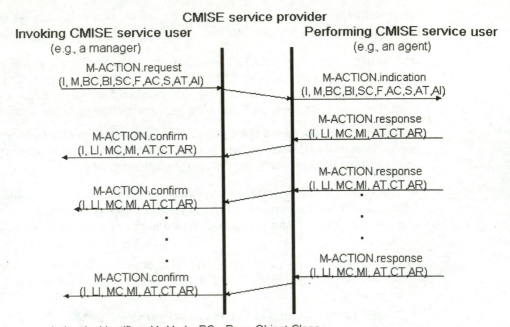

I-- Invoke Identifier, M--Mode, BC-- Base Object Class,
BI--Base Object Instance, SC--Scope, F-- Filter, AC--Access Control,
S--Synchronization   AT--Action Type, AI--Action Information.
LI--Linked Identifier, MC -- Managed Object Class, MI - Managed Object
Instance, AR -- Action Reply, CT -- Current Time.

Figure 5.11   The M-ACTION procedure.

TABLE 5.9   M-CREATE Parameters

Parameter name	Request/ indication	Response/ confirm
Invoke identifier	M	M(=)
Managed object class	M	U
Managed object instance	U	C
Superior object instance	U	—
Access control	U	—
Reference object instance	U	—
Attribute list	U	C
Current time	—	U
Errors	—	C

and values that shall be assigned to the new object instance. These values override those derived from either the reference object instance or the default values.

- In case of an error, the error parameter may take one of the following values: access denied, class instance conflict, duplicate invocation, duplicate managed object instance, invalid attribute value, missing attribute value, mistyped argument, no such attribute, no such object class, no such object instance, no such reference object, processing failure, resource limitation, or unrecognized operation. For definitions of these errors, see Table 5.5.

The procedure for the M-CREATE operation.   The procedure for the M-CREATE operation with error is shown in Fig. 5.12. Note that only one object instance can be created at a time and there are no multiple replies. In case of no error, the sequence of the protocol message exchanges is the same as the error case except that the error parameter is absent.

M-DELETE.   The M-DELETE service is for a CMISE service user to request a peer CMISE service user to delete one or more managed object instances and the associated identifiers from the MIB.

M-DELETE parameters.   The parameters for the M-DELETE operation are shown in Table 5.10. In addition to the generic definitions of the parameters, the following special provisions are defined for this service.

- If the scope parameter is absent, the default scope is the base object alone.

- The managed object instance parameter identifies the object to be deleted, and the managed object class parameter denotes the class the deleted object belongs to.

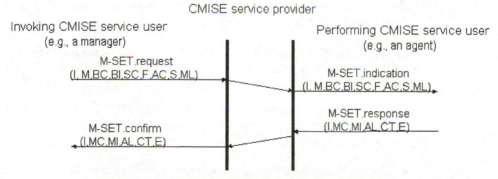

I-- Invoke Identifier, M--Mode, BC-- Base Object Class, BI--Base Object Instance, SC--Scope, F-- Filter, AC--Access Control, S--Synchronization, ML--Modification List, MC -- Managed Object Class, MI - Managed Object Instance, AL- Attribute List, CT--Current Time, E -- Errors.

Figure 5.12   Procedure for M-CREATE operation.

TABLE 5.10   M-DELETE Parameters

Parameter name	Request/ indication	Response/ confirm
Invoke identifier	M	M
Linked identifier	—	C
Base object class	M	—
Base object instance	M	—
Scope	U	—
Filter	U	—
Access control	U	—
Synchronization	U	—
Managed object class	—	C
Managed object instance	—	C
Current time	—	U
Errors	—	C

■ In case of an error, the error parameter may take one of the following values: access denied, class instance conflict, complexity limitation, duplicate invocation, invalid filter, invalid scope, mistyped argument, no such object class, no such object instance, processing failure, resource limitation, synchronization not supported, or unrecognized operation. For definitions of these errors, see Table 5.5.

M-DELETE procedure. Three different procedures for the M-DELETE service are possible for the following three cases: (1) multiple replies with no error, (2) single reply with no error, and (3) error case where the operation request is rejected due to an error. These procedures are the same as those of M-GET except for the parameters. The correct parameters can be found in Table 5.10 for the M-DELETE operation.

M-CANCEL-GET. The M-CANCEL-GET service is for a CMISE service user to request a peer CMISE service user to cancel a previously requested and currently outstanding M-GET service. It is always in confirmed mode.

M-CANCEL-GET parameters. The M-CANCEL-GET parameters are listed in Table 5.11. The service specific semantics of the M-CANCEL-GET parameters are explained below.

■ The get invoke identifier parameter specifies the identifier of the outstanding M-GET service request to be canceled.

■ The error parameter may indicate the following errors: duplicate invocation, mistyped operation, no such invoke identifier, resource limitation, and unrecognized operation.

M-CANCEL-GET procedure.   Note that there are no multiple replies for this operation and there are two cases, i.e., failure and success confirmation as shown in cases B and C in Fig. 5.13.

An application example.   Again, we use the example of managing a simple, hypothetical switch. This time, the scenario is that the active unit of the processing card is down. The agent needs to report this to the manager at once. Upon receiving the notification, the manager sends an instruction to the agent to switch to the standby unit. We show how CMIS services are used to accomplish the management task. Assume that at the system initiation time an association is already established between the agent and the manager applications.

TABLE 5.11   M-CANCEL-GET Parameters

Parameter name	Request/ indication	Response/ confirm
Invoke identifier	M	M(=)
Get invoke identifier	M	—
Errors	—	C

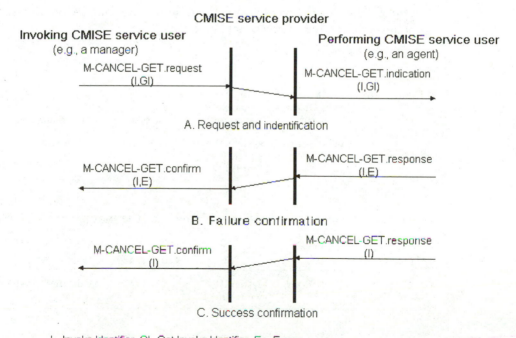

I-- Invoke Identifier, GI--Get Invoke Identifier, E -- Errors.

Figure 5.13   Procedures for M-CANCEL-GET.

Step 1. The agent sends notification to the manager with the following parameters:

M-NOTIFICATION.request

Invoke ID	Mode	MC (Managed-Object class)	MI (Managed-Object instance)	ET (Event type)	Event info
1	Confirmed	simpleSwitch	"systemID = 1, switchID = 1"	Major	Card down

Once the manager receives the notification, it sends a response like the following:

M-EVENT-REPORT.response

Invoke ID	MC	MI	ET	Current time	Event reply
2	simpleSwitch	"systemID = 1, switchID = 1"	Major	4/1,12:23:10	

Step 2. The manager requests the status information on the standby unit before making a decision to switch over.

M-GET.request

Invoke ID	BC (Basic Object class)	BI (Basic Object instance)	Scope	Synchronization	AL (Attribute List)
3	simpleSwitch	"systemID = 1, switchID = 1, cardId = 2"	Base alone	Best-effort	cardState

The agent sends the following reply back to the manager in M-GET.confirm:

Invoke ID	MC	MI	Current time	AL
4	simpleSwitch	"systemID = 1, switchID = 1"		cardDate = "in-service"

Step 3. Manager sends an M-ACTION message to the agent with the following parameters:

M-ACTION.request

Invoke ID	Mode	BC	BI	Scope	Filter	Sync	Action type
5	Confirmed	simple-Switch	"systemID = 1, switchID = 1"	Base alone	—	Best-effort	switchTo-backupUnit

The agent executes the action and returns the following in M-ACTION.confirm:

M-ACTION.confirm

Invoke ID	MC	BI	Action type	Action reply
6	simpleSwitch	"systemID = 1, switchID = 1"	switchTobackupUnit	—

Common Management Information Protocol

Essentially the CMIP has two tasks: defining the syntax for management information transmission and the procedure of the transmission. The former is accomplished by the PDUs, and the latter is defined by the protocol operations.

CMIP protocol data units

CMIP implements the CMIS services by conveying the CMIS service primitives in the PDUs between two management entities, e.g., a manager and an agent. The mappings between CMIP PDUs and CMIS service primitives are shown in Table 5.12. There are a total of 11 CMIP PDUs, implementing seven CMIS services, and the relationships between the CMIS operations and CMIP PUDs are shown in Fig. 5.14. For each CMIS service that can be in either the confirmed or nonconfirmed mode, there are two CMIP PDUs to implement both cases. Three CMIS services fall into this category: M-SET, M-ACTION, M-EVENT-REPORT. The m-linked-reply PDU is used to implement multiple replies of those CMIS services that have both the scope and linked identifier parameters. Four CMIS services are in this category: M-GET, M-SET, M-ACTION, and M-EVENT-REPORT. Note that ACSE and ROSE are omitted in Fig. 5.14 for brevity.

A CMIP PDU is characterized by a set of arguments supplied by the invoking protocol entity, and the results supplied by the responding protocol entity. Note that these arguments and results are the same as those used in CMIS services. The CMIP PDUs and the associated parameters are listed in Table 5.13.

CMIP procedures

How does CMIP implement the CMIS services? Conceptually, at the core of CMIP is a protocol machine called the CMIP machine (CMIPM) that exerts

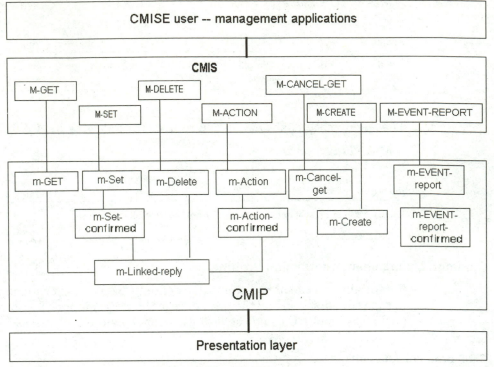

Figure 5.14 Mapping between CMIP PDUs and CMIS services.

control over the protocol operations. The CMIPM is responsible for receiving CMIS request primitives, mapping them into CMIP PDUs, and sending the PDUs over to the other end using lower-layer services. In the other direction, it converts CMIP PDUs into CMIS service primitives and sends them over to CMISE service users. The operations of the CMIPM are shown in Fig. 5.15.

An example of how CMIS M-GET is processed, adopted from Black (1995), will illustrate the CMIP procedures.

1. The CMIPM receives an M-GET request primitive from a CMISE service user such as a manager system.

2. The CMIPM constructs an m-Get *application protocol data unit* (APDU) that contains the parameters passed down in the M-GET request primitive.

3. The CMIPM uses the RO-INVOKE request service of the remote operation service element (ROSE) to send the PDU over to the other end.

4. The ROSE delivers the PDU to the responding CMIPM in an RO-INVOKE indication.

5. After ensuring there is no error in the received PDU, the CMIPM maps the received PDU to a CMIS primitive, in this case an M-GET indication.

6. The responding CMIPM receives a CMIS primitive M-GET response from the responding CMISE user.

7. The responding CMIPM constructs an m-Get APDU that contains the parameter passed down in the CMIS M-GET response.

8. The responding CMIPM sends the APDU back to the initiating CMIPM using the RO-RESULT request of the ROSE.

9. The ROSE delivers the APDU to the initiating CMIPM in an RO-RESULT indication.

10. The initiating CMIPM issues an M-GET confirmation to the initiating CMISE user.

In this example, ROSE services play a key role in conjunction with the CMIPM in implementing CMIS services and are explained as follows.

TABLE 5.12   Mapping Between CMIP PDUs and CMIS Primitives

CMIS primitive	Mode	Linked-ID	CMIP PDU
M-CANCEL-GET.req/ind	Confirmed	NA	m-Cancel-Get-Confirmed
M-CANCEL-GET.rsp/conf	NA	NA	m-Cancel-Get-Confirmed
M-EVENT-REPORT.req/ind	Non-confirmed	NA	m-EventReport
M-EVENT-REPORT.req/ind	Confirmed	NA	m-EventReport-Confirmed
M-EVENT-REPORT.rsp/conf	NA	NA	m-EventReport-Confirmed
M-GET.req/ind	Confirmed	NA	m-Get
M-GET.rsp/conf	NA	Absent	m-Get
M-GET.rsp/conf	NA	Present	m-Linked-Reply
M-SET.res/ind	Non-confirmed	NA	m-Set
M-SET.res/ind	Confirmed	NA	m-Set-Confirmed
M-SET.rsp/conf	NA	Absent	m-Set-Confirmed
M-SET.rsp/conf	NA	Present	m-Linked-Reply
M-ACTION.req/ind	Non-confirmed	NA	m-Action
M-ACTION.req/ind	Confirmed	NA	m-Action-confirmed
M-ACTION.rsp/conf	NA	Absent	m-Action-confirmed
M-ACTION.rsp/conf	NA	Present	m-Linked-Reply
M-CREATE.req/ind	Confirmed	NA	m-Create
M-CREATE.rsp/conf	NA	NA	m-Create
M-DELETE.req/ind	Confirmed	NA	m-Delete
M-DELETE.rsp/conf	NA	Absent	m-Delete
M-DELETE.rsp/conf	NA	Present	m-Linked-Reply

NOTE: NA = not applicable.
SOURCE: CCITT Recommendation X.711, 1991.

TABLE 5.13   CMIP PDUs

CMIP data unit	Arguments	Results	Error
m-Action	baseManagedObjectClass, baseManagedObjectInstance, accessControl, synchronization, scope, filter, actionInfo		
m-Action-Confirmed	baseManagedObjectClass, baseManagedObjectInstance, accessControl, synchronization, scope, filter, attributeIdList	Managed object class, managed object instance, current time, actionReply	
m-Cancel-Get-Confirmed	getInvokedId		
m-Create	baseManagedObjectClass, baseManagedObjectInstance, accessControl, referenceObject-Instance, attributeList	Managed object class, managed object instance, access control, reference object instance, attributeList	
m-Delete	baseManagedObjectClass, baseManagedObjectInstance, accessControl, synchronization, scope, filter	Base managed object class, base managed object instance, access control, synchronization, scope, filter	
m-EventReport	managedObjectClass, managedObjectInstance, eventTime, eventType, eventInfo	Managed object class, managed object instance, event time, event type, event info	
m-EventReport-Confirmed	ManagedObjectClass, managedObjectInstance, eventTime, eventType, enventInfo	Managed object class, managed object instance, event time, event type, event info	Managed object class, managed object instance, current time, event reply
m-Get	baseManagedObjectClass, baseManagedObjectInstance, accessControl, synchronization, scope, filter, attributeIdList		
m-Linked-Reply	getResult, getListError, setResult, setListError, actionResult, processingFailure, deleteResult, actionError, deleteError		
m-Set	baseManagedObjectClass, baseManagedObjectInstance, accessControl, synchronization, scope, filter, attributeIdList	Base managed object class, base managed object instance, access control, synchronization, scope, filter, modification list, modify operator, attribute, attribute identifier, attribute value	
m-Set-Confirmed	baseManagedObjectClass, baseManagedObjectInstance, accessControl, synchronization, scope, filter, modificationList	Base managed object class, base managed object instance, access control, synchronization, scope, filter, modification list, modify operator, attribute, attribute identifier, attribute value	Managed object class, managed object instance, current time, attribute list

SOURCE: Based on CCITT Recommendation X.711, 1991.

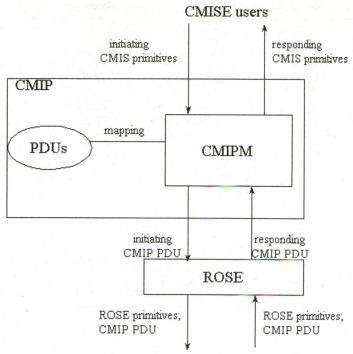

Figure 5.15 Illustration of the CMIP machine.

Use of remote operation service element

ROSE services enable an application entity (AE), called *invoker,* to invoke an operation to be performed by a remote application entity, called *performer*, and the performer to reply back with operation results. The CMIP uses the ROSE services to transfer the CMIP PDUs back and forth between two AEs.

There are five ROSE services as listed in Table 5.14. With the exception of the last one, each service has two service primitives that allow a request to be issued by the invoking application entity and an indication to be received by the performing application entity. The parameter in parentheses in Table 5.14, i.e., priority, is for a service request primitive only. Figure 5.16 presents an example of the RO-INVOKE service primitives, and the behavior of other ROSE service primitives is similar. The meanings of the parameters of the ROSE service primitives are as follows:

- *Operation value.* Identifies the operation to be invoked or performed, depending on the service primitive that contains it. The actual implementation of the operation is a local matter.

- *Operation class.* Indicates one of the five operation classes as described below.

- *Argument.* The argument(s) for the operation. The nature of these arguments is a local matter and not standardized.

TABLE 5.14   **ROSE Service Primitives, Parameters, and Descriptions**

ROSE service	Parameters	Description
RO-INVOKE.request RO-INVOKE.indication	Operation value, operation class, argument, invoke ID, linked ID, (priority)	Enable an invoking application entity to request an operation to be performed by the performing application entity. This is what the invoking end sends.
RO-RESULT.request RO-RESULT.indication	Operation value, result, invoke ID, (priority)	Enable a performing application entity to return results to the invoking application
RO-ERROR.request RO-ERROR.indication	Error value, error parameter, invoke ID, (priority)	Enable the performing application entity to return the negative reply of an operation failure to the invoking application entity
RO-REJECT-U.request RO-REJECT-U.indication	Reject reason, invoke ID, (priority)	Enable one application entity to reject the request or reply of another application entity if the ROSE user has detected a problem
RO-REJECT-P.indication	Invoke ID, returned parameter, reject reason	Enable the underlying ROSE provider to inform a ROSE user of a problem

SOURCE: Based on ITU-T Recommendation X.219 (1993).

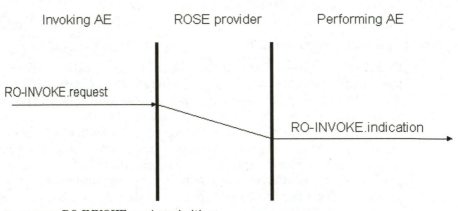

Figure 5.16   RO-INVOKE service primitives.

- *Invoke ID.*   Identifies a ROSE service request primitive and is subsequently used to match this request with the corresponding indication.

- *Linked ID.*   Identifies a parent operation of an invoked operation.

- *Priority.*   Identifies the priority of transferring this service primitive.

Please note that the ROSE is used not only for network management applications but for many other types of applications as well: it is a generic mechanism allowing one application to invoke an operation on another application, very much like the remote procedure call (RPC). There are two dimensions in classifying the operations: whether the performer reports the result of the requested operation and whether the operation is synchronous. An operation is said to be synchronous if the invoker has to wait for the reply from the performer before invoking the next operation. An operation is asynchronous when the invoker can invoke an operation while there are outstanding requests. As a result, there are five classes of ROSE operations:

- *Operation class 1.*   Synchronous, reporting success or failure
- *Operation class 2.*   Asynchronous, reporting success or failure
- *Operation class 3.*   Asynchronous, reporting failure only
- *Operation class 4.*   Asynchronous, reporting success only
- *Operation class 5.*   Asynchronous, no reporting

An application example

Again, we use the example of managing a simple, hypothetical switch. This time, the management scenario is that a manager wants to get status information on the central processing cards. For the purposes of this illustration, we assume that application level association has not yet been established. We use this application scenario to thread together the ACSE, the CMIS, the CMIP, and the ROSE.

First, the manager application initiates an application level association with the agent application. The major steps involved are as follows.

1. The manager side uses the ACSE's A-ASSOCIATE.request primitive to request that an application association be established.

2. As a result, the agent side receives an A-ASSOCIATE.indication and sends an A-ASSOCIATE.response to the manager side.

3. The manager side receives A-ASSOCIATE.confiirm and an application association has been established.

4. The manager-side AE issues an M-GET.request with the parameters indicating that the status attribute value of the central processing card object is to be retrieved and sends the request to the manager-side CMIPM.

5. The CMIPM converts the request into the CMIP PDU m-Get and constructs the CMIP PDU.

6. The CMIPM wraps the m-GET PDU into a ROSE RO-INVOKE.request and sends it across the network.

7. The agent side ROSE entity receives the request and converts it into RO-INVOKE.indication and sends it to its CMIPM.

8. The CMIPM unwraps the indication and forms an M-GET.indication from M-GET.request and passes the indication to the agent application program.

9. The agent application receives the indication, retrieves the requested status attribute value of the central processing card from its MIB, and sends an M-GET.response containing the retrieved attribute value to the agent-side CMIPM via the agent-side CMIS entity.

10. The CMIPM converts the M-GET.response into a CMIP PDU and sends it to the agent-side ROSE entity.

11. The ROSE entity wraps the CMIP PDU into an RO-RESULT.request and sends it over to the manager-side ROSE entity.

12. The manager-side ROSE entity converts it into an RO-RESULT.indication and sends it to the manager-side CMIPM.

13. The CMIPM unwraps the ROSE primitive and gets the CMIP PDU out; it converts it into a CMIS primitive before passing it to the manager-side CMIS entity.

14. The CMIS entity passes the retrieved attribute value to the manager application.

Summary

This chapter concludes the two-chapter series of discussions on the OSI network management model and discusses two topics in great detail: the CMIS and the CMIP. The CMIS in essence provides an interface to a management application in the form of seven CMIS services that allow the management application to use the OSI environment. The first issue discussed about the CMIS was how to select a desired set of managed objects to apply the management operations. Three aspects of object selection are discussed: scoping, filtering, and synchronization. The seven CMIS services, i.e., M-GET, M-SET, M-CREATE, M-DELETE, M-CANCEL-GET, M-ACTION, and M-EVENT-REPORT, are described in detail. Other than the different management operations these CMIS services enable a management application to perform on the remote managed objects, a CMIS service can also be characterized by whether it is performed on single or multiple objects and whether it is in a confirmed or unconfirmed mode.

The CMIP implements the CMIS services by conveying the CMIS service primitives in the CMIP PDUs back and forth between two communicating management applications. The CMIP is introduced from two aspects: CMIP PDUs and the CMIP operation procedure.

Exercises

5.1 Explain what scope and synchronization mean in selecting a set of objects on which to apply management operations.

5.2 Explain why the scoping is applicable to all seven CMIS services while filtering is applicable to only four of the seven.

5.3 Explain what filtering is in selecting a set of objects on which to apply management operations and list three types of filtering criteria.

5.4 How many CMIS services does an application program directly use and what are they?

5.5 Explain why there is no mode parameter in the parameter list for the M-GET operation.

5.6 Explain why the mode, filter, and synchronization parameters are absent in the M-CREATE parameter list.

5.7 Look at Table 5.7, which lists the M-SET parameters. Explain why the managed object class and managed object instance parameters are optional when the value of the scope parameter is base object alone.

5.8 Explain the meaning of the parameter managed object class of M-ACTION.

5.9 For M-CREATE, what does the managed object class parameter mean?

5.10 Discuss different ways a newly created object gets its initial values.

5.11 Explain the relationships between the CMIS and the CMIP. Does your management application directly use CMIP PDUs?

5.12 Explain the functionality of the CMIPM using a simple scenario.

5.13 Summarize the functionality of the ACSE, and explain under what situation A-ABORT is used.

5.14 Summarize the functionality of the ROSE, and explain the use of the two ROSE service primitives, i.e., request and indication, using the ROSE service RO-INVOKE.

Telecommunications Management Network

Functional, Physical, and Logical Layered Architectures

Outline

- The first section, "Introduction," provides a brief discussion on the motivations for telecommunications management network (TMN) and an overview of TMN history and standards.

- The second section, "TMN Functional Architecture," describes TMN functional blocks, functional components, and reference points between the blocks.

- The third section, "TMN Physical Architecture," introduces the five TMN nodes and four classes of interfaces between the nodes and concludes with a step-by-step methodology for defining a TMN interface.

- The last section, "TMN Logical Layered Architecture," presents the five TMN logical layers and the five functional areas that categorize the management tasks of each layer.

Introduction

This section provides an overview of TMN from two aspects, i.e., what TMN is and why TMN is needed. Certain TMN terminology may not be familiar to those not well versed in the area of telecommunications network management, but these terms will be explained by the end of this chapter.

What is TMN?

TMN is a set of international standards for managing telecommunications networks. First and foremost, as defined in ITU-T Recommendation M.3100 (1995), "TMN is conceptually a separate network that interfaces a telecommunication network at several different points to send/receive information to/from it and to control its operations." In other words, the principal idea of TMN is to use an independent management network to manage a telecommunications network by communicating via well-defined and standardized interfaces. Standards are required because telecommunications networks typically consist of network elements of different technologies supplied by multiple vendors. Figure 6.1 shows the relationship between a TMN and the telecommunications network it manages.

As shown in Fig. 6.1, a TMN consists of a set of operation systems (OSs), a data communication network, and portions of the managed network elements. It suffices now to state that the OSs implement management functions in coordination with the portion of the network elements that fall in a TMN domain (i.e., switches, transmission systems). The underlying data network a TMN uses to transport management information can be either the same network the TMN manages or a designated transport network.

A TMN is intended to manage not only a wide variety of telecommunications networks but also a wide range of the equipment and software that constitute the networks and the services carried on each network. For example, the managed networks could include any combination of the following technologies:

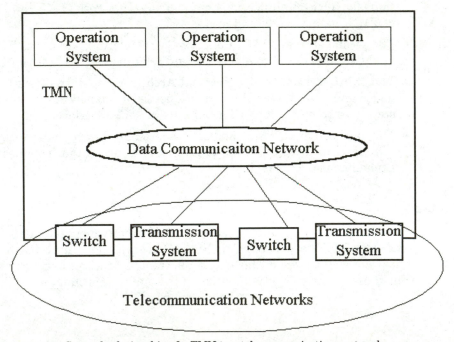

Figure 6.1   General relationship of a TMN to a telecommunications network.

- Circuit-switched or packet-switched networks
- Public switched telephone networks (PSTNs) or private enterprise networks
- Wireline or wireless networks
- Conventional PSTN or intelligent networks

Each of these types of networks is comprised of a variety of equipment. The managed network equipment includes devices such as transmission terminals, digital switches, and analog switches. Many of these devices contain software which must also be managed. As each of these types of networks provide many and varied services, service management is an important and integral part of the TMN management portfolio. Service management must cover both bearer services and the conventional telecommunications services offered to customers.

The rest of this section briefly describes the components of the TMN and the major TMN standards.

TMN components.   The major TMN standards were first published in the late 1980s and have continued to evolve. New component standards and new versions of existing standards are being added on a continuous basis. Now the TMN standards have become the most widely accepted standards governing operations in the telecommunications industry.

A TMN is an encompassing framework that covers every aspect of telecommunications network management, among which are the following main components:

- *A set of architectures* that collectively provide a formal architectural foundation for the management of telecommunications networks that up until their introduction had largely been ad hoc practice. These architectures include physical, functional, logical layered, and information architectures. The TMN functional architecture describes the appropriate distribution of functionality within the TMN. The TMN information architecture describes an object-oriented management information modeling methodology adopted from the OSI management framework for the TMN environment. The TMN physical architecture describes the TMN nodes and interfaces between the nodes that physically make up a TMN. The TMN logical layered architecture provides a framework to divide the management network into five logical layers.

- *An interface specification methodology* that provides recipelike steps for defining a TMN interface.

- *A set of TMN management services,* each of which is a relatively independent area of management activities, described from the user's viewpoint.

- *A set of management functions,* such as surveillance alarm and logging management, that can be used as building blocks to build a management application for a given TMN management service.

■ *A set of standard management information models* that can be categorized into three groups: generic, resource, and process. Generic information models represent managed network resources in generic, technology-independent terms. These object models are intended to be reused (inherited) to derive technology-specific information models. The second group is the derived set of managed object classes modeling the managed network resources in a particular technology area such as synchronous digital hierarchy (SDH) and Asynchronous Transfer Mode (ATM). The final group is a set of support object models that each implements a particular functional aspect of a management process. For example, the objects that model log control and traffic control processes all belong to this category.

■ *Protocol specifications* providing both communications protocols and management protocols for transferring management information (e.g., the management information models) across TMN interfaces. The communication protocols include standard OSI protocols, Integrated Services Digital Networks (ISDN) protocols, and Signaling System No. 7 (SS7) protocols. Management protocols include the CMIP and the SNMP.

TMN standards overview.   The TMN standards originated from the ITU-T's efforts in defining interfaces and interface protocols between OSs and transmission terminals. Soon TMN standards expanded to cover such areas as architectural frameworks, interface methodology, and application-specific interfaces. The major events in the evolution of the TMN standards are listed here.

■ *1982.*   Questions were proposed for the study of operation and maintenance (O&M) aspects of intelligent transmission terminals.

■ *1985.*   These questions were approved.

■ *1986.*   The TMN concept was formally proposed.

■ *1989.*   The TMN architecture document, ITU-T Recommendation M.3010, Principles for a Telecommunication Management Network (TMN), was first published. This recommendation has since gone through several revisions.

■ *1992.*   ITU-T Recommendation M.3100, Generic Network Management Information Model, and ITU-T Recommendations Q.811 and Q.812, which specify the $Q3$ interface protocol suite, were published.

The TMN standards result from the efforts of multiple study groups within ITU-T, whose responsibilities with regard to TMN are outlined below. For other ITU-T study groups that are also involved in defining TMN standards but are not listed below, see Sidor (1995).

■ Study group 4 is responsible for the overall architecture, methodology, functional requirements, and specific characteristics of *Q, X,* and *F* interfaces.

■ Study group 7 is responsible for collaboration with the ISO/International Electrotechnical Committee on OSI interfaces covering communication protocols, messages, and information modeling methodology. For more information on the ISO, ITU, and OSI, see App. 4B.

- Study group 11 is responsible for generic TMN information models, all aspects of TMN interfaces for the SS7 network, and management of Intelligent Networks using TMN methodology.

- Study group 15 is responsible for TMN interfaces for all transmission equipment such as SDH, pleisochronous digital hierarchy (PDH), and some portion of ATM.

TMN standards consist of two parts: the TMN recommendations and the reference recommendations. The reference recommendations can be viewed as input to or as the base of the TMN standards. As shown in Fig. 6.2, the TMN recommendations can be grouped along the line of the TMN components described earlier. For example, the TMN architectures are defined in ITU-T

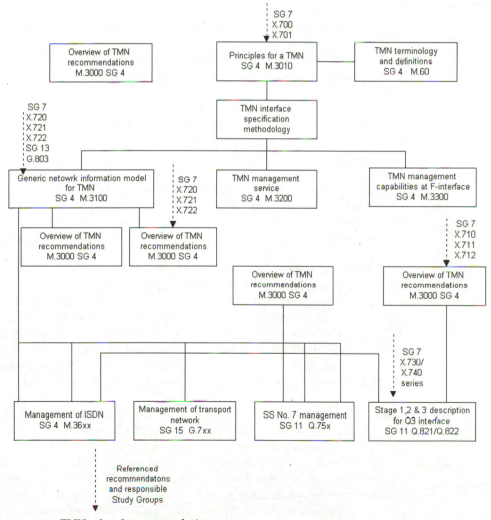

Figure 6.2 TMN-related recommendations.

Recommendation M.3010, which references CCITT Recommendations X.700 (1988) and X.701 (1992) defined by ITU study group 7. The TMN generic information models are defined in ITU-T Recommendation M.3100 (1995) by study group 4, and they reference CCITT Recommendations X.720 (1992), X.721 (1992), and X.722 (1992) by study group 7 and ITU-T Recommendation G.803 by study group 13. Figure 6.2 also illustrates the technology-specific or application-specific information models for three network technologies: ISDN, transport, and SS7. The TMN protocol suites are defined in ITU-T Recommendations Q.811 (1997) and Q.812 (1993) by study group 11.

An in-depth discussion of the TMN standardization process and related recommendations can be found in Sidor (1995) and Glitho and Hayes (1995 and 1996).

Motivations for TMN

Many factors prompted the development of TMN. Among them are

- Proprietary management systems
- Multivendor network operating environments
- Lack of integrated management
- Lack of automation and consequently the high operation cost
- New requirements posed by the new realities in the telecommunications industry

Proprietary environment. Telecommunications networks traditionally have been highly proprietary systems from inside out, as shown in Fig. 6.3. A direct consequence is the inability for one network's operation systems to interoperate with another network's proprietary operation systems.

Multivendor environment. Long gone are the days when homogeneity of a service provider's network was a normality, which had been achieved by using one vendor's equipment. It is now a rule rather than an exception that a service provider's network is made up of equipment from multiple vendors. Once coupled with proprietary operation systems, it becomes extremely difficult and inefficient to operate and manage such a multivendor network environment. Additionally, it is very expensive and time-consuming to modify these operation systems to accommodate new network services.

Lack of integration. One critical problem with the conventional telecommunications network is that a service is often tightly coupled with a special resource. Thus a management system was typically developed to cater to each service-resource pair. This resulted in multiple overlays of operation systems performing similar functions. Today's network operators prefer integrated systems providing a comprehensive view of services and resources in their network. But with the reality of the existing base of specialized component

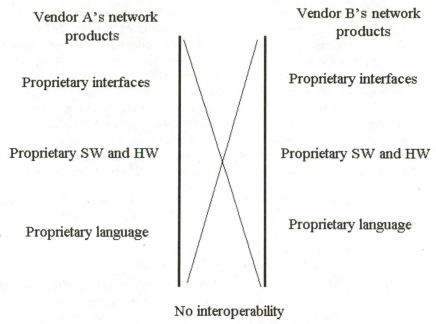

Figure 6.3 The proprietary aspects of telecommunications systems.

management systems coupled with a multivendor and proprietary environment, it becomes very difficult to integrate these management systems into a coherent, consistent one. The lack of integrated systems has resulted in overly large organizations operating the network, as each system has a separate group performing similar functions for each level of management and for each network component.

Lack of automation. Many network and service management processes are manual, partially due to the lack of standardization of the processes themselves. But even the exorbitant personnel costs usually don't justify automating a proprietary process for proprietary equipment when the process is likely to be outdated by the next equipment update.

One direct consequence of this is the high network operating cost. For example, some surveys estimate that the 50 largest carriers annually spend $100 million each for software related to operation support systems. The next 200 largest service providers spend approximately $10 million each (Roeckle 1997). It is becoming increasingly difficult to absorb such high operation costs in a deregulated, increasingly competitive market environment.

New reality of telecommunications industry. While the old practices of telecommunications network management are outdated by the new telecommunications market reality, telecommunications technological advances have created and will continue to create new requirements for network and service management. For example, because service processing is separated from the call processing and network resource functions in Intelligent Networks, the traditional scheme of coupling a service with a resource will not fare well. As the network intelligence, and thus the network management functions, becomes more distributed, the monolithic management system is no longer viable. As telecommunications networks are becoming larger and more complicated than ever, the management systems have to be able to scale up accordingly. As network services evolve at a more rapid pace, so must the management of new and modified services be available at a more rapid pace.

Among the many requirements for network and service management are the following three fundamental ones that TMN is intended to address:

- *Support of automation.*   The increase in the complexity of the management system and a deregulated worldwide telecommunications market makes it too expensive to continue manual operations. The prohibitive cost of in-house development of proprietary systems dictates that automation be accomplished via standards-based systems.

- *Support of interoperability.*   This is a key requirement imposed by market, regulation, and technological advances. For service providers, the multivendor environment replacing the single-vendor environment is a reality to deal with rather than a choice to be made. Interoperability between the network elements of different vendors is critical. Given a worldwide deregulation in telecommunications industries, the interoperability between different service providers is no longer a matter of a mutual agreement between two providers but is mandated by legislation such as the Telecommunications Act of 1996 in the United States. The convergence of multiple technologies [e.g., voice and data, Synchronous Optical Network (SONET) and ATM, wireless and wireline] and the merging of the markets (e.g., merging of Internet and PSTN services, cable and telephony business) make the interoperability between the diverse management systems more important than ever.

- *Support rapid technological evolution.*   Breathtaking technological advances, rapid market evolution, and a global market make it imperative for a network and service management paradigm to be able to adapt to this rapid pace of change.

TMN Functional Architecture

The TMN functional architecture consists of a set of functional blocks, a set of reference points, and a set of functional components. A *functional block* is a logical entity that performs a prescribed management function. A *reference point* separates a pair of functional blocks, and two functional blocks must communicate via a reference point. One or more *functional components* make up a functional block.

Before diving into the details of TMN functional architecture, we first present a network example as shown in Fig. 6.4 to illustrate the issues addressed by the TMN and TMN functional architecture.

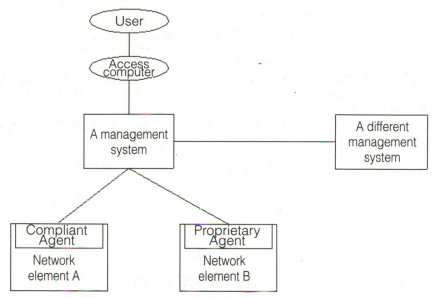

Figure 6.4 Illustration of a management system and its interfaces.

To illustrate the TMN functional architecture, Fig. 6.4 shows a possible configuration of a simple telecommunications network with a management system. A management system manages two network elements. Network element A has an agent with a standards-compliant interface, and the other element is from a different vendor with an agent that has a proprietary management protocol. A user can access the network management system via a remote-access computer. The management system is connected to another management system of yet another management protocol. This is not an unrealistic scenario in today's multivendor environment. The following basic management issues arise from this simple example:

- How can one provide a uniform model to represent the network management functionality as a management network?

- How can one provide the capability to manage a network that is made up of multivendor equipment?

This example will be revisited after the discussion of the TMN functional architecture to illustrate how the functional architecture addresses these issues.

TMN functional blocks

A TMN is to manage a telecommunications network consisting of diverse network elements from different vendors. To accomplish this management goal, the following components are involved:

- A set of management functions to monitor, control, and coordinate the network

- A set of managed network elements
- The capability for the TMN user to access the management operations and to get a presentation of the operation results

In addition, there shall be the capability to convert management information in a proprietary format into a standard TMN format to achieve the interoperability.

There are five functional blocks to accomplish the above described management:

1. Operation system function (OSF) block
2. Network element function (NEF) block
3. Workstation function (WSF) block
4. Mediation function (MF) block
5. Q-adaptor function (QAF) block

These functional blocks are illustrated in Fig. 6.5.

Operation system functions. Note that in Fig. 6.5, the OSF is one of the two functional blocks inside the TMN boundary and is responsible for providing the bulk of management functionality. It processes management information for the

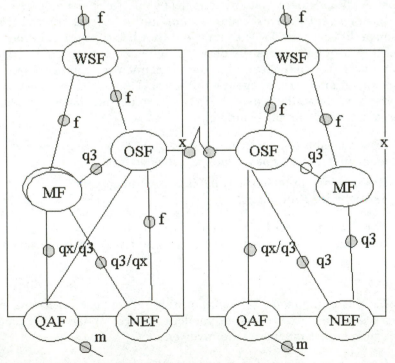

Figure 6.5 Functional blocks and reference points.

purpose of monitoring, coordinating, and controlling telecommunications networks. The OSF provides a set of core services for a TMN-based management system. Examples of these services include

- Application support in the areas of configuration, fault, performance, accounting, and security management
- Database functions to support network resources, configuration, topology, control states, and status
- User terminal support to provide human-machine interface capabilities
- Analysis programs providing the analytical capability in fault and performance analysis
- Data formatting and reporting to support the communication between two TMN functional entities or between a TMN functional block and an outside entity (a human user or another TMN)
- Analysis and decision support to provide the decision-making capabilities to the management applications

Network element function. The NEF has two sides: it presents a view of the managed network elements to the OSF, and it provides the network being managed with telecommunications and support functions. The telecommunications functions are the subject of management. This management is presented to the TMN via the support functions of the NEF. The primary responsibility of these support functions is to handle traffic, not management. It is the destination and origin of management supervision and control. Telecommunications functions are not a part of the TMN, whereas support functions are a part of the TMN itself.

Mediation function. The mediation function acts on information passing between an OSF and NEF. It provides storage, adaptation, filtering, thresholding, and condensing operations on the data received from the NEF. In addition, a mediation function routes and acts on information passing between a manager and an agent. Although the MF is completely inside the TMN domain, meaning that communication between the MF and OSF or NEF is through a standards-compliant interface, the information conversion is still expected. One reason for this is the differences in vendor implementations.

Q-adaptor function. The QAF provides the translation capability to connect a proprietary NEF or OSF to a TMN, or a non-TMN network element to a TMN. This is a crucial capability of TMN given that its mission is to integrate existing legacy networks into a TMN.

Workstation function. The WSF provides the functionality for interaction between a human user and the OSFs. The WSF can be viewed as a mediation function between a human user and the OSF. It converts information coming out of the OSF into a format that can be presented to a user and vice versa.

Functional components

A functional block consists of a set of functional components, and a functional component is the basic unit used to implement a TMN service. The hope is that each functional component is sufficiently self-contained with a well-defined interface to the outside world so that they can be implemented as off-the-shelf components. Consequently, this will enable a TMN management system to be assembled efficiently from off-the-shelf components. The identified functional components are described as follows.

Management application functions (MAFs).   MAFs are a group of components that implement the core TMN management services. Each of the components can be in either an agent role or a manager role depending on the system that contains the component. The TMN standards only describe the capabilities of each MAF, and the implementations of MAFs are not subject to standardization. One MAF is distinguished from another by a prefix of the functional block to which it belongs.

Mediation function–management application functions (MF-MAFs).   The MF-MAF components make up the mediation function blocks. Examples of the MF-MAF component functions include

- *Temporary storage.*   If the messages sent from a manager to an agent or from an agent to a manager exceed the receiving party's capacity, it is the responsibility of the FM-MAF to provide temporary storage for the message and any data that cannot be immediately processed.

- *Filtering.*   There can be multiple agents communicating with one manager or multiple agents communicating with multiple managers. The MF-MAF helps control and direct traffic and queues up the message in case there is congestion. Duplicate information from multiple agents or the same agent is filtered out.

- *Thresholding.*   The MF-MAF may monitor the information flow between an OSF and an NEF and set the thresholds to either prevent overflow or to detect an error condition (e.g., duplicates).

- *Concentration.*   The MF-MAF may perform preliminary processing on the information passed between managers and agents. One scenario is that data from multiple agents to one manager need to be gathered to form a complete unit of information to be processed by the manager.

- *Security.*   Both managers and agents may have different access control requirements. The MF-MAF helps administrate the security-related procedures.

Operation system function–management application functions (OSF-MA).   These management components are the core functions of OSFs. They may range from simple to more complicated functions such as

- Support of manager and agent roles in access to managed object information.

- Processing of raw information such as data concentration, alarm correlation, statistics, and performance analysis.

- Processing of information coming into the TMN from outside sources such as another TMN. One example would be the ability to pass trouble tickets between management systems.

Network element function–management application functions (NEF-MAFs). These are the component functions in the NEF supporting the agent role of a management system. Examples of functions in an agent role include

- Managing the objects that represent managed network resources to detect events that take place at the resources and executing the control action on the managed resources

- Interfacing MF or OSF in a manager role

- Maintaining the agent's management information base (MIB)

Q-adaptor function–management application function (QAF-MAF). These component functions are present in QAF to support both manager and agent roles. Examples of the specific functions of the QAF-MAF functional component may include

- Interfacing to the manager to pass the management information from the agent

- Interfacing to the agent to pass the management data and instructions from the manager

Information conversion function (ICF). The ICF is used in intermediate systems to translate the information model at one interface into the information model at the other interface. The translation can be done at both a syntactic and a semantic level. The ICF is a required component of an MF block.

Workstation support function (WSF). This component provides support for the WSF block and hides the existence of NEFs, OSFs, and MFs from the WSF users. The following specific functions have been identified for this functional component:

- *Data access and manipulation.* Provides the OSFs with WSF user data; processes and presents the data from OSF or NEFs via MF to the WSF user.

- *Invocation and confirmation of actions.* Invokes an action requested by the WSF user and sends the action results back to the user.

- *Transmission of notification.* Notifies the appropriate WSF user once an event has taken place at the OSFs, NEF, or MF that is of concern to the WSF user.

- *Administrative support for WSF.* Provides access control, logging service, and other administrative support.

User interface support function (UISF). The UISF has two major responsibilities:

- Translating the information represented in the form of a TMN information model to a displayable format for the human-machine interface
- Translating the user input to the TMN information model for communications in the other direction.

If required, the UISF is responsible for integrating information from multiple sessions with multiple OSFs or MFs to present the information to the user in a consistent and correct form.

Message communication function (MCF). Note that each functional component has a message communication function (MCF) that will allow connection to the data communication function (DCF). As mentioned earlier, except for NEF and OSF, little detail has been specified for the rest of the function blocks.

Directory system function (DSF). This functional component provides the directory system capability to both agent and manager. A directory system provides a hierarchical structure to organize and store the management information in the form of directory objects. This form of information representation and organization is independent of any given management protocol and thus is interoperable with many systems that support the directory service. One implementation option for DSF is to use a set of directory system agents, as described in the X.500 series of the ITU-T recommendations.

Directory access function (DAF). This functional component is used in all function blocks that need to access the directory. In essence, this is a general-purpose application program interface (API) to the directory which provides read, list, search, add, modify, and delete operations for a function block.

Security function (SF). The SF provides security services that can be used by a function block to implement its security policy and meet the security requirements. The security services are classified into five categories:

- Authentication
- Access control
- Data confidentiality
- Data integrity
- Nonrepudiation

See Chap. 12 for details on security management.

The relationship between function blocks and functional components is shown in Table 6.1, which identifies the functional components required by a function block.

TABLE 6.1   Relationship of Function Blocks to Functional Components

Function block	Functional components*
OSF	OSF-MAF, WSF, ICF, DSF, DAF, SF
WSF	UISF, DAF, SF
NEF	NEF-MAF, DSF, DAF, SF
MF	MF-MAF, ICF, WSF, DSF, DAF, SF
QAF	QAF-MAF, ICF, DSF, DAF, SF

*DAF—directory access function, DSF—directory system function, ICF—information conversion function, MCF—message communication function, MAF—management application function, SF—security function, UISF—user interface support function, WSF—workstation support function.
SOURCE: ITU-T Recommendation M.3010, 1996.

Reference points

The intention of reference points is to define service boundaries between two management function blocks and to identify the management information passing between function blocks. While a reference point is a logical entity, its implementation becomes an interface. As shown in Fig. 6.5, the TMN framework identifies the following five classes of reference points:

1. *q class.*   Reference point between NEF and OSF, QAF, or MF (either directly or via DCF). Within the class of *q* reference points:
   - *qx.*   The reference points between NEF and QAF, or MF, and between two MF
   - *q*3.   The reference points between NEF and OSF, MF, and OSF, and OSF and OSF
2. *f class.*   Reference point for attachment to a WSF
3. *x class.*   Between two OSFs belonging to different TMNs

In addition there are two more classes of non-TMN reference points:

1. *g point.*   Between the WSF and the operating personnel
2. *m point.*   Between a QAF and non-TMN managed entities (proprietary interface)

An implementation of a reference point is an interface. For example, a *Q*-class interface is an implementation of a *q*-class reference point. Though in most cases it is a one-to-one mapping between a reference point and an interface, it may not be true in some cases for reasons that will be explained later. The relationships between the logical functional blocks expressed as reference points are shown in Table 6.2.

TABLE 6.2   Relationships Between Logical Function Blocks Expressed as Reference Points

	NEF	OSF	MF	QAF$_{q3}$	QAF$_{qx}$	WSF	Non-TMN
NEF		q3	qx				
OSF	q3	q3,x	q3	q3		f	
MF	qx	q3	qx		qx	f	
QAF$_{q3}$		q3					m
QAF$_{qx}$			qx				m
WSF		f	f				g
Non-TMN				m	m	g	

The reference point is $q3$ between two OSFs if either one of the two is non-TMN OSF. The x reference point applies only when two OSFs are in separate TMNs. Both the x and g reference points are outside the TMN boundary.

An example.   Figure 6.6 illustrates a mapping of the TMN functional architecture onto the network example shown earlier in Fig. 6.4.

The management system maps to OSF, and two network elements with an agent map to two NEFs. The access computer is the WSF. In addition, two more function blocks are added: the mediation function and the Q-adaptor function. The flexibility of the TMN architecture is shown in its capability of providing a functional view without commitment to the actual implementation. The two additional function blocks can be independent physical nodes or functions within an existing node.

TMN Physical Architecture

While the TMN functional architecture presents a functional view in terms of the logical entities such as function blocks and reference points, the TMN physical architecture is defined in terms of various nodes in the network and the communications interfaces between the nodes. The nodes (e.g., OSs and network elements) and the links between the nodes can both be mapped to some software or hardware entities. A simple layout of an isolated TMN is shown in Fig. 6.7. At a conceptual level, a TMN consists of five different types of nodes and four types of links. Each node is characterized by the functionality provided by the node. Each link is characterized by the interface between the two nodes. The following sections will describe each type of node and each type of link.

TMN building blocks

First and foremost, the TMN is a network. As such, it has nodes, links, and interfaces, each of which is explained below. A node in the TMN can be a hardware system, a software application system, or a combination of the two.

Operation systems.   An OS is a system that performs the OSF as described in the TMN functional architecture. In general it processes information related to telecommunications management for the purpose of controlling, coordinating, and monitoring a telecommunications network. The OS provides the core capabilities of a TMN management system, providing the supervisory or control capabilities for a management application. One OS can be interconnected with other OSs either in the same TMN or a different one and thus form management hierarchies or other types of structures.

Functional configuration of an OS.   The configuration of an OS depends on the OSF configurations. OSFs can be categorized into several types: business OSF, service OSF, and network OSF, as shown in Fig. 6.8. Business OSFs are concerned with enterprise policy issues and carry out overall business-level coordination. A service OSF is concerned with the service aspects of the network and also performs the role of interfacing the customer. A network OSF is the network-based TMN application, responsible for supplying network-level information to the service OSFs. It also communicates with the NEF and/or the MF to carry out the management functions on network elements.

Physical configuration of an OS.   An OS physical architecture can be realized in either a distributed or a centralized fashion. A centralized OS has the complete set of OS functions concentrated in a single system. A distributed OS may have its functions distributed across a number of OSs. A number of factors affect the choice of a centralized or distributed implementation of an OS

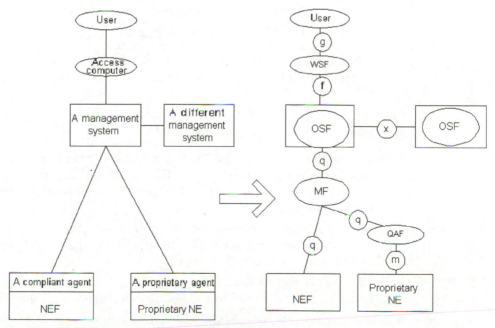

Figure 6.6   TMN functional architecture illustration.

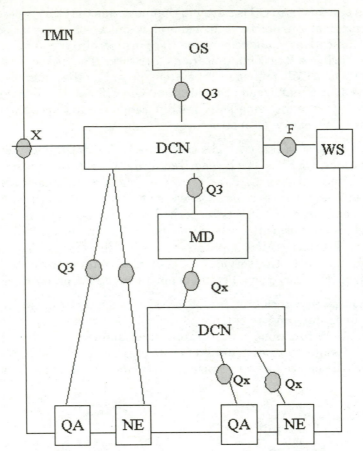

Figure 6.7 Simplified TMN physical architecture. (*From ITU-T Recommendation M.3010, 1996.*)

functional architecture, many of which are specific to the local environment. Examples of these considerations include

- *Real-time requirement.*   The TMN protocol selection is an important factor in the OS physical architecture. For example, the choice of hardware depends strongly on whether an OS provides real-time, near-real-time, or non-real-time service.

- *Management communication traffic.*   For a distributed OS, a network element needs to communicate with multiple OSs and management traffic load should be taken into consideration.

- *Fault tolerance requirement.*   For example, a distributed OS is less prone to a catastrophic failure caused by the failure of a single communication channel.

- *Administrative and organizational considerations.*   A distributed OS with well-defined functional boundaries may suit an organization well.

Given the prevalence of the distributed computing platforms, distributed OS implementation is being widely applied in the industry.

Mediation device (MD).   An MD implements a mediation function as defined in the TMN functional architecture. The declared mission of the mediation function is to process the information passing between an OS and a network element (NE) to ensure the information conforms to the agreed format and semantics. The function at this node may store, adapt, filter, threshold, and condense information.

Process of mediation.   The following list identifies five mediation processes corresponding to the responsibilities for the mediation function block as described in the TMN functional architecture:

- *Information conversion.*   An example of this type of process is translating between information models. Information conversion processes translate multiple information models to a generic information model, augmenting and enhancing information from a local MIB to be compliant with a generic information model.

- *Interworking.*   Interworking processes provide higher-order protocol interworking by establishing and negotiating connections and by maintaining the communication context.

- *Data handling.*   These processes provide concentration of data, collection of data, data formatting, and data translation.

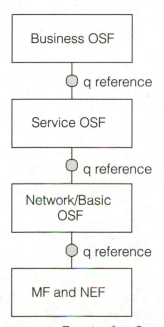

Figure 6.8 Functional configuration of an OSF.

- *Decision making.*   Decision-making processes include workstation access, thresholding, data communication backup, routing and rerouting of data, circuit selection and access for tests, and circuit test analysis.

- *Data storage.*   These processes include database storage, network configuration, equipment identification, and memory backup.

Configuration of mediation devices.   The mediation function can be implemented either as an independent mediation device, as logical entities distributed between the OS and the NE, or as a combination of the two. Fig. 6.9*a* shows an independent mediation device that serves multiple NEs, while in Figures 6.9*b* and 6.9*c*, the mediation function is distributed between the two NEs. In the stand-alone case, the interfaces toward the NEs, QAs, and OSs are one or more of the standard interfaces *Qx* and *Q*3. When mediation is part of an NE, only the interfaces toward the OSs are specified as one or more of the standard interfaces. Mediation that is part of an NE may also act as mediation for other NEs. In this case, standard interfaces (*Qx*) to these other NEs are required.

The mediation function can also be implemented as a hierarchy of cascaded mediation devices as shown in Fig. 6.10. One advantage of the cascading of mediation devices is greater flexibility in the TMN. The mediation device is regarded as the most ill-defined component of TMN (Glitho and Hayes, 1995), and few examples exist that strictly fit into the definition. In practice, what is often referred to as a mediation device is actually a Q adaptor.

Q adaptor.   A Q adaptor can be a hardware device, software, or a combination of the two. This Q adaptor implements the Q-adaptor function (QAF), which converts a non-TMN interface to a standard TMN interface. Specifically, a QAF converts a proprietary interface to a *Q*-class interface (*Q*3 or *Qx*). A Q adaptor can contain one or more QAFs.

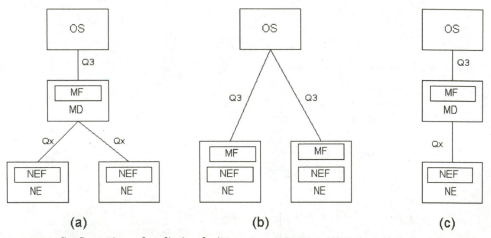

Figure 6.9   Configurations of mediation devices.

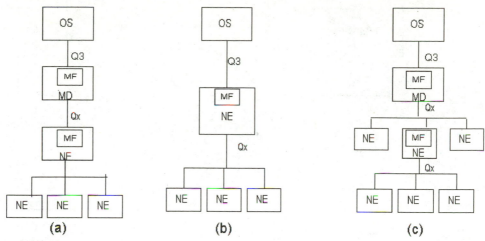

Figure 6.10 Different configurations of mediation devices.

The existence of the Q adaptor reflects concerns about the interworking of TMN and the preexisting systems. It has often proven difficult to build Q adaptors due to the difficulties in mapping between the TMN interfaces and the proprietary interfaces. Because each proprietary interface is unique, it is practically impossible to standardize the mapping between TMN and legacy interfaces.

Currently in the industry, many people use the term *mediation devices* to mean Q adaptor. In fact, such use is so prevalent that the term mediation device has practically overtaken the meaning of Q adaptor.

Configuration of Q adaptors. Q adaptation can be performed on two cases, as shown in Fig. 6.11: between a legacy network element and a TMN OS, and between a legacy OS and a higher-order TMN OS [such as a proprietary element-management-layer (EML) OS and a TMN network-management-layer (NML) OS]. Physical implementation of a Q adaptor can vary from system to system, and Fig. 6.12 shows three different configurations. Most of the industry's focus has been on the first case.

Q-adaptation process. A QAF performs two fundamental functions: information conversion and protocol conversion.

Information conversion. A QAF maps a TMN information model into a non-TMN model or vice versa. This requires the QAF to understand the syntax, semantics, and the MIB structure of both information models involved. As shown in Fig. 6.13, the proxy has the knowledge of both the SNMP and the CMIP information models.

Many TMN information models have been defined for *Q3* interfaces between network elements and EML OSs. They include

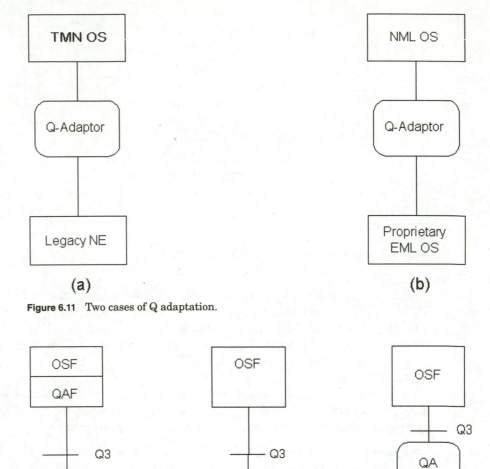

Figure 6.11 Two cases of Q adaptation.

Figure 6.12 Different configurations of a Q adaptor.

- The customer administration information model (ITU-T Recommendations Q.824.0 to Q.824.4, 1995) and the traffic management information model (ITU-T Recommendation Q.823, 1996)

- The information model for the message transfer part of the SS7 network (ITU-T Recommendation Q.751.1)

Commercially available are the Q adaptors that convert a range of non-TMN information models to TMN information models. The non-TMN information models include the SNMP models, a host of human-machine language–based models, and the Translation Language 1 (TL1) models. The TL1 models are widely used in the Bell operating companies in the United States. The Network Management Forum (NMF) has specified standard mapping from the OSI information models to SNMP information models and vice versa.

Protocol conversion. Protocol conversion, needed only when the Q adaptor is located in a separate node, maps non-TMN protocols into TMN protocols or vice versa. For all practical purposes, the OSI protocol stack has been designated as a TMN protocol, although some other lower-layer protocols have also been adopted (see ITU-T Recommendation Q.811, 1997) as part of the TMN protocol suite. Recently, other upper-layer protocols such as the SNMP have been endorsed by the ITU as part of TMN management protocols.

There are two types of protocol conversion, i.e., deep mapping and application level mapping between TMN and non-TMN protocols. A deep mapping converts both lower- and upper-layer non-TMN protocols into a whole stack of TMN protocols or vice versa. The application level mapping converts one management protocol to another. For example, it maps a non-CMIP management protocol such as the SNMP to the CMIP or vice versa. In practice, the deep mapping is rarely used because of the obvious difficulties in implementation.

A Q-adaptor example. Figure 6.13 is adopted from the Network Management Forum's integrated framework for interworking between different management protocol-based systems. This framework features a proxy, which in essence is a Q adaptor that converts between CMIP- and SNMP-based information models and management protocols.

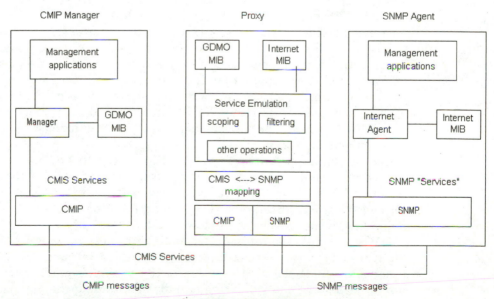

Figure 6.13 Interworking between the CMIP and the SNMP.

Data communication network (DCN). The DCN of a TMN implements the data communication function (DCF) of the TMN functional architecture and provides connectivity between TMN nodes. Specifically, a DCN connects network elements, Q adaptors, and mediation devices to the OSs for $Q3$ interfaces and connects mediation devices to network elements and Q adaptors for Qx interfaces as shown in Fig. 6.7.

There is little restriction on the component networks that make up a DCN. They might include packet-switched data networks, dedicated lines, public switched networks, or local area networks. The only requirement is to provide the transport capabilities between the TMN nodes. Though the DCN can be a separate network, in practice it often is the same network that is being managed by the TMN.

Network element (NE). An NE within TMN consists of telecommunications equipment and any support equipment that performs the network element function (NEF) of the TMN functional architecture. The NEF consists of the following two categories.

- Telecommunications functions such as switching and transmission
- Telecommunications support functions such as fault localization, billing, and alarm reporting

Besides the network element function, an NE may optionally contain other functions such as the MF, QAF, OSF, or WSF. An NE may provide an X interface or an F interface if it also contains one or more OSFs or WSFs.

Workstation (WS). A TMN workstation is considered to be a terminal connected through a data communication link to an OS or an MD. This terminal has the capabilities to translate the information at the f reference point represented in TMN information models to a displayable format for presentation to the human user at the g reference point or vice versa, as shown in Fig. 6.14. This implies that the terminal shall have data storage, data processing, and interface support. As shown in Fig. 6.14, part of the workstation falls within the TMN boundary and part remains outside of the TMN. A workstation implements two types of functions: the presentation function and the WSF.

The presentation function provides the user with physical input, output, and editing facilities to enter, display, and modify details of the information inside a TMN. It also provides support to the human-machine interface, termed the g reference point. The human-machine interface may be a command line, menu-driven, or window-based.

A WSF provides the user at a terminal with the general functions to handle input and output of data to and from the user's terminal. Examples of the functions include the secure access to the terminal; parsing and validating input; formatting and validating output; maintaining the database; and supporting the menu, screens, windows, and scrolling.

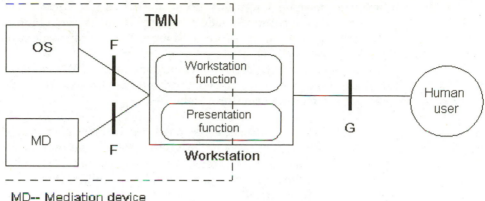

MD-- Mediation device
OS-- Operation system

Figure 6.14 A typical TMN workstation and its connections.

A workstation must have an *F* interface and shall not include any OSFs. If OSFs and WSFs are combined in one implementation, it is considered an OS. Note that a workstation as a TMN node does not convey the same meaning as "workstation" in the computer world.

TMN standard interfaces

TMN standard interfaces provide for the interconnections between NEs, QAs, OSs, MDs, and workstations through data communication networks. The ultimate goal of the interfaces is to ensure that the network equipment of different vendors can communicate with each other to accomplish management tasks. It is like providing a common language to the network devices of different vendors in order to achieve common management goals. Since TMN is interface-centric, detailed interface specifications are the essence of TMN standardization.

The emphasis on and the efforts in TMN interfaces are a direct reaction to the reality of the telecommunications industry. With worldwide deregulation and globalization in the telecommunications industry, most public network service providers purchase their NEs from multiple vendors. In many cases, OSs have been bought from the NE suppliers to ensure operational compatibility with each NE type. The common interfaces based on TMN standards provide a common language for equipment to communicate with a managing system and allow the suppliers to concentrate on innovative end-system applications. The common interfaces make it possible for the network equipment and the managing systems (OSs) to be developed separately. The equipment development is decoupled from the development of managing systems because each only needs to know the protocol and message it can expect from the other party.

An interface can be viewed as consisting of two basic components: information models and a communication protocol suite to carry the management information models. The information models can be further categorized into two types: the information models to support the management functions and those modeling network resources. Sidor (1995) presents a detailed discussion of TMN interfaces.

Three types of interfaces have been defined for TMN, namely, Q, F, and X interfaces, each of which is discussed from two aspects: the information models and the communication protocols.

X interface. The X interface is used to interconnect two TMNs or to interconnect a TMN with a non-TMN system which accommodates a TMN-like interface. In practice, this interface is used to connect the two TMNs that belong to two different service providers. For example, this interface is used to exchange management information between

- An access provider and a local exchange carrier that provides local phone service to an access customer
- An access provider and an interchange carrier that provides long-distance phone service

Information models. Only a few information models have been defined for the X interface. The information carried on this interface includes the data for service ordering, provisioning, billing, and trouble ticketing. One widely used information model is for trouble management. It allows the exchange of trouble ticket information between different service providers. A detailed discussion of this model can be found in Chap. 10.

Protocol suites. Few details have been specified for the X interface with regard to a protocol suite, though OSI seven-layer protocols have been used in some published applications (Weese et al. 1996). Recently there has been movement toward using generic distributed computing technology like Common Object Request Broker Architecture (CORBA) as the underlying communicating protocol for the X interface.

Issues related to the X interface. One issue important to the X interface is security. Since this interface is envisioned as a gateway of a TMN to the outside world, a higher level of access security is required. Another issue is the standard information model, or the lack of one. The difficulties lie in the fact that many aspects of the information exchanged at this interface are proprietary, and the service providers may have reservations in opening up this information to the outside world.

F interface. The F interface is intended to give a TMN user access to the management system. Through an F interface combined with a human-machine interface, a user can

- Define, request, view, and modify report parameters; schedule a report
- Specify real-time performance criteria, request performance statistics, modify performance data logging criteria, and retrieve the performance log

- Retrieve alarm reporting and logging criteria, acknowledge the receipt of an alarm report, schedule a test, request a test report, report a service problem, and check status information on a problem report

- Perform configuration management–related tasks such as processing (create, check status of, cancel, close, etc.) a service order and querying resource information

- Perform accounting management–related tasks such as creating, modifying, and listing billing criteria and querying billing information

- Perform security management–related tasks such as accessing usage, security event, and security alarm information for security auditing purpose and backing up files for intrusion recovery

Information models.   Little has been specified for the information model of the F interface. However, some efforts have been undertaken by organizations such as the TeleManagement Forum (formerly Network Management Forum) to specify a set of telecommunications network management–specific Graphic User Interface (GUI) objects.

Protocol suites.   Similarly, very little has been officially specified with regard to the protocol for the X interface. However, CORBA-based client-server architecture in combination with web-based tools (e.g., Java Applets) is being widely used in the industry.

Q interfaces.   The Q interface is applied at the q reference points as shown in Fig. 6.5. In order to provide implementation flexibility, the Q interface is further divided into the $Q3$ and Qx categories, where the interface Qx is applied at the qx reference point, and $Q3$ at the $q3$ reference point, respectively.

Q3 interface.   The $Q3$ interface is the flagship of the TMN interfaces. This is the only interface that has a substantial number of specifications and that has been widely implemented in practical applications. It is used to communicate between an operation system and a network element, a Q adaptor, or a mediation device, or between two OSs that belong to the same TMN. The bulk of the existing TMN specifications have been devoted to this interface.

Information models.   The resource information models of $Q3$ have been defined for generic network elements, ISDN, the CCS7 network, ATM network elements, and SONET network elements. A survey of the information models can be found in the following chapter.

Protocol suites.   The second part of the interface is the $Q3$ protocol suite which is specified in ITU-T Recommendations Q.811 (1997) and Q.812 (1993). The requirements of the $Q3$ protocol suite are to support bidirectional data communications and to provide the flexibility to make use of a variety of existing data communication protocols, both connectionless and connection-oriented.

The $Q3$ protocol suite consists of two portions: the lower-layer protocols and the upper-layer protocols. The lower layers are the OSI layers 1 through 3, i.e., physical layer, data link layer, and network layer. The upper layers are the

OSI layers 4 through 7, i.e., transport layer, session layer, presentation layer, and application layer.

The lower-layer protocol suite of the $Q3$ interface practically includes all the protocols in use. A total of seven protocol suites are identified in ITU-T Recommendation Q.811 (1997). They fall into two general categories: those used in telecommunications networks and those used in data networks.

One category of protocols is those traditionally used in data networks. It includes the protocols of both connection-oriented and connectionless modes. For example, one connection-mode packet-switched protocol suite uses Recommendation X.25 (1993) for the data-link and network layer protocols and one of a variety of protocols such as Recommendation X.21 for the physical layer. An example of the connectionless-mode protocol suite uses local area network (LAN) technologies such as Carrier Sense Multiple Access with Collision Detection (CSMA/CD).

A second category consists of the protocols traditionally used in telecommunications networks. Both SS7 and ISDN protocol stacks can provide lower-layer protocol suites. For example, the message transfer part (MTP) of the SS7 protocol stack is used to provide the layer-1 and layer-2 protocols, and the signaling connection control part (SCCP) of the SS7 provides the network layer protocol.

The upper-layer protocols are of two types, one providing transaction services and one providing file transfer services. The transaction-oriented services are widely used for a TMN manager to send a management directive to an agent or for an agent to send an event notification to a manager. As shown in Fig. 6.15, it is mandatory that for a connection-mode service, the transport layer protocol shall conform to ITU-T Recommendation X.214 (1993). The session layer uses the protocol specifications of ITU-T Recommendation X.225 (1994). The presentation layer shall conform to the specification of ITU-T Recommendation X.226 (1994) and uses ASN.1 and basic encoding rules (BER). Three component protocols are specified for the application layer: ACSE, ROSE, and CMISE, as shown in Fig. 6.15. Refer to Chap. 4 for a detailed discussion of the application layer protocols.

The protocol profile for file transfer services is identical to that of transaction services for layers 4 through 6, as shown in Fig. 6.16. One example of the application of file transfer services is software downloading into a network element.

Qx interface.   The Qx interface connects a mediation device to a Q adaptor or a network element.

Information models.   What primarily distinguishes this interface from $Q3$ is the information it carries. The management information carried on the Qx is shared between the mediation device and those NEs it supports, while the $Q3$ interface is characterized by the information model shared between the OS and the TMN nodes to which the OS directly interfaces.

Protocol suites.   The protocol suite of the Qx interface is the same as that of $Q3$ interface. With regards to function, the Qx can be a subset of $Q3$. It is supposed to be used when cost or efficiency precludes the use of $Q3$. However,

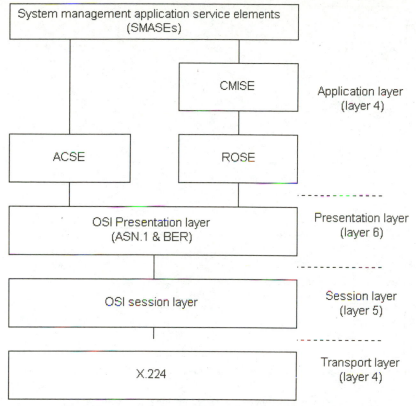

Figure 6.15 Upper-layer protocols of a *Q*3 interface using the transaction function.

since the mediation device is not a well-defined node in the TMN, the precise definition of the contents of a *Qx* interface has not been specified in the standards. As a result, it is largely up to the equipment vendors to make the decision on the scope of this interface. Also, the term *Qx interface* is not as widely used in the industries as *Q*3.

Relationship of a TMN interface to TMN building blocks. The relationships between the TMN building blocks and interfaces are summarized in Table 6.3. A comparison of the relationships listed in this table with those relationships between function blocks and reference points as discussed in the TMN functional architecture reveals a close correspondence between the two types of relationships. This may make the reader wonder what the motivation is behind the reference points and function blocks if they seem to just duplicate the interfaces and nodes of the physical architecture. The seeming repetition stems from the reality of current telecommunications networks. In cases where one entity provides both NE and OS functionality, how do we classify a node where the vendor has provided both NE and OS functionality? How do we refer to the management information passed between functional components of a

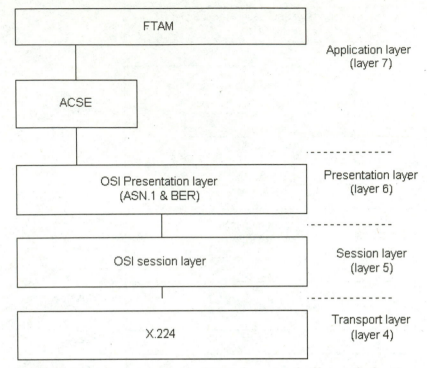

Figure 6.16 Upper-layer protocols of a *Q*3 interface using the file transfer function.

TABLE 6.3 **Relationships between a TMN Interface and Building Blocks**

	*Q*3	*Qx*	*X*	*F*
NE	One of the three interfaces must be present.			O
OS	One of the three interfaces must be present.			O
MD	One of the two interfaces must be present.		O	O
QA	One of the two interfaces must be present.			
WS	NA	NA	NA	M

M—mandatory, O—optional, NA—not applicable.
SOURCE: From ITU-T Recommendation M.3010 (1996).

single entity? To enable these distinctions, TMN uses the concept of function blocks and reference points to describe the functional architecture, since a reference point and function block may or may not map to a physical network entity (like a node or an interface).

Interface methodology

Assume that you are given the responsibility of defining a TMN interface between a management system and a network element. What are the major steps involved in defining such an interface? ITU-T Recommendation M.3020 describes a generic methodology for defining interfaces. While the order of the steps may be modified to adapt to a particular application on hand, ITU-T Recommendation M.3020 sheds valuable light on the steps necessary to implement a TMN interface. The steps are summarized in the following paragraphs.

Step 1.   Identify the TMN management services to be provided by the management system. Examples of these services include traffic management, alarm surveillance, and customer administration, as defined in the TMN functional architecture.

Step 2.   Describe the TMN management context. Several subtasks can be identified as follows:

- Define the management role such as manager, agent, or both for each management system entity.

- Clearly identify the network resources to be managed, since these resources will determine the information models to be defined later. Identify the TMN management functions that implement each TMN management service identified in the previous step.

- Finally, define the relationships between the management roles, network resources, and the TMN management functions.

Step 3.   Build the information model. This task can be broken into several subtasks.

- First, identify all the managed object classes required to support the TMN management functions identified in the previous step. The first place to look is the existing standards, as a majority of the needed classes may be found. It is likely that additional managed object classes need to be defined in order to fully support the TMN functions identified.

- Then modify the existing standardized object models to incorporate the newly created object classes. The new classes can be subclasses of the existing class or stand-alone classes.

- Finally, associate each identified TMN management function with one or more managed object classes to ensure that all the TMN functions are supported by the object classes. A function is supported when
  - The monitoring part of the function can obtain all the necessary information from the support objects
  - The controlling part of the function can exert influence over the support objects

If a TMN function is not supported by any of the support object classes, the previous subtasks of this step need to be executed again.

Step 4.   Define communication requirements for the most likely scenarios. For example, there may be requirements for simple transactions, file transactions, file transfer, file access, or a combination of all of these. Furthermore, there may be requirements for the performance of each transaction. Performance requirements might include throughput, reliability, or transit delay.

Step 5.   Define the protocol suite for the TMN interface. This task is broken into the following subtasks. First, analyze the message needs of the TMN services by analyzing the message needs of each TMN function identified above. The identified messages can be roughly mapped to the application layer protocol messages. Next, examine the standard TMN protocol suites to determine whether the message needs of each layer can be met by the existing standards, with emphasis on the application layer. If yes, the TMN interface definition is complete; if no, carry out the following step.

Step 6.   Define new protocols or amend existing protocols. For each layer for which no adequate protocol has been found to meet the message need, either define additional protocols or amend existing protocols. To do this, the following subtasks are performed.

- First, define the protocol requirements.

- Define appropriate new or amended layer $(N - 1)$ service to support layer N, and define or amend the corresponding protocol mechanism.

- Select the service requirement from layer $(N - 1)$ to layer N for layers 1 through 6. For the application layer, identify the application service elements necessary to support the specific TMN function.

- Finally, put all layer protocols together to form a protocol suite.

The whole process of defining a TMN interface can be viewed as two types of tasks: one that deals with application and one that deals with protocol. The first three steps fall in the category of application tasks, and the latter three are protocol tasks.

TMN Logical Layered Architecture

This section introduces TMN's five functional areas and five logical management layers. These architectural concepts have gained wide acceptance in the field of telecommunications network management. The concepts of five functional areas and five management layers are extensively used throughout this book.

Five functional areas

ITU-T adopts the framework that groups operation, administration, and maintenance (OA&M) functions for telecommunications network in the following five functional areas in its Recommendation M.3400 (1992):

- Fault management
- Configuration management
- Performance management

- Accounting management
- Security management

This section provides an overview of the management functions within each area, and the following five chapters will treat each of the five areas in a systematic manner.

Configuration management. Configuration management is responsible for the planning and installation of NEs, their interconnection into a network, and the establishment of customer services that use the network. According to CCITT Recommendation M.3400 (1992), configuration management is further divided into three general areas: provisioning, NE status and control, and NE installation.

Provisioning is a set of procedures that bring already installed equipment into service and is further divided into several function groups. One is NE configuration that provides a set of functions that manage the configuration of an NE. The tasks include reporting configuration information to an outside system and changing the configuration to accommodate addition, deletion, or modification of an NE's configuration. Another one is NE administration that involves system administration. The tasks include setting clock, tape backup, and process initiation and shutdown. Another function group is for NE database management that provides the capability to initialize, query, back up, and update the database.

NE status and control provides a set of functions to perform status reporting, to schedule status reports, and to allow the request of a status report. It can be the messaging network status, leased circuit network status, transmission network status, or a combination of the above.

NE installation supports the installation of equipment and the associated software that make up a telecommunications network. The tasks include determining a set of installation operations, plus sequencing and scheduling the operation to achieve maximum efficiency and minimum interference with ongoing operations.

Performance management. Performance management provides the operator with the capability to monitor and measure the utilization of network resources, to analyze these measurements, and to make adjustments to improve network performance. Performance management generally encompasses four groups of functions: performance monitoring, performance management control, performance analysis, and performance quality assurance.

Performance monitoring is further partitioned into traffic status monitoring and traffic performance monitoring. The traffic status monitoring functions provide the current status of the network, its elements, and services provided by the network. These functions report the service availability of NEs, the busy or idle status of circuit groups, the congestion status of exchanges, the receipt of congestion control signals, and the congestion status of the signaling network, among others. The traffic performance monitoring functions assess the performance of the network and the traffic being carried on the network. The functions report circuit group data and parameters on a scheduled basis, switch load measurements, switch congestion, and signaling network load measurements, among others.

Performance management control is further divided into traffic control and traffic administrative functions. The traffic control functions are concerned about establishing, modifying, or removing a manual or an automatic control over the network traffic. The traffic administrative functions involve activities such as establishing, changing, or removing a measurement schedule, a performance threshold, and schedules, and administrating a network management database.

Performance analysis involves analyzing the network and service performance data collected by the performance monitoring functions and reporting the analysis results.

Performance quality assurance involves the activities to ensure quality of service. The activities include monitoring and ensuring the connection quality, billing integrity, cooperation with fault management to establish the cause of a resource failure, and cooperation with configuration management to change routing and load control parameters.

Fault management. Fault management enables the detection, isolation, and correction of abnormal operation of the telecommunications network and its environment. It consists of the following subareas: alarm surveillance, fault localization, fault correction, testing, and trouble administration.

Alarm surveillance is mainly concerned with monitoring and detecting failures in an NE. The alarm surveillance functions provide capabilities to report alarms, summarize reported alarms, set alarm reporting criteria, enable and disable visual or audible alarm indications, and manage alarm logging.

Fault localization is concerned with locating the root cause of faults. Among others, the process involves scheduling a diagnosis test, setting test criteria, managing the auditing process, and arranging a program trace.

Fault correction is a process of restoring the network operation and service interrupted by the faults. It may involve reloading of software, replacement of faulty components, or switching to the standby unit.

Testing is a process that involves test configuration (e.g., test access, test circuit configuration), test control, test results and status reporting, and network control for testing purposes. Testing capabilities are often used in the fault localization process to determine the root cause of a fault.

Trouble administration is about maintaining a trouble information base, retrieving trouble history data as requested, and keeping track of a trouble report through its life cycle.

Accounting management. Accounting management primarily provides a set of functions to measure the use of the network resources and services and to determine the cost of such use. It provides facilities to collect accounting records and to set billing parameters for the usage of services and for periodic changes of access to the network. Based on CCITT Recommendation M.3400 (1992), the following functional groups make up accounting management: usage metering, tariffing, and billing.

Usage metering is the first stage of accounting management and is responsible for measuring and recording usage of network resources and services and for creating usage data records and sending the records to other systems for further processing.

Billing functions are mainly concerned about controlling billing data collection at NEs, setting the charging criteria for billing data collection, processing the billing data, and interfacing the customers to complete billing transactions.

Tariffing functions determine the amount of payment for network resource use and addresses issues such as creating tariff classes, determining tariff periods, and integrating diverse tariff data to create a single bill.

Security management. Security management provides for prevention and detection of the improper use of network resources and services, for the containment of and recovery from security breaches, and for security administration. The areas that security management covers include fraud prevention, fraud detection, fraud recovery, and security services.

Fraud prevention addresses the proactive measures to prevent frauds.

Fraud detection provides mechanisms to detect fraud promptly and efficiently once it has occurred.

Fraud recovery and containment determine how to contain and limit the damage incurred by frauds.

Security services introduce five different types of security services: authentication, integrity, confidentiality, access control, and nonrepudiation.

Five logical layers

The concept of layered architecture, well known for its application in OSI's seven-layer network architecture, is also used to describe TMN's functional architecture. One of the main advantages to this layered architecture is that it provides a framework for abstract network management functions at different levels. This approach also facilitates the organization and the development of network management systems in a modular and incremental fashion.

The five-layer TMN functional architecture, as shown in Fig. 6.17, has gained widespread popularity. Though the general characteristics of each layer are relatively well understood, the precise contents at each layer are still subject to different interpretation.

Business management layer (BML). There are two main tasks at the business management layer: defining business processes and setting management policy. A management operations process formally defines a sequence of steps

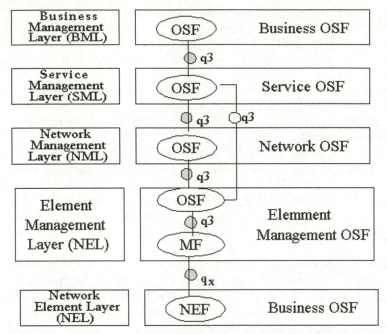

Figure 6.17   Reference model of TMN logical layered architecture.

and functions necessary to achieve a well-defined business goal. It is assumed that the operations processes used in various enterprises within the telecommunications industry share common characteristics, and therefore the standardization of the process is warranted. A standard management operations process provides a common terminology to facilitate the exchanges of information across the industry. More importantly, a standard management operations process provides a foundation for automating the process, which is a key goal for many telecommunications operators.

The TeleManagement Forum is an industrial consortium for defining telecommunications network and service management standards and agreements. It has defined general processes for service management, billing, service ordering, and others. Defining standard processes for the BML is not a trivial task. For example, the process of service ordering that involves two providers consists of many steps. The steps may include accepting service orders, determining preorder feasibility, preparing price estimates, developing order plans, performing credit checks, requesting customer deposits, initializing service installations, establishing service level agreement (SLA) terms, tracking order statuses, and finally notifying customers.

A wide range of policies with regard to service and network management need to be decided at the BML. For example, a provisioning policy may designate certain types of NEs as preequipped during initial installation to facilitate immediate service activation. Some routing criteria may be imposed for a variety of reasons, ranging from ensuring service reliability to enabling business agreements with another service provider. For example, a performance threshold policy needs to be decided to set the condition for taking corrective actions and to designate the type of action to be taken when performance degeneration crosses the threshold.

Service management layer (SML).   The SML has two sides: one facing the customer and one facing the network. As shown in Fig. 6.18, both are related to providing service. The customer-facing side provides an interface to customers and directly interacts with customers to receive information. This information from customers includes customer service trouble reports, service inquiries, and service orders. Through this interface, the SML identifies customer access to the network, reports service usage for billing, and maintains and reports quality of service data to customers.

The network-facing side converts the customer view of a service into a network view and implements and delivers the service to customers. Among the major functions of this interface are service provisioning, service problem resolution, and service quality management. Service provisioning is the process which turns a service order from a customer into an appropriate provisioning of network resources to implement the order. The service planning and development process is to plan and develop a service from an existing set of building blocks and then deploy them into the network. The service problem resolution process analyzes the network performance data and determines the

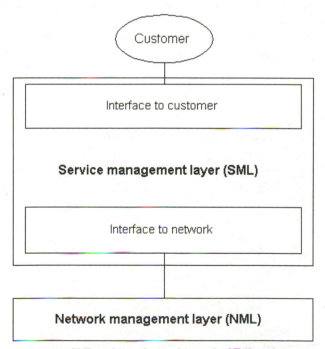

Figure 6.18 SML and its relationship to the NML and the customer.

cause of the customer service problem reported via the customer interface. This process also involves taking corrective, as well as preventive, actions. Service quality management is a process which ensures that the delivered service is consistent with the service agreed upon by the customer (i.e., quality of service) and takes actions in case of performance deterioration crossing a threshold.

At the SML resides the function for one service provider interface other service providers to provision a service. One application of such a function is the service configuration that involves multiple service providers to provision a service. Issues such as "What capacity is provided by whom?" and "How should tariffs be applied?" are resolved via information exchange between service providers.

One principle of the SML is that the management functions at this layer should be independent of underlying switching and transmission technologies.

Network management layer (NML).   The management tasks of this layer, which include all five functional areas as described above, are characterized by a networkwide view. The relationships between the NML and the adjacent layers are shown in Fig. 6.19.

To provide this view, the NML aggregates the management information from the component network elements and forms a network view. For example, tasks within the NML in the area of fault management include the correlation and filtering of networkwide fault data for networkwide alarm surveillance. In the area of performance management, the performance data and events are collected from all NEs of the network in order to detect a performance trend.

One of the responsibilities of the NML is to coordinate network elements in traffic control, event trending, performance analysis, and other management

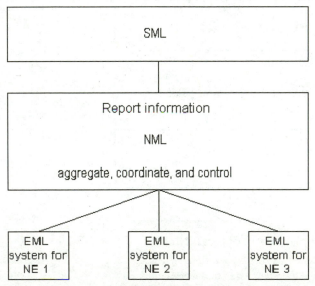

Figure 6.19   The two sides of the network management layer.

operations. For example, in the process of finding the root cause of a fault, circuit tests involving multiple NEs need to be coordinated. The same is true for network traffic control and fault recovery, where coordinated efforts are required for a management operation and the responsibility of coordination resides at the NML. It is also the responsibility of the NML to exert control over networkwide resources. The resources include networkwide circuits, switching facilities, and software.

Element management layer (EML). The mission of this management layer is to manage a single network element or in some cases, a group of network elements given that they are closely coupled together. The management scope includes all five management areas, as described earlier. An example of configuration management at the EML is keeping track of the states of all circuits and resources on a switch (an NE). An example of fault management at the EML is displaying all trunk alarms reported by that switch and forwarding them to the interested management systems in the NML.

The EML provides the NML with NE-specific management information. For example, an EML management entity may report the configuration of the NE resources (channel, circuits, trunks, etc.) to help an NML management entity to form a networkwide configuration. The NE fault correlation results are also needed by the NML to help determine the root cause of a fault that involves multiple NEs. The EML directly interfaces an NE or a group of NEs to collect the data of management interest from each NE.

Network element layer (NEL). The NEL represents the NE being managed by the TMN and presents those NE aspects of management interest to the upper layers. This layer is the ultimate destination of the management commands such as switching to a backup unit and running diagnostic tests. Also this layer originates management information that may result in eventual management actions. Alarms are detected at the NE and reported to the upper layers. Performance data and billing data are generated at an NE.

The five-logical-layer abstraction recognizes that the management of a telecommunication network deals with NEs, networks, services, and business matters. The upper four layers address management functions, and the lowest layer (the network element layer) represents the management view of the telecommunications functions provided by the network resource.

Figure 6.20 shows a logical layering of network management functions, in increasing levels of management abstraction. Corresponding to each layer, there is a set of open system functions (OSFs) that implement the management functions. Though not an exact one-to-one mapping, the functions in the top four layers generally correspond to the manager side's functionality, and the NEL corresponds to that of the agent side.

Relationships between the five functional areas and the five logical layers. The listing of five functional areas across the five functional layers can be expressed as a two-dimensional table, or TMN function matrix, as shown in

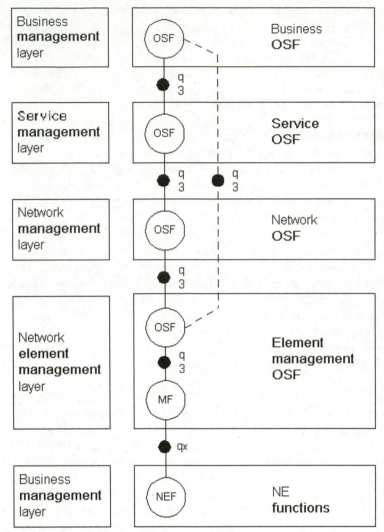

Figure 6.20   TMN functional blocks and TMN layers. (*From ITU-T Recommendation M.3010, 1996.*)

Table 6.4. It is conceivable that a set of function building blocks (FBBs) at each grid of the matrix can be defined, the interface of each FBB standardized, and the FBBs developed as off-the-shelf functional components. The saving in development effort to build network management applications and the time to market will be tremendous if this vision can be turned into reality. As alluded to, no standard FBBs have been defined yet, though some efforts in this direction are under way. The contents of Table 6.4, primarily based on CCITT Recommendation M.3400 (1992) and Bellcore GR-2869 (1995) provide an example set of possible FBBs.

TMN management services

A TMN management service is a relatively independent area of management activities defined by a TMN that can be viewed as a functional requirement from the management system user's perspective. A TMN management service

TABLE 6.4   Relationships Between Five Functional Areas and Five Layers

	Fault management	Configuration management	Accounting management	Performance management	Security management
BML	Alarm policy, management of repair process, test policy, trouble report policy	Material management policy, provisioning policy, procurement guideline, priority service policy	Management of the usage measurement process, pricing strategy	Network performance goal setting, performance monitoring policy, exception threshold policy	Investigation of revenue change pattern, security policy
SML	Service test, trouble reporting to customer, trouble information query, trouble ticket notification	Arrangement of installation with customer, customer need identification, customer service feature definition	Feature pricing, totaling usage charge, service usage correlation, usage validation, usage aggregation	Customer performance reporting, customer traffic performance summary	Customer profiling, customer usage pattern analysis, customer security administration, service intrusion recovery
NML	Network fault event correlation and filtering, network fault localization, selection of test suite, circuit selection for test, trouble ticket administration	Network installation administration, access circuit design, leased circuit design, network connection management	Network usage correlation, usage data storage	Data aggregation and trending, traffic control, traffic capacity analysis, network performance characterization	Internal traffic activity pattern analysis, network security alarm, severing internal connection, network intrusion recovery, network security alarm management
EML	NE alarm reporting, alarm summary, alarm indication management, NE fault localization, log control, NE test circuit configuration, NE fault correction	Loading NE software, NE installation administration, loading program for service features, NE database management, generate NE status and control	Usage data collection, NE usage data validation, usage accumulation	NE trend analysis, NE traffic performance analysis, NE performance control, NE performance analysis, NE traffic capacity analysis	NE security alarm management, NE audit trail management
NEL	Failure event detection and reporting	Parameter counting and reporting	Usage data generation	Detection of state change, storage of state information	

SOURCE: Based on ITU-T Recommendation M.3400 (1992) and Bellcore GR-2869 (1995).

can be implemented with a set of function building blocks as described. A list of TMN management services, while their implementations and requirements are not subject to standardization, is a collection of the services defined across various TMN standards. A total of 17 such services have been identified but only a few have detailed definitions, as shown in Table 6.5.

The following management services are identified, but little has been specified for them with regard to the service contents:

- Management of the security of TMN
- Management of the Intelligent Network
- Restoration and recovery

TABLE 6.5   TMN Management Services

TMN management service	Description
Customer administration	A management activity performed by the network operator to obtain customer data required to offer the requested services and to exchange the customer data with the network for it to produce the services
Routing and digit analysis administration	A management activity performed by the network operator to change the static routing information at run time
Traffic measurement and analysis administration	A management activity performed by the network operator to change the traffic measurement–related parameters
Tariff and charging administration	The management activities including creation, modification, and deletion of charging data and the management of the data collection process
Traffic management	The management of traffic associated with circuit-switched networks and transmission networks. It collects and processes traffic data, and modifies the operation of a switch and reconfigures the network as needed.
Management of customer access	The management of the customer access equipment, including network terminating units, digital access loop, and multiplex equipment
Management of transport network	Includes the management of PSTN circuits, leased circuits, and transmission paths and links. The management covers all functional aspects necessary for bringing a transport network into service.
Switching management	Management service covering all aspects of managing a digital switch, including performance, fault, and configuration management
Common channel signaling system management	Management service covering all aspects of managing a signaling network: configuration of the signaling network, availability and traffic conditions of signaling links, control of the signaling links, etc.

SOURCE: CCITT Recommendation M.3200 (1992).

- Quality of service and network performance administration and system installation administration

- Management of equipment in customer premises system installation administration

- Management of the TMN

- Materials management

- Staff work scheduling

At a conceptual level, a TMN service shall be implemented using a TMN management function group. A group consists of a set of functions as identified in the five TMN functional areas of this chapter. The relationships between TMN management services, the TMN management function group, and TMN management functions are demonstrated in Fig. 6.21.

Summary

TMN defines a network to manage a telecommunications network. At the core of TMN are four architectures: functional, physical, logical layered, and information. This chapter has discussed the first three architectures. The information architecture and related issues are the subjects of Chap. 7.

The functional architecture consists of five functional blocks and three classes of reference points. The five functional blocks are the operation system function (OSF), the network element function (NEF), the workstation function (WSF), the mediation function (MF), and the Q-adaptor function (QAF). Each

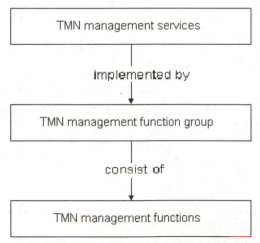

Figure 6.21 Relationships between TMN management services, the TMN management function group, and TMN management functions. (*From ITU-T Recommendation M.3020, 1995.*)

of the functional blocks represents a management function entity. The OSF can be viewed as the aggregation of manager functions, and the NEF as the aggregation of the agent functions. The MF smooths the information transfer between an NEF and an OSF. The QAF makes it possible for a TMN OSF to exchange management information with a non-TMN NEF. The WSF interprets and presents the TMN management information to the information user, most often the human operator.

The physical architecture represents the physical entities of a TMN. The TMN physical architecture consists of five nodes and four interfaces. The five nodes, closely correspond to the TMN function blocks: operations system (OS), network element (NE), mediation device (MD), workstation (WS), and Q adaptor. Each implements the corresponding function block. Similarly, the three classes and four interfaces implement the corresponding reference points. TMN has become well known primarily because of the $Q3$ interface, which indicates the interface-centric nature of TMN. $Q3$ interfaces, especially the object-oriented information models, have become part of TMN that are most often implemented.

The logical layered architecture presents TMN as a logically layered network. The five layers (BML, SML, NML, EML, and NEL) provide different levels of abstraction of management functionality and separation of concerns. On each layer, the management capabilities are categorized into five functional areas (fault, configuration, accounting, performance, and security management). This logical layered architectural concept has gained wide acceptance in the industry.

Exercises

6.1   Given the network configuration shown in Fig. 6.22, determine the interface and TMN nodes, based on the TMN physical architecture.

6.2   Given the configuration shown in Fig. 6.23, explain how the Q adaptor uses M-GETs issued by the TMN manager.

6.3   Determine which of the five functional areas [fault, configuration, accounting, performance, and security (FCAPS)] each of the following functions belongs to.
- Analyze traffic performance data
- Detect and report an alarm
- Authenticate a user
- Provision trunks and lines for a service
- Generate a bill for a customer

6.4   For each of the management tasks described below, identify which layer of the TMN layered architecture the functionality belongs to.
- Analyze network traffic conditions
- Collect billing information on a switch
- Monitor a switch performance
- Take customer service orders
- Provision the networkwide resources (trunks, lines, circuits)

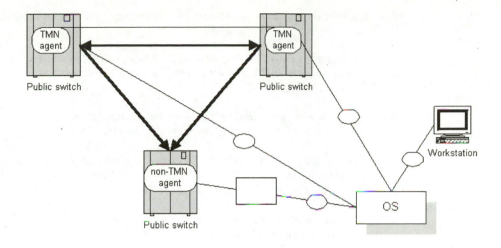

Voice/signaling link

datalink

Figure 6.22   A sample network configuration.

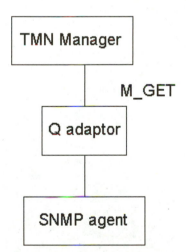

Figure 6.23   A sample configuration of a Q adaptor.

6.5   Given that the function blocks and reference points of the TMN functional architecture are very similar to the nodes and interfaces of the physical architecture, what is the main justification for having a functional architecture?

6.6   Explain the factors that prompted the development and subsequent wide acceptance of TMN.

6.7   If you develop an alarm surveillance system, explain which part of the system belongs to OSF and which to NEF of the TMN functional architecture.

6.8 Explain the differences between the Q-adaptor function and the mediation device function.

6.9 Assume that the OS is implemented as two separate systems, one at the network layer and one at the NE layer. Use an example to show the different tasks performed by the two systems.

6.10 It is rare for a mediation device to be implemented as an independent system. Give some reasons for combining an MD with an OS.

6.11 A Q adaptor, which has almost universally been called a mediation device in the industry, makes it possible for a TMN OS to communicate with a non-TMN NE. What are the major steps involved to accomplish this?

6.12 Explain which TMN interface should be used if two service providers desire to exchange management information, say, billing data.

6.13 Use an example to explain the usage of the *F* interface.

Telecommunications Management Network

Information Architecture and Generic Information Models

Outline

- The first section, "TMN Information Architecture," presents three aspects of the TMN information architecture: object-oriented management information modeling, the model for management information exchange, and the TMN naming and addressing scheme.

- The second section, "TMN Generic Management Information Model," introduces five groups of generic managed object classes that intend to represent management aspects of a telecommunications network: network, network element, switching and transmission, connection and cross connection, and termination points.

- The last section, "Survey of Q3 Information Models," provides a high-level description of the published information models for different types of telecommunications networks such as ATM, SONET, ISDN, and SS7.

TMN Information Architecture

The preceding chapter covered the TMN functional, physical, and logical layered architectures. This chapter concludes the discussion on TMN with an introduction to the TMN information architecture and TMN information models. This section focuses on the three aspects of the TMN information architecture:

- An object-oriented model for representing management information
- A model for exchanging management information
- The TMN naming and addressing scheme

Object-oriented management information modeling addresses the issue of how to represent the managed network resources and present the information to the management system. The *management information exchange model* defines the entities involved in the exchange of management information, the relationships between the entities, the context of the exchange (such as management and communication protocols), and the knowledge shared between entities. Note that two aspects of TMN's information architecture, the manager-agent concept and the premise of shared management knowledge [as presented in ITU-T Recommendation M.3010 (1996)], are combined into the model for management information exchange. The *naming and addressing scheme* outlines the application directory structure to support distributed TMNs and TMN interworking.

Object-oriented management information modeling

This section provides a high-level review of object-oriented information modeling concepts, describes how managed network resources are represented as managed objects, and briefly discusses modeling languages such as Guideline for Definition of Managed Objects (GDMO) and Unified Modeling Language (UML).

Basic concepts.   Since an introduction of object-oriented concepts can be found in Chap. 4, only a brief review is provided here. In an object-oriented approach, all managed network resources (such as switches, components in a switch, and software that drives the switch) are represented as managed objects. Not only the network resources themselves but also the relationships between the managed elements can be managed objects. In general, any entity, physical or conceptual, concrete or abstract, can be an object, as long as the entity is of interest from a management perspective.

An object is characterized by a set of attributes that represent properties of the entity and a set of operations that can be performed on the attributes. Each object has a state and an identity. Viewed from the outside, an object is an encapsulated entity whose attributes and internal implementation of behavior are hidden. The outside world interacts with an object through its published interface. An object class is an abstraction of a set of objects sharing the same attributes and common behavior. A class is like a type of object that can be instantiated. Sometimes the terms *type* and *class* are used interchangeably. An instantiated object class is called an *object instance,* or *object* for short.

One object class can inherit from another class if they have many behaviors and attributes in common. The inheritance means that the class inheriting from another class, often called the *subclass* or *child class,* has available to it the behavior and attributes of the class being inherited, often termed the *superclass* or *parent class.*

One object class may contain another class as part of its attributes. This is also referred to as the relationship "has a," as in "the containing class *has a* contained class as a component." The containment relationship models the real-world hierarchies of "is part of" relationship (e.g., an engine *is part of* a car) or real-world organizational hierarchies (e.g., a directory contains files, a file contains records, and a record contains fields).

Relationships between managed objects and managed resources

- What is the relationship between the managed objects and the managed network resources? The ultimate goal of a network management system is to manage the telecommunications network resources such as switches, software in the switches, and transmission systems. A managed object is a logical representation of a network resource or an aspect of a network resource. In addition, the following aspects of the relationship are mentioned in ITU-T Recommendation M.3010 (1996). There is not necessarily a one-to-one mapping between managed objects and real resources. A resource may be represented by one or more objects, and an object may represent all or part of one or more resources. When a resource is represented by multiple managed objects, each object provides a different abstraction of the resource.

- A managed object may provide an abstract view of resources that are represented by other managed objects.

- A managed object can be embedded; i.e., a managed object may represent a large resource that contains subresources that are represented by other objects.

- If a resource is not represented by a managed object, it cannot be managed across the network because it is invisible to the management system. It is the managed objects, the abstract representations of resources, not the resources themselves that are being manipulated by a management system.

These aspects of the relationship between managed object and network resources, described in the first items of the above list, are depicted in Fig. 7.1.

Language for modeling management information.   OSI GDMO has been the choice of modeling language for telecommunications network management, largely because that was the only choice available at the time. However, alternatives are emerging, in part as a result of adopting technologies of the general computing world, such as Common Object Request Broker Architecture (CORBA). In this subsection, a review of GDMO is followed by a brief discussion of alternative information modeling languages.

GDMO.   When it comes to the information modeling language, OSI's GDMO was the most logical choice at the time the TMN information architecture was first defined. For one reason, there were few alternatives at the time. Another

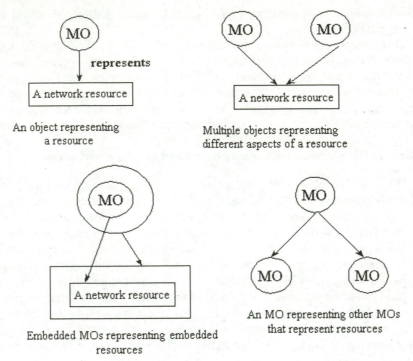

Figure 7.1 Relationships between managed objects and network resources.

obvious reason is that the TMN architecture had already adopted other major components of the OSI management framework (such as the object-oriented approach to the management information modeling and the CMIP management protocol). A detailed discussion of GDMO can be found in Chap. 4. A high-level review is provided here.

GDMO, though based on the object-oriented information modeling approach, is designed specifically for telecommunications applications and has its own unique syntax and semantics. The top-level structure of GDMO is the managed object class, which consists of a set of packages. A package contains attributes, attribute groups, actions, and notifications, as shown in Fig. 4.6. All these can be either mandatory or optional. All instances of a managed object class must have all the mandatory portions of the class. A brief description of each component provides an overall view of the managed object, while detailed syntax and semantics are given in the GDMO section.

A *package* is a logical grouping of attributes, attribute groups, actions, notifications, and behavior. Once encapsulated in a managed object, the package becomes an integral part of the managed object and becomes accessible only through the interface of the managed object.

An *attribute* represents a property of the network resource being modeled. Any access to an attribute has to go through the interface of the managed object that contains the attribute. The value of an attribute can be either retrieved or modified, again via the containing object interface. The permitted operations on an attribute are defined in the containing managed object class. An attribute can be single-valued or set-valued. If set-valued, each of the elements of the set is of the same data type. Additionally, the set can be empty. An attribute group is a logical grouping of a set of attributes for organizational and operational convenience.

The behaviors that a managed object may take on are expressed in the form of a set of actions and/or a set of notifications. An *action* is an operation that can be performed by a managed object. A *notification* is a report emitted by a managed object and sent to a manager via the associated agent.

Two types of relationships, inheritance and aggregation, are explicitly modeled between managed object classes by GDMO. The inheritance relationship is defined inside a managed object class and the containment relationship is defined by the *name binding* construct.

OMT, UML, and others.   While GDMO has been and still is widely used for specifying management information models, developments on several fronts have motivated the search for alternatives. On the front of computing technologies, one development is the convergence between the telecommunications technologies and general computing technologies. The general, less-expensive, standardized computing technologies are gradually replacing the proprietary and expensive computing technologies used exclusively in the telecommunications industry. Examples of computing technologies include design methodologies, operating systems, programming languages, and the database. Another development is that object-oriented technologies have matured considerably since the time when GDMO was first defined, and a standard object-oriented modeling language has begun to be accepted on a large scale.

In the meantime, some of the inherent characteristics of GDMO appear to limit its further acceptance. For one, the syntax and constructs of GDMO are closely tied to the OSI management protocol CMIP (this was the intent since the very beginning). The protocol dependency of GDMO will constrain its use when alternative management protocols are adopted.

Efforts are under way in the international standards bodies to search for management information modeling languages that are neutral to any particular management protocol and that have wide support in general computing industries. This is intended to be a complement to, rather than a replacement of, GDMO. The candidates include Object-oriented Modeling Technique (OMT), UML, and CORBA's Interface Definition Language (IDL).

OMT is an object-oriented modeling language with well-defined notations. It has achieved industrywide use since the early days of object-oriented technologies. UML stems from the efforts to unify and standardize diverse object-oriented methodologies and notations. It combines most of the widely used object-oriented (OO) methodologies such as Booch's method (Booch 1994), OMT (Rambaugh 1991), Object-Oriented Software Engineering (Jacobson et al. 1992), and other methods. UML was accepted by the industry consortium, the Object Management Group (OMG), as an industrial standard in late 1997. CORBA's IDL is emerging as another alternative for the information modeling language, largely because CORBA is being applied to telecommunications network management on an increasingly large scale. (Chapter 15 provides a detailed discussion on applications of CORBA and IDL to telecommunications network management.)

Model for management information exchange

A management information exchange model consists of an exchange paradigm and the context for the exchange. The paradigm defines the entities involved in the management information exchange. The context defines the conditions that enable the information exchange.

Manager-agent paradigm.   Management of a telecommunications environment is in essence an information-processing application. Because the network resources being managed are distributed, network management is a distributed application. The distributed application involves the exchange of management information between management processes for the purpose of monitoring and controlling the physical and logical resources of the network (such as switching and transmission hardware, and software).

TMN adopts the OSI manager-agent paradigm in which a management process can take on a manager role, an agent role, or both in a single management session, as shown in Fig. 7.2. These roles are characterized as follows:

- The manager role is the part of a distributed application that issues management operation directives and receives notifications.

- The agent role is the part of a distributed application that acts upon the management operation directives by manipulating managed objects. The agent also emits event notifications to the manager reflecting the behavior of the managed objects with which it is associated.

- A manager can be associated with multiple agents, and an agent can be associated with multiple managers.

In Fig. 7.2, the management process of system B takes on both manager and agent roles, while the management processes in systems A and C take on one role each.

There are two other entities involved in the management information exchange, i.e., managed objects and network resources. An agent effects the management operation on the managed objects (instead of the managed network resource itself). The managed object in turn propagates the effect onto the network resource, the ultimate destination of management operations. The relationships between all four entities involved in management information exchange are shown in Fig. 7.3. There are three types of management information exchange:

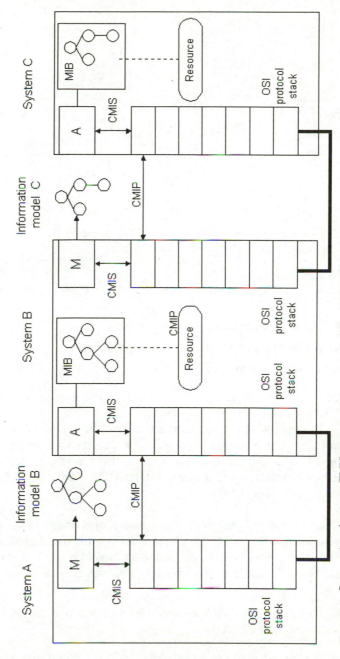

Figure 7.2 Interactions between TMN management processes. (*From ITU-T Recommendation M.3010, 1996.*)

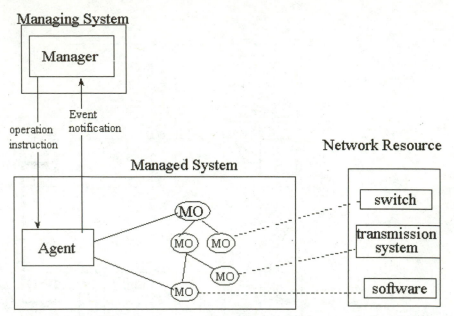

Figure 7.3 Relationships among manager, agent, managed objects, and network resources.

- Between a manager and an agent
- Between an agent and a managed object
- Between a managed object and a network resource

Only the first type of exchange, between a manager and an agent, is subject to TMN and OSI standardization.

A managed object can emit a notification to a manager via an associated agent and take certain actions according to the instruction by the manager through an agent. The information exchange between an agent and a managed object is a local matter, and so is the information exchange between a managed object and a network resource.

Context for management information exchange. The context refers to the management knowledge shared between a manager and an agent that make the management information exchange possible. *Context negotiation* is the process of establishing shared knowledge between two parties.

Shared management knowledge. In order for the management information exchange to take place between a manager and an agent, they must share the knowledge of the following areas.

- *Supported protocol capabilities.* The two parties must use the same management protocol (such as the CMIP). In case different protocols are used, a Q adaptor shall be present to convert one to the other. The limitations and restrictions of the Q adaptor, if any, must be known to both parties.

- *Supported management functions.* For example, the management functions that an agent is capable of performing must be known to the manager that issues the management directives to the agent.

- *Supported managed object classes and the object instances.* A manager shall have the static knowledge of what managed object classes are defined for a given agent and the dynamic knowledge of what object instances have been created during a particular session.

- *Authorized capabilities.* A manager and an agent must be aware of each other's authorized capabilities (such as filtering and scoping).

- *Containment relationship between objects.* A manager must know the containment relationships between managed objects in an agent's MIB in order to formulate a request.

Context negotiation. Context negotiation can be static or dynamic, depending on the time the negotiation takes place and the nature of shared management knowledge that is the subject of the negotiation.

Static context negotiation. The static process of context negotiation is to establish the shared management knowledge between a manager and an agent for a contractual period. The static process can take place off-line prior to the establishment of a session or on-line during a session between a manager and an agent.

An off-line process of static context negotiation occurs as two equipment vendors agree between themselves as to what management protocol to use and what management functions to provide. The examples include cases where a management system and an agent with associated network equipment are from two different vendors.

The on-line process of static context negotiation occurs at the beginning of a session. At the session establishment time, information is exchanged to allow two parties to come to a common understanding of what the shared management knowledge will be for the session. Shared management knowledge negotiated this way will remain in force for the duration of this session. For example, a manager and an agent may negotiate the scoping and filtering capabilities using ACSE messages as part of the process of establishing an association.

Dynamic context negotiation. A dynamic context negotiation is required if shared management knowledge is subject to change during a session. For example, the management functions may change, or a new managed object class or instance may be added during the session. One possible mechanism to support dynamic context negotiation would be to define a managed object class that supports a notification of any change to the shared management knowledge.

TMN naming and addressing

It is essential to have a naming and addressing scheme for identifying and locating the various communication objects within a TMN and between TMNs. The architectural requirements for such a naming and addressing scheme to support distributed and TMN interworking are described below. The TMN architecture with such a naming and addressing scheme should

- Allow for geographic distributed configuration of network management operations
- Facilitate service assistance and interaction with a customer
- Provide the fault-tolerant capability and security support for management functions
- Make it possible for parties other than the owner of the system, such as a customer or value-added service provider, to access the management functions
- Make it possible to have different or the same management services at different locations, even if it accesses the same network element
- Make the interworking between separately `managed networks possible so that the management services can be provided between TMNs

The directory service specified in ITU X.500 is being considered to provide the naming and addressing scheme. The X.500 directory service can provide the following naming-related services:

- General information service (also known as the "yellow pages" service) for TMN-related management information
- Global naming of managed objects
- Name and address resolution
- Representation of shared management knowledge as discussed above

In brief, the X.500 directory, as shown in Fig. 7.4, is a directory information base (DIB) that consists of one global root node, a set of directory system agents, and a set of directory objects (DOs). Together they provide unambiguous distinguished object names in a global environment. Remember that the managed objects (MOs) in the OSI MIB are uniquely named only in the local MIB environment. The X.500 directory provides the naming service that will make the managed objects visible in a global environment. This is an essential condition for TMN interworking. The relationships between the directory object and the managed object and between the MIB and the DIB are shown in Fig. 7.4.

Generic TMN Management Information Models

This section introduces the general TMN information models that consist of six groups of managed object classes. These groups of managed object classes are intended to model six different aspects of a generic telecommunications network that is devoid of any specific network technology and yet can be used to derive the information models for a specific network through inheritance.

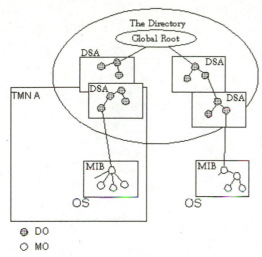

Figure 7.4 Integrated directory in a TMN information architecture.

The six aspects of the network [also called six "fragments" in ITU Recommendation M.3100(1995)] are network, management element, termination points, switching and transmission, cross connection, and functional areas.

Network fragment

The network fragment has only one managed object class, *network,* that represents a collection of interconnected telecommunications and management objects capable of exchanging management information. A network can be nested within another network, forming a containment hierarchy.

As shown in Fig. 7.5, the *network* class contains a managed element object class and the managed element object class in turn contains the managed object classes of other fragments.

Managed element fragment

This fragment consists of eight managed object classes: circuitPack, equipment, equipmentR1, equipmentHolder, managedElement, managedElementComplex, software, and softwareR1. Each managed object class in this fragment models an element of a network that is independently manageable and visible to the management system. The inheritance and containment relationships between the managed object classes of this fragment and others are shown in Fig. 7.8 later in the chapter.

The following list provides descriptions of the managed object classes:

- *managedElement.* Represents a telecommunications equipment entity or a TMN entity that performs managed element functions (such as providing

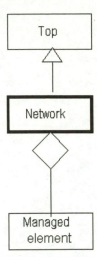

Figure 7.5 Containment and inheritance relationship for the network fragment and related classes.

support and service to subscribers). A managed element communicates with a manager over one or more TMN *Q* interfaces to be monitored and controlled by the manager.

- *managedElementComplex*.   Represents a set of network elements that are grouped together for management purposes.

- *circuitPack*.   Represents a plug-in replaceable unit that can be inserted into or removed from an equipment holder.

- *equipmentR1*.   A subclass of the object class equipment which is designed to provide equipment alarm monitoring capabilities.

- *equipment*.   Represents physical components of a managed element, including replaceable components. An instance of this object class is in a single geographic location. An equipment object may be nested within another equipment object.

- *equipmentHolder*.   Represents physical resources of a network element that are capable of holding other physical resources. Examples of the physical resources represented by the instance of this class are equipment bay, shelf, and slot.

- *software*.   Represents logical information stored at a network element, including programs and data. One software object may be contained within another software object.

- *softwareR1*.   A subclass of the software object class with additional attributes.

The following example of a switch will help illustrate the managed element fragment and its relationships to the network resources it models. A switch consists of a set of shelves that in turn contains a set of cards. A variety of cards can be in a shelf: central processing cards, line interface cards, trunk interface cards, timing cards, switching matrix cards. A card may have onboard software to drive it and perform the desired functions. The rough correspondence between the managed object classes and the represented components of switching equipment is listed in Table 7.1. A logical view of switching equipment is shown in Fig. 7.6.

The attributes, notifications, and actions of each object class in the managed element fragment are listed in Table 7.2. The definitions of the attributes are provided shortly. See the Managed Object Class Dictionary (App. D) for definitions of the generic notifications and actions.

Termination point fragment

A network element is connected to other network elements via logical links called *connections*. From the perspective of one network element, the concern is not a whole connection, but rather the point of the connection that interfaces

TABLE 7.1 Correspondence Between the Managed-Element-Fragment Managed Object Classes and the Represented Network Components

Managed object classes	Represented network components
managedElement	Network element, switch
circuitPack	Card
software, softwareR1	Software
equipmentHolder, managedElement-Complex	Shelf, frame

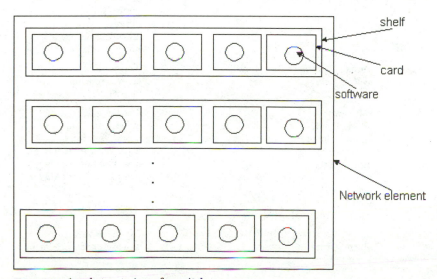

Figure 7.6 An abstract view of a switch.

TABLE 7.2   Managed Object Classes of the Managed Element Fragment

Class name	Attributes	Notifications/actions
circuitPack	circuitPackType, availabilityStatus, administrativeState, operationalState, alarmStatus, currentProblemList, alarmSeverityAssignmentProfilePointer	*Notifications:* objectCreation, objectDeletion, stateChange, equipmentAlarm
equipment	equipmentID, replaceable, administrativeState, operationalState, affectedObjectList, alarmStatus, currentProblemList, userLabel, vendorName, locationName, version	*Notifications:* objectCreation, objectDeletion, attributeValueChange, stateChange, equipmentAlarm, environmentalAlarm, communicationsAlarm, processingErrorAlarm
equipmentHolder	equipmentHolderType, equipmentHolderAddress, acceptableCircuitPackTypeList, holderStatus, subordinateCircuitPackSoftwareLoad	*Notifications:* environmentalAlarm, processingErrorAlarm
equipmentR1	alarmStatus,serialNumber, supportedByObjectList, alarmSeverityAssignmentProfilePointer	*Notifications:* equipmentAlarm
managedElement-	managedElementId, systemTitle, alarmStatus, administrativeState, operationalState, usageState, userLabel, vendorName, version, locationName, currentProblemList, externalTime, systemTimingSource	*Notifications:* environmentalAlarm, equipmentAlarm, communicationsAlarm, processingErrorAlarm, objectCreation, objectDeletion, attributeValueChange, stateChange *Actions:* allowAudibleVisualLocalAlarm, inhibitAudibleVisualAlarm
managedElement-Complex	managedElementComplexId, systemTitle	*Notifications:* objectCreation, objectDeletion
software	softwareId, administrativeState, operationalState, affectedObjectList, alarmStatus, userLabel, vendorName, version, currentProblemList	*Notifications:* objectCreation, objectDeletion, attributeValueChange, stateChange, processingErrorAlarm
softwareR1	alarmSeverityAssignmentProfilePointer, alarmStatus	*Notifications:* processingErrorAlarm

the element. This point of connection is referred to as a *termination point* (TP), as shown in Fig. 7.7. There are three aspects of a termination point:

- Where the termination point is used
- Who uses it
- Direction of the termination point

Depending on the purpose of the termination point, two generic types are defined: trail termination points and connection termination points. Based on the direction of the trail or connection associated with a TP, the TP can be a sink, source, or bidirectional TP. A sink TP is a point where a connection or trail terminates. A source TP is a point where a connection or trail originates. A bidirectional TP is a point where a connection or trail passes through.

The following list provides descriptions of the managed object classes of the segment:

- *CTPBidirectional.* Originates at one link connection and terminates at another link connection. All its attributes and behaviors are inherited from the object classes connectionTerminationPointSource and connectionTerminationPointSink. CTP is an acronym for connection termination point. This object class is bidirectional.

- *CTPSink.* Terminates at a link connection. The attribute downstreamConnectivityPointer points to the TP that receives information from this TP at the same layer.

- *CTPSource.* Originates a link connection. The attribute upstreamConnectivityPointer points to the TP managed object within the same managed element that sends information to this TP at the same layer.

- *terminationPoint.* Represents the point that terminates transport entities such as trails and connections. This is an abstract class from which the subclasses trailTerminationPoint and connectionTerminationPoint are derived.

- *TTPbidirectional.* Represents a trail termination point (TTP) where one trail is originated and another is terminated. This object class is bidirectional and is meant to be subclassed by the technology-specific (e.g., SONET, ATM) trail termination bidirectional object classes. All its attributes and behaviors are inherited from the object classes trailTerminationPointSource and trailTerminationPointSink.

- *TTPSink.* Represents a TP where a trail is terminated. It represents the access point in a layer network that is a focus of both the trail relationship and the client-server relationship.

- *TTPSource.* Represents a TP where a trail is originated.

The attributes, notifications, and actions of each object class in the TP fragment are listed in Table 7.3. The definitions of the attributes are provided shortly, and the definitions of the generic notifications can be found in the Managed

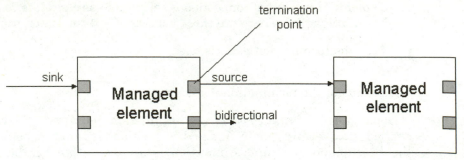

Figure 7.7   Illustration of termination points.

Object Class Dictionary (App. D). The relationships between the object classes of the TP fragment are shown in Fig. 7.8.

Switch and transmission fragment

This fragment of managed object classes represents the connections that link managed elements. Conceptually, a managed element is divided into different functional layers, each performing a distinct function. This is similar to the seven layers of the OSI protocol stack and the four layers of the Internet protocol. From the perspective of network management, there are three distinct types of connections: circuit, connection, and tail. A circuit, or a group of circuits, directly interconnects two network nodes and provides the physical connectivity between the two. A connection is responsible for the transparent transfer of information between connection termination points. A connection is a component of a trail. A sequence of one or more connections are linked together to form a trail. A connection can be either unidirectional or bidirectional.

Although a connection and a trail are distinct transport entities serving different functions, they share common characteristics and operations. Common characteristics include operational and administrative states. Common operations include sending alarm and state change notifications to a manager in the event an alarm or state change occurs.

The switch and transmission fragment consists of four object classes:

- *circuitEndPointSubgroup.*   Represents a set of circuit endpoints that interconnect one switch with another.

- *connectionR1.*   Responsible for transparent transfer of information between connection termination points.

- *pipe.*   Responsible for transferring information between two termination points. This is an abstract class to be inherited by the managed object class connection and trail that actually carry traffic.

- *trailR1.*   Responsible for transferring information from one end to another across one or more networks. A trail is composed of two trail termination points, connections, and the associated connection termination points. The relationships between the managed object classes are shown in Fig. 7.9.

The attributes, notifications, and actions of each object class in the switch and transmission fragment are listed in Table 7.4. The definitions of the attributes are provided shortly, and the definitions of the generic notification and action can be found in the Managed Object Class Dictionary (App. D).

Cross-connection fragment

The cross-connection fragment represents another type of connection that is between the termination points within a managed element like a switch. This type of connection is important for representing the routing capabilities and switching entities within a network element. A cross connection is represented by a pair of termination points, i.e., fromTerminationPoint (fromTP) and toTerminationPoint (toTP), as shown in Fig. 7.10. A fromTP represents a termination point where a cross connection originates, and a toTP represents a termination point where a cross connection terminates within a network element. A cross connection can be either point-to-point or point-to-multipoint. Multiple termination points grouped together for management purposes are treated as a single entity and are termed a *group termination point* (GTP).

The cross-connection fragment consists of seven object classes:

TABLE 7.3   Managed Object Classes of a Termination Point Fragment

Class name	Attributes	Notifications
CTPBidirectional	(Derives its attributes and behavior from both CTPSink and CTPSource)	
CTPSink	downstreamConnectivityPointer, CTPId, channelNumber	
CTPSource	upstreamConnectivityPointer, CTPId, channelNumber	
terminationPoint	supportByObjectList, characteristicInformation, administrativeState, operationalState, networkLevelPointer, crossConnectionObjectPointer, alarmStatus, currentProblemList, alarmSeverityAssignmentProfilePointer	*Notifications:* objectCreation, objectDeletion, attributeValueChange, stateChange, communicationAlarm
TTPBidirectional	(Derives its attributes and behavior from both TTPSink and TTPSource)	
TTPSink	upstreamConnectivityPointer, administrativeState, operationalState, supportableClientList, TTPId	
TTPSource	downstreamConnectivityPointer, supportableClientList, TTPId	

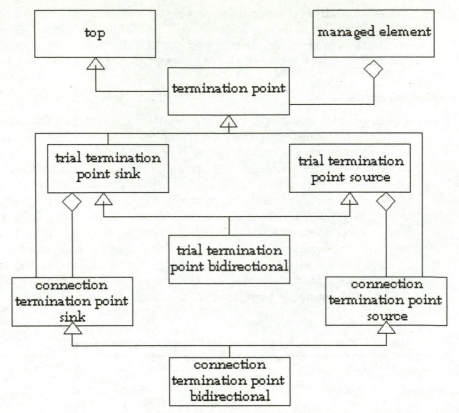

Figure 7.8 Inheritance and containment relationships for the termination point fragment.

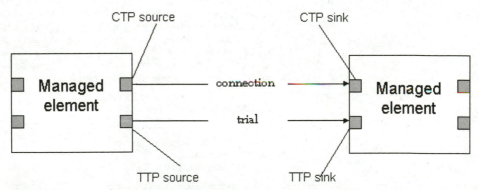

Figure 7.9 Illustration of switching and transmission managed object class fragment.

TABLE 7.4 Switching and Transmission Managed Object Class Fragment

Class name	Attributes	Notifications
circuitEndPointSubgroup	circuitEndPointSubgroupId, numberOfCircuits, labelOfFarEndExchange, signalingCapabilities, informationTransferCapabilities, circuitDirectionality, transmissionCharacteristics, userLabel	*Notifications:* attributeValueChange, objectCreation, objectDeletion
connectionR1	connectionId, serverTrailList, clientTrail	
pipe	directionality, operationalState, a-TPInstance, z-TPInstance, characteristicInformation, protected, alarmStatus, currentProblemList, alarmSeverityAssignmentProfilePointer, userLabel	*Notifications:* objectCreation, objectDeletion, attributeValueChange, stateChange communicationAlarm
trailR1	trailId, serverConnectionList, clientConnectionList	

- *crossConnection.* Represents an assignment relationship between the termination point of a GTP object listed in the fromTermination attribute and the termination point of GTP objects listed in the toTermination attribute of this managed object.

- *fabric.* Responsible for establishment and release of cross connections and the assignment of termination points to TPPools and GTPs.

- *groupTerminationPoint* (*GTP*). Represents a group of termination points that are treated as one unit for management purposes.

- *multipointCrossConnection* (*mpCrossConnection*). Represents an assignment relationship between the termination point or GTP object listed in the fromTermination attribute and the termination point or GTP objects listed in toTermination attributes of the contained crossConnection managed objects.

- *tpPool.* Represents a set of termination points or GTPs that are used for management purpose such as routing.

- *namedCrossConnection.* Represents a designated cross connection that is marked for special attention (e.g., for being sensitive to environment variation). This object class is a subclass of crossConnection.

- *namedMpCrossConnection.* Represents a designated multiple-point cross connection that is marked for special attention. This object class is a subclass of mpCrossConnection.

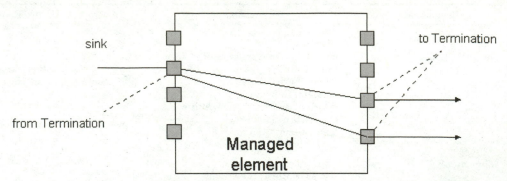

Figure 7.10 Illustration of a cross connection.

The attributes, notifications, and actions of each object class in the cross-connection fragment are listed in Table 7.5. The definitions of the attributes are provided shortly, and the definitions of the generic notification and action can be found in the Managed Object Class Dictionary (App. D).

Functional area fragment

This fragment of managed objects is different from the previous fragments. It does not model managed network resources, but instead provides a set of generic management functions that normally are required by any telecommunications network regardless of the network technology. For example, the capability of representing, setting, reporting, and controlling alarms is one of the first items to be considered for a network. There are three managed object classes for this capability: alarmRecord, alarmSeverityAssignment, and currentAlarmSummary. Generic event logging and filtering functions can be found in all telecommunications network management systems. Most of the managed object classes in this fragment are adopted from the OSI information models.

This fragment consists of thirteen object classes as listed below. Applications of most object classes in this fragment will be demonstrated in the following five chapters (configuration, performance, fault, accounting, and security management), and a detailed definition of each object class can be found in the Managed Object Class Dictionary (App. D). The thirteen classes are

- alarmRecord
- alarmSeverityAssignmentProfile
- attributeValueChangeRecord
- currentAlarmSummaryControl
- discriminator
- eventForwardingDiscriminator
- eventLogRecord

- log
- logRecord
- managementOperationsSchedule
- objectCreationRecord
- objectDeletionRecord
- stateChangeRecord

Attribute definitions of the TMN generic information models

The attributes used in the TMN generic information models (above) are described briefly below, in alphabetic order. The precise ASN.1 definitions of each attribute can be found in ITU-T Recommendation M.3100 (1995).

TABLE 7.5   Managed Object Classes of the Cross-Connection Fragment

Class name	Attributes	Actions
crossConnection	*Mandatory:* crossConnectionId, administrativeState, operationalState, signalType, fromTermination, toTermination, directionality	
fabric	administrativeState, operationalState, availabilityStatus, listOfCharacteristicInfo, supportedByObjectList	*Actions:* addTpsToGTP, removeTpsFromGTP, addTpsToTpPool, connect, disconnect *Action:* switchOver
fabricR1		
gtp (groupTerminationPoint)	gtpId, crossConnectionObjectPointer, signalType, tpsInGtpList	
mpCrossConnection (multipointCrossConnection)	mpCrossConnectionId, administrativeState, operationalState, availabilityStatus, signType, fromTermination	
namedCrossConnection	redline, crossConnectionName	
namedMultipointCrossConnection	redline, crossConnectionName	
tpPool	tpPoolId, tpsInTpPoolList, totalTpCount, connectedTpCount, idleTpCount	

a-TPInstance.   Identifies one of the two termination points of an instance of the connectivity object class or one of its subclasses.

acceptableCircuitPackTypeList.   Indicates the types of circuit packs that can be contained in an equipment holder object.

administrativeState.   Indicates the administrative state of a resource, with one of the following three values: locked, shutting down, and unlocked.

affectedObjectList.   Specifies a set of object instances that can be directly affected by a state change of a given managed object instance.

alarmSeverityAssignmentList.   Lists all abnormal conditions that may exist in instances of an object class and shows the assigned alarm severity information (major, minor, etc.) for each condition.

alarmSeverityAssignmentProfileId.   An attribute type whose distinguished value can be used as a relative distinguishing name (RDN) for naming an instance of the alarmSeverityAssignmentProfile object class.

alarmSeverityAssignmentProfilePointer.   Identifies an alarmSeverityAssignmentProfile object.

alarmStatus.   Indicates an alarm condition associated with an object and the status of the alarm with the following severity values (from highest to lowest):

- activeReportable-Critical
- activeReportable-Major
- activeReportable-Minor
- activeReportable-Indeterminate
- activeReportable-Warning
- activePending
- cleared

channelNumber.   An integer that represents a channel number.

characteristicInformation.   An object identifier used to indicate whether a connection or a trail can be connected to a termination point object.

circuitDirectionality.   Indicates the directionality of the circuits in a circuit group with one of the three values: one-way in, one-way out, two-way.

circuitEndPointSubgroupId.   An identifier that can be used as an RDN for naming an instance of the circuitEndPointSubgroup object class.

circuitPackType.   A printable string value that indicates the type of the circuit pack.

clientConnectionList.   A list of objects identifying the client connections served by a trail. These client connections may be either of a lower or the same transmission rate as the trail.

clientTrail.   Identifies a trail object instance in the same network layer that is served by a connection object.

connectedTpCount.   An indication of the total number of termination points associated with a tpPool that have been connected.

connectionId.   An identifier of a connection that can be used as an RDN for naming an instance of the connection object class.

crossConnectionId.   An identifier uniquely identifying a cross connection whose value can be used as an RDN for naming an instance of the crossConnection object class.

crossConnectionName.   A name of a character string for a crossConnection or multipointCrossConnection object.

crossConnectionObjectPointer.   A pointer to a managed object such as a crossConnection, a groupTerminationPoint, or a fabric object. When a termination point is neither connected nor reserved for connection, this pointer points to the fabric object responsible for establishing its connection.

cTPId.   An identifier uniquely identifying a connection termination point whose value can be either an integer or a graphic string.

currentProblemList.   A list identifying the current existing problems with severity indications that are associated with the managed object containing this attribute.

directionality.   Specifies whether the associated managed object is unidirectional or bidirectional.

downstreamConnectivityPointer.   Indicates that the downstream (i.e., outgoing) connectivity is either single, concatenated, broadcast, or broadcast concatenated. The difference between broadcast and broadcast concatenated connectivities is that the order of the recipients in the list is important for the concatenated connectivity.

equipmentHolderAddress.   Indicates the physical location of the resource represented by the equipmentHolder object. Since the equipmentHolder object can be nested, the address may be nested as well. For example, if a system has three levels of equipment holders representing bay, shelf, and slot, respectively, then the address can have a frame identification code for a bay, a bay shelf code for a shelf, and a position code for a slot.

equipmentHolderType.   Indicates the type of equipment holder using a character string. The possible value can be bay, shelf, drawer, slot, and rack.

equipmentId.   An identifier uniquely identifying an equipment object, whose value can be used as an RDN for naming the instance of the equipment object class.

externalTime.   Provides the time-of-day system time.

fabricId.   Uniquely identifies a fabric object whose distinguished value can be used as an RDN for naming an instance of the fabric object class.

fromTermination.   Identifies a trail termination point (source or bidirectional), a connection termination point (sink or bidirectional), or a group termination point composed of members of one of the above categories.

gtpId.   Uniquely identifies a group termination point whose value can be used as an RDN for naming an instance of the groupTerminationPoint object class.

holderStatus.   Indicates the status of a physical holder, i.e., empty, in the acceptable list, not in the acceptable list, or unknown.

idleTpCount.   The total number of termination points associated with a tpPool that are in the enabled and available operational state for cross connection.

informationTransferCapabilities.   Specifies different service types supported by the circuit group, such as speech, 64 kbits of unrestricted data, and 56 kbits of unrestricted data.

labelOfFarEndExchange.   A name for the far end switch where this circuit subgroup terminates.

listOfCharacteristicInfo.   Lists the characteristic information types that can be cross-connected by a fabric.

locationName.   A character string identifying a location.

managedElementId.   Uniquely identifies the managed element object whose distinguished value can be used as an RDN for naming an instance of the managed object class.

managedElementComplexId.   Uniquely identifies a managedElement-Complex object whose distinguished value can be used as an RDN for naming an instance of the managed object class.

mpCrossConnectionId.   Uniquely identifies a multipointCrossConnection object whose value can be used as an RDN for naming the instance of the managed object class.

networkId.   Uniquely identifies the network object whose value can be used as an RDN for naming the instance of the managed object class.

networkLevelPointer.   A pointer to a network managed object class instance.

numberOfCircuits.   Total number of circuits in a circuit subgroup.

operationalState.   Represents the operability of a resource with two possible values: enabled and disabled.

protected.   A Boolean value to indicate whether the associated managed object is protected or not.

redline.   Indicates whether the associated managed object is red-lined, i.e., a part of a sensitive circuit.

replaceable.   Indicates whether the associated managed object is replaceable or nonreplaceable.

serialNumber.   A serial number for a physical resource such as a network node.

serverConnectionList.   Identifies a list of connection objects within the same network layer that constitute the associated trail.

serverTrailList.   Identifies the trail objects in a network layer that may be used in parallel to serve a connection object.

signalingCapabilities.   Specifies the signaling types supported by the circuit subgroup, such as ISDN User Part (ISUP).

signalType.   Specifies the signal type of a cross connection, TP pool, or group termination point. The signal type can be simple, bundle, or complex. If the signal type is *simple,* it consists of a single type of characteristic information. If the signal type is *bundle,* it is made up of a number of signal types of the same characteristic information. If the signal type is *complex,* it consists of a sequence of bundle signal types.

softwareId.   Uniquely identifies a software system whose distinguished value can be used as an RDN for naming an instance of the software object class.

subordinateCircuitPackSoftwareLoad.   Indicates whether there is designated software to be loaded into the containing circuit pack whenever an automatic reload of software is needed. The value of this attribute can be either a printable string, a sequence of object instances, or NULL. The NULL choice is used when the contained circuit pack is not software loadable or no software load has been designated. The sequenceOfObjectInstances choice identifies an ordered set of software instances to be loaded. When the choice of printableString is used, the semantics is a local implementation issue.

supportableClientList.   A list of object classes representing the clients a particular managed object is capable of supporting.

supportedByObjectList.   A set of object instances that can affect the associated managed object. The object instances in the set can be objects representing physical or logical resources.

systemTimingSource.   It specifies the primary and secondary timing sources for synchronization between managed elements.

systemTitle.   A unique identifier for the system managed object class whose value can be used as an RDN for naming instances of the managed object class.

totalTpCount.   The total number of termination points associated with a termination point pool.

toTermination.   Identifies a source or bidirectional connection termination point, a sink, or bidirectional trail termination point, or a group termination point made up of a member of the above categories.

tpPoolId.   A unique identifier for the termination point pool managed object class, whose distinguished value can be used as an RDN for naming the instance of the managed object class.

tpsInGtpList.   A list of termination points in a group termination point.

tpsInTpPoolList.   A list of termination points that are in a TP pool.

trailId.   An identifier uniquely identifying a trail object class whose distinguished value can be used as an RDN for naming the instances of the managed object class.

transmissionCharacteristics.   Specifies whether different transmission characteristics (such as satellite or echo control) are supported by the circuit subgroup.

tTPId.   Uniquely identifies a trail termination point object class whose distinguished value can be used as an RDN for naming instances of the managed object class.

upstreamConnectivityPointer.   Indicates that the upstream (i.e., incoming) connectivity is either single, concatenated, broadcast, or broadcast concatenated. The difference between the broadcast and broadcast concatenated connectivities is that the order of the recipients in the list is important for the concatenated connectivity.

userLabel.   A name assigned to a managed object.

vendorName.   Identifies the vendor of the associated managed object, such as a network or circuit pack.

version.   Identifies the version of the associated managed object.

z-TPInstance.   Identifies one of the two termination points of an instance of the connectivity object class.

Survey of Q3 Information Models

The *Q*3 interface is by far the most mature and widely deployed among the TMN interfaces so far. This section briefly surveys the published *Q*3 information models for a variety of telecommunications networks. The *Q*3 information models are being defined by various international standards bodies on a continuous basis. This survey is not meant to be an exhaustive listing, but rather representative of those that are already stable and widely in use. Also this overview mainly covers the standards defined by ITU-T. Note that generic information models are covered in the preceding section, and only the network-specific or technology-specific ones are described here.

Subscriber administration

ITU-T Recommendation Q.824 (1995) defines information models for PSTN and ISDN service provisioning. Three aspects of service provisioning are modeled, i.e., service, network resources, and customer. The services include both PSTN and ISDN services.

The service provisioning–related network resources include access port, access port profile, directory number, and access channels. ITU-T Recommendation I.210 (1993) defines services that can be assigned to a customer. The services are grouped into bearer services, teleservices, and supplementary services. The customer is modeled by the customerProfile object class that is associated with a directory number and a set of access ports.

Traffic flow control and management

ITU-T Recommendation Q.823 provides an object model for network traffic control that allows the control to be exercised manually by specific input from an operation system or automatically in response to an internal or external stimulus. Manual controls are activated and deactivated by the creation and deletion of traffic control objects. Other object classes defined for manual control include

- cancelRoutedOverflow
- destination Code Control
- alternativeRoutingTo
- alternativeRoutingFrom

Automatic controls are modeled with trigger and control objects. A trigger object can be armed or disarmed (inhibited or disinhibited). Control objects specify the action and nature of the control to be applied, once triggered.

Q3 object models for SS7 network management

SS7 is the most widely used signaling standard in PSTN networks. Like layered OSI network reference models, it consists of four layers: message transfer part (MTP), ISUP, signaling connection control part (SCCP), and transfer capability application part (TCAP). Two object models are defined for MTP and SCCP layers in Q.751.1 and Q.751.2, respectively, and they provide an NE view only.

An SS7 network is made up of signaling points, which are connected by the signaling links. Various types of signaling points and links that make up signaling networks are modeled as MTP signaling point, signaling link termination point, signaling terminal, signaling link set termination point. The MTP layer is responsible for routing the signaling messages between signaling points. The signaling routes and routing capabilities are modeled by the managed object class signalingRouteSetNEPart.

The SCCP, similar in functionality to the network layer of the OSI network reference model, is responsible for call connection control. Various types of controls are modeled by the following object classes: sccpRoutingControl, sccpConnectionlessControl, sccpConnectionOrientedControl, and sccpManagement. For a more detailed discussion on the object models for SS7 network management, see Petermueller (1996).

Q3 object model for an ATM network

The ATM Forum, an industry consortium dedicated to the development of standard ATM technology, has defined several standard object information models or MIB, including the M4 MIB [ATM98], and provides an NE view only. To accommodate different management protocols, two different versions of the MIB, one for the SNMP and one for the CMIP, are defined [ATM97]. The managed object classes model a wide variety of components within an ATM NE, ranging from ATM cell layer interfaces trap log, and cell protocol monitoring interfaces to virtual-circuit–virtual-path traffic monitoring interfaces.

Q3 interface object model for an SDH network

ITU-T Recommendation G744 defines an information model for the synchronous digital hierarchy (SDH), which provides only a network element view of an SDH network. The model consists of over 80 managed object classes, representing various aspects of an SDH network element, including multiplex section connection termination points, optical interfaces, and regenerator section connection termination points. These object classes are specialized from the object classes of the TMN generic network information models as described above.

Summary

This chapter concludes the introduction to TMN. We use an application scenario to briefly sum up some of the major benefits of TMN. But before that, a brief summary of the issues discussed in this chapter is in order.

This chapter covers three topics: the TMN information architecture, generic TMN information models, and a survey of published Q3 information models. The TMN information architecture is described from three aspects: an object-oriented approach to network management information modeling, a model for management information exchange, and the naming and addressing scheme.

The object-oriented management information modeling subsection provides an overview on three topics: object-oriented concepts, representation of the managed network resources with objects, and the information modeling language used in the representation. The management information exchange model subsection describes the management entities involved in the management information exchanges (i.e., manager and agent), the relationships between the entities, the context of the exchange (such as management and communication protocols), and the knowledge shared between the entities that makes the management information exchange possible. The naming and addressing scheme within the TMN information architecture provides a way to name and address TMN management entities and information models in a global environment that supports distributed TMNs and TMN interworking.

The generic TMN information models section describes six sets of managed object classes that model the six aspects of a generic telecommunications network devoid of specifics of any particular network technology. These models are widely used to derive the information models for a technology-specific network through inheritance. The last section of this chapter provides a survey of the published $Q3$ information models for technology-specific networks such as ATM, SONET, and SS7.

Now we use a hypothetical network management scenario to illustrate the applications and benefits of TMN. Assume that a large local service provider has five different networks, each with a different operation system. These five network management systems cannot communicate with each other due to differences of interfaces.

The first aspect of the TMN application is to use TMN to integrate the five management systems into a centralized operation control center. One benefit is the reduction of the operation cost by consolidating five technical teams operating five systems into one team operating just one system. A side effect is that a single integrated user interface is easy to learn and use.

The second aspect of the application is to use the TMN $Q3$ interface to standardize the management information exchange between the management system and diverse types of network elements. The benefits include use of off-the-shelf products to achieve savings in development cost and time required for the system to be operational. A uniform interface across the entire network makes the network easy to operate and control.

A third aspect of the application is to use the TMN information models to model all network elements to be managed. The advantage is to have a uniform representation of network resources that are easy to develop and manage.

Exercises

7.1 List and discuss different kinds of shared management knowledge that are required in order for a manager and an agent to communicate.

7.2 Explain the difference between static and dynamic context negotiations and the mechanisms that make the dynamic negotiation possible.

7.3 Many TMN information models are defined using GDMO. Discuss the pros and cons of using OSI GDMO as a management information modeling language, as opposed to other alternatives such as CORBA IDL or UML.

7.4 The naming and addressing scheme is a relatively new addition to the TMN information architecture. Discuss why such a scheme is needed.

7.5 Briefly discuss how each of the ITU-T Recommendation X.500 (1993) directory services can be used in a distributed TMN.

7.6 The TMN generic information models as presented in ITU-T Recommendation M.3100 (1995) model a whole telecommunications network from six different aspects. Briefly explain each aspect. Can you think of any additional aspects for a technology-specific network?

Configuration Management

Outline

- The first section, "Network Planning and Engineering," has two parts: network planning and network design. The network planning part describes a generic network planning process and a model for network capacity planning. The network designing part describes network infrastructure design and routing design.

- The second section, "Installation and Software Management," discusses both hardware and software installations with an emphasis on the software management, which includes software installation.

- The third section, "Provisioning," discusses the network level and network element level provisioning.

- The last section, "NE Resource Status and Control," covers network element control operations and network element resource state and status management.

Introduction

Configuration management is a very complicated area, maybe more so than the four other functional areas of management (i.e., fault, accounting, performance, and security management). There is also less agreement on what constitutes configuration management. Four areas included in this chapter, i.e., network planning and engineering, installation, provisioning, and status and control, are based on the CCITT recommendations while taking into consideration other widely used documents such as the Bellcore generic requirement specification. Two areas are specified under configuration management in CCITT Recommendation M.3400: provisioning and network element status and control; both of these areas are covered at the network element level only. The Bellcore GR-2869 (1995) takes into account the other three TMN management layers (i.e., business management layer, service management layer, and network management layer) and identifies the following five areas of configuration management: network planning and engineering, installation, service planning and negotiation, provisioning, and status and control. Since the service planning and negotiation area is covered in Chap. 14, which is exclusively devoted to service management, this chapter covers the other four areas of configuration management.

We first describe a management scenario to provide a context as well as motivations for the discussion of configuration management. Imagine that you are part of a team that is assigned to the task of building a telecommunications network from scratch. The starting point is a set of requirements on the types and capacity of the network. What are the basic steps for building, configuring, and operating such a network? These essentially are the issues configuration management intends to address, which can be generalized into a process model as shown in Fig. 8.1. The four areas of the configuration management roughly form a linear process that starts from designing and planning a network all the way down to controlling the state and status of a network element.

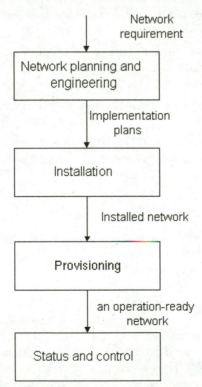

Figure 8.1 A configuration management process model.

Network Planning and Engineering

The primary goal of network planning and engineering is to turn a system requirement into a network system design and a plan for implementing the design, a largely manual process in the past. The major issues addressed include design of access infrastructure, facility infrastructure, and network infrastructure. Note that business-related issues such as demand forecasting and infrastructure planning are assumed to have been addressed already.

A model for network planning and design

This subsection presents a model for network planning and design in terms of the process for accomplishing the task. The process begins with a set of requirements for the network and concludes with an agreement with a potential customer on the network infrastructure, capacity, routing, and cost. This planning process involves a number of activities, ranging from network capacity forecasting, circuit number calculation, and traffic engineering. The goal of a network-planning model is to organize and sequence the planning activities. The discussion is based on a network-planning model defined in CCITT Recommendation E.175 that provides a guideline for network-planning engineers and specifies the steps and sequence of the steps of the planning process, plus the input and output of each planning step and activity.

Network planning is basically a modeling process, mapping a set of economic and commercial capacity requirements of a new network from a potential customer to a set of requirements on the network infrastructure and capabilities. The planning model consists of the following nine steps, which do not form a straight linear process and some of which may have to be repeated before processing to the next step. The starting point of the process is the assumption that the requirements of the new network have already been defined, at least at a high level.

Step 1. Agree on a modeling method. This is the very first step of the process because the method will determine the types and forms of the output and lays a foundation for communications between the involved parties. The following factors should be considered in determining the method.

- Manual or computer-based model
- The time and resources allocated to the planning tasks
- Method of financial comparison study (e.g., present value of an existing network or a competitor network)
- Capital cost and other economic parameters
- Type of input information required such as network financial and capacity requirements
- Form of results to be provided

Step 2.   Gather requirement input information. Once the modeling method is determined, the types and forms of input information should already have been defined. The information may include but is not limited to

- Network traffic forecasts
- Operation, administration, maintenance, and provisioning (OAM&P) requirements on fault tolerance, routing and service, and network restoration
- Infrastructure life span
- Infrastructure cost, capacity, and availability data
- Quality of service requirements

Step 3.   Determine and select network infrastructures based on the requirements. The infrastructure selected shall reflect the operational, technical, and commercial requirements of the potential customer with the operational requirements focusing on OAM&P capabilities. Technical requirements include the requirements on the network technology to be deployed and network capabilities. The commercial requirements define the projected cost, revenue, and profits.

Step 4.   Design the routes for the traffic on the proposed network. This can be done either manually or using an automated tool. The purpose of this step is to evaluate the network infrastructures defined in the previous step. If the network infrastructures prove inadequate, step 2 needs to be repeated and so does this step. Manually routing traffic on the proposed network can be done quickly and efficiently and is suitable for high-level estimation rather than detailed, precise evaluation. Conversely, the computer-based simulation provides a finely detailed evaluation but may require detailed analysis and precise parameters, which may be difficult to come by at the initial stage.

Step 5.   Determine and classify the network costs. The costs can be classified by the following categories so that different levels of partnership in building and operating the proposed network can be defined.

- Capital costs of the network infrastructure
- Estimated network maintenance costs by facility type
- Network extension costs

Step 6.   Check the results with the potential customer and obtain agreement. At this point, the result of the network planning process should be presented to the potential customer. This is to evaluate whether the proposed network has met the customer requirements and whether there are any changes in the requirements. When the results are deemed satisfactory, an agreement on the proposed network shall be obtained from the potential customer.

Step 7.   Repeat part or all of the planning process if necessary. The network requirements from a customer are rarely final and often evolve with the customer's needs and network technology. When changes in the requirements are considered major or the discrepancy between the requirements and the proposed network is nontrivial, the planning process may have to start all over again from step 1 or 2, as warranted by each individual case. In some cases, only a portion of the process needs to be revisited.

Some of the steps of the network planning and design process model, as shown in Fig. 8.2, are self-explanatory and some others deserve further elaboration. The following two subsections further expand on steps 2 to 4.

Network capacity projection

The fundamental task is to project future traffic demands and decide the network capacity that will adequately meet the demands. This is a process consisting of analytic modeling, heuristics, and human experience–based decision making, as shown in Fig. 8.3. The focus here is on defining a generic process, though maybe not the only process, in terms of the tasks involved and the component process for accomplishing each task.

In traffic data collection and aggregation, historical traffic data of a similar network is collected as a base for extrapolation into future traffic patterns. The traffic data may be network technology–specific. For example, the traffic on the public switched network might be measured in hundreds of call seconds per hour, the amount of busy hour that can be carried by a network element, types of services, and quality of service.

In market forecasting, a projection is made on the market demand of the services offered by the network. The forecasting data used to make the projec-

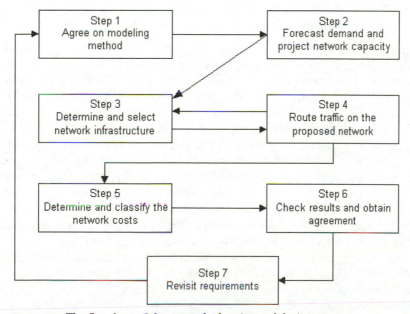

Figure 8.2 The flowchart of the network planning and design process.

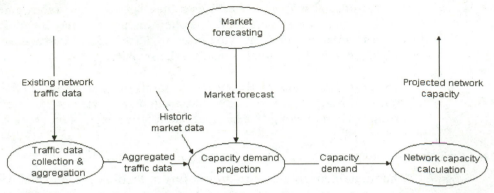

Figure 8.3 Network capacity planning process.

tion include projected population, service offering pattern, and service usage pattern in the target market.

Projected capacity demand, in terms of hundreds of call seconds per hour or other technology-specific traffic capacity measurements, is determined by projecting the aggregated traffic data onto the projected market.

In network capacity calculations, projected capacity demand is converted into network capacity, in terms of a network facility such as trunks, lines, and channels. Widely used methods such as the Erlang B traffic table have achieved standard status in the field of capacity planning. For more details, see Flood (1995). Details on the usage of the traffic table can be found in Chap. 9, "Performance Management."

Selection of network infrastructures

Once the network capacity is determined, the next step is to plan and design the network infrastructure, which traditionally has been a highly manual process. From a designer's point of view, a telecommunications network consists of three types of infrastructure: access infrastructure, core network infrastructure, and facility infrastructure, which includes the physical links between network components.

Access infrastructure, sometimes also called access network, constitutes the portion of a telecommunications network that interconnects customers with the switching network. One important part of access design is to design access circuits based on factors such as customer location, service feature, route, and demand forecast. For example, the planned service feature offering influences the choice of type of access circuit: primary rate Integrated Service Digital Network (ISDN), basic rate ISDN, or simple analog line.

Core network infrastructure consists of the network elements (e.g., switches, routers, bridges) and hardware and software in a network element. The task is to design the network infrastructure capacity, including both software and hardware, to meet the forecast demands, projected traffic load, planned service offering, and other factors that may be specific to each network.

Facility infrastructure includes power supply, repeaters, a building to house the network equipment, and other supporting facilities. Planning and design of facility infrastructure takes into consideration factors such as the physical location of network elements, route, demand forecast, and product line budgets to arrive at the overall facility requirements of the network. They may include specifications of quantity as well as location and functional characteristics of each type of facility. For example, the physical distances between network elements and demand forecasts dictate the type and quantity of physical links required, while the type of switching equipment may help determine the power supply requirement.

Routing traffic on the proposed network

Two issues related to network routing design are important: selection of a network link that physically connects two network nodes, and the routing scheme that directly influences how two nodes should be connected.

Selection of physical links.   Once network components and other infrastructure are in place, the next step is to connect them together. This process defines the physical links between network elements as well as the type and capacity of each link. Link selection must reflect the capacity, operation, and OAM&P requirements defined in the network design. In determining a link, the following factors shall be taken into consideration.

- Distance between a source and destination nodes or between any two directly linked nodes

- Forecasted traffic and possible congestion condition

- Alternative routing requirement for fault tolerance

- Customer-specific criteria such as guaranteed quality of service or security

- The chosen routing structure and routing scheme

One simple, yet effective approach is to route the projected traffic on the proposed network. In very general terms, a stepwise approach works as follows. For each pair of nodes, assign a link that satisfies the projected traffic demand and other requirements. Repeat the steps for all pairs of nodes. Then perform a networkwide optimization to eliminate redundancy and detect the holes and inconsistencies. This process can be repeated as many times as needed. One complicating factor is the choice of routing plan and routing scheme, which can significantly impact the network connection and consequently the traffic capacity.

Routing plans and routing schemes.   A *routing plan* defines how network nodes are connected to one another and how traffic flows and overflows from one node to the next and in what sequence. Generally, there can be a choice of hierarchical versus nonhierarchical and static versus dynamic routing plans. Note that the concept of hierarchical routing can be completely independent of the concept of a hierarchy of network nodes.

Fixed hierarchical routing structure.   Often static, the fixed hierarchical routing is based on the hierarchical network configuration often seen in public switched networks. As shown in Fig. 8.4, the first layer is the local network that consists of customer access lines, local central office switches, and regional tandem switches. On top of that is a national tandem switch network, often known as a long-distance network. Multiple national networks can form an international switching network through international gateways.

In such a hierarchical network, the routing paths are fixed in the sense that all alternative routes are predetermined. All traffic traveling on a given route always chooses the same set of alternative routes at a specific node in case of traffic overflow, irrespective of the routes already traversed. In case of traffic congestion, overflow traffic is directed to these predetermined alternative routes. The routes in the alternative route set will always be traversed in the same sequence. In other words, the routes are ordered in such a way that they are always tested in the same order in the search for an available route at a given node. The fixed hierarchical routing is a convenient way to design routing paths if the traffic pattern is very predictable and stable, with little deviation.

In defining a fixed hierarchical routing plan, a designer normally needs to specify

- How a pair of switches should be connected: by direct circuits or via a tandem switch

- How many tandem switches are needed and where they should be located to achieve maximum efficiency

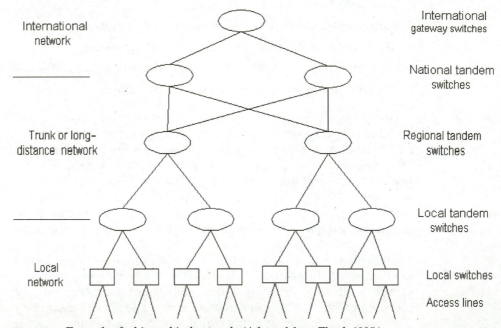

Figure 8.4   Example of a hierarchical network. (*Adapted from Flood, 1995.*)

- How many levels of hierarchical tandem switches there should be in a network for the forecasted volume of traffic

- What the alternative routes are for each given route, and under what traffic conditions the alternative routes should be used

Nonhierarchical dynamic routing scheme.   In recent years, advances in digital network technologies and a rapid increase of traffic volume have made the dynamic, nonhierarchical routing or combination of static and dynamic routing an attractive option in the PSTN. Contrary to the fixed hierarchical routing, which determines all routes statically, the dynamic routing determines the routes for a call on demand and breaks down the hierarchical barrier of the network nodes. This scheme routes calls away from a high-usage route or a route that has encountered fault conditions. The choice of routes is made based on the traffic conditions of the moment.

The modern SS7 network makes the dynamic routing practical. The out-of-band signaling network can transmit routing signals independent of the routes on which calls are actually carried, and the information on traffic conditions can be transmitted through the signaling network for making a routing decision. A simple, yet illustrative example works as follows. If a connection is needed between switch A and switch B and all circuits on the direct route between A and B are busy, then when A finds circuits available on the route between A and C, it tries to route the call to B through C. However, at the moment, the route $C–B$ is congested and C immediately notifies A of the congestion condition via the signaling network and an alternative route, say, $A–D–B$ is then chosen.

A number of factors should be considered in determining a routing structure. First is the projected traffic pattern. If the traffic pattern is highly predictable, the hierarchical routing structure with a fixed routing scheme may be more suitable than the alternatives because of the possibilities for a designer to optimize the resource utilization.

The projected amount of traffic is another major factor. It is economically more efficient to use a direct route between two switches if a large amount of traffic is expected. Also, large groups of circuits are more efficient than smaller groups because a large group has a lower probability of a busy circuit and thus a higher occupancy rate. Otherwise, for little traffic between two switches, it is more economical to combine this with traffic to other destinations to have a large amount of traffic traveling over a common route to a tandem switch. The cost of circuits relative to the traffic volume is an obvious factor as well.

The regulatory requirements are to be considered too. Because of the long history of heavy regulations in the telecommunications industry, the routing plan must be subject to the requirements from areas such as tariff and dialing plans (e.g., area code plan).

Installation and Software Management

This section discusses the installation of both hardware and software with an emphasis on software management, of which software installation is a part.

The bulk of this section is devoted to an introduction of the software management model defined in ITU-T Recommendation in X.744.

Hardware installation management

The main task of hardware installation management involves scheduling and managing the installation process. As shown in Fig. 8.5, the process consists of several steps. A preinstallation check ensures the compatibility of installed hardware components. Scheduling installation takes into consideration such factors as dependency between components and the order of pretests. Hardware installation management also needs to arrange a preservice test and to coordinate the associated software installation. Once the installation is completed, the database is updated to reflect the newly installed hardware.

Software management

Although software management is not entirely equivalent to software installation, the majority of the issues addressed in software management are related to software installation processes, such as installing, uninstalling, or upgrading a software system.

Figure 8.6 lists a set of the software management operations, which are normally requested by a managing system and whose results shall be returned to the requesting party. The subject of the operations is the software unit object that is a management view of a piece of software and may include data, control information, and executable instructions. The management operations are eventually carried out by the three managed object classes, software unit, executable software, and software distributor, as shown in Fig. 8.6.

This subsection first introduces the software management operations and then describes the three managed object classes in detail.

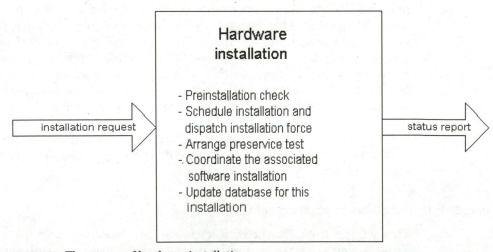

Figure 8.5 The process of hardware installation.

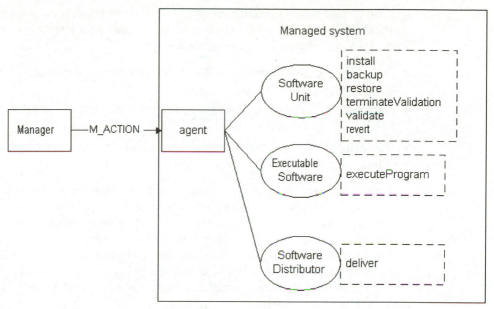

Figure 8.6 Software management operations.

Operations of software management

Create. This operation creates a software unit managed object by creating a new software unit or duplicating a reference object. A software unit may also be created as a result of the delivery operation on a software distributor managed object or as a result of some local action. The state of the created software unit depends on the context of the create operation, which can be in either the delivered state or the created state. Once created, a software unit managed object may emit an object creation notification.

Delete. This operation, which can be carried out by an agent as directed by a manager, removes a software unit and the associated resource from a managed system.

Deliver. A manager may request an agent to perform this delivery operation. The agent in turn sends the deliver action request to the software distributor object. As a result of the action by the software distributor object, a copy of a software unit is delivered to the target system in the delivered state.

Execute program. A manager can ask an agent to carry out this operation to initiate the execution of an executable software. The management of the software execution process and the results from the execution are a local matter.

Get. This is a query operation that can be requested by a manager and carried out by the agent. The information that can be queried includes what software is present and available for use and the relationships between software units.

Install. The install operation starts where the deliver operation ends. Once the deliver operation has delivered a set of software units to the target systems, it is the responsibility of the install operation to load the software units into appropriate memory or disk space and ensure they are ready for use. The install operation is also responsible for

- Coordinating the software delivery to maintain the order if the delivery order is significant
- Ensuring that there is no conflict between new and old versions of installed units
- Ensuring that the proper dependency relationships between the managed objects are maintained

Revert.   This operation causes an installed software unit managed object to revert back to be where it was before the install operation. This may include unloading the software unit from the disk, removing one or more applied patches, and resetting the state of the software unit.

Set administrative state.   This can be part of the revert, install, or deliver operation: after each operation, the administrative state can be set from the unavailable to the available state, or vice versa, or from the shutting down or locked to unlocked state, or vice versa.

Terminate validation.   A manager can request an agent to perform this operation to terminate a validation operation, possible prematurely.

Notification.   An agent sends a notification to the managing system of the results along with confirmation on a completed operation.

Validation.   A manager may request an agent to perform this operation to check the integrity (e.g., whether code or data are corrupted), state (e.g., whether a software unit is in a desired state), or the relationships (whether the dependent software units are in place) of a software unit.

Backup.   A manager may request that an agent make a copy of a target software unit without any direct effect on the source software unit.

Restore.   A manager may request that an agent restore a previously backed up software unit. Note that both the state (both operational and administrative) and availability status of the software unit shall be restored.

The CMIS services that can be used to implement the software management operations just described are shown in Table 8.1.

Managed object classes.   Four managed object classes related to software management are defined in ITU-T Recommendations M.3100 (1995) and X.744 (1996). The inheritance and containment relationships between these object classes are shown in Figs. 8.7 and 8.8. Notice that the containment relationships are primarily for storing and retrieving the objects in an OSI-based MIB. If another type of MIB is used, it is not necessary to enforce the relationships if not called for by the application.

Software.   The software object class represents logical information such as a data table and executable code, which is stored at a network element. The

**TABLE 8.1   Software Management Operations and
the Corresponding CMIS service**

Software management operations	Implemented by the CMIS service
Backup	M_ACTION
Create	M_CREATE
Delete	M_DELETE
Deliver	M_ACTION
Get	M_GET
Install	M_ACTION
Restore	M_ACTION
Revert	M_ACTION
Set administrative state	M_SET
Terminate validation	M_ACTION
Validation	M_ACTION

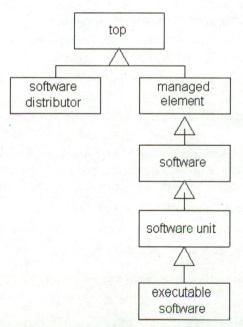

Figure 8.7   Inheritance hierarchy of software
management–related managed object classes.

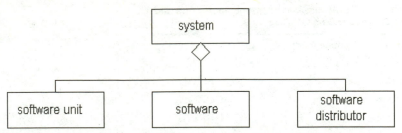

Figure 8.8 Containment relationships for software management–related object classes.

software managed object class is defined in ITU-T Recommendation M.3100 (1995) as part of the information model of the TMN information architecture. Because it is a direct or indirect superclass of other software management–related object classes to be introduced shortly, its attributes, notifications, and actions are briefly reviewed.

The software object class has only one mandatory attribute, namely, softwareId, which is used to uniquely identify a software object instance as a relative distinguishing name (RDN) in an OSI-based MIB.

Optionally, the software object class has the following attributes, notifications, and actions. The details of the following attributes can be found in the Managed Object Class Dictionary (App. D).

- *administrativeState.*   Indicates the administrative state of a software object instance. The state value, which is either locked, unlocked, or shutting down, is administratively set by a managing system.

- *affectedObjectList.*   A list of object instances that can be directly affected by the state change or deletion of a software object instance.

- *currentProblemList.*   A list of current existing problems with severity and associated object.

- *operationalState.*   Indicates whether a software object instance is enabled or disabled operationally.

- *alarmStatus.*   Indicates the alarm status, if any, associated with a software object instance. The status can be either cleared (i.e., no alarm at this moment) or active with a severity indicator. See App. D for more details.

- *userLabel.*   A character string identifying a software object.

- *vendorName.*   A name identifying the developer of the software object.

- *version.*   Identifies the version of the software object.

The following notifications allow the software managed object to notify a managing system:

- *objectCreation.*   This notification is sent to a managing system when the associated software object instance is created.

- *objectDeletion.* This notification is sent to a managing system to report the associated software object instance is deleted.

- *attributeValueChange.* This notification is sent to a managing system to report that a value of a specified set of attributes has changed.

- *stateChange.* This notification is sent to a managing system to report that either the operational or administrative state has changed.

- *processingErrorAlarm.* This notification is sent to a managing system to report a processing failure.

Software unit. An instance of the softwareUnit object class represents a software unit, which can be either a piece of executable code or a data table, and the operations that can be performed on the software unit. The information about a software unit represented in the object includes the file type (i.e., executable or nonexecutable), file location, and file size. Management operations on a software unit are through the corresponding softwareUnit object. This object class is a subclass derived from the software managed object class and thus inherits the attributes and notifications described above. It is a superclass of the executableSoftware managed object class, and the relationships between the classes are shown in Figs. 8.7 and 8.8.

The remainder of this subsection introduces the softwareUnit managed object class in terms of the attributes it contains, the notifications it is capable of emitting, and the actions it can perform, if any. The attributes, notifications, and actions are packed into a set of packages, a GDMO construct as has been discussed in Chap. 4, which is no more than a simple placeholder.

- *availabilityStatus.* Indicates the conditions that may affect the availability of the softwareUnit object, e.g., power off, off-line, in test.

- *proceduralStatus.* Indicates the procedure that the softwareUnit object has to go through in order to be operational, e.g., the procedure initialization.

The processing ErrorAlarmOnServicePackage provides a processingError-Alarm notification to notify a manager of an alarm in processing software such as an error in software downloading, switching over, or restoring. The notification takes a Boolean parameter to indicate whether the software management service has been affected by the alarm.

The appliedPatch package has an attribute appliedPatches, which represents a set of software patches, and each patch has a patch identification (ID) and a pointer to the actual software patch object.

The checkSum package has an attribute checkSum which is a bit string to ensure the integrity of software that might have been transferred across a network. To the outside world, this is a read-only attribute.

The fileInformation package contains a set of read-only attributes providing information on the software being managed:

- *dateOfCreation.* The time the softwareUnit managed object is created.

- *identityOfCreator.*   An identifier of the party that created this softwareUnit managed object instance.

- *dateOfLastModification.*   The time of the last or most recent modification to the information represented by the softwareUnit object instance such as file type or file location. A NULL value indicates no modification has occurred.

- *identityOfLastModifier.*   The party that made the most recent change to the information represented by this softwareUnit object instance.

- *dateDelivered.*   The time the software represented by this softwareUnit object instance is delivered to the managed system. A NULL value means that the software has not been delivered.

- *dateInstalled.*   The time that the software represented by this softwareUnit object instance was installed on the managed system. A NULL value means that the software has not been installed.

The informationAutoBackup package has the following two attributes and one notification for future automatic software backup.

- *futureAutoBackupTriggerThreshold.*   An integer threshold that will trigger an automatic backup of the information represented by this softwareUnit object instance. The integer value is the number of times that the information has been modified.

- *futureAutoBackupDestination.*   A destination where the information represented by this object instance will be backed up. The backup criterion is defined in the futureAutoBackupTriggerThreshold attribute above. The value of this attribute can be either a remote network address for a file transfer protocol such as FTAM (the OSI's equivalent of FTP), a local object instance of the same class, or an in-line indicator for the information to be included in a notification.

The autoBackupReport notification is sent to a managing system by this softwareUnit object instance once an automatic backup on the information represented by this object instance has been completed.

The informationAutoRestore package has the following two attributes and one notification for an automatic restore of software.

- *futureAutoRestoreSource.*   The source of information to be restored to this softwareUnit object instance. Like autoBackDestination, the source can be either a local object instance of the same class or a remote network address. The criteria for triggering an automatic restore are locally determined.

- *futureAutoRestoreAllowed.*   A Boolean attribute to indicate whether automatic restore of the information represented by this softwareUnit object instance is allowed.

The autoRestoreReport notification is sent by this object instance to a managing system once an automatic restore has occurred.

The informationBackup package has the following two attributes and an action to implement the information backup.

- *lastBackupTime.*   The last time the information represented by this software-Unit object instance was backed up.

- *lastBackupDestination.*   The destination to which the information was backed up the last time. A NULL string indicates that no backup took place. Otherwise, a destination can be a remote system or a local object instance of the same class.

The backup action is an operation implemented by this local object instance and used by a managing system to request the local system to perform a back-up operation on this softwareUnit object. The CMIS M-ACTION is used to implement this action. The managing system includes a backupArgument parameter in the operation request. The parameter specifies the destination of backup and optionally additional information when necessary. The destination can be either a local managed object instance of the same class, the managing system that sent the action request, or a remote system to which the information is transferred off-line. In the case where the managing system is the back-up destination, the information to be backed up is included in the action reply. The action reply the local system sends to the managing system has a backupReply, which indicates the result of the backup. For local or off-line backup, a NULL value indicates the backup is successful. For in-line backup, the backup information will be included in the action reply.

The informationRestore package has the following two attributes and one action to implement the restore operation.

- *lastRestoreTime.*   The time the most recent restore operation was performed.

- *lastRestoreSource.*   The source of the information to be restored to this softwareUnit object instance. Again, the source can be a remote system, a local object instance of the same class, or the managing system that requests the restore operation.

The *restore* action is an operation that is implemented by the local system with M-ACTION service and used by a managing system. The action information argument that the managing system sends to the local system with the operation request indicates the source from which the information will be restored. Again, like the lastRestoreSource attribute, it can be either a remote system, a local object instance of the same class, or the managing system that sends the request for the restore operation. In the case where the source is the managing system, the information to be restored is included in the action information argument. The action reply the local system sends to the managing system indicates the result of the operation, and a NULL value indicates the restore is successful.

The install package contains only an install action that is implemented with M-ACTION and used by a managing system to request the local system to install one or more software units. Each software unit is specified by a software ID and a pointer to the software object instance. The action information argument that is sent to the local system by the managing system with the operation request specifies the target software to be installed.

The noteField package contains only a noteField attribute that is a text holder to hold installation instructions, startup parameters, default values, and other useful information necessary to activate the feature of this softwareUnit object..

The revert package contains a revert action that is a service implemented by the local system with CMIS M-ACTION. A managing system can use this service to request the local system to revert one or a set of software patches represented by this softwareUnit object instance. The action information argument sent by the managing system contains one or more software path IDs and optionally additional information. The local system returns the reverted software patches in the action reply to the requesting system. If the operation fails because of one of the reasons other than those already defined by CMIS, the local system is responsible for defining the syntax of the error data.

The terminateValidation package contains only a terminateValidation action that is a service implemented by the local system with CMIS M-ACTION. A managing system can use this service to request the local system to terminate a current active validation operation on one or more objects. The action information argument sent by the managing system indicates the mode of termination, namely, cancel-mode or truncated-mode. For the truncated-mode argument, the validation will be terminated and the result of the partially completed validation, if any, is returned in the action reply. For the cancel-mode argument, the validation will be terminated and the result, if any, will be discarded. The action reply sent to the managing system indicates either noOutstandingValidation to be terminated, validationCancelled, or resultOfPartialValidation. In the last case, the result of partial validation is included in the reply.

The usageState package contains only a usageState attribute whose definition can be found in App. D.

The validation package contains only a validate action that is a service implemented by the local system with CMIS M-ACTION. This service can be used by a managing system to request the local system to perform a validation on this softwareUnit object instance. The action information argument sent by the management indicates the target of validation that can be either the current softwareUnit object instance by default, or a registered object instance, or a set of object instances specified by the local system. The action reply that is returned to the managing system can take on one of the following values: validationTerminated (by a terminateValidation action), passValidation, passValidationWithResult, failValidation, or failValidationWithResult. In the case where a result is returned (i.e., passValidationWithResult and failValidationWithResult), the result is included in the reply.

Executable software.   An instance of the executableSoftware object class provides a logical representation of an executable program and a function to execute the program. An executable program may be in the form of binary code or in a form that is not readable by the managing system. Any management operation on such a program is through an instance of the executableSoftware object class. ExecutableSoftware is a subclass derived from the softwareUnit object class and thus inherits all the attributes, actions, and notifications as described above.

Mandatory package.   The executableSoftware package contains only a usageState attribute to indicate whether there are any active executions of the program.

Conditional packages.   The conditional executeProgram package contains an action executeProgram, which is a service implemented by the local system with CMIP M-ACTION. This service can be used by a managing system to request the local system to execute the program represented by this instance of the executableSoftware object class. The action information argument sent by the managing system specifies the parameters necessary to invoke the program. Obviously, the parameters are program-specific. The action reply that is sent back to the managing system by the local system contains the ID and owner of the process executing the program, the start time of the execution, and optionally additional information.

Software distributor.   The softwareDistributor managed object class is responsible for distributing one or more software units to a target managed system when it receives a deliver instruction from a managing system. This managed object notifies the managing system of the result of distribution when the operation is terminated. The softwareDistributor object class is a subclass of the top object class.

Included in the softwareDistributor package are three attributes, one action, and four notifications. The attributes represent the state information of a software distributor object.

- *softwareDistributorId.*   A unique identifier for an instance of the software-Distributor management object class.

- *administrativeState.*   Indicates the administrative state of the software-Distributor object instance whose value can be either locked, unlocked, or shutting down. The administrative state value is set by a managing system. When in the locked state, the softwareDistributor cannot accept any request to distribute software. Conversely, when in the unlocked state, it can carry on the normal operation given that the object is not in the disabled operational state. See App. D for more details.

- *operationalState.*   Indicates the operational state of a softwareDistributor object instance whose value can be either enabled or disabled. The value of this state is determined by the conditions of the object instance and cannot be set by a managing system. See App. D for more details.

The four notifications are

- *deliverResultNotification.* This notification is sent to the managing system once a deliver operation is terminated, either normally or abnormally. Included in the notification are a deliver ID, i.e., an ID for the deliver operation, a result of the deliver operation, and optionally additional information. The deliverResult field indicates one of the following outcomes:
  - *pass.* Deliver is successful.
  - *communicationError.* Deliver failed because of a communication error such as connection down.
  - *equipmentError.* Error condition associated with the equipment, such as target system not responding.
  - *qosError.* Predefined quality-of-service criteria, such as rate of deliver operation, are violated.
  - *accessDenied.* Cannot access the target system.
  - *notFound.* Target system not found.
  - *insufficientSpace.* Not enough space on the target system to hold the software to be delivered.
  - *alreadyDelivered.* The same software is already in the target system.
  - *inProgress.* Delivery is in progress.
  - *unknown.* Any other reason.

- *objectCreation.* This notification is sent to a managing system once an instance of the softwareDistributor object class is created.

- *objectDeletion.* This notification is sent to a managing system once the associated instance of the softwareDistributor object class is deleted.

- *stateChange.* This notification is sent to a managing system when a change of either administrative or operational state has occurred.

The deliver action is a service implemented by the local system using the CMIS M-ACTION. The service is used by a managing system to deliver one or more software units to a target managed system. The argument for the action sent by the managing system specifies the following information:

- *deliverId.* A unique ID for each outstanding deliver action

- *targetSoftware.* A set of distributedSoftware, which each has a distributedSoftwareId and pointer pointing to a distributed software

- *transferInfo.* Specifies a transfer protocol and the information required by the protocol

- *additionalInformation.* Optionally additional information as required by the local system

An application scenario. We use an application scenario to illustrate the use of the managed objects described above in the software management process. Assume that a base station of a wireless network has just been installed along with the OAM&P software. Included in the OAM&P software is a software management system consisting of the components shown in Fig. 8.6. The next step is for the management system to deliver, install, and perform a test run on the application software at the base station. The following sequence of steps provides an illustrative description of the software management process using CMIS services, assuming that no error occurs in the process.

1. The managing system issues a CMIS M-CREATE requests to the agent on the base station to create an instance of the softwareDistributor object class.

2. The agent creates the requested object instances and returns the object instance IDs along with other information to the managing system.

3. After updating its view of the MIB to reflect the existence of the newly created object instance, the managing system issues an M-ACTION request to the agent to invoke the deliver action on the identified software-Distributor object instance. Included in the action request are the following:
   - Deliver type
   - Deliver ID
   - Target software
   - Target system
   - Transfer information (file transfer protocol, protocol parameters, etc.)

4. After ensuring that the request is valid (e.g., do the requested action and the associated object instance exist? Are the required parameters present?), the agent invokes the deliver action on the newly created softwareDistributor object instance.

5. The specified softwareDistributor object executes the deliver action by delivering the specified software to the target, a local disk of the base station, from the source specified. Once the deliver operation is completed, the softwareDistributor object sends an action reply via the agent to the managing system, and the reply includes the following information:
   - Deliver ID
   - Deliver result (i.e., whether deliver is successful and, if there is an error, the error type)

6. After receiving the action reply, the managing system sends an M-CREATE request to the agent to create a softwareUnit object instance for the newly delivered software.

7. The agent creates a softwareUnit object to be associated with the delivered software with the following information, some of which is passed down from the managing system in the M-CREATE request:
   - *File type.*   Executable
   - *File location.*   Directory
   - *File size.*   Size in bytes
   - *Date of creation.*   The time it was created

- *Identity of creator.*   The identity of the managing system
- *Date of last modification.*   Same as the date of creation
- *Identity of the last modifier.*   Same as the identity of the creator
- *Date delivered.*   The time the corresponding software is delivered
- *Date installed.*   NULL (not installed yet)
- *Future auto backup destination.*   Specified by the managing system
- *Future auto backup trigger threshold.*   Specified by the managing system
- *Future auto restore allowed.*   Specified by the managing system
- *Future auto restore source.*   Specified by the managing system
- *Last backup destination.*   NULL
- *Last backup time.*   NULL
- *Last restore source.*   NULL
- *Last restore time.*   NULL

8. After receiving an M-CREATE reply from the agent and modifying the MIB view to reflect the newly created object, the managing system sends an M-ACTION request to the agent to invoke the install action on the newly created softwareUnit object. Included in the request is the target software to be installed.

9. The agent, after ensuring the action request is valid, invokes the action on the newly created softwareUnit object.

10. The specified softwareUnit object executes the install action by installing (or loading) the specified software into memory from the specified location on the local disk. Once the install operation is completed, the softwareUnit object sends an action reply via the agent to the managing system to indicate the result of the install action.

11. After receiving the action reply, the managing system sends an M-CREATE request to the agent to create an executableSoftware object instance for the newly installed software.

12. The agent creates an executableSoftware object to be associated with the installed software with the following information:

- *Usage state.*   Idle (no active execution of the software)

13. After receiving an M-CREATE reply from the agent and modifying the MIB view to reflect the newly created object, the managing system sends an M-ACTION request to the agent to invoke the executeProgram action on the newly created executableSoftware object. Included in the request is target software to be executed.

14. The agent, after ensuring the action request is valid, invokes the action executeProgram on the newly created executableSoftware object.

15. The specified executableSoftware object executes the executeProgram action by invoking the specified executable and sends back to the managing system an action reply via the agent with the following information:

- ID of the process associated with the program

- Process owner
- Start time of the execution

In summary, the software management is accomplished via a sequence of exchanges of management instructions and information in the form of managed objects between the managing system and the managed system.

Provisioning

Provisioning can take place at different levels: service, network, and network element. *Service provisioning* is to bring a service offering to a subscriber and prepare all necessary data and resources to realize a subscribed service for the subscriber. *Network provisioning* is to bring a network into service, including provisioning of each network element. Provisioning at the network element level is to bring a network element into service. This includes design and connection of resources to realize the service, loading of software logic, and setting service features. Service provisioning is discussed in detail in Chap. 14, and provisioning at the network and element level is the focus of this section.

Network level provisioning

The goal of network level provisioning is to ensure adequate network level resources are configured to meet the needs of services offered by the network. For example, a service provider is to offer a lease line service. The management tasks involved in provisioning network resources to meet the service needs are the focus of this section.

Note that network level provisioning is an area for which very few standards, if any, have been defined. Part of the reason for this is that network level provisioning is closely tied to the specific network technologies [e.g., Intelligent Networks (INs), wireless networks and PSTNs], and all these network technologies are changing rapidly. Normally the standards of a technology emerge only after the technology has achieved wide acceptance and thus stability. Nevertheless, some regional efforts have been made in this direction. One such effort is Bellcore's GR-2869 (1995). Based on these efforts and today's common practices, four areas of network level provisioning are discussed: network topology management, network connection management, and network resource data management, as shown in Fig. 8.9G.

A generic provisioning functional model.   For the convenience of discussion, the network level provisioning can be viewed as consisting of the following basic steps. First, the necessary networkwide connections must be designed and set up in order to implement a service. The networkwide connection management is responsible for establishing, maintaining, and monitoring the connections between NEs. Then networkwide service data shall be provisioned and network topology shall be maintained, in order for a service to operate. These are the responsibilities of network topology and resource data management.

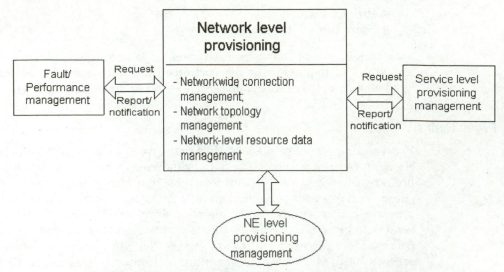

Figure 8.9 Network level provisioning and its relationships with other management applications.

Network connection management. Configuration management is responsible for establishing connections across NE boundaries to implement a circuit design. It accomplishes this by requesting a connection be established at one or more network elements (i.e., switches). This includes the connection of ports and facilities and the setting of features and parameters associated with those connections. This requires a networkwide view to coordinate the actions among multiple NEs to ensure end-to-end networkwide connectivity.

A related task is to manage the pending network changes in network connections. Networkwide end-to-end connectivity and associated parameters sometime require changes at a specific future time for the purpose of performance or addition or deletion of network resources as the network evolves. The network layer configuration management is responsible for coordinating the changes across multiple network nodes and detecting any inconsistencies. It is also responsible for sending notifications to the interested parties on the changes and updating the networkwide resource database.

Network topology management. Network topology is a fundamental part of the network configuration management. Two dimensions are associated with a network topology: modeling aspects and data aspects. The modeling aspects are discussed first.

Network topology is required for numerous network management applications. For example, a network topology is a necessary tool to present an overall view of networkwide resources, both logical and physical for the purpose of network planning and design. Topology information is needed for determining network routing and networkwide performance monitoring. The centralized dynamic routing as discussed earlier requires a view of overall network traffic conditions in order to make routing decisions that are optimal for the network as a whole.

Network topological information is often required for fault detection that involves multiple network components. It is rather difficult to determine the cause of a fault if the fault is considered in isolation. Some condition in a neighboring network component that itself is not considered a fault may be the culprit of a fault in this node. A network topology certainly facilitates the fault event correlation and filtering. For example, a failure in a connection between two nodes may cause faults in both connected nodes, and investigation of the nodes themselves will prove rather futile.

It is a common practice to represent a network topology as graphs based on well-established graph theory. Several characteristics of graphs make them well suited for representing a telecommunications network. First, a network component is modeled as a node of a graph and connections between network components can be represented as links between graph nodes. A link can have direction, and this is important for modeling routing and relationships where directionality is significant. The connection between two network components can have associated properties such as length, capacity, type, and other types of information as appropriate for an application on hand.

Network topology data can be of a static or dynamic nature. If there is a requirement that network topology reflect the up-to-the-minute conditions of the network, the topology data are said to be dynamic. Otherwise, the topology data are static. The nature of topology data is determined by the intended use of network topology. For example, if the topology is for dynamic routing as discussed earlier, the real-time updating of topology is necessary. In contrast, if the topology is used to represent the static network resources, such as network nodes, capabilities of the nodes, and the positions of the nodes relative to each other, then real-time updating may not be so critical.

Often the original topology is to be maintained for comparison purposes. Especially when topology is used for performance monitoring and it can be dynamically changed, a base topology serves as an "original" to detect the deviation.

Networkwide resource data management. There are both logical and physical resources that can be provisioned at the network level. Examples of the physical resources include switching fabrics (switches), lines, trunks, circuits, and access devices. The logical resources include special resources such as announcements, routing data, signaling data, protocols, logs, application programs, and network services. Provisioning of these resources involves operations ranging from loading a program into the right database to setting an appropriate attribute. The resource data management is responsible for

- Providing the capability for circuit inventory data management, for automatic notification of changes in status and states of network resources, and for access to information about the current status of network resources

- Sending notifications to the interested parties on the changes and updating the networkwide resource database

- Maintaining the states and status of the network resources, relationships between the resources, and the information on the physical characteristics of the resources

NE level provisioning

The goals of NE level provisioning are to ensure that adequate resources at the NE are deployed and configured to implement the services subscribed to by customers. For example, if a central office switch is to provide ISDN service to the area subscribers, the required physical resources, such as the interface card, and logical resources, such as software modules, should be in place before the service offering. The tasks involved in the resource provisioning at the NE vary from NE to NE, depending on the type of NE and services it provides. In listing TMN functions, CCITT Recommendation M.3400 (1992) identifies three general areas of NE provisioning: NE configuration, administrative functions, and database management.

Figure 8.10 shows the three functional areas of NE provisioning and the relationships between NE provisioning and other management applications. NE level provisioning reports NE resource configuration information to other management applications such as fault or performance management usually in response to a request. The NE level provisioning directly interfaces the objects that represent the resources on the NE to receive notification on the state or status change of a resource.

NE resource configuration management.   A variety of network resources at an NE, such as a switch, fall into three categories: infrastructure resource, access

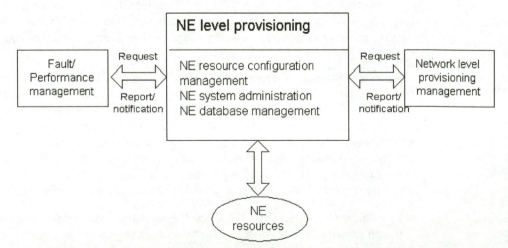

Figure 8.10   NE level provisioning and its relationship with other management applications.

resource, and logical resource. The access resource provides the means for an NE to connect to other NEs or users; examples of such access resources include trunks, lines, circuits, access ports, digit sender and receiver, tone generator and receiver, and interface cards. The infrastructure resources are internal to the NE and provide services to other NE resources. Examples include processors, input-output (I/O) devices, and power supply equipment. An NE logical resource can be associated with either access or infrastructure resources; examples include software modules, signaling data, NE routing data, and directory number (i.e., your telephone number).

NE resource configuration reporting. NE resource configuration reporting informs other management applications of the changes in the resource configuration or simply the overall view of the NE resources. There are two types of reporting: solicited and unsolicited. The unsolicited reporting is initiated by the NE configuration management and often performed on a periodic basis. The second type of reporting is solicited by another management application, such as a network-level configuration management application. Scheduling either solicited or unsolicited configuration information reports is another management task. In the OSI framework, the reporting can be accomplished by the event report mechanism of the CMIP.

Operation on resources. NE configuration management is responsible for a range of operations on the resources themselves as listed below.

- Load the resources into the database or memory as required by the services. For example, loading a service logic program into an NE database, or a plug-in card into a slot, is part of NE provisioning. The control and decision of software loading can be made either by the NE provisioning or the network-level provisioning, depending on individual situations. For instance, the decision to load software to activate service-specific features in an NE component (e.g., downloading of a software feature into a line card) at the time of service activation can be made by the NE provisioning application. Such software installment includes testing the software and backing out of the software load in case of loading failure.

- Select and assign a particular resource in response to a request from a higher-level management application and report the assignment to the requesting party.

- Cross-connect two channels on the NE. This is normally requested by network-level provisioning (e.g., connection management) or performance management applications. In addition to making the connection, the NE provisioning must also update the state and status of the involved channels.

- Remove the interconnection between two channels as requested by other management applications and update the status and states of the channels.

- Add or remove a channel from the available channel pool as new channels become available or existing channels are being taken out of service.

■ Set the default service or system parameters and update them as required. Examples of the NE-wide parameters include the number of maximum retries allowed in case of no response, default service parameters such as an announcement for an unavailable resource, and a default route in case of traffic congestion.

■ Administrate tests on circuits, cross connections, or other resources on the NE either as scheduled or as directed by another management application. The tasks include initializing and terminating tests, and setting and tuning the testing parameters.

NE resource configuration maintenance. Any of the above operations on the NE resource will result in a change in either the resource state, status, or relationship with another resource, and the changes need to be reflected in the NE resources configuration. The configuration, often represented as table, tree, list, or other form, can grow or shrink and needs constant attention. When a resource is disconnected, is newly installed, is deleted, or has a state or status change, the effects of the operation must be reflected in the resource configuration.

NE system administration. With very few exceptions, an NE in essence is a computing system. There is a set of administration tasks that the NE configuration management is responsible for.

■ Initialization and shutdown of the native computing system on the NE.

■ Management of the system clock.

■ Management of the application processes. The management tasks include initializing and terminating processes, allocating and deallocating application processes, and monitoring the performance of active processes, using services provided by the operating system.

NE database management.   The NE database is a vital part of the NE resources. Service data like service logic programs, customer data like subscriber profiles, and NE resource configuration and inventory data as described above all rely on the NE database management.

Normally NE database management is required to provide three types of capabilities:

■ To store, back up, restore, and maintain the NE resource configuration and connectivity data

■ To create, read, update, duplicate, and delete database records

■ To initialize and reinitialize the database as required

Real-time database performance issues have to be considered as well. Many NEs of a telecommunications network are based on an embedded real-time system (e.g., switch or transmission system) and have stringent performance requirements. For example, the subscriber information is queried at run time by the call processing software during a call and the response time must be up to speed. It is the responsibility of NE configuration management to determine what data to cache into memory for maximum real-time performance.

NE Resource Status and Control

One of the main tasks of NE configuration management is to manage and control the numerous logical and physical resources at run time on a network element. Trunks and lines are being constantly allocated, deallocated, tested, or suspended; a card is going out of or coming back into service; a software unit runs into an execution error; an alarm on the central processor is bringing down a whole array of resources; and on and on. This section describes the basic control operations on the network resources and the most important aspects of the information that must be maintained and updated constantly on the network resources in order to guarantee the normal operation of a network or network element. Though the operation control-related information can be numerous in type and large in quantity, the focus of this section is on the resource states and status, the two most critical aspects that are expected to be common to all network resources.

Note the main difference between the issues addressed by this section and those by the NE provisioning: while the mission of NE provisioning is to get the NE resources; ready for operation, NE status and control are concerned with the operations of NE resources themselves.

NE control operations

NE configuration management is responsible for controlling, maintaining, and monitoring all the resources on the NE, often on a real-time basis.

Resource activation and deactivation. Hardware and software initialization and shutdown are part of the resource activation and deactivation. For example, when a switch is first brought up, a whole range of resources needs to be activated. Part of the resource activation and deactivation is the resource state management that will be discussed shortly. Other tasks include resource parameter initialization and notification to all concerned parties of the activation or deactivation.

Switch-back. Switch-back is a control action to revert to a preferred resource whenever it becomes available for service, e.g., after recovering from a failure or first-time initialization. Using a preferred resource whenever possible is called *revertive protection policy,* and it is used mainly for the purpose of system reliability. For example, in a duplex configuration, one resource unit can be designated as primary and the other as secondary. The primary resource is active whenever possible.

Switch-over.   Switch-over is a control action to switch services from an active resource to the backup one because of a failure, routine maintenance, or scheduled control changeover. This is similar to the switch-back, apart from the fact that there is no primary or secondary distinction between the resources. To varying degrees, all the control operations involve the resource state and status management, which are discussed next.

Resource state management

As shown in Fig. 8.11, a management state is an attribute of a managed object that represents an instantaneous condition of the associated network resource, a switch in this case. The purpose of the state attribute is to provide information to the management system on the conditions of the managed object.

The number of state attributes for different management object classes can be large, and some of them are specific to a network resource. Out of the possibly many states, ITU-T has standardized three states in CCITT Recommendation X.731 (1992) with the expectation that they be common to most of the managed object classes.

- *Operability.*   Whether or not the network resource is in working condition
- *Usage.*   Whether or not the resource is in use and has the spare capacity to provide for additional users

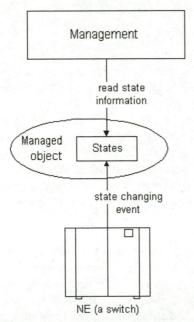

Figure 8.11 State of a managed object in relationship to the NE and management application.

■ *Administration.* Whether or not the resource is permitted by the management system for use.

Operational state. This state attribute represents the operability of the network resource and has two possible values: disabled and enabled. The operational state is set to *unknown* when the actual state is unknown. Of course, it does not reflect the actual operational state.

The *disabled* operational state indicates that the network resource is completely inoperable. As shown in Fig. 8.12, the events that render a resource completely inoperable include power or processor failure and will cause the transition into the disabled state. A special case of the disabled state is when the network resource has ceased to exist but the managed object still maintains the operational state for one reason or another. Thus the presence of disabled operational state may mean the absence of the resource.

The *enabled* operational state indicates that the network resource is either partially or fully operable. Those events that render the network resources operable cause the transition to the enabled operational state. However, some managed objects may always display an enabled operational state, regardless of the events that took place. This is when these managed objects have no visible dependencies on other resources and no components that can exhibit defects.

Note that it is the network events that change the operational state value, not the management system. Network management can only read the value of the operational state of a managed object and cannot set the value, again because this state is intended to reflect whether a resource is capable of normal operation at a given moment.

Usage state. A managed object can have a usage state attribute to describe the usage of the network resource. The usage state is single-valued (versus set-valued) and read-only for a manager with three possible values: idle, active, and busy. The *idle* usage state means that the resource currently is not in use

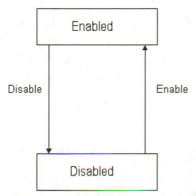

Figure 8.12 Operational state diagram. (*From CCITT Recommendation X.731, 1992.*)

and can accept users. The *busy* state indicates that the resource is in use and does not have the spare capacity to accept additional users. The *active* state means that the resource is in use but has sufficient capacity to provide for additional users.

The events that can cause a transition from one usage state to another include

- Addition of a new user when the resource usage state is active or idle and the operational state is enabled
- User quitting when the resource usage state is busy or active
- Capacity increases when the resource usage state is busy
- Capacity decreases when the resource usage state is active

The effects of the events on the state transitions are shown in Fig. 8.13. For example, the user quitting event can cause one of the two possible state transitions: if this is the last user, the state changes from active to idle; and the state becomes or remains active otherwise.

Administrative state.   The administrative state indicates whether or not it is permitted by the management to use a network resource, independent of its usage and operational states. The administrative state attribute of a managed object is single-valued, and a manager can read or write the state value. The administrative state can have one of the three values, not all of which are applicable to every managed object class:

- *locked.*   The resource is administratively prohibited from providing services to its users.
- *unlocked.*   The resource is administratively allowed to provide services to its users.

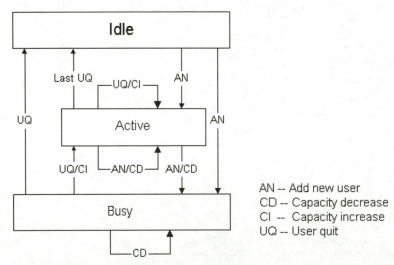

Figure 8.13 Usage state diagram. (*From CCITT Recommendation X.731, 1992.*)

- *shuttingDown.*   The resource is allowed to provide service only to the users with previously granted permission but not to a new user.

As shown in Fig. 8.14, the following administrative operations or events can cause the transition from one administrative state to another:

- *unlock.*   An administrative operation that can only be performed on a managed object whose administrative state is either locked or shutting down to allow the associated network resource to provide services to its user.

- *lock.*   An administrative operation that can only be performed on a managed object whose administrative state is either unlocked or shutting down to prohibit the resource from providing services.

- *shut down.*   An administrative operation that can only be performed on a managed object whose administrative state is unlocked. This operation will result in the transition to the shutting down state if the resource has existing users and to the locked state otherwise.

- *user quit.*   An event that can occur only if the managed object is in the unlocked or shutting down state. If the resource has existing users after the event, the administrative state will remain as shutting down; otherwise, it will result in the transition to the locked state.

State change notification.   The change of a network resource state often warrants a notification to be sent to the management application. It is the designer's responsibility to decide whether or not a state change triggers a notification. In the OSI framework, the CMIS M-EVENT, as you recall from Chap. 4, is the mechanism used to report a state change.

As noted earlier, not every managed object class will have all three state attributes, and it is the application designer's responsibility to determine

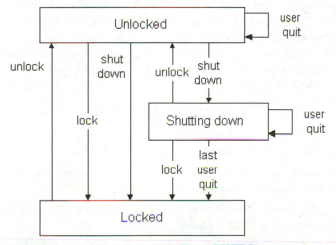

Figure 8.14   Administrative state. (*From CCITT Recommendation X.731, 1992.*)

which state attribute is applicable. It is also possible that the three generic state attributes are insufficient for some managed object classes and additional attributes are called for.

Bellcore has defined a different state model with emphasis on modeling telecommunications-specific services and equipment and on using the terminology commonly seen in the industry. This state model has two sets of state values: primary and secondary. The primary state value is either in-service or out-of-service. Associated with the primary state values are nine mutually exclusive primary state qualifiers, each further defining the condition and context of the state value. Four primary state qualifiers are associated with the in-service state and the other five with the out-of-service state. For example, the qualifier *restricted* associated with the in-service state indicates that a network element is capable of performing all its provisioned functions but is intentionally suspended from performing part of, but not all, these functions. The secondary state values are an extensive list, used to further describe the operational state or availability status of a network resource. Examples of the secondary state values include active, blocked, and busy. For more details, see Bellcore GR-1093.

NE resource status management

While a management state represents an instantaneous condition of availability, operability, and usability of a resource, a status of the resource provides further qualification for the state. The value of a status attribute can represent the presence or absence of a particular condition, providing further information on the state attribute. The following status attributes, viewed as common to many types of resources, have been defined in CCITT Recommendation X.731 (1992).

Alarm status attribute.   The status attribute indicates the status of an alarm associated with a resource: its severity, the effect of the alarm on the resource, and the status of the alarm resolution. The alarm status attribute, whose value can be read or written by a management application, is set-valued. This means that the attribute values are not mutually exclusive. It can have zero or more of the following values, of course, not all of which are applicable to every network resource. When the attribute value is an empty set, it means that none of the conditions below are present.

- *Under repair.*   The resource is currently being repaired.
- *Critical.*   One or more critical alarms indicating a fault have been reported in the resource and have not been cleared.
- *Major.*   One or more alarms indicating a fault have been detected in the resource and have not been cleared.
- *Minor.*   One or more alarms indicating a fault have been detected in the resource and have not been cleared.
- *Alarm outstanding.*   One or more alarms have been detected in the resource and have not been cleared for certain periods of time.

If a resource is in a disabled operational state and has a major or critical alarm status attribute value, the latter may well be the cause of the disabled operational state provided there is an absence of any other cause.

Procedural status attribute. This status attribute indicates a specific procedure required of a resource for it to operate, or indicates the phase of the procedure it is in. This status attribute is applicable to those network resources that normally go through some well-defined sequence of phases in their operations. One example is software units that go through a test process. Or a procedure must reach a certain phase before a resource can be operational and available for use. This status attribute is set-valued, and an empty set means that none of the status conditions described below are present.

- *Initialization required.* The resource requires initialization before it can perform its normal functions, and the initialization has not been done yet.

- *Not initialized.* The resource requires initialization before it can perform its normal functions, and the initialization has yet to be completed.

- *Initializing.* The resource requires initialization before it can perform its functions, and this procedure has been initiated and is not yet complete.

- *Reporting.* The resource has completed some processing operation and is in the middle of notifying about the results of the operation.

- *Terminating.* The resource is in a terminating phase. For example, a software unit has just finished testing.

One such resource that needs a procedural status attribute is a software system on an NE. If the initialization related status attribute value is present and the software system is not reinitialized, this can be sufficient to put a software system in a disabled operational state.

Availability status attribute. The availability status attribute indicates the conditions that may affect the availability of a resource, ranging from service degradation to maximum log size being exceeded. This attribute is set-valued and read-only for a management application. It can have zero or more of the following values, not all of which are applicable to every resource. An empty set means none of the status conditions described below are present.

- *In test.* The resource is undergoing a test. If the administrative state is locked or shutting down, the normal users are prohibited from using the resource and the control status attribute has the value reserved for test. For example, an interface card can be in test, and if the test does not preclude the additional user, the operational state can still be enabled.

- *Failed.* The resource has an internal fault that prevents it from functioning normally. For example, the interface card failed and is in the disabled operational state.

- *Power off.* The resource is not powered on and is in the disabled operational state.

- *Off-line.* The resource is taken off the available resource pool and requires a routine operation to be performed to make it available for use. This can result in the disabled operational state for the resource.

- *Off duty.* The resource has been inactivated by an internal control process according to a predetermined time schedule. Under normal conditions the control process is expected to activate the resource at a scheduled time. For example, a backup unit of a pool of interface cards can be taken off duty as part of a rotation. The operational state can be enabled or disabled depending on the local context.

- *Dependency.* The resource is not available or operable because other resources it depends on are unavailable. For example, a software unit is inoperable because the processor it runs on is down.

- *Degraded.* The service a resource provides is degraded in some aspects such as congestion level or processing speed. Despite the degradation, the operational state of the resource can remain enabled because some other services are still satisfactory.

- *Not installed.* The resource is either not present or its installation is not complete yet. For example, a software module is not loaded, a card is not installed, or a plug-in module is missing.

- *Log full.* The maximum log size has been exceeded and continued operation runs the risk of losing log records. The operational state of the resource can be enabled or disabled, depending on the local context.

The standardized availability status attribute values are not meant to be exhaustive or encompass every possible situation. As an application designer, you may need additional status attribute values for a specific application.

Control status attribute. The control status attributes indicate a control action that has been performed and its effect on a resource. The attribute is set-valued and read-write for a management system. It can have zero or more of the following values, not all of which are applicable to every resource. An empty set means that none of the conditions are present.

- *Subject to test.* The resource is available to normal users, but testing may be conducted at unpredictable times that may result in uncharacteristic behavior and affect users.

- *Part of service locked.* It indicates whether access to and use of a resource is restricted by the manager. Examples are incoming service barred, outgoing service barred, and read or write locked.

- *Reserved for test.* The resource becomes unavailable by some administrative control action because the resource is undergoing a test procedure. In this case, the administrative state is locked.

- *Suspended.* The service of a resource has been suspended for the users and will remain so until the suspension is revoked.

Standby status attribute. The standby status attribute indicates the type of standby condition a resource is in. This attribute is meaningful only when a resource can play a backup role in duplex resource configuration. The attribute is read-only for a management system and single-valued, meaning that the following standby conditions are mutually exclusive.

- *Hot standby.* The resource is not providing service but is operating in synchronization with the active resource it is backing up. A resource with a hot standby status will be able to immediately take over the role of the active resource without the need of going through initialization and state information transfer.

- *Cold standby.* A resource with a cold standby status is to back up a resource in an active role and is not synchronized with the active resource. A cold standby resource cannot immediately take over the role of active resource without going through an initialization process.

- *Providing service.* A resource is playing a backup role to another active resource and is providing services at the same time.

In summary, this section discusses the resource state and status management in the context of real-time NE control operations, although the state and status management can also be applicable in the context of non-real-time management operation such as resource provisioning as discussed earlier. The states and statuses are defined for generic applications, and as an application designer, you may find the need to augment or not use some of the states or statuses.

Summary

This chapter discusses configuration management, maybe the most complicated area with the least agreement on its scope. The scope, as discussed in this chapter, covers the whole process ranging from network planning and design, called network provisioning, to the normal operations and controls of a network element.

The network planning and engineering defines the network infrastructure based on the forecasted traffic volume, the choice of routing structure, and customer-specific requirements. The issues addressed in this area range from the business management layer to the network management layer. The issues at the business management layer include customer requirement specification, market demand forecast, and network cost classification and determination. At the network management layer, the choice of network routing structure and scheme is discussed.

Software management presents a model defined in ITU-T Recommendation X.744 (1996) for managing software based on the OSI model, and the ultimate goal is to automate the often complicated, tedious process. A range of management operations are supported, including create, delete, deliver, install, execute, backup, and restore. The bulk of the management functionality is embodied in the managed object softwareUnit. This software management model will be very useful even if a designer chooses not to use the OSI model: the object model is quite comprehensive and can be easily adapted to an implementation based on a non-OSI model.

The next section covers both network and network element layer provisioning. At the network management layer, the tasks of provisioning cover management of networkwide connections, network topology, and networkwide resource data. The goal of network element level provisioning is to prime a network element and ready it for providing services. Three areas discussed are resource configuration management, administrative functions, and database management. The resource configuration management deals with the issues related to reporting NE configuration to a higher-layer management entity and managing the changes to resource configuration. The system administrative functions cover normal system maintenance issues such as system initialization and process management. The database management touches on maintenance issues such as database backup, restoration, and real-time database operations.

The final section of the chapter focuses on the management of resource state, status, and control operations in a real-time environment. From the perspective of network operations, the information most important to a management system is the state of a resource. A state represents an instantaneous condition of the associated network resource. Three states that are expected to be common to all resources are operational, usage, and administrative. An operational state indicates to the outside whether or not a resource is in working order. A usage state indicates whether or not the resource is being used and can accept new service requests. An administrative state indicates whether access to a resource is permitted by the management system. A set of resource statuses such as alarm, procedural, and control are defined to provide qualifying information to a state.

Exercises

8.1 Explain some of the major factors that make the dynamic routing practical in modern public switched networks.

8.2 If the operational state of a resource has the value *enabled,* usage state value is set to *idle,* but administrative state is *locked,* can this resource be used by an application?

8.3 As presented in this chapter, the software management is realized via the message exchange between a manager and an agent. If we use a framework other than the OSI model, say, CORBA-based client-server architecture, how will the management operations like diver, execute, or revert be implemented? Discuss this in general terms.

8.4 What is the difference between the software management operations *deliver* and *install* as discussed in this chapter?

8.5   Describe the overall goal of network level provisioning and the major areas discussed in this chapter.

8.6   Explain the difference between a state and a status as discussed in this chapter.

8.7   Assume a resource has all three states. List all states whose values can be changed by an event initiated by the resource. List all the states that a manager cannot change.

8.8   Please offer some explanation as to why the values of all resource statuses can be changed by a manager.

Performance Management

Outline

- The first section, "Performance Monitoring," introduces a general performance monitoring model, describes a set of managed objects performing summarization on the performance data, and presents a model for logging and reporting performance data.

- The second section, "Performance Analysis," covers two aspects, i.e., performance analysis for circuit-switched networks and performance analysis for packet-switched networks. Methods for computing common network performance measures such as probability of congestion and delay and grade of services are presented.

- The last section, "Performance Management Control," discusses the network traffic control and performance administration. Traffic control is realized through routing pattern manipulation. A set of routing methods is introduced, and a range of tasks for performance administration is described.

Introduction

The primary goals of performance management are to monitor and evaluate the behavior and effectiveness of a network and the quality of services offered to its users on a constant basis and to take either preventive or corrective actions to maintain the desired performance level. The main tasks of performance management include collecting statistical data, analyzing the data, and exerting performance maintenance–related control actions. The organization of this chapter is based on the three functional areas of performance management identified in CCITT Recommendation M.3400 (1992): performance monitoring, performance analysis, and performance management control.

Performance Monitoring

Performance monitoring is used for observing and collecting data on a specified set of attributes associated with network resources to measure the network performance. This section first presents a performance monitoring model as defined in ITU-T Recommendation Q.822 that focuses on the performance data collection process and the managed object classes used to represent performance data. Then the discussion shifts to performance data trending and summarization that apply statistical algorithms to the collected performance data. Finally OSI-model-based performance data logging and reporting are described.

A performance monitoring model

The performance monitoring model presented here is based on ITU-T Recommendation Q.822. An introduction of the model is followed by a detailed description of the managed object classes of the model. This subsection concludes with an application scenario that illustrates the model behavior.

The model description.   Figure 9.1 presents an object model for performance monitoring that features major managed objects or object groups involved in performance monitoring. A high-level description of the involved objects to present an overall view of the model is followed by an attribute level description of each of the managed object classes.

This model presents the managed object classes within a managed system (agent side) that can be viewed as consisting of two categories: those representing performance data and those providing support functions such as scheduling and control and other functions for performance monitoring. The first category of classes includes currentData, historyData, and scanReportRecord. Among the second category of managed object classes are scanner, summarization, and thresholdData.

The monitored object represents the resource being monitored from which the performance measurements are being collected. For example, if the interest of the managing system is in the total signaling messages received during a specified time interval, a managed object can be a signaling interface object.

The *currentData* object class is a placeholder containing the performance measurements collected for the resource being monitored for the specified time interval. One way to associate the data object currentData with a monitored object is to define the monitored object class as a subclass of currentData so that the data attributes of currentData can represent the attributes of the monitored resource. For example, the signaling interface class can be defined

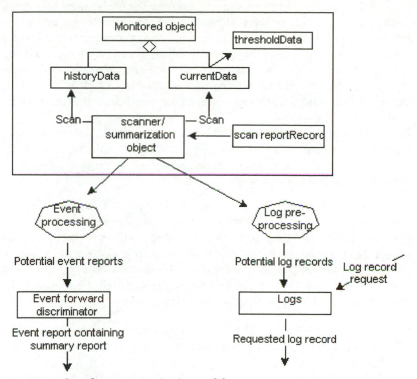

Figure 9.1 A performance monitoring model.

as a subclass of the currentData object class and the attributes and functionality defined in the currentData object class will be automatically available to the signaling interface object. Another way of making the association is defining a currentData object as a data attribute of the monitored object.

A currentData object may contain a pointer to a thresholdData object. If any of the thresholds in the referenced thresholdData object are violated, a quality of service alarm notification is emitted by the currentData object, and the resulting alarm record may be logged. At the end of each specified interval, a current objectData may do two things: emit a scanReport notification that will result in a corresponding event report being sent to a managing system or cause a historyData object to be created.

The *historyData* object contains a copy of the performance measurement and other attributes that are in the currentData object at the end of the same interval for the purpose of performance data logging.

The *thresholdData* object specifies the performance monitoring criteria using a set of threshold values for one or more classes of currentData. The threshold values define a set of performance-related attributes and performance threshold values for each of the attributes. The threshold values can only be referenced by a set of currentData objects, and thus it is the responsibility of the associated currentData objects to monitor the threshold value crossing by comparing the collected measurements against the corresponding threshold values.

The *scanner* objects have three basic responsibilities: scan the contents of either currentData or historyData objects; aggregate the scanned performance measurements by applying algorithms to them; and send either the simple scanned measurements or aggregated measurements in scanReport to a managing system. A detailed description of this object class and its subclasses is deferred until the discussion of performance data summarization.

Description of managed object classes. Described in this subsection are the attribute level details of the key managed objects in the performance-monitoring model. For each object, a list of attributes and optionally some notifications and actions are presented.

The currentData managed object class. This object class is a subclass of the scanner object class and contains the current performance measurement parameters to be collected. One way this object is used is to include it in a managed object that represents the network resource. The attributes found in the currentData object class generally fall into two categories: the attributes of performance measurements and the attributes that are related to control of the performance monitoring, as described below.

- *suspectIntervalFlag.*   If set, this attribute indicates that the performance data for the current period may not be reliable for one of the following reasons.
  - The network performing data collection detects unreliable data.
  - Either the administrative state has the value locked or the operational state is disabled when the data are collected.
  - The performance counter was reset during the interval.
  - The data collection was inhibited by the scheduler during the interval.

- *elapsedTime.*   The time elapsed since the previous collection of data.

- *discriminatorConstruct.*   Controls the creation of a historyData object at the end of the specified interval. A discriminator constructor specifies an attribute to be evaluated, called *criterion,* and an implicit evaluation function. No historyData object is created if the criterion is evaluated to FALSE. An empty discriminator constructor defaults to TRUE.

- *historyRetention.*   Specifies how long the new historyData object instance shall be retained at the NE once it is created.

- *maxSuppressedIntervals.*   Specifies the maximum number of intervals for which the creation of historyData shall be suppressed, if the discriminator constructor is evaluated to FALSE.

- *measurementList.*   Contains a set of performance parameters for the monitored object. Each measurement has an identifier and its value. A measurement is included only if it is not represented already as an explicit attribute of an instance of a subclass of the currentData class or an instance of a subclass of the historyData class.

- *numSuppressedIntervals.*   The number of consecutive intervals for which the current performance data collection has been suppressed from being sent to the managing system or for which historyData object instances have not been created. It reflects performance measurement up to, but not including, the current interval.

- *observedObjectClass.*   The object classes of the monitored object instances for which this currentData object instance is created.

- *observedObjectInstance.*   The object instances for which this currentData object is created to gather performance measurements.

- *scanAttributeIdList.*   A set of attributes of the observedObjectClass managed object class that are observed for performance monitoring data.

- *numericAttributeIdArray.*   Numeric attributes of the type counter or gauge that are observed to collect performance monitoring data.

- *onceReportedAttributeIdList.*   A list of attributes whose value will remain the same across all object instances. Those attribute values are reported only once for efficiency purposes.

- *reportAllAttributes.*   A Boolean value to indicate whether all the measurement attributes in this currentData object instance are returned to the managing system in case of a quality of service alarm.

- *suppressAdditionalThresholds.*   A Boolean value to indicate whether additional notification on threshold crossing will be sent to the managing system before the end of this interval.

- *ThresholdDataInstance.*   A pointer to an instance of the thresholdData object class which contains the threshold settings for this currentData object. This attribute is used to report threshold violation on performance data contained in this currentData object.

The currentData object can emit notification under different circumstances. For one, once the monitoring interval is up, it will send a notification to the associated managing system to report the collected performance measures.

- *scanReport.*   This is the report sent out by the currentData object upon the completion of the performance data monitoring for the current interval. The report may include scanAttributeIdList, numericAttributeIdArray, and onceReportedAttributeIdList, as described above, depending on how the control attributes such as reportAllAttributes are set.

- *qualityOfServiceAlarm.*   The currentData object sends an alarm notification for violation of thresholds specified in thresholdDataInstance.

The historyData managed object class. At the end of each performance monitoring interval, a historyData object instance is created out of a currentData object, if its creation has not been suppressed inside the currentData object. The attributes found in the historyData object class include

- *historyDataId.* A unique identifier for this historyData object instance.

- *periodEndTime.* The time at the end of the interval.

- *granularityPeriod.* The time interval between two successive observations.

- *measurementList.* A mirror image of the measurementList in the associated currentData object.

- *suspectIntervalFlag.* If set, it indicates that the one or more suspected performance measurements collected in the measurement list during the same interval may not be reliable for one of the reasons explained above.

- *numSuppressedIntervals.* The number of consecutive intervals for which the current performance data has been suppressed from being sent to the managing system or for which historyData object instances have not been created. It reflects performance measurement up to, but not including, the current interval.

- *observedObjectClass.* The object class for which this historyData object is created to gather performance measurements.

- *observedObjectInstance.* The object instance for which this historyData object is created.

An objectDeletion notification will be sent to the associated managing system when a historyData object is deleted as a result of the fact that the specified historyRetention period has expired. Included in the notification is a list of attributes such as historyRetention period and observedObjectInstance.

The thresholdData managed object class. As the name of the object suggests, this object class contains various threshold settings for performance management parameters. This managed object class has only one mandatory attribute, *thresholdId,* and its behaviors are defined by the following attributes in the conditional packages.

- *gaugeThresholdList.* A set of threshold settings for performance attributes of the gauge type. Each threshold setting consists of the attribute identifier, a threshold value, and an optional severity level for the event of crossing this threshold.

- *counterThresholdList.* A set of threshold settings for performance attributes of the counter type. Each threshold setting consists of the attribute identifier, a threshold value, and an optional severity level for the event of crossing this threshold.

- *monitoredEntityType.* A set of objects to be monitored for the performance data. This attribute together with the next attribute, granularityPeriod, allows the agent to validate that the threshold settings defined in counterThresholdList and gaugeThresholdList are reasonable. Since each monitored object has its permissible threshold values and granularity period, the agent can compare the permissible values against the ones in the two lists set by the manager. When a manager sends a threshold setting request to a thresholdData object in an M-SET request, the agent may reject the request by returning an error reply of setListError if the requested value is found inappropriate for the attribute type and monitoring interval. In this case, the next lowest threshold value supported by the managed system is used and the manager side is notified of the value in the reply to the M-SET request.

- *granularityPeriod.* The time interval between two successive observations.

The thresholdData object can emit the generic notifications under the circumstances described below, and the parameters of the notification can be found in the Managed Object Class Dictionary in App. D.

- *objectCreation.* This notification is sent to the managing system when this thresholdData object is first created.

- *objectDeletion.* This notification is sent to the managing system when this thresholdData object is deleted.

- *attributeValueChange.* This notification is sent when any of the performance thresholds such as gaugeThresholdList and counterThresholdList are modified.

An application scenario. We use a scenario to illustrate the behavior of the performance monitoring model discussed so far. Assume that a managing system is interested in knowing the traffic conditions at a signaling network node and queries the total number of signal messages transmitted at the signaling interface. Assume that there is a managed object signalingInterface that is responsible for receiving and transmitting signaling messages at the network node. The following steps provide an illustrative description of the behavior of the performance monitoring model described above.

1. The managing system sends a CMIS M-CREATE to the agent requesting that a currentData object be instantiated as an attribute of the signalingInterface object. Among the initial attribute values the managing system provides are

   - historyRetention = TRUE
   - discriminatorConstructor = default (TRUE)
   - observedClass = SignalingInterface class
   - numericAttributeIdArray = total number of call signaling message received; number of dropped signaling messages; number of signaling messages in error

2. The agent creates the currentData object as requested; a notification is sent back to the managing system to report the success of the object creation operation.

3. The managing system sends another M-CREATE request to the agent to instantiate a thresholdData object instance as the attribute threshold-DataInstance of the currentData object just created. Among the initial attribute values provided by the managing system for the thresholdData object are

- counterThresholdList = number of dropped signaling messages no more than 1.5 percent; number of signaling messages in error no more than 2.5 percent
- monitoredEntityType = signalingInterface

4. The agent system creates the thresholdData object as requested and binds it to the currentData object created earlier; a notification is sent back to the managing system to report the success of the object creation operation.

5. At the end of the observation period specified by the attribute granularity period, the agent instantiates a historyData object, assuming that no threshold violation occurred during the period.

6. The agent creates a scanReportRecord object from the historyData object and sends the scan report to the logging subsystem to be logged.

Performance data summarization

In the general performance monitoring model presented earlier, currentData and historyData objects collect performance measurements. Note that both objects are capable of sending the collected raw performance data to the managing system and other interested parties by notifications. Alternatively, the performance data can be aggregated and condensed before being sent out. Summarization is a process of aggregating and optionally applying algorithms to observed and collected performance data to produce summary information. This subsection presents a performance data summarization model as defined in ITU-T Recommendation X.738 (1993).

As part of the service performance monitoring process, the summarization function is shown in Fig. 9.2.

Again we use a simple scenario to illustrate the issues addressed here. Assume that a managing system is monitoring the congestion condition of a signaling interface card, which is modeled as a signaling interface managed object. In order to reduce the management information traffic between an agent and a manager, instead of sending raw performance measurements (e.g., a notification per interface message received), the agent sends summary information to the manager. The summary information includes the mean value, peak hour variance, and the minimum and maximum of the number of signaling messages received over a specified period of time.

The summarization objects, as derived from the scanner managed object class, are capable of scanning and retrieving the performance measurements from the monitored objects. Once the raw performance measurements are scanned and collected, summarization objects aggregate the data, apply required algorithms, and produce summary performance information. The

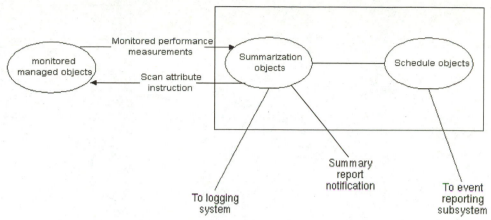

Figure 9.2 Summarization function model. (*From ITU-T Recommendation X.738, 1993*)

summary information may be sent to a managing system via event notification or logged for further analysis.

The summarization managed object classes. An inheritance hierarchy of the summarization managed object classes is shown in Fig. 9.3. All summarization object classes inherit from the scanner class, which provides the capability of scanning and retrieving the attribute values of monitored objects. Three types of scanner object classes, i.e., homogeneous scanner, heterogeneous scanner, and buffered scanner, all derived from the scanner class, represent different categories of the aggregation and summarization capabilities. The dynamic scanner class provides the flexibility for a managing system to dynamically specify the scanning criteria.

The homogeneous scanner is a general category of scanners that scans the same set of attributes across a selected set of monitored objects. The homogeneous scanner object class provides to its two subclasses and their respective subclasses the time stamping support and the basic capability to select a set of managed objects to monitor. The two subclasses of the homogeneous class, the simple scanner and ensemble statistical scanner, represent two specific types of homogeneous scanners: one simply scans and collects the performance data; and the other scans, collects, and applies statistical algorithms to the performance data.

The other two object classes at the same level of the inheritance hierarchy as the homogeneous scanner are the heterogeneous scanner and buffered scanner. A heterogeneous scanner object can scan possibly different sets of attributes for a set of explicitly named objects, as opposed to the same set of attributes of the homogeneous scanner. A buffered scanner object differs from other scanner objects in that it can buffer the scanned attribute values instead of reporting them at the end of each observation period. The scanner object classes will be introduced next in more detail in a top-down fashion along the inheritance hierarchy shown in Fig. 9.3.

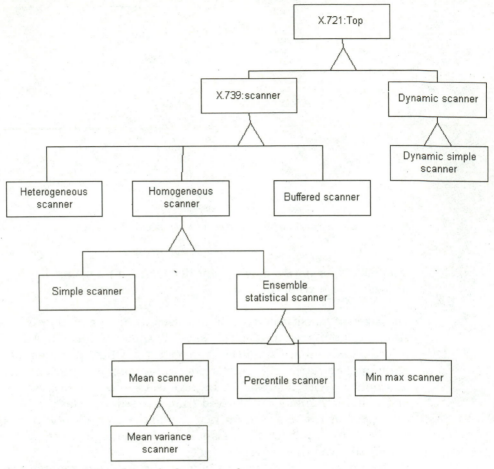

Figure 9.3 Inheritance hierarchy for scanner classes.

The scanner managed object class. The *scanner* object class is an abstract (i.e., non-instantiable) object class to provide to its direct and indirect subclasses the capability of periodic scanning and retrieving attribute values of monitored objects. In addition, it provides the capability to schedule the data scanning and retrieval on the basis of a specified interval, such as daily, weekly, or a more extensive period. In object-oriented terminology, the attributes and behavior of the scanner object class are inherited by all its subclasses, direct and indirect alike.

The mandatory attributes found in the scanner managed object class include

- *scannerId.* A unique identifier for an instance of the scanner managed object class

- *OperationalState.* Indicates whether the scanner object is enabled or disabled

- *AdministrativeState.* Indicates whether the managing system has administratively prevented the scanner object from operating

- *GranularityPeriod.* The time interval between two successive scans or the intended elapsed time of a scan

In addition, the scanner object class has the following optional attributes that specify the various scheduling, notification, and synchronization capabilities.

- *availabilityStatus.* The status of the scanner, e.g., in testing or off-line.

- *startTime.* The start time for the performance data scanning operation.

- *stopTime.* The stop time for scanning.

- *intervalsOfDay.* A set of time intervals for which the scanner object is turned on if the current day is one of the days selected in the weekMask attribute.

- *weekMask.* A mask to select a set of days of the week when the scanner object shall be turned on.

- *globalTime.* Specifies the synchronization time for repeating periods. The start for each period is at a time that is an integral number of periods before or after this attribute value. This attribute provides the reference point to which the granularity period shall be synchronized.

The scanner object has the capability to send the following notifications to the managing system.

- *attributeValueChange.* The scanner object emits this notification when any value of a specified set of attributes is modified. The choice of the set of attributes can be application-dependent, but generally those attributes are critical to the normal operations of the scanner object such as operationalState and administrativeState.

- *objectCreation.* This notification is emitted when the scanner object instance is created.

- *objectDeletion.* This notification is emitted when the scanner object instance is deleted.

- *stateValueChange.* This notification is emitted when an operational or administrative state change occurs.

The homogeneousScanner object class. The homogeneousScanner class is designed to select a set of managed objects and scan the same set of attributes across these managed objects. The homogeneousScanner object class, derived from the scanner class, is an abstract object class itself, to be inherited by other scanner object classes. For example, one instance of a subclass of the homogeneousScanner can observe one attribute of all selected signaling interface objects, say, signaling messages counter.

In addition to the attributes and behavior inherited from the scanner object class, the homogeneousScanner object class has the mandatory attribute scanAttributeList, which is a set of identifiers of the attributes to be observed and reported.

- *scanAttributeList.* Specifies a list of attributes to be observed for performance data.

- *timeStampReportMode.* Takes one of the three modes for time stamping: stamp off, global time stamp only, or individual time stamping.

- *baseManagedObject.* The base object for the scoping operation, i.e., the starting point of selection of a set of objects to be monitored.

- *scope.* Specifies the level of object selection in the MIB. This is the same scoping concept as discussed in Chap. 5.

- *scanningFilter.* The criteria in the format of logical expression used to filter the scoped managed object.

- *beginTimeOffset.* Used together with the attribute endTimeOffset to provide a time window relative to the current time for selecting managed objects.

- *endTimeOffset.* The end of the time window.

- *timeAttributeIdentifier.* Identifies the attribute whose value will be used as part of the criteria for selecting the managed object. All selected attributes are of the ASN.1 type GeneralizedTime.

- *objectList.* Identifies a list of objects to be observed for the performance data.

One important responsibility of the homogeneousScanner class is selecting an appropriate set of objects and attributes of the objects to scan for performance data. The objects are selected using a scoping and filtering mechanism similar to that used in the CMIS, as discussed in Chap. 5. Put simply, three attributes are used together to perform the selection. The base managed object attribute specifies the starting point of selection. The scope attribute identifies the levels in an MIB naming hierarchy. The filter specifies the criteria for the object filtering.

A set of attributes specifies the timing criteria for selecting managed objects to be observed. The two attributes, beginTimeOffset and endTimeOffset, together define a time window relative to the current time for selecting managed objects. Some managed objects have time-related attributes such as arrival and send-off time of a signaling message. An object is selected only if its time attribute falls in the specified time window. A third attribute, timeAttributeIdentifier, specifies the attribute of a managed object to which the time window shall be applied.

The timeStampMode attribute indicates whether a time stamp should be included in a summary report. The mode of global-time-stamp-only means that only the scan initiation time shall be included in the summary report. The mode of individual-time-stamping means that the scan initiation time, as well as the offset of each value, shall be included in the summary report.

The homogeneous scanner has two subclasses: simple scanner and ensemble statistic scanner, as described below.

SimpleScanner object class. A simpleScanner object simply scans the requested attribute of the selected objects and reports the attribute values without applying any algorithm to the scanned data. The attributes found in the simpleScanner class include

- *numericAttributeIdArray.* An ordered set of scanned attribute values without the associated attribute identifiers for efficient reporting.

- *suppressObjectInstance.* A Boolean-valued attribute, if set to the value TRUE, to indicate that the object instance parameter shall not be present in the summary report. This attribute is set to the value TRUE only when either the object names are not required by the receiving system or the same information is already available somewhere else.

- *onceReportAttributeIdList.* Same as above—a list for those attributes whose values remain the same across all selected objects and thus need to be reported only once.

The simpleScanner object has an action activateScanReport that allows the simpleScanner object to be activated by a managing system via an action request, and the scanned results are returned via an action reply. The simpleScanner object may send the scan results to the managing system unsolicited via the scanReport notification.

EnsembleStatisticScanner and subclasses. The ensembleStatisticScanner, a noninstantiable class along with a set of subclasses, provides statistic algorithms to be applied to the same set of attributes across the selected objects. The ensemble statistics scanner has two attributes, i.e., numericAttributeIdList and suppressObjectInstance, to scan the same set of attribute values across all selected objects and to suppress the object identifiers in the summary report as needed. As for the simpleScanner object described above, the ensembleStatisticScanner object has the activateStatisticalReport action to allow scan reporting to be activated by an action request from a managing system and the scanned results to be returned as an action reply. In addition, the statisticalReport notification allows the statistical scanner to report the scan results by emitting a notification.

The ensembleStatisticScanner object class, a noninstantiable abstract class is inherited directly or indirectly by four specific statistic scanner classes, i.e., mean scanner, mean variance scanner, minimum-maximum (min-max) scanner, and percentile scanner.

The meanScanner object class. The meanScanner is a class that scans the same set of attributes across all selected managed object instances and applies the algorithm to derive the mean value out of the sample population. The attribute values and the mean are reported back in a numericAttributeIdList with the number of samples as the first element of the array and the sample mean as the second element.

The meanVarianceScanner object class. The meanVarianceScanner is a subclass of meanScanner and is a special kind of mean scanner. It scans the same set of attributes across all selected managed object instances. The scanned data reported in a numericAttributeIdList with the first element of the array being the number of samples, the second element being the sample mean, and the third element being the mean variance. The sample variance S is defined as

$$S = \frac{\sum_{i=1}^{N} (X_j - X)}{N - 1}$$

or equivalently as

$$S = \frac{\sum_{j-1}^{N} x_j^2 - \frac{1}{N}\left[\sum_{j-1}^{N} x_j\right]^2}{N - 1}$$

The percentileScanner object class. The percentileScanner class is a subclass of the ensembleStatisticScanner class as discussed above. Its main responsibility is to calculate and report the percentile of a set of attribute values of the same type across selected managed objects. This object class has one attribute, configurablePCT, or configurable percentile, that has an integer value between 1 and 49, used for calculating the percentile.

A percentile P_j indicates the percent of a distribution that is equal to or below P_j. A set of collected attribute values is represented as $\{X_1, X_2, X_3, ..., X_n\}$, where the values are ordered from smallest to largest. Assume that q is the largest integer less than or equal to Q, and Q is defined as

$$Q = \frac{j(N+1)}{100}$$

Then,

$$P_j = X_1 \qquad \text{if } Q < 1$$

$$P_j = X_n \qquad \text{if } Q > N$$

$$P_j = X_q + (X_q + 1 - X_q)(Q - q) \qquad \text{if } 1 \leq Q \leq N$$

The result percentileScanner is returned in the attribute numericAttributeIdList, which is inherited from the ensembleStatisticScanner object class, with the following arrangement of elements in the list:

- *First element:* Number of samples
- *Second element:* Sample minimum
- *Third element:* Sample *j*th percentile
- *Fourth element:* Sample median
- *Fifth element:* Sample $(100 - j)$th percentile
- *Sixth element:* Sample maximum
- *Seventh element:* Sample mean

The MinMaxScanner managed object class. The minMaxScanner object class is a subclass of the ensembleStatisticScanner class, and its main responsibility is to identify and report the minimum and maximum of a set of observed attribute values of the same type. The results of the observation are returned in the attribute numericAttributeIdList with the maximum value as the first element and the minimum value as the second element of the list.

The heterogeneousScanner managed object class. As opposed to observing the same set of attributes across a selected set of managed object instances, the heterogeneousScanner object can scan possibly different sets of attributes for a selected set of objects.

The attribute observationList identifies a set of managed object instances and the associated attributes to be monitored. In the attribute is the scanAttributeIdentifierList field to specify the associated attribute list and the numericAttributeIdArray field to provide a more efficient way to report numeric attributes in an array without the associated attribute identifiers.

The attribute onceReportAttributeIdList is used to avoid multiple reporting of the same values of an attribute across a selected set of objects, as described above.

The bufferedScanner managed object class. The bufferedScanner managed object class, a subclass of the scanner object class, has a behavior similar to the other subclasses of the scanner object class except that it retains the scanned performance data instead of sending them to the managing system. The bufferedScanner attribute values are sent out in a notification at the end of each report period instead of at the end of a granularity period (i.e., observation period). Each report period can consist of an arbitrary number of granularity periods, depending on the specification of the application designer. The report periods can be scheduled using the scheduling capabilities inherited from the scanner class.

The buffered scanner has the following attributes, in addition to the ones inherited from the scanner class.

- *reportPeriod.* The number of granularity periods for a report period. The buffered attribute values are reported at the end of each report period. The report period begins and ends on granularity period boundaries and includes the scan that is initiated at the end of the report period.

- *BufferedObservationIdList.* Identifies a set of objects to be observed along with the relevant information about each object in the list. The information includes
    - *Observed object.* Identifier of the object instance whose attribute value is to be buffered.
    - *Scan attribute identifier list.* A list of attributes to be observed. The attribute value can be of an arbitrary type.

■ *Numeric attribute identifier array.* An ordered set of attribute values without the associated attribute identifiers for efficient reporting.

■ *Report time attribute identifier list.* A list of attributes that are scanned and reported only at the end of each report period.

■ *SuppressObjectInstance.* A Boolean variable to indicate, if set to TRUE, that all object instance names shall be suppressed in all summary reports.

■ *timeStampReportMode.* As defined above, this attribute takes one of the three modes for time stamping: stamp off, global time stamp only, or individual time stamping.

The bufferedScanner object will emit notifications at the end of each report period if the reporting schedule is activated. Alternatively, a managing system can explicitly request the bufferedScanner object to report previously scanned attribute values by sending an M-GET request, and the summary reports are returned in the reply.

The dynamicScanner managed object class. DynamicScanner, a noninstantiable superclass, allows a managing system to dynamically specify the criteria for selecting objects to scan. The criteria are specified in the action-information field of the CMIS M-ACTION request sent from a manager to an agent. The results are reported back in the action reply of the CMIS message.

Only three attributes are included in the dynamic scanner class, i.e., scannerID, operationalStates, and administrativeState. The detailed behaviors are specified in the subclass dynamicSimpleScanner.

The dynamicSimpleScanner managed object class. An instance of this object class is activated by an action request from a managing system to scan and report the requested attribute values across the selected managed object instance as long as its administrative state is "unlocked." The administrative state, inherited from the superclass dynamic scanner, is used to suspend or resume the scanning or reporting.

The attributes to be scanned are identified in the scan attribute identifier list and the numeric attribute identifier array.

An application example of summary report. We continue with the application example of monitoring the performance of a signaling interface as described in the preceding subsection. Assume that the managing system is interested in the mean variance and mean of the signaling message drop rate over n observation periods. The following sequence of events provides an illustrative rather than a prescriptive description of how the summary report objects work together to perform the task.

1. The managing system sends a CMIS M-CREATE request to the agent, requesting that a bufferedScanner object be instantiated as an attribute of the signalingInterface object. Among the initial attribute values the managing system provides are

- reportPeriod = n, the desired observation periods
- numericAttributeIdArray = total number of call signaling messages received; number of dropped signaling messages; number of signaling messages in error

2. The agent creates the bufferedScanner object as requested; a notification is sent back to the managing system to report the success of the object creation operation.

3. The managing system sends another M-CREATE request to the agent to instantiate a meanVariableScanner object with the following initial attribute values:

- scanAttributeIdList = number of dropped signaling messages
- timeStampReportMode = global time stamp

4. The agent system creates the meanVarianceScanner object as requested; a notification is sent back to the managing system to report the success of the object creation operation.

5. The bufferedScanner object scans the specified attribute values and buffers them at the end of each granularity period.

6. At the end of the nth observation period, the meanVarianceScanner object applies its algorithm and computes the mean and variance of dropped signaling messages. The result is returned in the scanAttributeIdList with the sample size as the first element of the list, and the computed mean and variance values as the second and third elements, respectively.

Performance data logging

Collected performance data can be reported to a managing system, logged or both. This subsection describes a generic log control function as defined in CCITT Recommendation X.735 (1992) that can be used in performance monitoring to log performance data. The log control model can be used to log the performance data as well as data such as alarm, security session, or any information that is worth logging.

A *log* is a repository for log records, and a *log record* contains information that is logged. The log control model provides the ability for a managing system to create, delete, modify, and retrieve log records; the ability to suspend, resume, and modify the criteria of logging; and the ability to determine whether a log record has been modified or lost.

A log management model. A log management model, as shown in Fig. 9.4, is very similar to the event model that will be presented in Chap. 10. Logging essentially is a special event, and thus log management shares the same characteristics of the event management. The log management model involves three components, i.e., managed objects, log preprocessing, and logs.

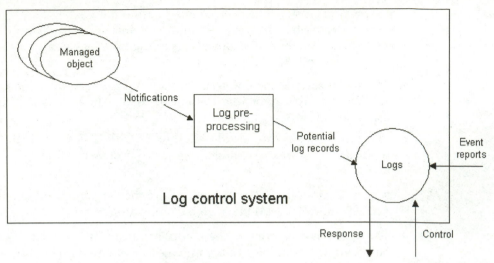

Figure 9.4   Log management model.

We continue to use the scenario of a signaling interface to describe the behavior of the log management model. Assume that the interest of a managing system is in monitoring the traffic condition at a signaling interface to detect signs of congestion or degradation of quality of service the managed object in the model represents the information to be logged, the a signalingInterface object in this case. As described before, the signalingInterface object has the scanner object as its attribute to scan and collect performance data. The log preprocessing component formats the performance data into a potential log record. The log object functions very much like a controller of the logging process. It is responsible for applying logging criteria to potential log records and determines which record to log. It is responsible for storing the log records and responding to a logging-related instruction from a managing system to create, delete, modify, and retrieve log records; to suspend, resume, and modify the criteria of logging; and to determine whether a log record has been modified or lost. In the case of performance data at the signaling interface, the log managed object may receive a log record either for each signaling message or for signaling messages of a specified interval, and the criteria of logging is up to each specific application. The log object also adds additional information to form a log record: log ID, time stamp, etc. The log may store log records at the local NE and also sends them to a managing system on a scheduled basis. The attribute level details of the key object classes will further explain the log model.

Managed object classes of the log model.   Two managed object classes, log and logRecord, are described in detail.

The log managed object class.   The numerous attributes of the log managed object class can be put into three different categories: log state or status related, log control action, and log scheduling. The first category has three attributes: operationalState, administrativeState, and availabilityStatus. The log control actions include maximum log size, log threshold alarm, and log full action. A third category of attributes allows a managing system to schedule logging activities on a daily, weekly, or more extensive period basis.

In general, the values of the state and status attributes such as operational and availability status are not modifiable by a manager. The nonstatus attribute values may be modified, though restrictions may apply. For instance, the value of the maximumLogSize attribute cannot be changed to a value smaller than that of the currentLogSize attribute. The logId attribute, once assigned, is not resettable.

The log managed object class. This is a support managed object class that controls logging activities for performance data by defining criteria for controlling the logging of information in an open system.

- *logId.*   The distinguishing identifier for log object.

- *discriminatorConstruct.*   Criteria and test to be applied to the potential log records for deciding whether the information is logged.

- *administrativeState.*   Indicates whether the log object is administratively prevented from functioning.

- *operationalState.*   Indicates whether the log object is operational.

- *availabilityStatus.*   Indicates whether or not the managed object that is the target of logging is available.

- *logFullAction.*   Specifies the action to be taken when the log is at its maximum size. Two basic options are available: *halt* or *wrap*. The action of wrapping means that the oldest log records in log, as determined by the log record identifier, are deleted to make room for new log records.

- *maxLogSize.*   Maximum number of log records that can be logged.

- *currentLogSize.*   Current number of log records that have been logged.

- *numberOfRecords.*   The number of records in the log.

- *capacityAlarmThreshold.*   Specifies a percentage point of maximum log size at which a notification is sent warning the managing system of the approaching log-full condition.

- *startTime.*   Start time for logging operation.

- *stopTime.*   Stop time for logging operation.

- *intervalsOfDay.*   A list of intervals for which the log will be turned on.

- *schedulerName.*   The scheduler used for the logging activity.

- *weekMask.*   A list of days during a week for which the log can be turned on.

The logRecord managed object class. The logRecord managed object class is used to define the records contained in a managed object log. Log records contain the information to be logged and are created as a result of the receipt of an event report or notification.

There are only two attributes in the logRecord managed object class: logRecordId and loggingTime. Then, where are the attributes that represent the information to be logged? The specific attributes representing the logged information are simply too diverse to be standardized and are left to each application.

Performance Analysis

The next phase of performance management is performance analysis, which analyzes the performance data collected during the performance monitoring phase and provides information on whether network congestion is occurring and helps the management system to determine whether performance is adequate for existing services and planned new services. Based on this information, a management system can make performance control decisions.

This section discusses two aspects of performance analysis: performance analysis of circuit-switched networks and of packet-switched networks. The performance analysis of circuit-switched networks introduces the commonly used methods for calculating the important performance measures such as network traffic capacity, call drop rate, or congestion probability. The performance analysis of packet-switched networks introduces the methods for computing common performance measures such as probability of delay and mean number of messages in a queue.

Note that the performance analysis as discussed in this section focuses on the network and network element management layers of the TMN logic layered architecture. Performance analysis on the other layers, the service management layer in particular, is discussed in Chap. 14.

Performance analysis of circuit-switched networks

This subsection introduces the commonly used traffic measurement and the calculation of congestion probability for circuit-switched networks (e.g., conventional PSTNs). The emphasis is on the application of the analysis of generic methods. For details behind the formula, the reader is referred to numerous other books listed in the References.

This is a network level functional unit that describes, analyzes, and assesses end-to-end performance and service characteristics of a dedicated digital network with the focus on traffic.

Traffic measurement. First introduced are some basic concepts related to traffic measurement for circuit-switched networks. Network capacity is measured by the number of trunks needed. A *trunk,* the basic traffic-carrying entity of circuit-switched networks, is a circuit or a communication line between two switching systems. A trunk group is a group of trunks sharing the same characteristics and linking two switching systems.

Network traffic is measured by the number of calls in progress. The time a trunk is busy is called the *holding time* because the trunk cannot be shared while carrying a conversation.

The unit of telecommunications traffic is called the *Erlang* (E), named after A. K. Erlang, the Danish pioneer of telephone traffic engineering. One Erlang is equal to one full hour of the use of a trunk, or $60 \times 60 = 3600$ seconds (s) of telephone conversation. For example, on average, if a company as a whole makes 100 outgoing calls of 3-minute (min) average duration and receives 120 incoming calls of 2-min average duration, the total traffic carried on the trunk group leased by the company can be computed as $[100 \times (3/60)] + [120 \times (2/60)] = 5 + 4 = 9$ E.

In North America, traffic is often expressed in terms of hundreds of call seconds per hour or centi call seconds (CCS). Thus, 1 Erlang = 36 CCS, derived from $(60 \times 60)/100$. The traffic carried by a group of trunks is derived from the Erlang definition as follows:

$$A = \frac{Ch}{T}$$

where A = traffic in Erlang E
  C = average number of call arrivals during time T
  h = average call holding time, minutes

Grade of service. It is prohibitively costly for a service provider to have as many trunks in a network to accommodate all subscribers making calls simultaneously. The usual practice is to have the number of trunks just sufficient to handle peak-hour traffic and to use skewed pricing to discourage deviation from the normal traffic pattern (e.g., premium rate for peak-hour calls and discount for off-peak-hour calls). The situation may arise where all trunks in a trunk group are busy, and it can accept no more calls. This situation is called *congestion*.

For a circuit-switched network like a telephone network, new calls have to be dropped when the network is congested. Thus this type of network is also called a *lost-call system*. In contrast, new messages can be queued up as they arrive at a packet-switched system even when the network is congested. Thus this is called a *delaying system* or *queuing system*.

In a lost-call system, the traffic actually carried is less than the total traffic that actually arrives at the network. An important performance measure is the grade of service (GoS) which can be measured by the proportion of calls lost due to congestion:

$$Q = \frac{\text{number of calls lost}}{\text{total number of calls received}}$$

If A E of traffic is received at a trunk group with an average GoS Q, then traffic lost is $A*Q$ and the actual carried traffic is $A*(1-Q)$.

Probability of congestion. In order to carry out basic performance analysis for lost-call systems, a few basic assumptions are introduced first.

The concept of *memoryless traffic* assumes that call arrivals and terminations are random events, completely independent of each other. This refers to the behavior observed over a large number of calls by a large number of people, instead of all the calls by an individual. This sometimes is also called *pure chance traffic.*

The concept of *statistical equilibrium* means that the generation of traffic is a stationary random process. In other words, the probability of a call arrival at one point is the same as that at another point during the period being considered.

Assume that a lost-call system like a switching system has N outgoing trunks and A E of received traffic. The probability of having x calls in progress is described by the following formula:

$$P(x) = \frac{A^x/x!}{\sum\limits_{k=0}^{N} A^k/k} \; !$$

This is also called the first Erlang distribution. Of particular interest to performance analysis is $P(N)$, because this is the probability of congestion, i.e., the probability of a lost call. This is also the grade of service. Expressed in terms of grade of service, the probability of lost calls, denoted as $E_{1,N}(A)$, for a trunk group of N trunks with received traffic A E, is computed as follows:

$$B = E_{1,N}(A) = \frac{A^N/N_1}{\sum\limits_{k=0}^{N} A_k/k!}$$

The formula above allows for the computations of $E_{1,N}(A)$ for all values of N. Tables of $E_{1,N}(A)$ have been published in Beckman (1968). A traffic table is published in Flood (1995) that rearranges the data computed using the above formula and enables an engineer to determine parameters other than the GoS: network capacity or expected traffic load. Appendix 9A presents the traffic table for N from 1 to 50.

Performance analysis for packet-switched systems

Packet-switched networks are fast becoming an indistinguishable part of the telecommunications scene that had been exclusively dominated by the circuit-switched network not long ago. This section introduces some of the most basic performance analysis tools, most from queuing theory. Included are the second Erlang distribution, delay probability, and a delay table.

The main issue the second Erlang distribution addresses is how to determine the probability of encountering delay when traffic load A is received at a queuing system, such as a packet-based switching system, with N trunks. In queuing theory, *trunks* are also called *servers,* for the fact that queuing theory is also used in fields other than telecommunications traffic engineering. The second Erlang distribution depends on the following assumptions:

■ *Random traffic,* meaning that traffic events such as arrivals and terminations of messages are completely independent of each other.

- *Statistical equilibrium,* meaning that the probability of an event occurring remains the same throughout the period T.
- *Full availability,* meaning that each arriving packet can be connected to an outgoing trunk if there is one free and there is no delay in switching and connecting the packet.
- Calls (data calls, or packets) which encounter congestion are queued up until a trunk becomes available.

Note that the only difference between a packet-switched and a circuit-switched system is the queuing capability of a packet-switched system, as far as the performance analysis is concerned.

Probability of delay. The probability of delay, an important measure of quality of services, provides the statistical likelihood that a customer will have to wait before the service request can be processed by a server (e.g., connecting to a network). Assume a packet-switched system with a traffic load of A erlangs, N servers, and x data calls in progress; the probability of delay is given by

$$P(x \geq z) = \frac{N^N}{N!} \left(\frac{A}{N}\right)^z \frac{N}{N - A} P(0)$$

where $z \geq N$ and $P(0)$ is the probability of no call in the queue and is given by

$$P(0) = \left[\frac{NA^N}{N!(N - A)} + \sum_{x=0}^{N-1} \frac{A^x}{x!} \right]^{-1}$$

This is also called second Erlang distribution and provides a base to derive other useful performance measures.

Mean number of data calls. If the mean number of data calls over time a network shall expect can be calculated using the historical traffic load, a management system will be able to plan the network capacity and determine the routing pattern accordingly. When there is delay, the mean number of data calls in a packet-switched system is defined by

$$\overline{X}' = \frac{A}{N - A} + N$$

Averaging over all time, the mean number of calls is given by

$$\overline{X}' = \frac{A}{N - A} - E_{z,+N}(A) + A$$

Mean delay time with a first-in–first-out queue. If the first-come–first-serve queue discipline is used, the mean delay time is calculated as

$$\overline{T}' = h(N - A)$$

where h is the mean holding time. Averaging over all time, the mean delay, T, is given by

$$\overline{T}' = E_{z,N}{}^{(A)h\setminus(N-A)}$$

Performance Management Control

Performance management control is used to control the operations of a network to either maintain the desired level of performance or to prevent certain conditions (e.g., congestion) from developing. Performance management control uses both the results of performance analysis and the performance data collected during the performance monitoring. Specifically, performance management control has two aspects, according to Bellcore GR-2869 (1995) and CCITT Recommendation M.3400 (1992): traffic control and performance administration. Traffic control focuses on creating and modifying routing patterns, and selectively shedding traffic if necessary, to either prevent or relieve network congestion. Performance administration is responsible for management of schedules, thresholds, and other performance-related data.

Traffic control

The main task of traffic control is to maintain and manage the network routing pattern. Routing is an important part of ongoing maintenance of network performance. As the network load changes, it may be necessary to redesign the network to accommodate the change in the load by moving traffic from one route to another. Various routing methods can be used for traffic control, and this subsection surveys several of the commonly used ones.

Flooding.   Flooding is one of the simplest routing methods according to which each node in the network, except for the node to which the message is destined, sends a copy of any message it receives to each of its neighbors. This method relieves a network node of the burden of having to know the topology of the network. It does not need to know the nodes, links of the network, relationships between them, or the states of each network element. All it needs to know is its local topology, i.e., its neighboring nodes and links.

Flooding has the advantages of simplicity and speed in determining a route. Obviously it also has the disadvantages of duplicating a message many times and using more network resources than another method would have. Usually a node will send multiple copies of the same message and an infinite loop of a message may occur.

Some mechanism must be in place to prevent a message from being sent forever between nodes, even after it has been delivered to the destination. There are two commonly used methods. One is to have each message carry a hop count, the number of nodes it has already traveled. Each node that forwards the message increments the count, and when the count reaches a predefined limit, the message is discarded. The second method is to assign an ID to each message and to have each node keep a record of the message it has forwarded. When the same message comes around for second delivery, it is discarded. The destination node keeps track of messages it has received as well, in order to be able to discard the duplicate copies.

In the traditional telecommunications networks, especially PSTNs, it is difficult to imagine using flooding to route signaling messages. However, this seemingly simple and even trivial method is widely used in other types of networks, where the network topology changes rapidly, such as mobile networks.

Explicit routing.   The so-called explicit routing is widely used in conventional telecommunications networks; a routing table containing explicit, predefined routes for each source and destination is stored at each network node. Sometimes, multiple routes can be defined for fault tolerance purposes or different routes are defined for different classes of traffic. Also sometimes special routes can be reserved to accommodate special requirements.

The explicit routing has the advantages of not requiring any computation of routes or exchange of network topology information between two nodes. In addition, it provides the flexibility for the network designer to optimize the network performance by taking advantage of the fact that the network topology and traffic conditions will be fairly predictable and stable.

There are two major disadvantages associated with this routing method. First, the routing cannot be changed dynamically in response to the occurring network conditions and a message has to travel the predetermined route even when the route is rendered inefficient by the network conditions. Secondly, the burden of creating and maintaining the often very large routing lists falls on the shoulders of the network managing staff. This tends to be a very tedious and time-consuming process. It also proves to be a problem to maintain net-workwide consistency of routing tables, because routing tables are often very large and a network node usually does not have a view of the entire network. In addition, the errors stemming from the manual process only make the sit-uation worse. In some cases, it is difficult to synchronize the cross-network updating of the routing tables, and this can result in inconsistency between nodes.

Shortest-path routing.   The general idea of the shortest-path routing is simple: a length or weight is associated with each link, and traffic is routed over the shortest path based on the accumulated weight or length from the source to the destination. The computation of node length can be expressed as

$$D_j = D_i + L_i + L_{ij}$$

where D_i = length of path up to node i
   L_i = length assigned to node i
   L_{ij} = length of link from i to j

For a more detailed discussion on this topic, see Kershenbaum (1993).

There are several variants of shortest-path routing, based on the way the link length is assigned. The simplest case is where a length is assigned to each link. This is also known as *min-hop routing,* a widely used routing method because of its simplicity. It is simple to compute the accumulated length. Min-hop routing tends to produce ties because there are often multiple alternate paths with the same number of hops. One simple way to break the tie is to randomly choose one path, and this tends to work out well in practice.

The next variant is to assign different lengths and weights to different hops and nodes in order to encourage or discourage traffic on a particular path for purposes such as balancing the load and relieving congestion. In this case, the link length can be assigned based on the node capacity to encourage flow on high-capacity links and discourage the traffic on the low-capacity links. This is achieved by assigning smaller lengths to high-capacity links and larger lengths to low-capacity links. The link length can be adjusted periodically to accommodate the change in traffic patterns. It is the responsibility of performance management to assign the link lengths and adjust them as needed.

Another variant of shortest-path routing is called *least-cost routing.* This routing method assigns a cost or weight to each link, based on criteria such as the distance of a hop or monetary cost. While this method does not minimize the delay, it conserves the capacity on the more expensive or high-priority links.

A small variation to the above method is adding a constant to the length of each link. Paths with a small number of hops are favored over paths with a larger number of hops, because the constant weighs more when multiplied many times for a path with a large number of links. This is another means of controlling and directing traffic flow toward a desired direction.

An important advantage of the shortest-path routing is that it relieves the network craftperson of the burden of maintaining large routing tables. Once link lengths are defined, the routes are implicitly determined. When new links and nodes are added, a few link lengths need to be added and a few existing link lengths may be modified to adapt to the new traffic flow caused by the new links. In addition, the link length–based routing makes it easier for the performance management to automatically adapt to link failure or node congestion by changing the link lengths.

A disadvantage of shortest-path routing is its lack of flexibility in dealing with traffic with special requirements. In min-hop routing, once a message reaches a node, the path it takes to reach the destination has nothing to do with its origin. Traffic with special requirements cannot be directed toward a particular desired path if the path is not the shortest one. In shortest-path routing, all traffic with the same characteristics and bound for the same destination follows the same path. In contrast, the routing-table–based routing can assign a route based on a specific traffic requirement.

Distributed routing. Distributed routing is a kind of dynamic routing that can dynamically change the routing behavior. According to this routing method, each node makes a decision only about the next hop for the message. The node at the other end of the link then makes the decision about its next hop. Since it does not require global updating of congestion information, the overhead and delay is reduced but more coordination among the nodes is required.

Distributed routing, like shortest-path routing, requires minimal human intervention in the performance management process. It dynamically adapts to the network traffic conditions and tends to route around traffic-heavy or congested areas.

Performance administration

The performance administration is responsible for coordinating performance management activities such as performance monitoring and performance analysis to implement a performance management policy. For example, when a segment of the network requires special attention due to heavier than normal traffic volume, the performance administration sets up performance monitoring parameters and defines threshold values according to the general network performance requirement and the local context. It also arranges the input to the performance analysis functions and uses the analysis result to help make performance control decisions.

The performance administration manages performance-related scheduling activities. For example, it may establish, change, and remove a schedule according to the need of a performance management function. Also it can initialize, suspend, resume, and shut down a performance-related scheduler, as requested by another management component.

The performance administration is responsible for executing the control actions such as enacting a new route, taking down a route, or rerouting traffic in a certain direction, based on the routing decision from the traffic control component. In addition, it also interacts with the configuration management application to manage the routing data and table, performing tasks such as accessing and reporting routing data and updating the routing table as instructed by the traffic control and other management components.

The performance administration interacts with other management applications such as fault management, configuration management, or security management to exchange management information. For example, it may generate a quality-of-service alarm in case of network congestion and send the alarm information to the fault management.

Summary

This chapter presents three aspects of performance management: performance monitoring, performance analysis, and performance management control. The performance monitoring covers three aspects. First, a generic performance monitoring model and major managed object classes are introduced. Second, a set of scanner managed objects that can summarize the collected performance data in a variety of ways, with focus on the statistic algorithms that can be applied to the collected data is presented. Each scanner is introduced from two aspects: how data are summarized and how data are reported. Third, the performance data logging subsection introduces the ITU generic logging model in the context of performance data logging.

The performance analysis section discusses the performance analysis from two different aspects: performance analysis for circuit-switched networks and performance analysis for packet-switched networks. The focus of this section is on the methods for computing common performance measures for the two types of networks, such as probability of network congestion and delay, grade of services, and mean number of messages.

The performance management control covers two aspects: traffic control and performance administration. The traffic control is realized through manipulation of traffic routing patterns. Several commonly seen routing methods are introduced. The performance administration is responsible for a variety of tasks, ranging from coordinating various performance management components and managing scheduling activities to interacting with other management applications to exchange management information.

Exercises

9.1 Discuss two ways that a scanner object can be associated with a monitored object. *Hint:* Consider the two basic types of relationships between two objects in object-oriented design.

9.2 What are the relationships between the currentData object and the thresholdData object as defined in the generic performance monitoring model?

9.3 What is the main difference between a homogeneous scanner and a heterogeneous scanner?

9.4 What is the main difference between a loss-system and a queuing system as far as performance analysis is concerned?

9.5 Explain how the probability of congestion on a telephone network is calculated.

9.6 Explain how to interpret the traffic table as listed in App. 9A.

9.7 In a wireless network, a mobile station can move from cell to cell, thus making it difficult for a management system to send a message to it. Discuss in general terms why the flooding method might be a good way to route a message to a mobile station.

9.8 Contrast the distributed routing with the explicit routing in terms of involvement of a human operator and flexibility to adapt to changing traffic conditions.

Appendix 9A: Traffic Table

An engineer can find expected quality of service (QoS) from a given network capacity or expected traffic load. For example, if the desired QoS is 1 call loss per 1000 calls and the expected traffic load is 20 E, 35 trunks are required. Or, if 50 trunks are available, then no more than a 32-E traffic load shall be received in order to maintain the desired QoS of 1 call loss per 1000 calls. Table A.1 is the traffic-capacity table for N from 1 to 100.

TABLE 9A.1   Traffic-Capacity Table for Full-Availability Groups

Number of trunks	Traffic capacity (E) for 1 lost call in			
	50 (0.02)	100 (0.01)	200 (0.005)	1000 (0.001)
1	0.020	0.010	0.005	0.001
2	0.22	0.15	0.105	0.046
3	0.60	0.45	0.35	0.19
4	1.1	0.9	0.7	0.44
5	1.7	1.4	1.1	0.8
6	2.3	1.9	1.6	1.1
7	2.9	2.5	2.2	1.6
8	3.6	3.2	2.7	2.1
9	4.3	3.8	3.3	2.6
10	5.1	4.5	4.0	3.1
11	5.8	5.2	4.6	3.6
12	6.6	5.9	5.3	4.2
13	7.4	6.6	6.0	4.8
14	8.2	7.4	6.6	5.4
15	9.0	8.1	7.4	6.1
16	9.8	8.9	8.1	6.7
17	10.7	9.6	8.8	7.4
18	11.5	10.4	9.6	8.0
19	12.3	11.2	10.3	8.7
20	13.2	12.0	11.1	9.4
21	14.0	12.8	11.9	10.1
22	14.9	13.7	12.6	10.8
23	15.7	14.5	13.4	11.5
24	16.6	15.3	14.2	12.2
25	17.5	16.1	15.0	13.0
26	18.4	16.9	15.8	13.7

TABLE 9A.1   *(Continued)*

Number of trunks	Traffic capacity (E) for 1 lost call in			
	50 (0.02)	100 (0.01)	200 (0.005)	1000 (0.001)
27	19.3	17.7	16.6	14.4
28	20.2	18.6	17.4	15.2
29	21.1	19.5	18.2	15.9
30	22.0	20.4	19.0	16.7
31	22.9	21.2	19.8	17.4
32	23.8	22.1	20.6	18.2
33	24.7	23.0	21.4	18.9
34	25.6	23.8	22.3	19.7
35	26.5	24.6	23.1	20.5
36	27.4	25.5	23.9	21.3
37	28.3	26.4	24.8	22.1
38	29.3	27.3	25.6	22.9
39	30.1	28.2	26.5	23.7
40	31.0	29.0	27.3	24.5
41	32.0	29.9	28.2	25.3
42	32.9	30.8	29.0	26.1
43	33.8	31.7	29.9	26.9
44	34.7	32.6	30.8	27.7
45	35.6	33.4	31.6	28.5
46	36.6	34.3	32.5	29.3
47	37.5	35.2	33.3	30.1
48	38.4	36.1	34.2	30.9
49	39.4	37.0	35.1	31.7
50	40.3	37.9	35.9	32.5
51	41.2	38.8	36.8	33.4
52	42.1	39.7	37.6	34.2
53	43.1	40.6	38.5	35.0
54	44.0	41.5	39.4	35.8
55	45.0	42.4	40.3	36.7
56	45.9	43.3	41.2	37.5
57	46.9	44.2	42.1	38.3
58	47.8	45.1	43.0	39.1
59	48.7	46.0	43.9	40.0
60	49.7	46.9	44.7	40.8

TABLE 9A.1 *(Continued)*

Number of trunks	Traffic capacity (E) for 1 lost call in			
	50 (0.02)	100 (0.01)	200 (0.005)	1000 (0.001)
61	50.6	47.9	45.6	41.6
62	51.6	48.8	46.5	42.5
63	52.5	49.7	47.4	43.4
64	53.4	50.6	48.3	44.1
65	54.4	51.5	49.2	45.0
66	55.3	52.4	50.1	45.8
67	56.3	53.3	51.0	46.6
68	57.2	54.2	51.9	47.5
69	58.2	55.1	52.8	48.3
70	59.1	56.0	53.7	49.2
71	60.1	57.0	54.6	50.1
72	61.0	58.0	55.5	50.9
73	62.0	58.9	56.4	51.8
74	62.9	59.8	57.3	52.6
75	63.9	60.7	58.2	53.5
76	64.8	61.7	59.1	54.3
77	65.8	62.6	60.0	55.2
78	66.7	63.6	60.9	56.1
79	67.7	64.5	61.8	56.9
80	68.6	65.4	62.7	58.7
81	69.6	66.3	63.6	58.7
82	70.5	67.2	64.5	59.5
83	71.5	68.1	65.4	60.4
84	72.4	69.1	66.3	61.3
85	73.4	70.1	67.2	62.1
86	74.4	71.0	68.1	63.0
87	75.4	71.9	69.0	63.9
88	76.3	72.8	69.9	64.8

TABLE A.1   *(Continued)*

Number of trunks	Traffic capacity (E) for 1 lost call in			
	50 (0.02)	100 (0.01)	200 (0.005)	1000 (0.001)
89	77.2	73.7	70.8	65.6
90	78.2	74.7	71.8	66.6
91	79.2	75.6	72.7	67.4
92	80.1	76.6	73.6	68.3
93	81.0	77.5	74.3	69.1
94	81.9	78.4	75.4	70.0
95	82.9	79.3	76.3	70.9
96	83.8	80.3	77.2	71.8
97	84.8	81.2	78.2	72.6
98	85.7	82.2	79.1	73.5
99	86.7	83.2	80.0	74.4
100	87.6	84.0	80.9	75.3

10

Fault Management

Outline

- The first section, "Alarm Surveillance," discusses the process of detecting alarms and subsequent activities. It describes a generic model of detecting alarms, the detection mechanisms, the objects involved in detecting alarms, and subsequent alarm reporting and logging.

- The second section, "Fault Localization," describes a generic process for finding the root cause of a fault and each of the major steps in the process. The major steps include diagnostic testing and alarm correlation and analysis. A set of well-known methods for alarm correlation and analysis is introduced.

- The third section, "Test Management," first describes a diagnostic test model and the major objects involved in the test process. Then a set of six categories of diagnostic tests is introduced, which includes the connection test, data integrity test, loop-back test, and protocol integrity test.

- The fourth section, "Fault Correction and Service Restoration," describes a general process of correcting faults and restoring services once the proper root cause of a fault has been correctly diagnosed.

- The last section, "Trouble Administration," is responsible for keeping track of a fault for its entire life cycle in a network. The focus of the discussion is on the trouble report management, i.e., its creation, tracking, management, and closure.

Introduction

A *fault* is an abnormal condition that adversely affects services in a telecommunications network. It can be caused by an event such as equipment failure or software error. The ultimate goal of fault management is to minimize the impact of a fault on the service with a minimum amount of interference to the normal network operation. Functionally, fault management consists of a set of functions which enables the detection, isolation, and correction of abnormal conditions affecting the operation of the telecommunications network and its environment.

Life cycle of fault management

The life cycle of fault management goes through several stages in a relatively linear fashion. First a fault has to be detected and reported promptly. Once the management system receives a fault report, the root cause of the fault is determined in order to decide on the subsequent actions to take. Diagnostic tests are often conducted in the process of determining the root cause of a fault. Once the root cause of the fault is determined, corrective measures are taken to rectify the situation or minimize the harm caused by the fault. In this process, there is a large amount of information that is passed between the stages that needs to be managed. The whole fault management process, which is shown in Fig. 10.1, can be modeled as five distinct phases and is largely based on Bellcore GR-2869 (1995) and CCITT Recommendation M.3400 (1992).

- *Alarm surveillance.*   Responsible for detecting a failure which is defined as the onset of a fault, while a fault is defined as a persistent condition in a system unit that prevents the unit from performing its function.

- *Fault localization.*   Responsible for determining the root cause of a fault

- *Test management.*   Responsible for designing and conducting tests in order to determine the root cause of a fault

- *Fault correction.*   Responsible for taking appropriate action to correct the fault once the root cause has been identified

- *Trouble administration.*   Responsible for overall record keeping of a trouble report originated by the customer or the network management system itself

As shown in Fig. 10.1, a fault report can be initiated either by an alarm surveillance system or by a customer who reports a service problem.

The five sections in this chapter may be viewed as the five functional areas within fault management, maybe for lack of a better term. Some of the areas may be described from multiple management layers, i.e., the service management layer (SML), the network management layer (NML), and the element management layer (EML), while others may primarily involve one layer.

Before diving into the topics of this chapter, we use a fault scenario to assist our discussion of fault management and illustrate the issues addressed by fault management. The scenario is about a line card in a switch that lost its incoming signaling messages. It is required that the signal loss be detected and reported to the managing system as soon as possible. Once the managing system receives the alarm, it needs to find out what caused the fault and conduct the necessary tests if needed. In addition, it needs to keep track of the fault report, trouble ticket, and the states of the alarm and finally correct the fault and restore services. These are the basic responsibilities of a fault management system for a telecommunications network.

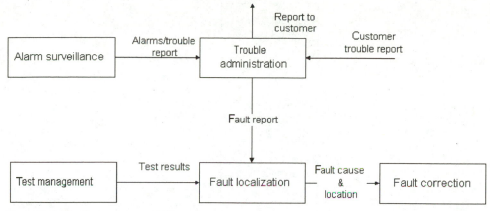

Figure 10.1   Fault management life cycle.

Alarm Surveillance

The alarm surveillance centers on detecting a fault event and generating and reporting an alarm to a managing system promptly. Various aspects of alarm surveillance are summarized in Fig. 10.2, where three steps form a general alarm surveillance process. First, the alarm must be detected as soon as possible. Next, the detected fault events go through preliminary processing to filter out the duplicate, unwarranted, or dubious information. Subsequently alarms are generated that represent the fault events in a predefined form. Then, alarms are promptly reported to the managing system and may also be logged if so required. The focus of the discussion will be on the latter two parts, i.e., alarm filtering and reporting, for two reasons. First, alarm reporting is an area in which there is great potential for automation and for interoperability between management systems of different vendors. Second, alarm detection is highly specific to each network device and system and thus difficult to generalize.

We first present some general information on alarm management to provide a background for further discussion.

General alarm information

An alarm can be characterized from many different perspectives, for example, its severity, type, and probable causes. Some of those aspects are standardized in that they have been defined in CCITT recommendations and are widely accepted in practice. These common aspects of alarms are introduced first to provide a context since they will be used throughout the chapter.

Alarm category.   Alarms are classified into five general categories according to the causes of alarms as follows:.

- *Communications alarms* are those primarily associated with the procedures or processes in transporting information from one system to another (e.g., loss of signals or frames).

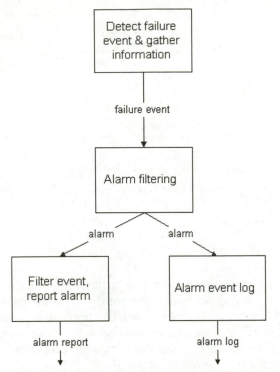

Figure 10.2   An example of the alarm surveillance process.

- *Quality-of-service alarms* are those caused by degradation in the quality of service that have violated certain predetermined thresholds (e.g., call drop rate exceeding the limit).

- *Process alarms* are those that are primarily associated with software or processing faults [e.g., such as central processing unit (CPU) response time exceeding limits and queue size too long].

- *Equipment alarms* are those that are primarily associated with equipment faults (e.g., transmitter failure and multiplexer problem).

- *Environmental alarms* are those that are primarily associated with conditions of the environment in which the equipment operates (e.g., fire alarm, temperature, and humidity of the room where equipment resides).

Perceived severity.   This is a parameter that indicates how much an alarm is perceived to have affected the service of a network or the capabilities of a managed object. Four severity levels representing service-affecting conditions are ordered from the most severe to the least severe: critical, major, minor, and warning. Two additional levels are defined to complement the four severity levels: indeterminate and cleared.

- *Critical.*   Indicates that a severe service-affecting condition has occurred and immediate corrective action is required. For example, when an important object is totally out of service, the immediate restoration of its capabilities is required.

- *Major.* Indicates that a service-affecting condition has developed and immediate attention is required. For example, a severe degradation of the service quality may warrant an alarm of this category.

- *Minor.* Indicates that the existence of a non-service-affecting condition has occurred and action should be taken to prevent it from developing into a major fault condition. For example, a threshold crossing may indicate that the network service either is not being affected at the moment or is being affected to a lesser extent. But the service will be severely impacted if the condition is allowed to continue.

- *Warning.* Indicates an impending or potential fault condition that calls for further diagnosis and possible corrective action to prevent it from developing into a major service-affecting condition.

- *Cleared.* Indicates the clearing of one or more previously reported alarms. An alarm with the cleared severity level clears all previous alarms for a particular managed object that has the same alarm type, probable cause, and specific problems.

- *Indeterminate.* The severity level cannot be determined for one reason or another.

Probable causes. A set of probable causes is identified in CCITT Recommendation X.733 (1992) and listed in App. 10A. Please note that this is not an exhaustive listing and new causes of alarm will be added as network technologies progress.

Alarm detection

Alarm event criteria and indication management. What constitutes an alarm and what does not may vary from system to system and application to application. One of the first tasks of an alarm surveillance system is to set the alarm event criteria by specifying the alarm attributes (e.g., call drop) and assigning the threshold values (e.g., x number of call drops per second). The alarm criteria should be set and organized at an NE in such a way that it can be easily applied to the occurrence of an event to determine whether it is an alarm and that it can be easily reset and queried by a manager.

Another task is to prepare and set the mechanism to indicate an alarm once it occurs. An alarm enters the management domain only after its occurrence is made known to the management application. The agent application at NE must decide the indication mechanism to use, such as audible (e.g., speaker), visual (e.g., alarm lamp), printer, screen display, and notification message, and must set the alarm indicator in the desired mode (i.e., on/off). Note that the notification message of an alarm is different from the alarm reporting from an agent side to a manager, which will be discussed shortly. This is a message from a resource to the managed object representing the resource to indicate the occurrence of an event at the resource. For example, a processor going down may set the alarm indicator by sending a message to the management application. For the purpose of management, all alarm indicators must be resettable by the management application.

Alarm detection mechanisms.   An alarm is often signaled by the transition of a resource state from an enabled to a disabled state or a threshold crossing of counts taken over a designated time period. There are perhaps as many methods and ways to detect alarms as alarms themselves. Therefore, alarm detection mechanisms can best be discussed only in generic terms. Two often-used alarm detection mechanisms are built-in hardware test and check and built-in software test and check. The examples of hardware test and check include sensors and test circuits. Software checks include the software exception handling mechanism.

Alarm reporting function

Alarm reporting is an important management area of alarm surveillance and involves reporting alarms by an agent to a manager application for analysis and decision making. This section introduces the ITU-T defined alarm reporting model and managed object classes for alarm reporting.

Generic event reporting model.   This section describes a generic event reporting model defined in ITU-T Recommendation X.734 (1992). The goal of event reporting is to provide the capability for an agent system on a managed NE to report events that take place at the NE to the interested parties, such as a management application, on demand, on schedule, or as a combination of the two. Specifically, the following functional requirements of the model are specified in the standard:.

- It shall provide the capability for an outside application (e.g., a manager) to register for one or more events.

- It shall allow an outside application to schedule the event reports on both a short- and long-term basis.

- It shall allow an outside application to initiate, terminate, suspend, and resume event reporting (i.e., event forwarding).

- It shall allow an outside application to define and modify the criteria of event forwarding.

- Of course, it shall allow an outside application to specify and modify the destination of event forwarding.

The model description   As shown in Fig. 10.3, the event model consists of three components: a set of managed objects, an event preprocessing unit, and a set of event forwarding discriminators. A managed object is responsible for generating notifications for the events that happen at the network resource represented by the object. The event preprocessing unit is responsible for detecting and receiving the event notifications emitted by the managed objects and forming potential event reports as input to the event forwarding discriminator.

An event forwarding discriminator (EFD) is itself a managed object that provides the bulk of the management capabilities of the event reporting model. For example, it is the EFD that allows an outside application to schedule event forwarding and to specify and modify event forwarding criteria and a destination.

How the event forwarding model works is best illustrated by walking through a scenario. Assume that when the rate of signaling message loss has exceeded a predetermined threshold an alarm is generated. The event forwarding model will behave as follows in generating an alarm:

1. A managed object representing the part of the system responsible for signaling messages, say, a signaling agent object, detects the crossing of the signaling message loss rate threshold and sends a notification to the event preprocessing unit.

2. The event preprocessing unit, after receiving the notification, creates an internal event report out of the notification and sends it on to all the EFDs in the local system by default. An EFD, like any other managed object with state and status attributes, can be in a state that prevents it from receiving an event report.

3. An EFD, after receiving the event report, first checks who has registered for this event and then checks the event forwarding criteria. The event is forwarded to a destination only if the criteria specified by that application are met. Assume that two outside applications have registered for this event, i.e., a call processing and a service management application. Assume further that the criterion for the call processing application is that the signaling

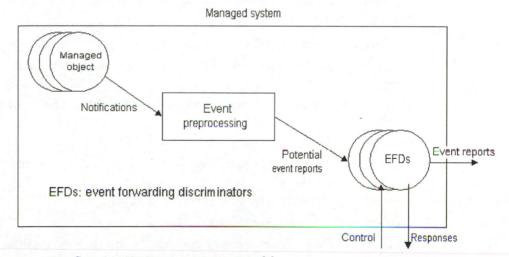

Figure 10.3 Generic event report management model.

messages must be for a call setup and the criterion for the service management application is that the same alarm report has not been reported in the last x number of minutes. Again, assume both conditions happen to evaluate to TRUE. The EFD forwards the alarm report to both applications using the CMIS M-EVENT-REPORT mechanism.

4. Then the EFD performs housekeeping chores: resetting its own state and status if necessary, incrementing the counter on the forwarded event of this particular type, etc.

Note that before this scenario takes place, there are steps for registering events, defining criteria, supplying the destination address, and other steps that must take place first.

Managed object classes.   Out of the three components of the event forwarding model, the managed objects that represent managed network resources vary from resource to resource. The functionality of the event preprocessing unit is relatively simple, and this unit is normally implemented either as part of the managed object or part of the EFD. The focus of this subsection is on the two managed object classes, the discriminator and the event forwarding discriminator.

The discriminator managed object class.   This is the basic superclass of the event forwarding model. It provides the following capabilities to be inherited by the subclasses for event forwarding.

- To allow an outside application to schedule event forwarding
- To allow an outside application to set criteria for event forwarding
- To allow an outside application to control the EFD by setting the administrative state of the discriminator object

Specifically, the discriminator object class has the following attributes to realize the described functions.

- *DiscriminatorId.*   A unique identifier for this discriminator object instance.
- *discriminatorConstruct.*   A filter that specifies the criteria to determine forwarding events.
- *administrativeState.*   Indicates whether the EFD is administratively permitted to operate.
- *operationalState.*   Indicates whether the EFD is operable.
- *availabilityStatus.*   Indicates the status of the EFD in terms of its being ready for operation.
- *startTime.*   The time for the EFD to start operation.
- *stopTime.*   The time for the EFD to cease operation.
- *intervalsOfDay.*   A list of days for which the discriminator will be turned on.

- *weekMask.* Defines a set of mask components each of which specifies a set of time intervals for a day on a 24-hour clock. Together with itnervalsOfDay, this attribute specifies the days of the week and intervals of each day when the discriminator object is scheduled to be operational.

- *schedulerName.* The name of the scheduler that allows the EDF to schedule the event forwarding.

A discriminator object is capable of sending notifications to a managing system. The events that can trigger a notification are

- *stateChange.* When the operational or administrative state value changes

- *attributeValueChange.* When the values of the specified attribute such as discriminator Construct change

- *objectCreation.* When a discriminator object is created

- *objectDeletion.* When a discriminator object is deleted

The EventForwardingDiscriminator (EFD) managed object class. The EFD managed object class is a subclass of the discriminator class and thus inherits all the attributes and notifications discussed above. In addition, it also defines the conditions that shall be satisfied by potential event reports before the event report is forwarded to a particular destination. The additional attributes include

- *Destination.* One or more management entities that will receive forwarded events.

- *activeDestination.* Optionally an active destination, and backup destinations may be specified.

- *backupDestinationList.* This attribute together with activeDestination provides backup destinations for event forwarding in case the original destination is not available for whatever reason.

- *confirmedMode.* A Boolean value to indicate whether this EFD requires a confirmation from the receiver of the event.

Alarm reporting model. An alarm reporting model, based on the event forwarding model just described, is defined in CCITT Recommendations X.733 (1992) and Q.812 (1993).

The focus of the alarm reporting model is on the agent side where there are three components of the reporting model: alarms, alarm information, and the EFD, as shown in Fig. 10.4. Alarms are the managed objects representing alarm conditions that are associated with a network resource. Alarm information, closely mirroring the event processing unit of the event forwarding model, is a logical entity that is responsible for processing alarms before having them sent to the EFD. The examples of processing alarms include assigning alarm severity, setting appropriate alarm states, and even performing some basic filtering to eliminate duplicates.

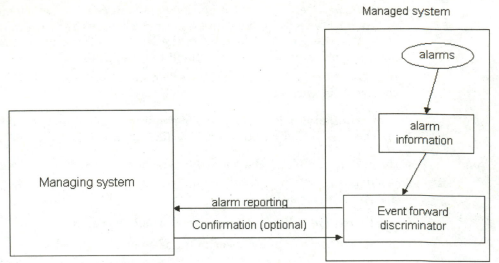

Figure 10.4 Alarm reporting process model.

The EFD is the control center where the bulk of alarm reporting–related functionality resides. It interfaces with the alarm information unit to receive alarms, to set the criteria for alarm preprocessing, and to query for an alarm state or other information. Optionally the alarm information logical unit can also generate the alarm reports. The mechanisms for generating alarm reporting are internal to the managed system and not subject to standardization. It is through the discriminator object that the outside managing systems can schedule alarm reports, set the alarm reporting criteria, and modify and delete the reporting schedule or criteria.

Again we use a scenario to illustrate how the alarm reporting model works. Assume that a line card on a switching system is down and an alarm shall be sent to the managing system as promptly as possible.

1. When the condition that the line card is down is detected, an internal notification is sent to the alarm information logical unit by the line card object represented as the alarm in the model.

2. The alarm-information logical entity collects all the information available to it, which may include alarm type, possible cause, and alarm status, and create an alarm report. There may be multiple EFDs within the managed system. By default, the alarm reporting is passed on to all the discriminators known to the alarm-information logical entity unless a discriminator is configured to reject any alarm report.

3. An EFD, after receiving an alarm report, first checks which managing system has registered for this alarm report and then checks the event forwarding criteria. The alarm is forwarded to a destination only if the criteria specified by that application is met. Assume that two outside applications have registered for this event, i.e., a configuration management application

and a service management application. Assume further that the criterion for forwarding an alarm report is that the alarm severity level is *critical*. In this case, that the line card is down is a critical alarm and the criteria evaluate to TRUE. The EFD forwards the alarm report to both applications using CMIS M-EVENT-REPORT mechanism.

4. Then the EFD performs housekeeping chores: resetting its own state and status if necessary, incrementing the counter on the forwarded event of this particular type, etc.

5. Once a managing system, say, a network configuration management application, receives the alarm report, it makes appropriate management actions such as modifying its resource configuration map to reflect the changes and switching to the backup unit.

It is required that alarm reporting be available on demand and on a scheduled basis. The behavior of the alarm reporting model described above is sufficient for the on-demand reporting. The schedule-based reporting will require the scheduling capabilities described in the following subsection.

Alarm reporting control and scheduling. A managing system shall be able to control alarm reporting by setting the event forwarding criteria and the values of the state and status attributes of the EFD managed object. As shown in Fig. 10.5, the control information is exchanged in the same way as the alarm reports are.

At a minimum, a managing system shall be able to suspend and resume alarm reporting. In addition, it may be desired to allow a managing system to initiate and terminate alarm reporting and to set or retrieve the EFD managed object. A managing system initiates alarm reporting by the following general steps.

- Create an EFD object and initialize the appropriate state values such as administrative state.

- Set the event forwarding criteria.

- Set the destination attributes of the created EFD object to the address of this managing system.

In the other direction, a managing system can terminate the alarm reporting by deleting the associated EFD if it is not shared by other managing systems. Otherwise, the same effect can be achieved by removing the address of this managing system from the destination list of the associated EFD object.

Again, we use a simple scenario to describe the behavior of the model in Fig. 10.4 and to illustrate how a managing system suspends the alarm reporting.

1. The managing system sends the CMIS service primitive M-SET to the managed system to request the attribute administrative state of the EFD object be set to *locked*.

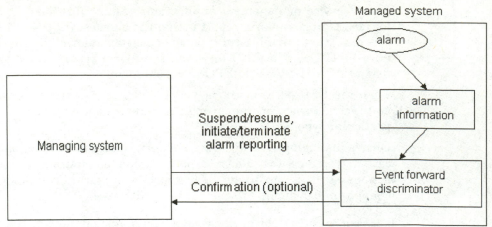

Figure 10.5   Alarm reporting control functions.

2. The managed system sets the attribute value as requested and then sends back a confirmation to the managing system to report the success of the operation.

The effect of the operation is that the associated EFD will reject any alarm reports sent to it by other function units like an alarm-information logical entity within the managed system.

Alarm summary reporting.   A large volume of alarm information can be a heavy burden on the connection between a managing system and a managed system. Thus it is desirable that a managing system receive a summary of alarm reports rather than the bulky reports themselves. Some additional capabilities are required to accomplish this. First, a managing system and a managed system must agree on the summary report format and contents, and additional capability is required of a managed system to generate an alarm report summary. Second, an additional set of capabilities is required to control and schedule summary reports.

Model description.   As shown in Fig. 10.6, in place of the alarm EFD is a currentAlarmSummaryControl object and replacing the alarm-information logical unit is a managementOperationSchedule object. The currentAlarmSummaryControl object, as the object name suggests, is responsible for generating summary reports according to the criteria specified by the managing system. The summary report contains a set of object instances that provide summary information on the alarm reports of interest to the managing system. Included also are the probable causes, alarm status, and perceived severity of each alarm included in the summary report. The other managed object, managementOperationsSchedule, allows a managing system to schedule the summary report.

The currentAlarmSummaryReporting managed object class.   As in the case of alarm reporting, it is a requirement that a managing system should be able to control the alarm summary reporting: initiate, terminate, suspend, resume, and route the report to a particular destination. All these are achieved via the managed object current AlarmSummaryControl.

The currentAlarmSummaryControl/managed object class. CurrentAlarmSummaryControl is a support object class that allows a managing system to specify three different categories of criteria for selecting object instances with an outstanding alarm to be in the current alarm summary report. This object class has the following attributes:

- *currentAlarmSummaryControlId.* An identifier to uniquely identify an instance of the CurrentAlarmSummaryControl object class.

- *alarmStatusList.* Specifies criteria for an alarm object to be included in a current alarm summary report. When this list is set to null, it means that the alarm status is not a criterion for object inclusion.

- *objectList.* A set of object instances that are included in the report.

- *perceivedSeverityList.* Also specifies criteria for an object with an alarm to be included in a current alarm summary report. It consists of a set of possible perceived severities as defined earlier. An object must have an outstanding alarm with the perceived severity level matching one of those specified in this list. If this list has a null value, it means that perceivedSeverityList is not a criterion.

- *probableCauseList.* Specifies the criteria for an object with an alarm to be included in a current alarm summary report. It contains a set of probable causes as defined earlier that are used as a filter to screen out any object whose alarm's probable cause does not match any of the ones in this list. If this list has a null value, it means that the probable cause list is not used as a filtering criterion.

This current alarm summary object class has the notification currentAlarmSummaryReport, and the action retrieveCurrentAlarmSummary, which are described as follows.

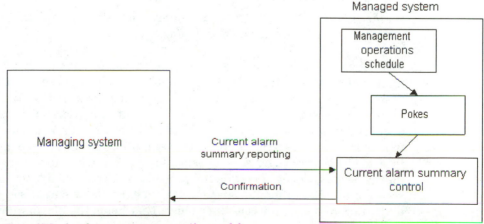

Figure 10.6 An alarm summary reporting model.

The notification consists of a set of object instances that have outstanding alarms and meet the selection criteria just described. The object instances are sent to a managing system along with their alarm information such as alarm types, perceived severity levels, and probable causes. The notification can be sent on a demand basis or a scheduled basis.

The action retrieveCurrentAlarmSummary is to support a managing system to retrieve current alarm summary reports as opposed to passively receiving a report from a managed system. The action allows a managing system to specify a set of criteria for selecting the alarm summary reports.

The ManagementOperationsSchedule managed object class.   Management-OperationsSchedule is a support managed object class that allows a managing system to schedule management operations such as reporting the current alarm summary report and to specify the destination of an alarm report. It has the following attributes:

- *administrativeState.*   Allows a managing system to suspend or resume the scheduling operation. See the Managed Object Class Dictionary (App. D) for a detailed definition of this attribute.

- *affectedObjectClass.*   Identifies the object class that is affected by this scheduled management operation.

- *affectedObjectInstances.*   Identifies the object instances to be affected by the scheduled management operation.

- *beginTime.*   Indicates the starting time for the management operation.

- *destinationAddress.*   Identifies the destination to which selected alarm reports will be sent. The address can be a management application entity name or an address such as a network address.

- *endTime.*   The termination time of the management operation.

- *interval.*   The time interval of the management operation.

- *operationalState.*   The operational state of this managementOpeations-Schedule object instance. See the Managed Object Class Dictionary (App. D) for a detailed definition.

- *scheduleId.*   An identifier uniquely identifying an instance of the managementOperationsSchedule object class.

Alarm logging.   A managing system may choose to have alarms logged at the managed system instead of or in addition to having them sent to the managing system. A general alarm log control model is shown in Fig. 10.7. Inside the managed system are the following components for the alarm logging: event, event information, log, and alarm record. The event represents the source from which an alarm is generated. The event information is responsible for the intermediate processing that collects and filters alarm information and creates alarm records. The log object is the control center responsible for logging alarm records: initiate, schedule, suspend, resume, or set size of alarm records. The alarmRecord object is a container object that holds the alarm information.

We use a scenario to describe the behavior of the logging model. Assume that a line card is down and the alarm generated for this event is to be logged as an alarm record. In general terms, the alarm-logging model behaves as follows.

1. A network resource object (e.g., line card or a network interface) generates an alarm in response to the occurrence of a fault condition and sends the alarm to the event information.

2. The event information, as a logical entity whose implementation is a local matter, assigns the alarm type, severity level, and probable causes; filters out duplicates; and performs other processing as required. Optionally, this entity may create an alarm record using the collected alarm information.

3. Then it is the responsibility of the log object that decides whether an alarm record meets the logging criteria and, if yes, when the alarm record shall be logged. In addition, the log object also decides what action to take when the maximum log size is reached and what size the alarm log record should be. The log-full action, maximum log size, and logging criteria should be settable by a managing system either at initialization time or during the alarm recording logging.

4. The log object logs the alarm records either on schedule or on a demand basis. A notification is sent to an appropriate managing system, if so desired.

The alarmRecord managed object class. The two managed object classes for the alarm record logging are alarmRecord and log. Since the log managed object class has already been introduced in Chap. 9, this subsection describes the alarmRecord managed object.

The managed object class alarmRecord represents logged information that resulted from alarm notifications or event reports from a network resource

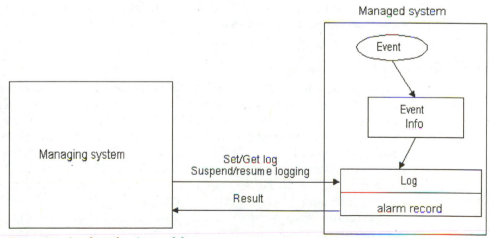

Figure 10.7 An alarm logging model.

object. Note that this object does not have any notifications or actions since this is a data object. The attributes found in this object class include

- *probableCause.*   One of the probable causes for the alarm, as defined in the alarm information subsection earlier.

- *perceivedSeverity.*   The perceived severity level: critical, major, minor, or other as discussed earlier.

- *specificProblems.*   A set of specific problems that have caused an alarm, each of which is represented as an object identifier. This information augments the probable cause of the alarm by providing more detailed information.

- *backedUpStatus.*   A Boolean value to indicate whether to back up the alarmRecord object.

- *backupObject.*   The object instances to be backed up.

- *trendIndication.*   Indicates the direction of the change of the alarm, for better or worse, compared to the previous status of the alarm: lessSevere, noChange, or moreSevere.

- *thresholdInfo.*   Provides the specific information on the threshold, e.g., threshold level, attribute IDs, and observed values.

- *stateChangeDefinition.*   The attribute change that caused the alarm, i.e., attribute ID, the old value and the new value of the attribute.

- *monitoredAttributes.*   A set of attributes that were monitored for potential faults.

- *proposedRepairActions.*   A set of identifiers for the proposed repair actions.

In summary, this section first provides some general information on alarms, i.e., alarm categories, alarm severity, and probable causes of alarms. Then alarm detection is discussed in general terms. The bulk of this section is devoted to alarm reporting, including the ITU T–defined event reporting model, alarm reporting model, alarm summary reporting model, and alarm logging.

Fault Localization

The responsibility of fault localization is to analyze filtered alarms, diagnostics, and other conditions of a fault to identify the root causes. This assumes that the initial fault or failure information is insufficient to determine the cause of the failure. Additional internal or external tests may be conducted to obtain additional information. Although each individual fault may have some unique condition requiring special processing on the part of the fault management system, it is the combination of fault analysis and diagnostic testing that is the generic approach to fault localization.

The relationship between alarm analysis and test management that is responsible for diagnostic tests is shown in Fig. 10.8. Alarms are input into the fault analysis from the alarm surveillance. The results of diagnostic

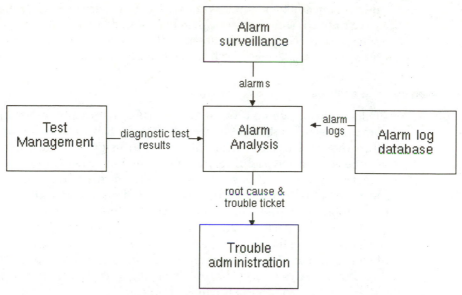

Figure 10.8 A logical view of the fault localization process.

testing and the alarm log database feed into alarm analysis. The root cause of a fault is directed to the fault administration for determining the repair action.

Diagnostic tests

This subsection describes diagnostic tests in generic terms, and a detailed discussion on the test model and six different types of diagnostic tests is given in the following section. Diagnostic tests can be of a variety of types and forms. In general, they fall into hardware equipment diagnostic tests, software checks, or a combination of the two. Hardware testing examples include circuit testing, segment testing, and path selecting. Equipment tests are often intrusive, meaning that the equipment being tested cannot perform its normal function while the test is being conducted.

Examples of software-based checks include audits, checksums, operational tests, and program traps. An audit may compare redundant copies of data within a system, or it may compare copies of data in separate systems. An audit failure may discover the point of fault occurrence. An audit can run on schedule or by request. A checksum is an integer whose value depends on the total contents of a physical unit of software. If a program is patched or data are altered, the checksum changes. The checksum is used to detect any change to the data or program received. An operational test can be applied to programs. A program is executed and a controlled set of inputs is supplied to the program. The output is observed and compared against the expected values. A program trap helps to locate a fault in the design of a program by reporting the path taken through a branch point.

A decision needs to be made whether the fault affects services and what services are impacted, once a root cause has been determined. If it is decided that a service is impacted, service status information needs to be updated in the customer database, the subscription database, etc.

Alarm correlation and analysis

Analysis is performed on filtered alarms, diagnostics, and other information of a fault to determine the root cause. Often a fault is not isolated at one NE but is interconnected with multiple alarms that originated at other NEs. A networkwide correlation and filtering of alarms is needed to remove redundancy and narrow the range of possible causes. Alarm correlation is a means to filter out a large volume of alarm messages and eliminate marginally meaningful, duplicate, or unrelated information. Knowledge from other functional units may be needed to perform this function. This will also prevent such disasters from happening as flooding the network with often prohibitively large volumes of alarm messages. This is a critical step in determining the root cause of a fault. Alarm analysis uses the result of alarm correlation in order to reason about and determine the root cause of a fault.

Several commonly used alarm correlation and analysis methods are introduced below. This is not an exhaustive listing.

Causal relationship model approach. This is a commonly used approach to alarm correlation that consists of four key elements: a network configuration information repository, a causal relationship model, an observed alarm repository, and a correlation engine, as shown in Fig. 10.9. The network configuration repository contains information about managed network objects representing managed network resources and relationships between the objects. Information in the repository can be dynamically updated as changes in network configuration take place. Feeding off the network configuration information is a causal relationship model that describes various possible faults in the system and the causal relationships of the faults with reference to the network configuration information. The causal relationships can be represented in rule, tree, or graph structures or as a finite state machine. The correlation engine applies the causal relationship to the collected alarms stored in the observed alarm repository and arrives at a decision as to the root cause of a fault. For more details on this approach, see Gardner and Harle (1996).

Codebook approach. This is a simplified variation of the above causal model approach. One potential problem with the causal relationship model approach is that the number of potential network faults can be very large, which thus renders the causal relationship model extremely complicated and inefficient to implement. According to the codebook approach, a minimum set of possible network faults is encoded called *codebooks*. The codebooks are kept as small as possible without losing the ability to distinguish the network faults. The

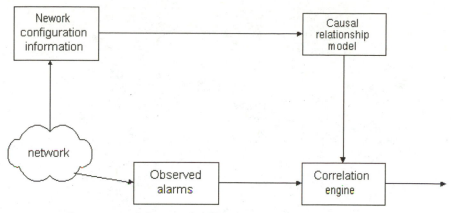

Figure 10.9 Components of the causal model approach to alarm correlation.

observed network alarms, on the other hand, are converted into a vector for each access and efficient processing. The correlation engine reasons the root cause of a fault by referencing the codebook. For an application example, see Kliger and Yemini (1994).

Case-based reasoning approach. Case-based reasoning is a technique developed in the field of artificial intelligence for problem solving. The basic idea is to use past experience to solve a current problem, that is, to determine the root cause of a fault in our case. Each of the past problem-solving experiences, including a problem description, solution, and the reasoning process, is stored as a problem-solving episode or a case. A current problem is solved by finding a stored case that most closely matches the current problem and then adapting its solution to the new problem. An application example of the case-based reasoning approach for alarm correlation and analysis is reported in Gardner and Harle (1996).

A case-based reasoning system for alarm correlation and analysis, or for that matter, a generic case-based reasoning problem-solving system has the following four basic steps that constitute a cyclic process, as shown in Fig. 10.10. Normally a current alarm is input into the system. The step *retrieve* is to retrieve a set of previous cases that closely match the current problem based on certain matching criteria. The next step is to *reuse* a case to suggest a solution as to the root cause of the current fault. If the suggested solution fits the current problem perfectly, skip over the next step. Otherwise, the next step is to *revise* the suggested solution to better match the fault at hand. Once the root cause of the current fault has been determined, it becomes a new case to be *retained* in the knowledge base for future use, the last step in the problem-solving process.

Rule-based approach. Rule-based systems are widely used in fault management and alarm correlation in particular. There are two components generally seen in a rule-based correlation and analysis system: a knowledge base and an inference

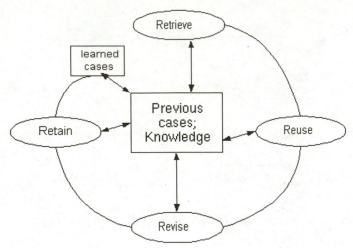

Figure 10.10 Case-based reasoning approach to alarm correlation and analysis.

engine. The knowledge base is a repository for all knowledge about the fault represented as rules. A rule is expressed in the form of an IF part and a THEN part. The IF part expresses the condition or conditions to be met in order for the rule to be executed or *fired*. The THEN part specifies the actions to be carried out when the rule is fired. Only one condition and one action are specified in the IF and THEN part of the example rule shown here:.

IF the alarm severity level is Major,

THEN check the state of the card that generated the alarm.

Multiple rules can be chained together to provide inference capability for determining the root cause of a fault. The inference engine contains the knowledge specific to the alarm correlation and analysis that provide inference control functions: determining which rule to be fired and in what order. It normally uses a heuristics approach to search the knowledge and decides which rule to fire, in what order, and whether an acceptable solution has been reached.

Test Management

Testing is an important part of general network operation and maintenance as well as fault management. This section discusses diagnostic tests with respect to fault localization.

Levels of testing

Testing management can be broken down into service, network, and network element levels. Each level has different tasks. This subsection provides an overview of test management at each management layer.

The tasks of service management layer testing include determining the testing strategy and designing a service feature combination for testing in order to determine whether a service feature or a set of service features is working properly. To design service feature combinations, knowledge of the dependencies and interactions between services is essential. This is a very complicated, tedious, and largely manual process that is yet to be automated and standardized.

Once a request for testing is received from either a network operator or other fault management functional unit like fault localization, a test is arranged. There are several steps involved in testing at the network management layer:

- A systematic test of segments of a circuit is arranged to determine which segment caused the fault.

- An appropriate test suite is selected for the test required.

- Results from each segment test are correlated and reconciled if necessary to derive an overall conclusion.

Like the service level testing, few standards exist for network level testing. The bulk of ITU-T recommendations has been devoted to test management at the network element layer, and that is the focus of the remainder of this chapter.

Generic test model

An ITU-T test management framework, as defined in ITU-T Recommendation X.745 (1993), specifies a set of test models, support managed object classes, and test functions. The goal of the test models is to automate the test process by the remote test control, using the OSI management framework. The test models mainly apply to the network element level testing.

Test model. The simplest test model, shown in Fig. 10.11, involves two application processes, i.e., a test conductor and a test performer. A test conductor is a managing process that initiates a test, controls and monitors the test operations, and determines the conditions to terminate the test. A test performer is an agent process that performs actual test operations requested by the test conductor.

The following terms are important for understanding the test models to be described.

- *Managed object referring to test (MORT).* Those managed objects representing the network resources being tested, or test target objects.

- *Test action request receiver (TARR).* A capability or functionality to receive a test action request and distribute the request to the concerned managed objects to perform the requested test operation. Either embedding this capability into another managed object or modeling it as an exclusive managed object is a local matter.

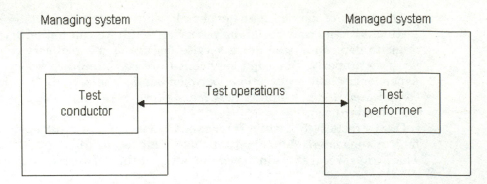

Figure 10.11 Generic test model.

- *Test conductor.* As said, this is a functional entity at the manager side responsible for initiating, controlling, and monitoring a test remotely. The implementation of this entity, which is a local matter, can take on a variety of forms: a single managed object class, a group of managed object classes, or a subsystem, all depending on the complexity of the test management tasks involved.

- *Test invocation.* One test instance. Each test invocation has a unique identifier for identifying associated test objects, test results, and other related information. Either a test conductor or a test performer can assign the test invocation identifier.

- *Test object (TO).* A managed object that exists exclusively for a controlled test invocation. A managed object has attributes, operations, and notifications pertaining to that instance of testing. It is mainly responsible for the arrangement of the MORTs to execute a test and hold the test results returned from the MORTs.

- *Test performer.* The agent-side application that is responsible for receiving the test request and initializing a test invocation.

A set of test-related terminology can be found in Table 10.1.

Uncontrolled tests. An uncontrolled test, as modeled in Fig. 10.12, is not subject to monitoring and control during the course of the test. Once invoked, the test runs its predetermined course and reports the test results at the completion of the test. An uncontrolled test is modeled by using one or more MORTs and a managed object containing TARR, as shown in Fig. 10.12.

A general course of events for an uncontrolled test is listed as follows:

1. The test conductor on the manager side sends a request to the test performer at the agent side to test a software module using CMIS M-ACTION. The following test input information is included in the request:
   - The identification of MORT(s)
   - The type of test to be performed

- A timeout period
- The identity of one or more support objects
- A test session identifier

2. The test performer directs the request to the managed object with TARR capability (managed object with TARR).

3. The managed object with TARR has the knowledge of which managed objects are relevant to this test and in what way and invokes a test suite appropriate for the object to be tested.

4. Once the test is completed, the test performer sends either the test result or status to the test conductor via a CMIS notification mechanism by instantiating an appropriate EFD in the test performer. The information returned to the test conductor for a successful test includes the test invocation identifier that identifies the particular test. A failure response will include a failure indication and information about causes of the failure.

The results of the test are reported in one or more confirmations to the test request. The final confirmation indicates that the test was completed by including the test result. The test result may take on one of the following values: pass, fail, inclusive, timed-out, or premature termination. An uncontrolled test cannot be suspended and resumed by a request from the test conductor and can only terminate spontaneously, i.e., at its completion or by a fault.

Controlled tests. The major difference between a controlled test and an uncontrolled test is that the controlled test is subject to controlling and monitoring actions initiated by a test conductor while the test is running. An example of a controlled test is shown in Fig. 10.13, and a possible sequence of events for a controlled test will further illustrate the point.

1. The test conductor sends a test request to the test performer using CMIS M-ACTION. The following test input information is included in the request.

- The identification of MORT(s)
- The type of test to be performed

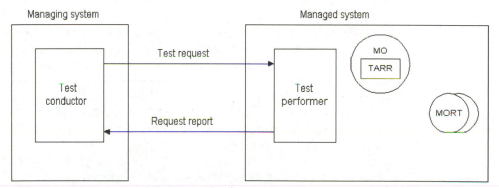

Figure 10.12 An example of an uncontrolled test.

- A time window in which the test conductor wishes the test to be performed
- The identity of one or more support objects
- A test session identifier
- Indication of whether one or more tests are being requested
- The class and, optionally, names of required TO(s)
- The initial attribute values of the TO(s)

2. The test performer creates a test instance by creating a unique test invocation identifier and instantiating a set of TOs, for this test invocation, each of which is identified by the test invocation identifier and its own unique object identifier. The test performer will provide the TOs with the start and stop time and initial parameter values.

3. During the course of the test, the test conductor sends controlling or monitoring instructions (e.g., collecting data, suspending or resuming test) and the instructions are directed by the test performer to the managed object with TARR capability.

4. The managed object with TARR distributes a specific action request to a TO, based on the responsibilities of the TOs.

5. The TOs can direct the MORT in question to carry out the action. For example, if the action request is to collect data on a circuit under test, the TO can ask the circuit object representing the circuit for the data.

6. The results of the test are stored as attributes of the TO. The results can be sent to the test conductor via CMIS notification by instantiating an appropriate EFD in the test performer. The test results, either intermediate or final, may also be stored as a log in the test performer. This is a local decision for the application developer. The information returned to the test conductor for a successful test also includes

- Distinguished names of the TO(s)
- Test invocation identifier(s)
- Optional initial attribute values of the TO(s)

A failure response will include a failure indication and information about causes of the failure.

For a controlled test, the test result reporting can be either solicited or unsolicited. The results may be requested of the TO by the test conductor (solicited) or emitted as notification from the TO (unsolicited). In the case of unsolicited reporting, the results are provided by one or more notifications from a TO. The final notification from a TO is indicated by the presence of the test outcome parameter in the notification. Attributes of a TO may be defined to hold test results in place of or in addition to emitting notifications of test results to the test conductor.

A controlled test can be suspended or resumed by a request from the test conductor. Suspending or resuming a test means suspending or resuming all TOs associated with the test. The test invocation identifier or a test session

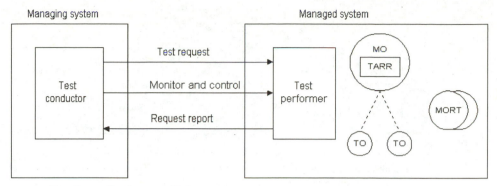

Figure 10.13 Example of a controlled test.

identifier identifies the TOs associated with a test. If a test session or invocation identifier is provided in the suspense or resume request, all TOs identified by the session or invocation identifier must be suspended or resumed; otherwise an error is returned.

A controlled test can terminate spontaneously or by a request from the test conductor. If a test session or invocation identifier is provided in the test terminate request, all TOs identified by the session or invocation identifier must be terminated; otherwise an error is returned.

Compound tests. A test conductor may request that multiple test invocations be initiated concurrently, sequentially, or in a combination of the two in order to meet a user requirement. Figures 10.14 and 10.15 show recursive and parallel uses of the test models, respectively.

Test information model. This subsection describes the test-related managed object classes just mentioned in more detail. As mentioned before, no distinction between mandatory and conditional attributes, notifications, and actions will be made, largely because this is viewed as a decision by the application designer.

The testActionPerformer managed object class. This is the object with TARR capability and is responsible for carrying out the test operations on the agent side.

- *testActionPerformerId.* A unique identifier for an object instance and the distinguishing attribute for this object class.

- *supportedTOClasses.* A set of object identifiers that specifies the TO classes supported by this object in case of a controlled test.

- *supportedUncontrolledTests.* A set of object identifiers that specifies the specific uncontrolled test supported by this object.

This object class models the test performer as described above. The core of the test performer is a set of actions for controlling and performing a test, as described below.

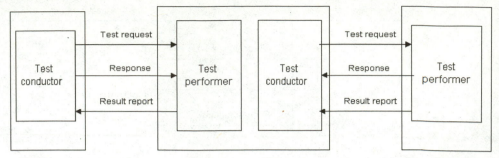

Figure 10.14   Recursive use of the test model. (*From ITU-T Recommendation X.745,* 1993.)

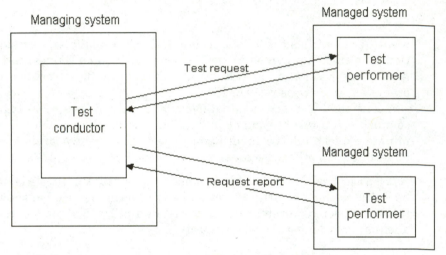

Figure 10.15   Parallel use of the test model (*From ITU-T Recommendation X.745,* 1993.)

- *testRequestUncontrolledAction.* This test performer invokes an uncontrolled test after receiving a request for this action. Included in the action request are test session ID, to-be-tested MORTs, the associated objects (AOs) for this test, and test timeout period. The action reply should contain the test outcome, tested MORTs, proposed repair action, and additional information specific to this test action.

- *testRequestControlledAction.* The test performer invokes a controlled test after receiving a request for this action. Included in the action request are the controlled test request type (i.e., independent or related), test session ID, to-be-tested MORTs, the AOs, the test timeout period for this test, and a list of TOs. The action reply contains (???)

- *testSuspendResumeAction.* The test performer either suspends or resumes the indicated test after receiving a request for this action, depending on the indicated choice (i.e., suspend or resume). The action reply contains the invocation ID of the suspended and resumed test and the states of the associated test objects.

■ *testTerminateAction.*   When this action request arrives from the test conductor, this testActionPerformer object shall terminate the test or tests specified in the request. The associated TOs, if any, may be returned in the action reply to the test conductor. The action reply contains the invocation ID of the terminated test.

The testObject managed object class.   The testObject class is defined for a controlled test, and instances of this object class are created to hold information related to such a test. A request to resume, suspend, or terminate a test is directed to the object with TARR functionality. However, a request to abort a test may be directed to the associated TOs, because the TO is in possession of the information needed to reply back to the test conductor and can emit a notification to report the test result. The information found in a TO includes the test invocation ID, the operational and procedural state of the test, the test session ID, the MORTs of the test, the start and stop time of the test, and the test outcome.

The schedulingConflictRecord managed object class.   This object class is a subclass of eventLogRecord, i.e., a type of log record that represents the information logged as a result of receiving a test scheduling conflict event report. The information included in the record ranges from test invocation ID, scheduled start and end time of the test, test session ID, and actual start and stop times.

The testResultsRecord managed object class.   This managed object class represents the information logged as the result of receiving a test results event report. Included in the result record are test invocation ID, the MORTs of the test, test session ID, and test outcome. The test outcome has five possible values: pass, fail, time-out, premature termination, or inconclusive.

Note that the managed objects only model the test management functionality on the agent side, and the test-related object on the manager side, which is relatively straightforward, is left to the application developer. Table 10.1 lists the commonly used terms in test management that are important for understanding the ITU-T defined test management model.

Diagnostic tests

The diagnostic test is based on the test model defined in ITU-T Recommendation X.745 (1993) that has just been discussed above. In brief, there are three entities involved in a test at the side of a managed system. Once a test request arrives from a managing system, first there is a test performer that is responsible for receiving the request, conducting the test, and returning the test results. These capabilities of the test performer are referred to as TARR functionality. The second entity is the managed objects that are subjected to a test, or MORT. A third entity, which may be optional, depending on the choice of test type, is one or more TOs. There is a choice of either a controlled test or an uncontrolled test. In a controlled test, TOs are created to monitor and control the test as it proceeds. In an uncontrolled test, no control and monitoring are exercised while the test is in progress, and the test result is returned at the end of the test.

TABLE 10.1   Test Management–Related Terms

Term	Definition
Associated object (AO)	Managed objects, distinct from an MORT, TO, and managed object with TARR capability, that are involved in a test
Controlled test	A test that is subject to control and monitoring by a test conductor before the test terminates. The control and monitoring are carried out by one or more special-purpose managed objects called test objects (TOs)
Managed object referring to test (MORT)	A managed object that serves as a target for testing the specified functionality
Test action request receiver (TARR)	An ability of a managed object to act upon a test request. It is often embedded in a managed object as part of the object behavior
Test conductor	A manager that controls the test procedures and initiates test operations
Test object (TO)	A managed object that exists only for a controlled test invocation. A TO is needed only in a controlled test
Test pattern	The signals or data to be applied to a communication path
Test performer	An agent that receives and carries out the test operations
Uncontrolled test	A test that is not subject to monitoring and control once the test commences

Diagnostic tests are important means for determining the root cause of a fault. Six categories of commonly seen diagnostic tests have been identified in ITU-T Recommendation X.737 (1995).

- *Connection test.*   To verify the connection between two known endpoints
- *Connectivity test.*   To verify the connectivity between two entities
- *Data integrity test.*   To determine whether or not the data exchanged between two entities is corrupted
- *Loop-back test.*   To determine the time it takes for two entities to exchange data over a given connection
- *Protocol integrity test.*   To determine whether or not two entities can exchange data using a specific protocol
- *Resource boundary test.*   To test the behavior of a network resource by observing and controlling the interactions between the resource and its boundaries

Two other categories of tests, also defined in ITU-T Recommendation X.737 (1995), are relatively simple and thus are omitted from the following discussion. They are the resource self test to determine whether or not a resource can perform its normal function and the test infrastructure test to verify the capability of a managed system to respond to a test request.

Correspondingly, six managed object classes are defined, for the six types of services, all of which are derived from the testObject as described above. All these test objects have the capability to emit a notification once the test is completed to return the test results. Generally, the parameters that are returned in the test result include the identity of the MORT, i.e., the managed object that has been tested, the test outcome (success, fail, premature termination, requested termination, or inconclusive), and information specific to each type of test.

Connection test. The connection test tests a communication path, and thus the MORT(s) represent the path to be tested. The two AOs represent the resources at the ends of the communication path that drive signals into and receive signals from the communication path. The TARR functionality is implemented in the AOs. The test patterns represent the signals or data to be applied to the communication path.

The connection test can be intrusive and nonintrusive to both the test subjects (MORTs) and to the AOs. If the test is intrusive, the administrativeState and availabilityStatus may need to be supported and to be set to a proper state before the test may be initiated.

A connection test may optionally have a direction applied to the MORT, a test pattern, test duration, test threshold, and reporting interval time. A test pattern is test data to be applied to the communication connections, and the test direction indicates the direction of data flow from the test subject MORT. The test threshold defines the error criteria for terminating the test.

The events that can trigger a test result reporting include completion of the test plan, expiration of the specified test interval period, a time-out, and reception of a request from a test object if it is a controlled test. Similarly, the events that can trigger termination of a test include the completion of the test plan, failure of the test due to a fault condition, expiration of test interval period, and reception of a termination request from the managing system if it is a controlled test.

The attributes found in the managed object class dataIntegrityTestObject include

- *mORTs.*   A set of object instances representing the network resource to be tested.

- *associatedObjects.*   A set of objects that are not directly targeted by the test but whose involvement is required for the test to be carried out.

- *testPattern.*   The data to be directly applied to MORTs for testing purposes.

- *connectionTestResult.*   The test results to be returned to the requester, which may optionally include test duration, error ratio in the test, and test direction made on the MORT.

- *timeOutPeriod.*   The value for the time-out period, measured from the initiation of the test. When the execution of a test exceeds the time-out period, the test is terminated.

The ConnectivityTestObject managed object class.   The goal of the connectivity test is to verify that connectivity can be established between two entities that are represented by a MORT and an AO. There are two ways of verifying connectivity. One way is to establish a connection between a MORT and an AO in case a connection-oriented protocol is used. The other way is to exchange data units between the two managed objects when a connectionless protocol is used. The time it takes to establish connectivity, called the establishment time, is measured from the issuing of a request by a MORT to the reception of a response from the corresponding AO. A time-out period may be set for the test, and if the establishment time exceeds the specified time-out period, the test is considered to have failed.

The connectivityTestObject managed object is responsible for controlling the connectivity test. Among its attributes, the MORT represents the entity that initiates the establishment of the connection with the associated object. The connectivityTestResults and connectivityType must be included as part of the parameter of the testResultNotification that reports the test result back to the managing system. If the timeoutPeriod attribute is present and expires before the test outcome can be concluded, the timeoutPeriod will be returned in the test result with the test outcome set to TIMEOUT.

Data integrity test.   The data integrity test verifies the integrity of the data exchanged between two entities and measures the time for the exchange. In this test, a MORT represents a managed object that can transmit data and an AO is the managed object that receives the data from MORTs and may respond by sending the reply data. The test is configured as such that the MORT transmits data to an AO and the AO, upon receipt of the data, sends the copy of the same data back to the MORT. The MORT compares the received copy against the original one to detect any data corruption. The MORT measures the time interval between the time the data is first sent out and the time when the copy of the data loops back to the MORT. The data can be transmitted in either a connection-oriented or connectionless fashion.

The data integrity test is intrusive to both the MORT and the AO: the normal operations of the objects must be interrupted to perform the test. The data integrity test will abort if the MORT or AO is not in the appropriate state or cannot be placed in the appropriate state for testing. In addition to the parameters that are required for all tests, the other required parameters include the identities and any related information of the AO and MORT, and the time-out period.

The test result reporting depends on the choice of the controlled or uncontrolled test model. In the case of the uncontrolled model, the reporting can be accomplished by means of one or more CMIS action replies while the CMIS M-ACTION is used by a managing system to invoke the test action on the managed system. When the controlled test model is used, solicited reporting is accomplished by means of the CMIS M-GET request while the unsolicited result reporting is realized via the rest result notification.

The attributes found in the managed object class dataIntegrityTestObject include

- *mORTs*.   The managed object that transmits data to the associated object for testing the integrity of the data.

- *associatedObjects*.   The managed objects that receive the data from MORTs and may respond by sending the reply data.

- *dataIntegrityResult*.   The result of the integrity test that may optionally include connection establishment time, original data and corrupted data, test period, and test threshold.

- *actualStartTime*.   The absolute time when the test commences.

- *actualStopTime*.   The absolute time when the test terminates.

- *timeoutPeriod*.   A time-out period set for a test and measured from the initiation of the test. When the test exceeds the specified time-out period, the test is terminated and the time-out is returned as part of the test result.

- *dataIntegrityType*.   Indicates the type of network connection, i.e., connection-oriented or connectionless, on which the data integrity test is carried out.

- *testThreshold*.   The error threshold for terminating the test.

- *testPattern*.   the data to be applied to the test target objects or MORTs.

Loop-back test.   The loop-back test is used to verify that data can be sent and received over a communication path within a specified loop-back time-out period with an acceptable error rate. The loop-back can be done in a variety of ways: physical loop-back, transparent physical loop-back, payload physical loop-back, or echo data back, etc. A loop-back test may be performed over a communication path that is connection-oriented or connectionless. In general terms, a loop-back test operates as follows.

1. The test conductor sends the test performer a request (e.g., M-ACTION) that includes the specification of the MORTs for the test. The MORTs represent the resources that are being tested by the loop-back test, i.e., the part of a communication path. Also specified are the AOs that represent the network resources that provide the loop-back points or locations. Included in the request are the loop-back parameters such as data type for the test, test start time, reporting interval, error interval, error threshold, and test time duration.

2. For controlled testing, the object with TARR functionality may create a TO, one for each loop-back test.

3. The test performer starts the test after the setup time, if any. The test conductor may specify the setup time for establishing the test environment and setting up the test itself.

4. The intermediate test results may be sent to the test conductor if so required.

5. The test performer causes the test to execute. Failure to receive the loop-back data within the loop-back time-out period or an error rate that exceeds the specified threshold will cause the test to fail.

The loop-back test can be intrusive or nonintrusive to both the MORTs and AOs. In the nonintrusive mode, the network resources are used as the MORTs and AOs whenever they are available and happen to be in the state appropriate for the test. In the intrusive mode, the loop-back test will abort if the MORTs or AOs are not in the appropriate state or if either of these cannot be placed in the appropriate state for testing.

In reporting the test results, the loopbackResults, loopbackType, and testPattern attributes must be included as part of the monitoredAttributes parameter of the testResultNotification attribute. The testThreshold attribute should be included if it is present in the loop-back test object. The errorRatioReportType attribute indicates in what form the test conductor desires the error ratio be reported, whether in the form of a number of error bits or in percentage of error seconds.

The attributes found in the managed object class loopbackTestObject include

- *mORTs.*   Identifies a part of the communications path that is to be tested.

- *associatedObjects.*   Identifies the loop-back points.

- *loopbackResults.*   The result of the loop-back test that may contain the duration of the test, the data units used in the test, the error ratio, and the error cause if any error was experienced.

- *loopbackType.*   The type of loop-back, for example, physical loop-back, echo, analog, digital, nontransparent physical loop-back, transparent physical loop-back, payload physical loop-back, and echo data back.

- *testPatterns.*   The test data used in the loop-back test.

- *errorRatioReportType.*   Specifies the type of error measurement such as number of error bits and percentage of error seconds.

- *timeoutPeriod.*   The time-out period set for the test measured from the initiation of the test. When the test exceeds the specified time-out period, the test is terminated and the time-out is returned as part of the test result.

- *testConditions.*   Specifies the conditions under which the resources should be Allocated to the test. For example, it indicates whether the loop-back test should start, when the test subject such as a connection is busy, and whether the user of the MORT can cause the loop-back test to abort.

- *dataUnits.*   Specifies the type and quantity of test data to be sent during the test. This is only settable by a managing system.

- *resultInterval.*   Specifies the result reporting interval. A new report is generated for every specified time interval after the test is initiated.

- *loopbackTimeout.*   A time-out period for transmission delay. It specifies a time-out period that measures from the start of transmission of each test data to the start of the receipt of the corresponding reply data. When the loop-back transmission delay exceeds the loop-back time-out period, the loop-back test is terminated and the result must contain the loop-back time-out period as part of the monitoredAttribute parameter in the testResult notification.

- *testThreshold.*   The error threshold for terminating the loop-back test.

Protocol integrity test.   The protocol integrity test is used to verify that the MORT can exchange protocol messages with a specified AO. In this case, the MORT represents a designated entity that is capable of transmitting the protocol message. The AO exhibits its normal behavior in reaction to the protocol messages received. The responses received by the MORT may be compared to the expected responses. The MORT may be required to transmit protocol messages in a particular sequence.

The protocol integrity test in general is nonintrusive to both the MORTs and AOs. The test protocol messages can be transmitted in the same way other protocol messages are transmitted. However, the MORTs and AOs are required to be in an appropriate state for testing or the test will be aborted.

As part of the test initiation for the protocol integrity test, the following parameters must be specified:

- Identity of the MORT and AO
- PDUs to be transmitted by the MORT and to be received by the AOs
- The condition to be met before a PDU is sent, which can be either a fixed time interval or the receipt of an expected reply PDU
- The interval for test result reporting if the intermediate test result reporting is required

The test may be terminated as a result of one of several events. One is the reception of a termination request from a managing system in a controlled test mode. Another is completion of the specified sequence of protocol messages. The failure to meet the condition set forth for the test to continue to the next protocol PDU can also cause test termination.

The result report of the protocol integrity test can be issued in various ways. In an uncontrolled test mode, the report can be sent back as a CMIS action reply. In a controlled test mode, the report can be sent back to the managing system as a response to a test result request. Otherwise, an unsolicited report can be sent as a result report notification. A result report must include the identities of the MORTs and AOs. The additional optional parameters may include the test outcome if this is the final report and the time from the test initiation to the completion of the test. If the test outcome is fail, more detailed information related to failure may be included in the report, for example, the incorrect PDU that caused the failure. A test may have one of five possible outcomes: pass, fail, time-out, terminate request, and inconclusive. While the first two outcomes are obvious, both the terminate request and time-out are premature terminations caused by a request from the managing system and violation of the specified time-out period, respectively. An outcome other than any of the first four is considered inclusive.

The protocolIntegrityTestObject managed object, as other test objects, is a subclass of testObject. In addition to the inherited attributes and behavior, it has the following attributes specific to the protocol integrity test.

- *mORTs.*   The managed objects that transmit protocol messages to the AOs for the protocol integrity test.

- *associatedObjects.*   The managed object that receives the test protocol message and responds in kind.

- *actualStartTime.*   The actual time when the test begins.

- *actualStopTime.*   The actual time when the test terminates.

- *protocolIntegrityResults.*   The result to be returned back to the requesting system. It may include the PDU sequence that has been performed; the received PDU with PDU type, value, and response time-out; the time interval for a MORT to receive the protocol message back; and the start and stop times.

- *pDUSequence.*   A sequence of PDUs to be performed. The information on each test PDU includes PDU type, parameter, the condition to be met for each PDU test to be viewed successful, and the wait duration before proceeding to the next test PDU.

- *timeOutPeriod.*   Specified time-out period for the time interval between the emission of a protocol message by the MORT to the reception of the reply protocol message.

- *waitingInterval.*   The waiting interval before the transmission of the next protocol message.

- *pDUReception.*   The expected PDU information to be compared against the received PDU information.

- *testCondition.*   An optional parameter that may be included in the parameter list for test initiation and that specifies the condition system under which the test may start.

Resource boundary test.   The resource boundary test is used to verify that a component resource of a system functions appropriately. From the perspective of system composition, a system consists of many resources or parts. It is important for fault detection to distinguish a part that malfunctions from the rest of the system.

This test inspects the behavior of a component resource by controlling and observing the interactions between the resource and its environment. The point of control and observation (PCO) will be located at the boundaries of the observed resource. The test inserts the test signals into the resource under test at the PCO and compares the behavior of the resource against the expected behavior of the resource. Test signals are different types of information that are exchanged between the resources under test.

The resource boundary test is configured as follows. The MORTs represent the resources that are being tested. A resource can recursively contain other resources. Several adjacent resources can be tested in one test, which is termed *multiresource testing,* and each resource is represented as a separate MORT. As PCOs are located at the boundaries of resources, each of them may have a connection with several resources. Each PCO is represented by an AO. At each PCO, signals can be observed and inserted. The relationships between PCOs and AOs, and between MORTs and resources, are shown in Fig. 10.16.

In general terms, the resource boundary test will operate as follows.

1. The test conductor sends the test performer a test request that includes a sequence of test events to be executed in the specified order.

2. The test performer first instantiates the resource boundary test object and appropriate MORTs and AOs.

3. The test performer executes each test event as specified in the test request. Each event specifies an action of either sending or receiving one or more signals to or from the MORT at the specified PCO. The signal type and value are specified as well. Sending a signal to the MORT means the signal is inserted into the MORT, and receiving a signal means the signal is emitted by the MORT.

4. Based on the value of the parameter resultReportIndicator that is included in the test request, the test performer decides whether or not to send the test results to the test conductor once the sequence of the test events is completed. Alternatively, the test conductor can request the test results intermittently.

The resourceBoundaryTestObject managed object, as other test objects, is a subclass of testObject. In addition to the inherited attributes and behavior, it has the following attributes specific to the protocol integrity test.

- *mORTs.* Represent the resource to be tested.

- *associatedObjects.* Identify the PCOs at which the signals of the resource under test are observed and inserted.

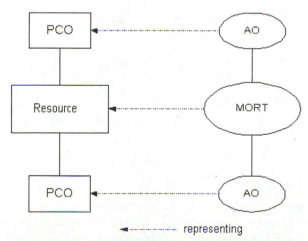

Figure 10.16 A possible configuration of a MORT and AOs for resource boundary testing. (*From ITU-T Recommendation X.737, 1995.*)

- *resultReportIndicator.*   This parameter is sent from the test conductor to the test performer to indicate whether a result report is required when a sequence of test events is passed.

- *sequenceOfEvents.*   This parameter is sent by the test conductor to the test performer to specify a sequence of test events that have to be performed sequentially by the test performer. Each event involves sending or receiving the specified test signal by the specified MORT. Specifically, a test event consists of the following information:

  - *eventId.*   A unique identifier for an event in the sequence
  - *signalType.*   The type of information to be inserted into the MORT for the resource boundary test
  - *signalValue.*   The information value to be inserted
  - *signalDirection.*   Sending or receiving the test signals to or from the specified MORT
  - *mORTs.*   The target MORTs for this test event
  - *associatedObjects.*   The involved PCOs
  - *waitDuration.*   The wait duration before proceeding to the next event in the sequence

- *resourceBoundaryTestResults.*   The test result.

- *timeoutPeriod.*   Specifies the time-out period for the time interval between the time of sending a signal to the MORT and the time of receiving a reply at the associated PCO.

- *testCondition.*   An optional parameter that may be included in the parameter list for test initiation and that specifies the condition under which the test may start.

The notification testResultNotification is used by the test performer to report the result of a resource boundary test to the test performer. Included in the result report are

- Test outcome, which can be pass, fail, etc.
- Signal type received
- Signal value of the signal type
- Event ID if the test event failed
- Identities of the MORTs from which the signal was received
- Identities of AOs where the signal was received

Fault Correction and Service Restoration

Fault correction and service restoration is an important area of telecommunications network management. However, because the method and procedure of fault recovery and service restoration can be highly specific to a network technology, there are few standards in this area. The discussions in this section, focusing on the network and network element level fault corrections, are largely based on the published literature and on reported common practices in the industry.

Fault correction and service restoration at the network element level

The fault correction and service restoration is concerned with correcting faults and restoring service on a single network element, and the tasks range from fault correction procedures to scheduling and administration of repair activities.

General fault recovery methods.   The methods are described in general terms. Common fault correction strategies include.

- *Replacement of faulty resources.*   This method of fault recovery is very applicable to a wide range of electronic components, which normally have well-defined interfaces to other components and modular functionality.

- *Fault isolation.*   This *cut-loss* strategy is used to isolate the resource that contains faults and prevent the faults from affecting other resources. The goal is to have the remaining resources continue in service. This strategy is used when the removal of a faulty resource from service does not affect over-all service quality or when additional resources are available.

- *Switching over to the standby unit.*   It is also called *hot standby procedure* meaning that services are switched from a faulty resource to a standby, redundant unit.

- *System reload.*   Many system faults, particularly when software is involved, can be fixed by a reload of the system software and starting afresh.

- *Installing another resource to specifically counter a fault.*   One example is a software patch.

- *Rearrange services* over complex routes to make room for a new route for a service.

Switching system fault recovery.   Switching systems are the nerve centers of modern telecommunications networks, and the impact of a total system failure is enormous. This points to the need and great importance of a fault recovery approach that aims at keeping the effects of switching system faults to a minimum (Nakamura et al. 1994).

Partial initialization is one such technique used at a switching system to recover from a software fault and to minimize the effect of a fault when it does occur. The basic idea is very simple: pinpoint the faulty process and then reinitialize only the faulty portion to avoid shutting down the whole switching system and thus minimize the loss. This method assumes the existence of advanced functions at the operating system level to determine the exact location of a faulty process.

This approach is based on the assumption that there is little chance that the problem will recur once the faulty process has been reinitialized. This is because the reinitialization does not address the root cause of the fault, for example, a software bug or a hardware error such as electric noise. The reason this approach works in many cases is that a survey of past faults indicates that the vast majority of software faults occurred under special circumstances such as rerouting because of traffic congestion and the chance is small for the same fault to recur under normal circumstances.

This method is effective only when it can be safely assumed that the effects of the fault do not extend to other processes. However, it is very difficult to predetermine whether and how much the effect of a faulty process will spread to other processes. Thus it is important to monitor the whole switching system after partial initialization to detect any spreading of the effect of the faulty process and to suspend any affected processes promptly.

Fault correction and service restoration at the network level

It is more complicated to perform a network level fault correction and service recovery because it requires a networkwide view and coordination among distributed network nodes. This subsection introduces the general concept of a self-healing network and a scheme for networkwide fault recovery and service restoration.

Self-healing network. The idea of a self-healing network has gained considerable attention in both the academic world and industry in recent years. The basic idea of a self-healing network, like many similar approaches, is to reserve redundant network resources to cope with possible failures or congestion. In a self-healing network, the spare capacity is calculated and assigned in the design phase. When a failure occurs, the network is reconfigured and the reserved resources are put into service to take over the part that has been affected by the failure. What characterizes the self-healing network is the manner in which the network is reconfigured. The reconfiguration is done in a distributed fashion, meaning that each local node that is affected by the failure makes its own decision, in the absence of the global view. Many algorithms have been proposed to accomplish the goal of self-healing. The loop-expanding method is one of the methods that have proposed, and it provides a concrete example of the self-healing network.

Loop expanding method. Loop expanding, as reported in Lee et al. (1998), is one particular algorithm of a self-healing network as applied to a transport network such as an ATM network. At the center of the issue is to find an alternative path when failure occurs on an original path. This method is based on a simple observation that the original path and any alternative path together form a loop. These loops are identified in advance for the purpose of fault recovery and service restoration. When a link failure occurs that results in a broken loop, the original loop is expanded to circumvent the failed node or link. In general terms, the procedure for expanding a broken loop works as follows. When a failure occurs on either a node or a link, each of the neighboring nodes expands the broken loop by connecting itself to the nearest neighboring loop. If no new, healthy loop results from the expanding, due to link or node failure in the new loop, the loop expanding continues until either a complete loop is found or the whole network is exhausted.

Trouble Administration

Trouble in a communication network is a problem that has adverse effects on the quality of service as perceived by network users. When trouble is detected, a trouble report is initiated either by a customer who reports the problem or by the network management system that first detects the problem.

As defined in ITU-T Recommendation X.790 (1995), trouble management is realized in the familiar manager-agent paradigm where the agent is responsible for carrying out activities to resolve trouble and the manager is responsible for keeping track of trouble reports and for interfacing customers for answering trouble-related inquiries. This section presents an introduction to the trouble management model defined in ITU-T Recommendations X.790.

Trouble administration requirements

The life cycle of trouble management is categorized into four different areas as shown in Fig. 10.17. It is evident that trouble administration centers on the management of trouble reports. There is a set of functional requirements for each of the four areas of trouble management, as discussed below. In general, a trouble report provides a means for customers to report a service problem and for a management system to record all information related to the trouble and to keep track of the progress of the trouble resolution.

Trouble report creation. A trouble report may be created on a resource such as a circuit or a dialed number or on a service offered to customers such as wireless call forwarding. Creation of a trouble report may be triggered by several entities. First, a management entity, acting in an agent role, can create a trouble report after receiving an event such as an alarm notification or threshold crossing. A person in the manager's organization can report trouble to the management system that creates the trouble report in response.

A trouble report shall contain a wide variety of information. Examples of the type of information found in a trouble report include

- The originator of the trouble report
- The contact person or point for this trouble report
- One or more aliases for the trouble report
- Comments about a trouble report
- The identity of the resource or network service on which the report is created
- A time stamp of the report

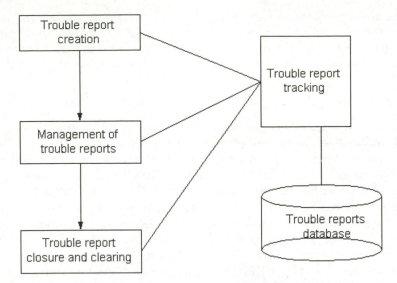

Figure 10.17   A process model of trouble report management.

The management entity creating the trouble report should have the flexibility to select the format of the trouble report for a particular problem.

Trouble report tracking.   Trouble management should allow a manager to track the progress of trouble report resolution and other information related to a trouble report. As the state and status of a trouble report change, the manager should be notified of the change if it so desires. The time stamps of any change in state, status, or other attributes of a trouble report shall be maintained.

The trouble report should maintain a record of all information related to a trouble: trouble diagnosis, test and repair, the repair type, length of time spent on each activity, whether an activity is billable or not, and the equipment involved in each activity. Some of the information is provided by the agent and some by the manager.

Trouble management should allow a manager to view a specific trouble report, part of a report, or historical records of trouble reports. Trouble reports should be viewable by customer and by service type.

Management of trouble reports.   It is largely the responsibility of a manager to manage trouble reports, and the following requirements outline the responsibilities:.

- While the agent shall be allowed to allocate resources such as the person or other system that is required to resolve trouble, the manager should be able to request escalation of a trouble report under certain circumstances.

- The manager should be able to request cancellation of a trouble report that is not closed yet.

- Access control should be used so that the manager can only update certain designated attributes of a trouble report. For example, the manager should be allowed to update a trouble report with new information such as a contact change.

- The manager should be able to query the agent for information about the time when a trouble report is created, deleted, or updated.

- It is the responsibility of the manager to authorize the repair activities for trouble that are requested by the agent.

Trouble management report closure and clearing.   Once the agent carries out the repair activities and the trouble is resolved, the agent is responsible for closing the trouble report. Before the agent proceeds to the closure, the manager needs to verify that the trouble is indeed resolved. This may involve testing or querying for additional information from the agent.

Once a trouble report is closed, there are some postclosure chores requiring the attention of both the manager and agent. The service outage caused by the trouble shall be computed for billing or other purposes. The trouble report should be updated with the information on the cause found and the solution applied. Then the trouble report may be logged for record keeping, if it is so required, for a specified period of time.

An object model of trouble report management

This model describes in more detail the conceptual model of trouble management in terms of managed objects involved. The management entity, acting in an agent role, and the managing system, acting in a manager role, may fall into two different jurisdictions. For example, the former can be a system of service provider *A,* while the latter belongs to the domain of service provider *B.* It is crucial that the interface used between the two for the exchange of management information is either standardized or, at a minimum, mutually agreed upon. Although the CMIP is used to illustrate the interface, as shown in Fig. 10.18, other implementation technologies have been used as well, CORBA in particular. The terms *manager* and *agent,* as used in this context, have a more generic denotation than when used elsewhere in the book. The agent system could be at the service management layer, network management layer, or element management layer, instead of being exclusively at the network element layer, as in most cases described in this book. This is particularly the case when the agent and manager are in the domain of two different service providers.

The telecommunicationsTroubleReport managed object class.   At the core of the model are the two managed object classes, i.e., telecommunicationsTroubleReport and serviceProviderTroubleReport. Both are subclasses of the troubleReport class. The telecommunicationsTroubleReport object class represents customer-reported troubles on telecommunications services or resources. It

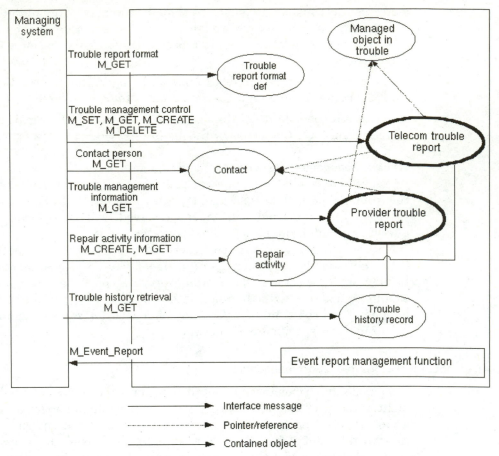

Figure 10.18   Conceptual model for trouble management (*From ITU-T Recommendation X.790, 1995.*)

describes the nature of the problem and provides a means for the management system to keep track of the status of the problem resolution. In essence, it acts as a container for the bulk of trouble report information. The large number of attributes that are found in the telecommunicationsTroubleReport can be broken down into different groups based on who can create and set values of the attribute. When an instance of this managed object class is instantiated, some of the attributes will be mandatory and some optional. The following five groups of attributes are found in the telecommunicationsTrouble Report object class.

The first group is the attributes that can only be supplied by the manager and generally are related to the interaction with the customer and the definitions (e.g., type and format) of trouble reports.

- *managedObjectInstance.*   The object that represents the network resource or service that has the trouble.

- *suspectObjectList.*   The object instances that may be associated with the cause of the trouble.

- *troubleType.*   The type of trouble, over one hundred of which have been identified. See App. 10B for details.

- *customerWorkCenter.*   The manager work center from which the trouble is entered.

- *custTroubleTickNum.*   A unique number assigned to the trouble for the customer to reference.

- *troubleReportFormatObjectPtr.*   A pointer to an instance of the troubleReportFormat Definition object class that is used for this trouble report.

The second group of attributes are those that can only be supplied and updated by the manager and are usually concerned with trouble administrative information such as trouble report search keys, detection time, severity, assigned priority, and contact person.

- *aLocationAccessAddress.*   The *A* address for which the associated *A* location access hours are valid.

- *zLocationAccessAddress.*   The *Z* address for which the associated *Z* location access hours are valid. See the Managed Object Class Dictionary (App. D) for the definitions of the *A* and *Z* locations.

- *aLocationAccessHours.*   The hours for each day of the week during which access to the *A* location is available.

- *zLocationAccessHours.*   The hours for each day of the week during which access to the *Z* location is available.

- *aLocationAccessPerson.*   The person at the *A* location responsible for this trouble report.

- *zLocationAccessPerson.*   The person at the *Z* location responsible for this trouble report.

- *additionalTroubleInfoList.*   Provides further information on the trouble type in a textual form.

- *alternateManagerContactPerson.*   An alternative contact person responsible for this trouble report.

- *alternateManagerContactObjectPtr.*   A pointer to a contact object that identifies an alternative individual that can be contacted for this trouble report.

- *callBackInfoList.*   The call-back types requested by the manager and the person responsible for the trouble report on behalf of a customer. The call-back types identified are as follows:
  - *escalation.*   The customer requested a call back from the higher officials of the service provider with regard to this trouble report.
  - *before-auto-test.*   The customer requests a call back before any automatic test is performed for this trouble report.

- *after-cleared.*   The customer requests a call back when the trouble is cleared.
- *commitmentTimeRequest.*   The trouble clearance time requested by a customer.
- *managedObjectAccessHours.*   The hours of each day of the week during which access to the managed object is available.
- *managedObjectAccessFromTime.*   The beginning of the time period during which the managed object can be accessed.
- *managedObjectAccessToTime.*   The end of the time period during which the managed object can be accessed.
- *managerContactPerson.*   An individual in a manager's organization responsible for this trouble report.
- *managerContactObjectPtr.*   A pointer to a contact object that identifies an individual in the manager's organization who can be contacted on this trouble report.
- *managerSearchKey1.*   A single-valued key that enables the manager to filter trouble reports by some attributes such as customerId.
- *managerSearchKey2.*   Another single-valued key to allow the manager to filter trouble reports.
- *managerSearchKey3.*   Another single-valued key to allow the manager to filter trouble reports.
- *managerSearchKeyList.*   A set of attributes that can be used to filter and scope trouble reports by the manager.
- *perceivedTroubleSeverity.*   Indicates the effect of the trouble on the managed object being report.
- *preferredPriority.*   Defines the perceived urgency associated with the trouble report.
- *troubleDetectionTime.*   The time at which the trouble was detected.

- *troubleReportStatusWindow.*   Specifies a sliding window during which a troubleReportProgress notification is expected by the manager from the agent.

The next group of attributes can only be supplied and/or updated by the agent. They are concerned with the detailed information on a trouble report that is only available at the agent side:

- *initiatingMode.*   Specifies the mode of initiation of the trouble report by the agent.

- *receivedTime.*   The date and time when the trouble report was entered.

- *troubleReportID.*   A unique identifier for this trouble report object instance.

- *activityDuration.*   The time spent on billable and nonbillable activities for the trouble report.

- *additionalTroubleStatusInfo.*   Further describes the value of the trouble report status attribute.

- *agentContactPerson.*   An individual in the agent's organization responsible for this trouble report.

- *agentContactObjectPtr.* A pointer to the contact object that identifies a contact person in the agent's organization responsible for this trouble report.

- *alarmRecordPtrList.* A set of pointers to instances of the alarmRecord object class that are available in the agent system. This attribute is present only if the trouble report is created as a result of one or more alarms.

- *commitmentTime.* The time given to a customer for trouble closure.

- *lastUpdateTime.* The time and date of the most recent update on the trouble report by either the manager or the agent.

- *responsiblePersonPtr.* Identifies an individual with the overall responsibility for resolving this trouble report.

- *troubleLocation.* Indicates where the trouble is.

- *troubleReportNumberList.* A list of internal trouble report alias identifiers in the agent system for this manager's trouble report.

- *troubleReportState.* The current state of the trouble report.

- *troubleReportStatus.* The current status of the trouble report. See App. 10B for detailed definitions of the status values.

- *troubleReportStatusTime.* The time of the most recent update of the trouble report status.

The following attributes can be set to default values by agents and only updated by the agent:

- *closeOutNarr.* A narrative supplied by the person who resolved the trouble report to provide additional information regarding the trouble closure

- *handOffCenter.* The service provider's control center to which a trouble report has been referred to

- *handOffLocation.* The location within a service provider control center to which the trouble report has been referred

- *handOffPersonName.* Identifies the person responsible for the hand-off center at the agent side where the trouble report has been referred

- *handOffPersonPtr.* A pointer that identifies the person responsible for the hand-off center at the agent side where the trouble report has been referred

- *handOffTime.* The time at which the trouble report was referred to the hand-off center

- *maintenanceOrgContactName.* The company or organization responsible for maintaining managed object instances that represent the network resource or service that are in trouble.

- *maintenanceOrgContactPtr.* A pointer to the company or organization described above.

- *maintenanceOrgContactTime.* The time at which the maintenance organization as described above was contacted by the agent to repair the trouble.

- *maintServiceCharge.*   Indicates whether the customer will be charged for the repair activities

- *outageDuration.*   The time interval between the trouble report creation and the trouble report clearing

- *repairActivityList.*   Information on the repair activities performed for a trouble report, i.e., activity types, who performed them, and when they were performed

- *restoredTime.*   The time when the trouble report was cleared and the affected service was restored

- *troubleFound.*   A numeric code identifying the problem that has been resolved

There are some attributes that can be set to default values by the agent and only updated by the manager:

- *afterHrsRepairAuth.*   A Boolean value indicating whether the customer has given permission to repair the service outside the normal business hours

- *cancelRequestedByManager.*   A Boolean value indicating whether the manager has initiated cancellation of the trouble report

- *closeOutVerification.*   A Boolean value indicating whether the manager has verified the completion of the problem resolution

- *troubleClearancePerson.*   The contact person for information related to the closure of the trouble report

The final group of attributes can only be supplied by the manager and updated by either the manager or the agent.

- *authorizationList.*   Indicates whether authorization is requested by the agent and granted by the manager, what type of activities are authorized, and optionally the authorizing person and time

- *dialog.*   A free-form text to record information on trouble resolution

- *escalationList.*   Indicates whether the escalation of the trouble report is requested by the manager and granted by the agent

- *repeatReport.*   Indicates whether there has been a trouble report on the same managed object in the specified past period, say, the past 30 days

The providerTroubleReport object classes.   The service providerTroubleReport managed object class is a subclass of the troubleReport managed object class. The instance of this object class is created by an agent to notify the manager of a planned maintenance that will render all or part of the services, resources, network, or system inaccessible during the specified time period. The object class contains the following attributes:.

- *beginTime.*   The beginning of the time period when the service will be unavailable

- *endTime.* The end of the time period when the services will be unavailable
- *locationPtr.* The location of the managed object instance for which the trouble report is created
- *unavailableServicePtr.* Identifies the services that will be affected

The troubleReport managed object class. The troubleReport object class is the superclass of the telecommunicationsTroubleReport and serviceProviderTroubleReport. As such, it supplies the attributes common to all derived subclasses of troubleReport. The common attributes include troubleReportId, troubleReportState, troubleReportStatus, troubleReportStatusTime, troubleType, repairActivityList, and trouble ReportFormatObjectPointer. See App. 10B for specified trouble types.

The troubleReportFormatDefinition managed object class. This managed object class provides a flexible scheme for a service provider to define a trouble report format. It also provides the flexibility to dynamically specify trouble report formats for service or a resource object on an object class basis. This is important because there is a variety of types of trouble reports for a wide range of resources and services. Instances of the troubleReportFormatDefinition object class are locally created and updated by the agent system.

Each instance of the troubleReportFormatDefinition managed object specifies what the conditional packages are and which of the packages may be included in a telecommunications trouble report. Each instance of the troubleReportFormat definition object is associated with one instance of the telecommunicationsTroubleReport object. A trouble report format is specified via a set of attribute lists as follows.

- *applicableManagedObjectClassList.* The managed object classes to which this trouble report format definition applies
- *applicableManagedObjectInstanceList.* The instances of the managed object classes identified above to which this trouble report format definition applies
- *tRConstrainedToSingleValueAttrIdList.* The set-values attributes in a trouble report that are constrained to a single value by the agent
- *tRMayBePresentAttrIdList.* A list of attributes in conditional packages of a troubleReportObject class that may be present if this particular trouble report format definition is used
- *tRMustBePresentAttrIdList.* A list of attributes in conditional packages of a trouble report object class that must be present if this particular trouble report format definition is used

The creation and deletion of an instance of the troubleReportFormatDefinition object class and any change in its attribute values will cause a notification to be sent to the manager to report the event.

The repairActivity managed object class. A repairActivity object instance is created for each repair activity required for the trouble resolution. Included in the object is the information on the details of repair actions, who has performed them, and when they were performed.

A repairActivity object is associated with one trouble report, and when the trouble report is deleted, so is the associated repairActivity object. The repair activities are carried out at the agent side, and thus the attributes of this object are supplied and updated by the agent while the manager can only query the information of this object. The attributes of this object class include

- *repairActivityId.* The distinguishing attribute for this object class
- *entryTime.* The time when the repair activity started
- *activityInfo.* A text string describing what repair activities are being carried out
- *activityPerson.* The person responsible for the repair activity
- *activityCode.* A code for a category of repair activities

The troubleHistoryReport managed object class. The troubleHistoryReport object class is a subclass of the logRecord object class defined in CCITT Recommendation X.721 (1992) and is used to log the troubleHistoryEvent notifications emitted from the troubleReport object and its subclasses such as telecommunicationsTroubleReport and serviceProviderTrouble Report. The instances of the latter two object classes generate a troubleHistoryEvent notification whenever the trouble report status attribute value changes to a final closed-out.

Contained in the troubleHistoryRecord object are the attributes found in the telecommunicationsTroubleReport and serviceProviderTroubleReport object classes. These attributes represent the information that falls into one of the following three categories: the trouble report and repair activity, information for or about the customer, and trouble report closure. The attributes related to the trouble report and the associated repair activities include troubleType, activityDuration, perceivedSeverity, and repairAuthorizationList. The attributes related to the customer include trouble RepairCommitmentTimeRequested, and customerTroubleTicketNumber. The third category of attributes, concerning the trouble ticket closure, are the most numerous and include troubleReportClearancePerson, restorationTime, and troubleReportCloseOutVerification.

The account managed object class. The account managed object class contains information describing a customer account from the perspective of trouble management with the following attributes:.

- *accountContactList.* A list of persons in the manager's organization who can be contacted with regard to this customer account

- *contactObjectPtrList.* A list of pointers to the instances of the contact object classes associated with this customer account

The contact managed object class. The contact managed object class represents a person or an organization that can be contacted for resolution of the associated trouble reports. It has attributes such as contactName, functionOfContact (e.g., repair, field service, testing), contactType (e.g., equipment-related activities and service-related activities), and the meansOfContact (e.g., phone, fax, and e-mail). An instance of this object class is associated with one instance of the troubleReport object.

A state transition diagram for trouble reports. A trouble report may go through any of the six states defined for trouble reports, as shown in Fig. 10.19. The disabled state is different from the rest of the six states: the semantics for a trouble report to transition into this state is a local matter and left to each application.

The queued state. A trouble report is in the queued state when it has been instantiated but the trouble resolution process has not begun yet. A manager can cancel a trouble report in this state, and the agent receiving such a request will close the trouble report if it is allowed to.

The open/active state. A trouble report transitions into the open/active state when appropriate actions to resolve the trouble are initiated. A manager may cancel a trouble report in this state, and the agent receiving this request will attempt to close out the trouble report.

The deferred state. This state indicates that corrective action to resolve the trouble has been postponed because the repair activity cannot be carried out for one reason or another. Trouble in this state can either directly transition into the open/active state if the condition hindering the repair activity no longer exists or move directly to the closed state if the trouble is cancelled.

The cleared state A trouble report moves into the cleared state by an agent when it concludes that the trouble has been resolved.

The closed state When a trouble report is in the closed state, the trouble report resolution for this trouble report is completed. Upon closure, the trouble report attributes are captured in a historical event generated at the trouble report closure that may be stored in a log of trouble history records.

The disabled state The disabled state indicates that trouble report information cannot be updated due to some local conditions. In this state, only read operations are allowed on the attributes of the managed object class.

An application scenario of the trouble report management model. We use a scenario to illustrate the trouble report administration model we have been discussing so far. As described earlier, a trouble report can be initiated by a manager in response to a complaint about service quality from a customer or by an agent in response to the detection of a fault. The scenario here is of the

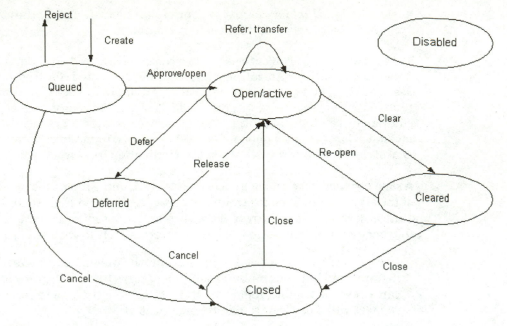

Figure 10.19 State transition diagram for trouble reports.

latter category: loss of signals from a line card caused an alarm and in turn resulted in the creation of a trouble report. The following sequence of events, in general terms, demonstrates the behavior of the trouble administration model.

1. Once the alarm for the event arrives, the agent generates an instance of the telecommunicationsTroubleReport class. In addition, the agent also instantiates instances of other object classes, such as contact and troubleReportFormatDefinition. The agent supplies the attribute and attribute values of the telecommunicationsTroubleReport object that must be assigned by the agent. As a result of the object creation, a notification is sent to the manager using M-EVENT-NOTIFICATION. Among the attributes that are supplied by the agent are

    - initiatingMode=alarmOriginated (the trouble report is generated as a result of the alarm)
    - receivedTime=date x, time y
    - troubleReportId = N + 1 (a unique number for all outstanding trouble reports)
    - activityDuration = 0 (the total amount of billable time is unknown at this point)
    - agentContactPerson = Joe (the contact person at the agent side for this trouble report)
    - troubleReportState = queued
    - troubleReportStatus = pendingTest (further testing is required to determine the repair action)

- troubleLocation = null (the exact location of trouble is not known yet)
- troubleType = noSignal

2. The manager creates a local copy of the trouble report for tracking the report and may provide the trouble report information to customers when queried. The manager uses M-SET to set the values of a set of attributes of the telecommunicationsTroubleReport object, among which are

    - custTroubleTickNum = N (the trouble ticket number given to the customer for the purpose of reference and trouble ticket tracking)
    - commitmentTimeRequest = date and hour (the requested time for the trouble to be resolved)
    - managerContactPerson = Henry (the contact person on the manager side for this trouble report)
    - preferred priority = major (requiring immediate attention)
    - perceivedTroubleSeverity = outOfService (highest level of severity)

3. The agent sets the attribute values as required by the manager without incident.

4. The manager also performs certain analyses to help decide the cause of the trouble and appropriate repair activities. For example, the manager may act as a test conductor and instructs the agent to perform certain diagnostic tests.

5. After performing diagnostic tests, the agent determines the cause of the problem and appropriate repair action. At this point, the agent decides to instantiate a repairActivity object instance to represent the repair activity. The state of the trouble report is set to open/active as a result. The state value change causes a notification to be sent to the manager.

6. The manager updates the state attribute value of the troubleReport object. In addition, the manager decides to get a more comprehensive picture on the trouble report by retrieving the information from the agent on the contact and repair activity using M-GET.

7. The agent carries out the repair activities, and as a result the fault is eliminated. The state of the trouble report is moved to cleared, and a notification is sent to the manager.

8. The manager needs to verify the clearance of the trouble report. Once the manager is satisfied that the trouble is indeed resolved, the manager indicates this to the agent by setting the closeOutVerification attribute value of the troubleReport object.

9. The agent sets the state of the troubleReport object to closed and creates a troubleReportLog object for this particular trouble report.

Summary

Fault management is a critical part of network maintenance and operation. This chapter covers the whole life cycle of fault management, from fault detection to fault recovery to trouble report management.

First discussed is the alarm surveillance that is responsible for detecting, reporting, and providing a summary of the trouble report. Fault detection is largely specific to each network system component, either hardware or software, and thus the discussion focuses on the generic mechanisms of fault detection. The bulk of the discussion is devoted to alarm reporting. Alarm reporting is a particular application of event reporting. A generic event reporting model is introduced first. At the core of the model is the eventForwardingDiscriminator (EFD) managed object class that allows a managing system to register events of interest and specify event forwarding criteria. Alarm reporting is strictly based on this event report model, which can be used for reporting many other kinds of events as well. Alarm summary reporting is a special case of generic alarm reporting that allows a managing system to retrieve a summary report of alarms instead of bulky alarm data that may not all be relevant.

The next step in the fault management life cycle is to determine the cause of a fault, or fault localization. Though the specifics of each fault shall be taken into consideration, the commonly seen approach is the combination of diagnostic tests and fault analysis. Since the diagnostic test is the topic for the following section, the focus of this section is on the introduction of several fault analysis methods, i.e., causal model, codebook approach, case-based reason, and rule-based method.

Test management, a huge topic that deserves a chapter of its own, is approached from the perspective of fault management. First, a generic test model is introduced, which automates the test process by allowing a managing system to remotely set up test parameters and schedule the test. At the core of the model is a set of managed object classes at the managed system, i.e., testPerformer, testObject (TO), and managedObjectReferringToTest (MORT). Two variations of the generic test models, control and uncontrolled test models, are introduced and contrasted. Then six categories of diagnostic tests, which are based on the generic test model, are introduced: connection, connectivity, data integrity, protocol integrity, loop-back, and resource boundary tests.

Fault correction and service restoration are discussed from two different perspectives, that is, the fault correction at the network element and networkwide fault recovery. At the network element level, several commonly seen fault recovery procedures and mechanisms are described, such as switching over and software reload. It is more complicated to attempt a networkwide fault recovery. The concept of a self-healing network is introduced. One particular scheme of self-healing–based fault recovery, the loop-expanding approach, is presented.

The last section is devoted to trouble administration. Trouble in a communication network is a condition that has an adverse effect on the quality of service as perceived by the network users. Initiation of a trouble report may be triggered by an alarm or simply by a customer complaint. This is an area that calls for the development of standard interfaces between various parties involved in resolving a trouble report. It is likely in the current telecommunications environment that more than one service provider is involved in resolving a service problem. This section introduces a trouble administration model as defined in ITU-T Recommendation X.790 (1995). At the core of the trouble administration model is the management of trouble reports. Maybe one of the most significant parts of the model is the definition of a set of managed object classes, in particular the telecommunicationsTroubleReport, serviceProviderTroubleReport, and their superclass troubleReport. These managed object classes with their well-defined attributes, provide a good foundation for achieving interoperability between different service providers.

Exercises

10.1 Explain how a managing system can suspend or resume alarm reporting.

10.2 Explain the basic procedure for a managing system to initiate the part of the managed system for alarm reporting.

10.3 Describe the criteria used to select managed object instances to be included in a current alarm summary report.

10.4 Describe the five alarm categories and present two probable causes for each type of alarm.

10.5 Compare the controlled test model with the uncontrolled test model and discuss the similarities and differences between the two.

10.6 Discuss the differences between the connection test and the connectivity test in terms of test purposes.

10.7 Discuss the differences between the data integrity test and the protocol integrity test in terms of test goals and test procedures.

10.8 Describe the basic idea behind self-healing networks. In particular, when a service-affecting fault occurs, a self-healing network reconfigures itself. Why is the reconfiguration achieved in a distributed fashion?

10.9 Discuss the semantics of the six states of a trouble report in the trouble report life cycle. In particular, what causes the transition from the queued state to the open/active state?

10.10 In the trouble management model as described in this chapter, how does a manager know when a trouble resolution is completed?

10.11 What does the escalation of a trouble report mean? Specifically, which attributes of a TelecommunicationsTroubleReport object are changed as a result of the escalation?

10.12 Assume the scenario where a trouble report management involves two management systems of two different jurisdictions, one acting in a manager role and the other acting in an agent role. What are the possible security issues involved? Refer to Chap. 12 for the security-related issues.

Appendix 10A: Probable Causes of Alarms

A *communication alarm* may be caused by one or more of the following events:

- *Loss of signal.* An error condition in which no data are present on a communication circuit or channel.
- *Loss of frame.* An inability to locate the information that delimits the bit grouping within a continuous stream of bits
- *Framing error.* An error in the information that delimits the bit groups
- *Local node transmission error.* An error that occurs on a communications channel between the local node and an adjacent node
- *Remote node transmission error.* An error that occurs on a communications channel beyond the adjacent node
- *Call establishment error.* An error that occurs while attempting to establish a connection
- *Degraded signal.* The quality or reliability of transmitted data has decreased
- *Communication protocol error* An error that stems from communications protocol-related operations.
- *Communication subsystem failure.* A failure in a subsystem that supports communications over telecommunications links, these may be implemented via a leased telephone line, by X.25 networks, token-ring LAN, or otherwise
- *LAN error.* An error that has been detected on a local area network
- *Interface error.* An error that occurs on a communications interface such as on an interface card

The *quality-of-service alarms* are related to a computing process that may be caused by one or more of the following events:

- *Response time excessive.* The elapsed time for a query has gone beyond the acceptable limits.
- *Queue size exceeded.* The number of items to be processed has exceeded the allowable maximum.
- *Bandwidth reduced.* The available transmission bandwidth has decreased.
- *Retransmission rate excessive.* The number of repeat transmissions is outside of the acceptable limits.
- *Threshold crossed.* A limit has been exceeded.
- *Performance degraded.* The service agreements or service limits are outside of acceptable limits.
- *Congestion.* A system or network component has reached its capacity or is approaching it.

- *Resource at or nearing capacity.*   The usage of a resource is at or nearing the maximum allowable capacity.

A *processing alarm* may be caused by one or more of the following events:

- *Storage capacity problem.*   A storage device has very little or no space available to store additional data.

- *Version mismatch.*   There is a conflict in the functionality of versions of two or more communicating entities which may affect any processing involving those entities.

- *Corrupt data.*   An error has caused data to be incorrect and thus unreliable.

- *CPU cycles limit exceeded.*   A CPU has issued an unacceptable number of instructions to accomplish a task.

- *Software error.*   A software error has occurred for which no more specific probable causes can be identified.

- *Software program error.*   An error has occurred within a software program that has caused incorrect results.

- *Software program abnormally terminated.*   A software program has abnormally terminated due to some unrecoverable error condition.

- *File error.*   The format of a file is incorrect and thus cannot be used reliably.

- *Out of memory.*   There is no more program addressable storage available.

- *Underlying resource unavailable.*   An entity upon which the reporting object depends has become unavailable.

- *Application subsystem failure.*   A failure in an application subsystem has occurred.

- *Configuration or customization error.*   A system or device generation or customization parameter has been specified incorrectly or is inconsistent with the actual configuration.

An *equipment alarm* may be caused by one or more of the following events:

- *Power problem.*   There is a problem with the power supply for one or more resources

- *Timing problem.*   A process that requires timed execution and/or coordination cannot be completed or has been completed but cannot be considered reliable.

- *Processor problem.*   An internal machine error has occurred on a CPU.

- *Dataset or modem error.*   An internal error has occurred on a dataset or modem.

- *Multiplexer problem.*   An error has occurred while multiplexing communications signals.

■ *Receiver failure.*   An error has occurred on the network device that receives incoming messages.

■ *Transmitter failure.*   An error has occurred on the network device that transmits messages.

■ *Output device error.*   An error has occurred on the output device.

■ *Input device error.*   An error has occurred on the input device.

■ *Equipment malfunction.*   An internal machine error has occurred for which no more specific probable cause has been identified.

An *environment alarm* may be caused by one or more of the following events:

■ *Temperature unacceptable.*   A temperature is not within acceptable limits.

■ *Humidity unacceptable.*   The humidity is not within acceptable limits.

■ *Heating / ventilation / cooling system problem*

■ *Fire detected*   A fire has been detected in the equipment room.

■ *Flood detected*   A flood has been detected in the equipment room.

■ *Toxic leak detected*   A toxic leak has been detected in the equipment room.

■ *Leak detected*   A water leak or other kind of leak has been detected in the equipment room.

■ *Pressure unacceptable.*   A fluid or gas pressure is not within acceptable limits.

■ *Excessive vibration.*   Vibratory or seismic limits have been exceeded.

■ *Material supply exhausted.*   A supply of needed material has been used up.

■ *Pump failure.*   A failure of a mechanism that transports a fluid by inducing pressure differentials within the fluid.

■ *Enclosure door open*   The enclosure door of the equipment room is found open.

Both the alarm categories and probable causes are not meant to be exhaustive. In fact, it is unrealistic to expect that possible causes of alarms can be exhausted. As telecommunications network technologies advance, it is conceivable that new categories may be added. As an application demands, additional causes of alarms must be defined.

Appendix 10B: Attribute Values for Trouble Administration

The troubleReportStatus attribute

The troubleReportStatus attribute as listed in Table 10B.1, indicates the status of an active trouble report. It provides information on the type of trouble repair activity that is currently being undertaken and the stage of the activity. The assigned code for each status value is for the purpose of achieving interoperability between different service providers that exchange trouble report status information. The thirty-five attribute values are not an exhaustive listing, and an application designer may choose a subset or to add additional attributes as needed.

TABLE 10B.1   TroubleReportStatus

Trouble status	Assigned code
screening	1
testing	2
dispatchedIn	3
dispatchedOut	4
preassignedOut	5
bulkDispatchedOut	6
startRepair	7
pendingTest	8
pendingDispatch	9
requestRepair	10
referMtceCenter	11
referVendor	12
noAccessOther	13
startNoAccess	14
stopNoAccess	15
startDelayedMtce	16
stopDelayedMtce	17
troubleEscalated	18
craftDispatched	19
temporaryOK	20
cableFailure	21
originatingEquipFailure	22
backOrder	23
clearedCustNotAdviced	24
clearedCustAdvised	25
clearedAwaitingCustVerification	26
closedOut	27
closeOutByCustReq	28
closeOutcustVerified	29
closeOutCustDenied	30
canceledPendingWorkInProgress	31
canceledPendingTestCompletion	32
canceledPendingDispatchCompl	33
techOnSite	34
techLeftSite	35

TABLE 10B.2

Trouble type	Assigned code
noDialToneGroup	100
noDialTone	101
slowDialTone	102
circuitDead	103
canNotCallOutGroup	200
canNotCallOut	201
canNotBreakDialTone	203
dialToneAfterDialing	204
highAndDry	205
canNotRaise	206
allAccessBusy	207
canNotCallOut2	208
canNotCallLongDistance	209
canNotCallOverseas	210
speedCall	211
canNotBeCalledGroup	300
canNotBeCalled	301
canNotBeCalledBusy	302
doNotGetCalled	303
canNotTripRing	304
falseRings	305
doNotAnswer	306
reachRecording	307
canNotRaiseAStation	308
canNotRaiseADrop	309
canNotRaiseACircuitLocation	310
ringNoAnswer	311
reorder	312
alwaysBusy	313
bellDoesNotRing	314
bellDoesNotRing2	315
bellRingsCanNotAnswer	316
bellRingsAfterAnswer	317
noRingNoAnswer	318
otherRingTrouble	319
receivesCallsForWrongNumber	320

TABLE 10B.2 *(Continued)*

Trouble type	Assigned code
recordingOnLine	321
canNotBeHeardGroup	400
canNotBeHeard	401
canNotHear	402
fading	403
distant	404
reachedWrongNumberGroup	500
wrongNumber	501
circuitOperationGroup	600
open	601
falseDisconnect	602
grounded	603
canNotBeSignaled	604
canNotSignal	605
permanentSignal	606
improperSupervision	607
supervision	608
canNotMeet	609
hangUp	611
noWinkStart	612
noSF	613
lowSF	614
noContinuity	615
cutCable	616
openToDEMARC	617
noRingGenerator	618
badERL	619
echo	620
hollow	621
circuitDead	622
circuitDown	623
failingCircuit	624
noSignal	625
seizureOnCircuit	626
lossEPSCSorSwitchedServices	627
monitorCircuit	628

TABLE 10B.2 *(Continued)*

Trouble type	Assigned code
newServiceNotWorking	629
openEPSCSorSwitchedServices	630
otherVoiceDescribeAdditInfo	631
cutOffsGroup	700
cutsOff	701
noiseProblemGroup	800
intermittentNoise	801
noisy	802
foreignTone	803
clipping	804
crossTalk	805
staticOnLine	806
groundHum	807
hearsOtherOnLine	808
HumOnLine	809
clicking	810
noiceEPSCSorSwitchedServices	811
levelTroublesGroup	900
lowLevels	901
highLevels	902
longLevels	903
hotLevels	904
highEndRollOff	905
lowEndRollOff	906
needsEqualized	907
lineLoss	908
doesNotPassFreqResponse	909
miscellaneousTroubleGroup	1000
hiCapDown	1001
carrierDown	1002
biPolarViolations	1003
frameErrorsHiCap	1004
outOfFrame	1005
lossOfSync	1006
frameSlips	1007
noLoopback	1008

TABLE 10B.2   *(Continued)*

Trouble type	Assigned code
canNotLoopbackDEMARC	1009
recordingOnCircuit	1010
lineNeedTagging	1011
outwatsRingingIn	1012
remoteAccess	1013
other	1014
alarm	1015
memoryServiceProblemGroup	1100
dataTroubleGroup	1200
canNotReceiveData	1201
cantNotTransmitCanNotReceive	1203
noResponse	1205
delay	1206
impulseNoise	1207
phaseJitter	1208
harmonicDistortion	1209
highDistortion	1210
noDataLoopback	1211
noCarrier	1212
notPolling	1213
dataFramingErrors	1214
dropOuts	1215
hits	1216
noAnswerBack	1217
steamer	1218
outOfSpecification	1219
canNotRunToCSU	1220
canNotRunToOSU	1221
deadDataCircuit	1222
circuitInLoopback	1223
errors	1224
garbledData	1225
invalidData	1226
crossModulation	1227
slowResponse	1228
otherDataDescribeAdditInfo	1229

TABLE 10B.2   *(Continued)*

Trouble type	Assigned code
gettingAllOnes	1230
slip	1231
stationTroubleGroup	1300
voiceEquipment	1301
dataEquipment	1302
videoEquipment	1303
otherEquipmnet	1304
stationWiring	1305
physicalTroubleGroup	1400
lightBurnedOut	1401
dataset	1402
ttySet	1403
highSpeedPrinter	1404
aNI	1405
aLI	1406
canNotActivatePC	1407
modem	1408
cathodeRayTube	1409
looseJack	1410
offHook	1411
physicalProblem	1412
processorDead	1413
wiringProblem	1414
wiringBrokeSetBrokePoleDown	1415
noRegister	1416
stuckSender	1417
otherStationTrouble	1418
otherCaseGroup	1500
callTransferProblem	1501
callwaitingProblem	1502
customerCallFeatureDoNotWork	1503
information	1504
threeWayCallingProblem	1505
orderWork	1506
ReleaseCktRequestedByIC	1507
ReleaseCktRequestedByEC	1508

TABLE 10B.2 *(Continued)*

Trouble type	Assigned code
ReleaseFacilityRequestedByIC	1509
ReleaseFacilityRequestedByEC	1510
RequestForRoutine	1511
Release	1512
RequestMonitorOfCircuit	1512
LostTimerReport	1516
HistoricalReports	1517
SwitchOrTrunkRelated	1518
TestAssist	1519

SOURCE: ITU-T Recommendation X.790, 1995.

The troubleType attribute

The identified trouble types as listed in Table 10B.2 cover a wide range of problems, including auido, connection, connectivity, and user equipment. The assigned code for each trouble type is for the purpose of achieving interoperability between different service providers that exchange trouble reports. The long list of trouble types is not exhaustive, and an application designer may choose a subset, to add additional attributes, or both.

Accounting Management

Outline

- The first section, "Accounting Management Process," introduces two complementary accounting processes, ITU's accounting process model and Bellcore's
  AMA data network system. A combination of the two process models yields
  three major processes of accounting management: usage metering, metering
  data processing, and charging and billing process.

- The second section, "Usage Metering and Data Collection," first introduces
  two metering methods, multimetering and automatic message accounting
  (AMA). It then presents a metering model as defined in ITU Recommendation
  X.742 (1995), discusses several practical issues involved in the AMA metering
  process, and concludes with a survey of sample AMA record fields for several
  commonly seen categories of services.

- The third section, "AMA Data Processing," addresses the issues related to
  AMA data processing at the network level that include service usage correlation, AMA records formatting and conversion, and AMA data aggregation
  and usage surveillance.

- The last section, "Charging and Billing," discusses the process of charging for
  network service usage by a customer and generating a final bill. The issues discussed include the steps for processing customer service usage, rating, and
  information required for billing such as service definition and pricing policies.

Introduction

Accounting management of telecommunications networks starts with collecting data of network resource usage from a network element such as a switch
and ends with sending out bills to customers for use of the telecommunications
services. In between is a whole range of tasks such as sending the collected

data out of the network element, validating and aggregating the usage data, and applying rating policy and tariffs. Traditionally this has been a highly proprietary process, including proprietary data format, noninteroperable interfaces for transferring data, and ways usage data are validated and aggregated that are specific to each service provider. It is not until fairly recently that some standard activities were initiated largely in response to the new realities of the telecommunications industry. The lack of standards is obvious from the terminology: different terminology has been used to describe the same process and the same terminology also means different things.

The new competitive telecommunications marketplace requires that multiple service providers interface with each other in delivering a service. The differences in interfaces and lack of interoperability between the billing systems can become a major obstacle to interservice provider operations.

The variety of service offerings has increased by leaps and bounds: PSTN services, wireless services, IN services, IP services, and broadband services (video on demand, telemedicine, etc.) and more are in pipe. New services and new practices in service offerings will continue to challenge the conventional accounting practices. New demands and requirements continue to rise. Real-time billing, customized billing, integrated billing with the capability to process large volumes of usage data, and many other new billing requirements continue to confront the accounting management.

This chapter describes a general accounting process that is common in today's telecommunications practices, introduces the existing standards in this area, and discusses related issues such as future trends.

Accounting Management Process

This section introduces two process models for accounting management, one defined in ITU-T Recommendation X.742 (1995) and one from Bellcore that is common in North America. These models present a high-level architectural view of the accounting management process, and the remainder of the chapter will discuss each component of the process in detail. Definitions of the commonly used terminology for accounting management can be found in Table 11.1.

ITU model

The process model for accounting as shown in Fig. 11.1 consists of three subprocesses. The *usage metering process* is responsible for generating usage metering records. Such a record contains a sequence of occurrences of accountable events and attributes associated with the events. Call setup completion and taking down of a call are two examples of such events. Among the attributes associated with the call setup are the time of call setup, the signaling channel that carries the setup message, and the type of call. The number of accountable events in a metering record is a local matter determined by each service provider. The metering process is also responsible for logging the metering records.

TABLE 11.1   Common Terminology for Accounting Management

Term	Definition
Automatic message accounting (AMA)	AMA is one of the most widely used records for recording detailed call-related data that can be automatically generated by an application system.
Billing process	Responsible for collecting service transaction records, selecting from these the ones that pertain to a particular service subscriber over a particular time period, and producing the bill from these records
Charging process	Responsible for collecting the usage metering records which pertain to a particular service transaction in order to combine them into a service transaction record. Pricing information is added to the service transaction record.
Multimetering	A simplistic metering method that takes snapshots of a call instead of recording the whole life cycle of events of a call
Rating	A process of associating a rate with a charge type to determine a charge for service usage
Usage metering	A process of measuring resource usage and collecting the usage data
Usage metering record	A data record to hold usage metering data

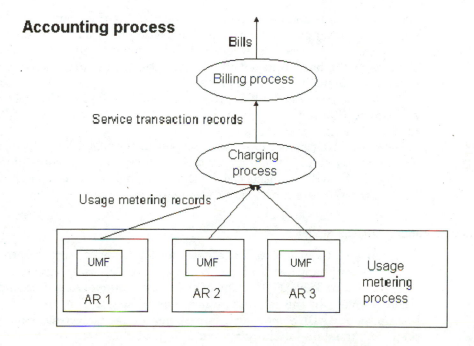

UMF-- Usage Metering Function; AR -- Accountable Resource

Figure 11.1 An accounting management process model. *(From ITU-T Recommendation X.742, 1995.)*

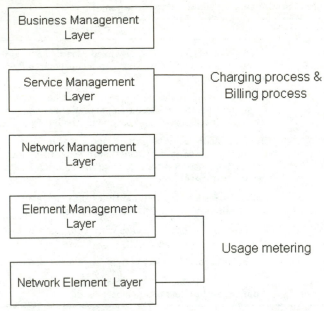

Figure 11.2 Accounting process model mapped to the TMN
logical layered architecture.

The *charging process* is responsible for collecting the usage metering records
associated with a particular service transaction in order to combine them into
service transaction records. A service transaction, i.e., an execution of a service
instance, is the basic unit for billing. The price information is added to transaction records in the charging process.

The *billing process,* the last process, produces bills for customers from service transaction records. The transaction records are combined, filtered, and
cut for a fixed time period.

The accounting model is shown mapped to the TMN logical layered architecture in Fig. 11.2. The usage metering process mostly takes place at NEL
and EML, as will be discussed in the following section. The charging and
billing processes are at the service management layer.

Bellcore accounting process model

A logical architecture of Bellcore's system for usage metering data collection
and processing is shown in Fig. 11.3. Automatic Message Accounting (AMA),
as explained in Table 11.1, is one of the most commonly used accounting data
record formats in the telecommunications industry. The AMA data network
system, or AMADNS for short, is responsible for collecting usage data and
preparing the data for charging and billing applications.

An AMADNS consists of two major components and a set of interfaces. The
first component is the data server. A data server, which can be implemented
either externally or internally to a generating system, after receiving AMA data,

processes the AMA records and stores them in a new AMA file. A data server can serve multiple generating systems. The AMA file then is transferred through standard protocol for file transfer from a data server to a Data Processing Management System (DPMS), the second component of the AMADNS.

A DPMS processes the AMA records contained in the AMA file according to the requirements of the application systems. For example, the format of AMA data may not be recognizable to an application system that desires to use the AMA data, and it is the responsibility of the DPMS to perform the necessary conversion or alteration and to prepare the AMA data for application systems. Again the AMA records are put in AMA files to be transferred to an application system and the AMA files are transferred using standard file transfer protocol and network protocol.

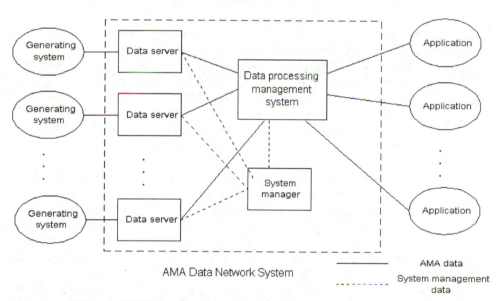

Figure 11.3   Bellcore's AMA data network system.

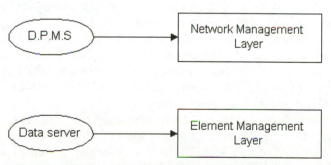

Figure 11.4   Mapping of the AMADNS to the TMN five-layer architecture.

The AMADNS interfaces two outside systems. At the front end it interfaces the generating system, and at the back end it interacts with billing and charging applications. The generating system is responsible for generating the AMA data that is input into the AMADNS. In today's telecommunications network, a generating system can be a switching node, service control point (SCP), or link monitoring system. The application systems in those network elements generate raw metering data, or AMA data. The AMA data-generating systems fall outside of the AMADNS and are responsible for sending the raw AMA data, whose format is normally network node (e.g., switch or SCP) specific, to the AMADNS.

At the back end, the AMADNS sends the processed AMA data to the application systems such as a charging system or billing system. Tariff rate, discount, and other rating information are applied to the AMA data, and customer bills are generated by the application systems. But this is out of the scope of the AMADNS.

So far we have described two accounting process models, one from ITU and one from Bellcore, for a reason: they complement each other, and together they constitute a complete accounting process. As shown in Fig. 11.4 (the figure shows mapping between the TMN five-layer architecture and the accounting model), the ITU model covers two ends of the process, mainly at the NEL and SML. The Bellcore model mostly deals with the middle part (the EML and NML) as shown in Fig. 11.4. The remainder of this chapter discusses an accounting process that is a combination of the two process models.

A scenario.   We use a scenario to describe the major steps of the accounting process to be described in the rest of the chapter. As shown in Fig. 11.5, the systems related to the accounting management include a switching node, a usage data-processing center, and a billing center. Assume that you made a long-distance call and used one other service such as call forwarding or call waiting as well. What are the major steps of the process that starts with recording your call and ends with determining the charge for the call?

Once a call attempt (you picked up the phone) is detected, the application system within the switching node recorded a sequence of events associated with the call along with a lot of other information. The events include call attempt detection, call setup completion, called party answering, and call completion. Among the examples of associated information are calling party directory number, called party directory number, starting time of each event, and duration of each event. An accounting metering data generation function extracts the accounting-related information and creates AMA records. Some preliminary processing may take place before a batch of AMA data records are shipped out to the usage data-processing center.

At the data-processing center, AMA records may be validated, to ensure the correctness of the data. A trivial example may be to check whether the value of an attribute is out of range. AMA data may be correlated and aggregated. For example, part of the data on your call may be collected on the switching node and part may be collected on a different node, say, an SCP. Data are aggregated for a single call. The billing center downstream may have special

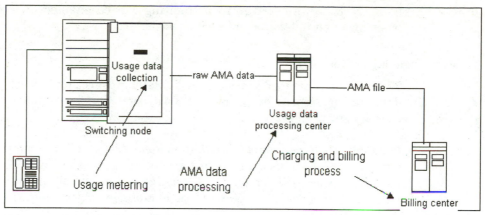

Figure 11.5   A sample accounting process.

requirements on the format of AMA records, which is different from that of AMA data generated by the switching node. Reformatting and conversion may be required in this case.

Once the processed AMA records arrive at the billing center, the rating policies, discounts, and other information such as tax are applied to the data to calculate the charge. Finally a bill is generated according to a predefined format.

Usage Metering and Data Collection

Usage metering is the first stage of accounting management, responsible for measuring and recording usage of network resources and services and for creating usage data records and sending the records to other systems for further processing. This section introduces the usage metering model defined in ITU Recommendation X.742 (1995), discusses more detailed steps of AMA usage metering, and finally presents sample AMA data record formats.

There are two general metering methods in use, one called multimetering (MM) and one called AMA. The multimetering method is an old, less-sophisticated scheme for usage metering that collects usage data at periodic intervals. In the old days, computing power and storage space were both very limited and only limited amounts of usage data could be collected and processed. This method is analogous to taking a snapshot of a call instead of continuously recording all the events in the lifetime of a call. For example, for a long-distance call, a snapshot is taken at the beginning of the call and one is taken at the end of the call. Many events in between such as call setup completion or call attempt cannot be recorded. This results in bulk billing of a subscriber's calls over a given period without much information on the details of a call. This method is still used in some parts of the world where crude billing is still acceptable.

In contrast, the AMA method is analogous to the continuous filming of a call throughout its lifetime with all major events of a call and other information being recorded. Modern AMA accounting can be enabled on a whole switch basis or on an individual subscriber basis.

ITU usage metering model

This metering model defines a generic way for collecting and controlling usage data and for representing usage information that is sufficiently general for subsequent processing.

Description of the usage metering model. Modeled as objects, the major components of the usage metering model are shown in Fig. 11.6. The focus of our discussion will be on the usageMeteringControl, usageMeteringData, usageMeteringRecord, and accountable managed object classes.

The accountable managed object class. Let's start with accountable objects that represent the network resources for which the usage data are gathered. Accountable resources can be logical or physical. For example, logical resources such as the communication line, trunk, access ports, and service features (e.g., call forwarding, call waiting) are all accountable resources whose usage shall be charged to generate revenue.

The usageMeteringData managed object class. The usageMeteringData object represents the accounted use of an accountable resource with information identifying the user, the service provided, type of resources, quantitative measures of usage, and many other relevant pieces of information. Specifically, this object has the following attributes:

accountableObjectReference. The accountable object for which this usageMeteringData object is created.

auditInfo. This attribute provides information on the source of the usage metering data, which may be required by the system processing this metering data. Examples include file or record numbers for the source data from which the usageMeteringData object is derived.

controlStatus. This attribute indicates the state of the usageMeteringData object. For example, a suspended controlStatus signifies that the usage metering is not taking place at the moment.

proceduralStatus. This attribute indicates the stage of usage metering. For example, the *terminating* proceduralStatus value indicates that the data object has stopped metering as a result of a delete request from a managing system.

dataErrors. This attribute is used to indicate whether the usage data are believed to be in error.

dataObjectId. An identifier for this usageMeteringData object.

providerId. An identifier of the provider of resources or services such as a phone company.

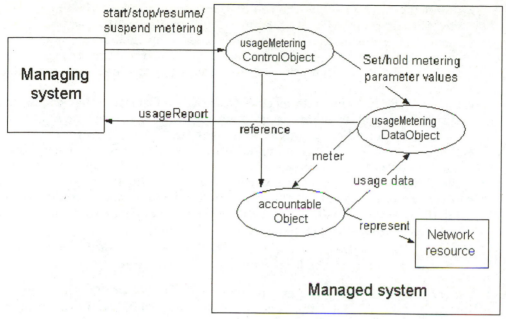

Figure 11.6   The relationships between the managed object classes for usage metering and metering data collection.

usageInfo.   This is a complex attribute that represents the core of resource usage data and is equivalent to the AMA data in a common accounting management system. The usage information is represented in terms of a set of accountable events while each event has a set of associated attributes that provide further detailed usage information related to each accountable event. Seven such accountable events are specified as follows, and each has a data block containing the information associated with the event.

- *Registration.*   It represents the detection of a user. One example is phone off hook as a caller attempts to make a call. The information contained in the block shall include the identifier of the user and a time stamp of the call stamp.
- *Request.*   This data block represents any form of input generated by the service-resource user such as dialed or key destination, input account number, or feature activation code. The information in the block may include the identity of remote parties like a called number, service-specific information, and metering data.
- *Accept.*   It represents any resulting response to the request such as called party answering the phone, establishment of a connection to a remote computer application, or activation of a service feature. Information associated with this event may include the identity of remote parties, service-specific information, and metering data.

■ *Complete.*   It represents the completion of one instance of using a service such as completion of a phone call or completion of a remote application. The information in this data block may include a completion cause such as a hang-up by the called party, a reporting trigger that specifies the event that triggers the generation of a usage report, and metering data.

■ *Corresponding.*   This data block contains the information that enables a system to correlate many usage metering records to provide a service transaction record, for example, a message ID.

■ *Bulk.*   This data block contains all other non-event-related usage measurements that are independent of an individual event. One example is the volume usage over a given period on a leased circuit.

■ *Interruption.*   The data in this data block indicate abnormal events during the accounting metering activities, such as clock change during service utilization or resetting of metering measurements.

The usageMeteringData object class has the following notifications that can be optional depending on the local context.

usageReport.   A usage report is generated according to the control instruction specified by the reportingTrigger attribute of the associated usageMeteringControl object that is discussed next.

objectCreation.   A notification is sent to the associated managing system to report the creation of this usageMeteringData object.

objectDeletion.   A notification is sent to the associated managing system to report the deletion of this usageMeteringData object.

The usageMeteringControl managed object class.   The usageMeteringControl object, which can be implemented either as a single, multiple, or embedded object (its implementation is a local matter), provides core control functions of usage metering. Its attributes are described first, followed by the notifications and actions.

controlObjectId.   The identifier for the usageMeteringControl object assuming that usage metering control is implemented as an independent object.

operationalState.   It indicates the operability of the usageMeteringControl object, that is, enabled or disabled.

reportingTrigger.   It specifies the events that will cause usage information to be reported. The events are categorized into three groups: periodically scheduled, instructed by a managing system, and some other external stimulus.

accountableObjectReferenceList.   It is a read-only list of accountable objects whose usage metering is controlled by this usageMeteringControl object.

dataObjectReferenceList.   It is a read-only list of metering data objects for which this usageMeteringControl object is provided.

The usageMeteringControl object has the following notifications, some of which may be optional, depending on the local context.

attributeValueChange.   A notification is sent to the associated managing systems to report the change of values of specified attributes such as reportingTrigger or operationalState.

objectCreation.   A notification is sent to the associated managing system reporting the creation of a new instance of the usageMeteringControl object.

objectDeletion.   A notification is sent to the associated managing system reporting the deletion of this usageMeteringControl object instance.

stateChange.   A notification is sent to the associated managing system reporting the change of the operationalState value.

meteringStarted.   A notification is sent to the associated managing system reporting the fact that usage metering has started with the metering control information.

meteringSuspended.   A notification is sent to the associated managing system reporting the fact that usage metering has been suspended.

meteringResumed.   A notification is sent to the associated managing system reporting the fact that usage metering has resumed.

The usageMeteringControl object has the following actions that may be optional, depending on the local context.

startMetering.   This action enables a managing system to start or restart usage metering, recording, and reporting of the collected usage data. The effect of starting usage metering is that the usage parameter values for all identified, related metering data objects are initialized or reinitialized.

suspendMetering.   This action enables a managing system to suspend active usage metering on specified usageMeteringData objects by holding the values of the usage parameters of the metering data objects.

resumeMetering.   This action enables a managing system to resume the usage metering on an identified set of metering data objects so their usage parameter values may be recorded.

The usageMeteringRecord managed object class.   This is a subclass of the eventLogRecord object class, which is described in Chap. 9. Usage data, in addition to being sent out for further processing, is also logged locally. In addition to the characteristics inherited from the eventLogRecord object class, the usageMeteringRecord object has the following attributes as well: accountableObjectReference, usageInfo, providerId, auditInfo, and dataErrors. The definitions of these attributes are the same as those in the usageMeteringData object class.

Behavior of the usage metering model.   The relationships between the usage metering–related managed objects are shown in Fig. 11.6. We use a scenario to illustrate the behavior of the usage metering model just described. Assume that the required managed objects such as usageMeteringData and usage-MeteringControl are already created. The managing system is interested in collecting the usage data on a switching node. The usage metering model defined by ITU assumes the OSI management framework and the CMIP management protocol. The following provides a description of high-level behavior of the model.

1. The managing system sends an instruction to the managed system to start the usage metering on a specified usageMeteringControl object and a set of usageMeteringData objects, using the CMIS service M-ACTION. Assume that in this case, the managing system is interested in the usage of call-forwarding services.

2. The managed system invokes the startMetering action on the usage-MeteringControl object. In turn, the usageMeteringControl object sets the attribute operationalState for the usageMeteringData object of the call-forwarding service to enabled so that it can start recording the usage information. The usageMeteringControl object also sets the reporting- Trigger according to the instruction from the managing system. In this scenario, the scheduled reporting type of the trigger is assumed.

3. The usageMeteringData object directly extracts the usage information from the associated call-forwarding service accountable object. Blocks of usage data are created one for each event such as a customer initiating a call, call setup completion, call termination, and total number of calls in the specified period of time.

4. At the end of the scheduled reporting period, the usageMetering Data object sends the usage report to the managing system using CMIS M-EVENT-REPORT.

In summary, the ITU usage metering model presents a high-level structure for usage metering, using the OSI management framework and an object-oriented approach to metering data modeling, in hopes of standardizing the usage metering data and usage metering control. Generally speaking, this model provides a generic reference, and many detailed issues related to usage metering are left for local implementation.

Practical issues with metering data generation

This subsection describes some issues or practices commonly seen in today's metering data-generating systems in telecommunications networks. Figure 11.7 features the issues in a metering data generation process model.

Usage metering system configuration. The configurations of a usage metering system are just as diverse as the network nodes are proprietary. Different configurations fall into three categories: configured internally within a network element like a switching node or SCP, as an independent unit with an interface to the network element from which it receives usage measurements, or a hybrid of the two.

Usage data receiver. The initial, unformatted AMA data are normally generated by application software or a dedicated software module embedded inside the application software. This portion of the metering system is highly proprietary. The usage metering system has an interface to an application program to receive unformatted data from the application.

Usage data validation. The usage metering system will validate the received unformatted usage data to ensure that the syntax is correct, the recorded value is within a specified range, and the usage data are collected in a timely fashion from the application program. Once an error is detected, the action to be taken is largely a matter of local implementation, and the action may include a simple error correction, discarding the data with error, or generating an alarm.

AMA record generation. The usage metering system generates AMA records from the unformatted data received from the application according to the predefined AMA format. Special flags are needed in the AMA record to indicate any special cases when AMA data are transmitted. For example, the service state is in the suspended state when the metering data are generated on this service, or the switching node is going through a test procedure at the time the AMA data are collected.

AMA data fault monitoring. The usage metering system is responsible for monitoring the local AMA data storage and generates an alarm if a certain threshold is crossed. For example, a threshold can be set for a mass storage space that is 85 percent full. AMA data loss can result from abnormally ended recording or other conditions that can also result in alarms being generated.

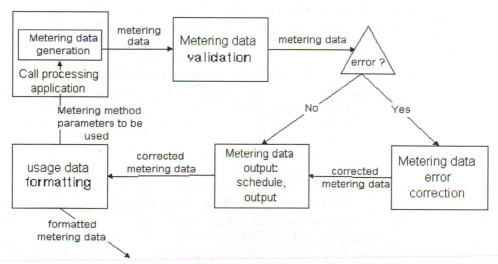

Figure 11.7 Preliminary processing of usage metering data.

AMA data output function. An interface to an outside system is defined that includes an information transfer protocol. Often a packet-switched data network such as that defined in ITU-T Recommendation X.25 (1993) is used for transferring AMA data. The output can be scheduled, or the usage metering system can be polled by the usage data-processing system.

Sample AMA record parameters

Since AMA data record formats are almost universally proprietary for each service provider or switching node, we instead present some sample fields of AMA data structure that are sufficiently general to illustrate the types of data that are generated in the usage metering process. Other than a few common header fields, the AMA structures highly depend on the services or network resource.

The sample AMA parameters described below are based on several commonly seen AMA record formats [for instance, the Bellcore AMA format (BAF) is one], but those fields specific to a service provider are stripped away in order to provide a generic view. In practice, the AMA structure is based on the types of services, such as ATM switched virtual circuit (SVC) and packet-switched data services. A predefined structure for each type of service is applied to a call once the call type or service type is determined. The granularity of service classification presented here is much coarser than those used in actual billing records, again for the purpose of presenting a generic view.

Sample AMA parameters for basic PSTN services. Table 11.2 presents some sample AMA parameters for the basic PSTN services such as line-to-line calls, both local and long distance without international long distance.

Sample AMA parameters for AIN or IN services. AIN or IN services are those provided by an AIN or IN that separates service logic from switching logic. Examples of AIN services include free phone, credit card call, and calling card calls. (See Table 11.3.).

Sample AMA parameters for wireless services. Although there is nothing to distinguish a wireless service per se from its wireline counterpart other than the fact that the services are carried on a wireless network, there are some fields uniquely identified for wireless network resources for billing purposes. (See Table 11.4.).

Sample AMA parameters for packet-switched data services. Considered here are only the data calls carried on public switched data networks. The field data network identification code (DNIC) is from the ITU data network numbering plan, which is explained in detail in Chap. 16. Note that the billable unit is the data segment count, and it is conceivable that finer granularity (e.g., bits) may be used in future billing practice when computing power and bandwidth constraint become a lesser issue. (See Table 11.5.).

TABLE 11.2   Sample AMA Fields for PSTN Services

Field name	Description
Calling party category	Category such as residential or business
Calling party number	The directory number that originated the call
Called party number	The directory number of the called party
Called party off-hook indicator	Indicates whether called party off-hook is detected
Elapsed time	The time interval of the call, often in minutes, seconds, and tenths of seconds
Service feature	Indicates a special service such as call forwarding, three-way call, centrex service
Connect time	The time connection is established
Recording office identification	The central office identification number

AMA Data Processing

The AMA data-processing system is responsible for intermediary processing of AMA data before the data are sent to downstream applications such as charging and billing. This system is characterized by

- A networkwide view of usage metering data
- Responsibility for networkwide error checking and validation
- Responsibility for providing a consistent view of the usage data on a per-service or per-customer basis
- Responsibility to provide AMA data in the format required by the downstream applications

Figure 11.8 provides a functional view of the AMA data-processing system. The AMA data management functional component provides database capabilities for managing the AMA records received from the usage metering system. The capabilities to store and retrieve AMA records are essential for processing AMA data. The system management component provides capabilities to manage fault and security for the AMA data-processing system. The system fault includes database full, hardware failure, and storage write failure. Among the issues related to security management are control of access to AMA data and authentication of the user. The focus of this section is on the rest of the functional components, for which a high-level description is provided. For more details, see Bellcore GR-1343 (1996).

Interfaces

The interface to the usage metering system is primarily for transfer of AMA data records. The AMA data are transported in files rather than in individual data records for efficiency purposes. Standard file transfer protocols such as

TABLE 11.3 Sample AMA Fields for AIN/IN Services

Field name	Description
AIN service type	The service types such as free phone and debit card call
Service feature	The code for service features such as call forwarding on busy, three-way call, and chargeable quotation. The default is the AIN service.
Service observed/ traffic sampled	Whether service observing or traffic sampling has been applied to the service
Recording office type	The type of central office switch that recorded the billing data, such as a particular vendor's central office switch (e.g., AT&T 4 ESS switch, Nortel DMS-100, Ericsson AXE switch)
Recording office identification	The identification of the central office that recorded the billing information
Date of query processing	The date when a query to the SCP by the SSP of an AIN network was made
Time of query processing	The time when a query to the global database by the SSP of an AIN was made
Service logic identification	The identification number of the service logic program inside the SCP of an AIN. This identification number is created by a service provider for administrative purposes and may not be applicable to another service provider of another country.
Start time	The time the call started
Elapsed time	The elapsed time of the call
Calling number	Directory number of the calling party
Called number	Directory number of the called party
Query originator	The type and identification of the service provider that originated the query to the global database
Service key	The service key associated with the service logic program
Response code	Code for the response from an AIN SCP

file transfer protocol (FTP) or file transfer access management (FTAM) can be used. This interface is responsible for receiving the AMA files; unpacking AMA records from files; and managing file status, file naming, and file integrity control and sessions for file transfer. One AMA data-processing system can be connected to multiple usage metering systems, and interface requirements may be different from one usage metering system to another.

The AMA data-processing system's interface to charging and billing systems is very similar to the one described above. The processed AMA data records are packaged into AMA files and transferred over to the applications such as billing and charging. It is responsible for creating the AMA file from the AMA records and managing the file status, naming convention, file integrity, and session for file transfer.

TABLE 11.4   Sample AMA Fields for Wireless Services

Field name	Description
Call type	The type of call. There are over 100 types of calls defined, such as ISDN permanent virtual circuit, public packet-switched network (PPSN) aggregate, ISDN virtual circuit aggregate, and wireless.
Calling party category	Category such as residential or business
Calling number	The directory number of the calling party
Called number	The directory number of the called party
Originating ESN	The electronic serial number (ESN), or the identification number, of the handset that originates the call
Terminating ESN	The electronic serial number (ESN), or the identification number, of the handset on which the call terminates
Call direction	Land to mobile, mobile to land, or mobile to mobile
Calling party initial BSC&BTS number	The identification number of the base station controller and the base station transceiver system that originated the call
Called party initial BSC&BTS number	The identification number of the base station controller and the base station transceiver system that first received the call
Calling party final BSC&BTS number	The identification number of the base station controller and the base station transceiver where the call terminates
Intra-MSC hand-off count	Total number of hand-offs within the same MSC
Inter-MSC hand-off counter	Total number of hand-offs between MSCs

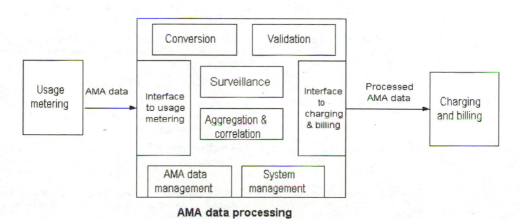

AMA data processing

Figure 11.8 Functional components of the AMA data-processing system.

TABLE 11.5 Sample AMA Fields for Data Services

Field name	Description
Call type	The type of call. There is a large number of types of calls, such as ISDN permanent virtual circuit, public packet-switched network (PPSN) aggregate, and ISDN virtual circuit aggregate.
Packet service type	Either virtual circuit or permanent virtual circuit
Calling party number	The data terminal number that originates the data call
Called party number	The data terminal number that receives the data call
Recording office type	The type of central office switch that recorded the billing data, such as AT&T 4 ESS switch, Nortel DMS-100, and Ericsson AXE switch
Recording office identification	The central office identification number
Present date	The date the billing record is created
Present time	The time stamp of the billing record
Connecting DNIC-INIC	Data network identification code (DNIC) or ISDN network identification code (INIC) that is from the ITU-T's international numbering plan for the public data network. See Chap. 16 for details.
Connection X.75 interface ID	The ID of the gateway interface between public data networks that were used for this account
Chargeable DNPA	Chargeable data numbering plan area, the equivalent of the numbering plan area (NPA) or area code of a phone number, but for the data network instead of voice network
Chargeable DCO-EPN	Chargeable data central office–endpoint number, the equivalent of the NXX of a phone number, or central office code for a data network. See Chap. 16 for details on the numbering plan, including NXX.
Chargeable ISDN channel number	The ISDN bearer or signaling channel used
Chargeable logical channel number	Identifier of a logical channel or the permanent virtual circuit (PVC) used
Rate period segment count	The total count of the packet segments sent and received from the beginning of the recording period to the generation of the billing record
Segment size	The size of the segments counted in the rate period
Auxiliary network identifier	The identity of the network element that contains the interface such as the X.75 interface that is being recorded. This field is used for auditing purposes on the network.
Count of call setups	The total number of call setups, made by all switched virtual circuits during the recording interval. It is up to the operating company to decide whether the failed attempts also count.
Count of packet segments	The total number of all segments sent and received for all circuits associated with a particular interface during the recording interval
Call holding time	The sum of the call holding time for all switched virtual circuits during the recorded interval

AMA data validation

Usage validation is the inspection of the contents of AMA data records to ensure that the records meet specific integrity checks. The validation functional component is responsible for reporting an error to other system components once an error is detected. It looks into data elements of each AMA record, verifying that the syntax is correct, the recorded value is within specified range, and the usage data are collected in a timely fashion from scheduled network elements. It also examines an entire AMA record, cross checking between fields in a record to ensure that all required information is included. Note that this functional component is similar to the one discussed earlier. The main difference is that this functional component works at a network level, while the earlier one is for one network element only.

Service usage correlation

In certain situations, usage data are collected across a network from multiple usage metering data generating systems for the same service instance. The data need to be correlated to produce a consistent view of a service instance. For example, two broadband switching systems can generate AMA records for the same connection, and it is the responsibility of this functional component to recognize that multiple AMA records represent a single connection rather than two different ones. In other cases, various components of usage data of the same service instance are spread across different AMA records and need to be associated with the service instance to provide a complete view according to the requirements of downstream applications.

Formatting, format conversion, and modification of the AMA records

An AMA data-processing system may receive AMA data records from multiple, different usage metering systems, some of which are unformatted records. In this case, the AMA data-processing system must format the data according to the requirements of the downstream application systems.

Another scenario is the case where the received AMA data records need to be transformed from one form to another suitable for the downstream applications. This is likely because the formats required by the downstream applications (e.g., charging and billing) are often different from that generated by the usage metering system due to the fact that these systems were developed either by different vendors or at different times.

Sometimes, due to the requirement from downstream applications, it is necessary to create multiple, unique AMA records from AMA data contained within a single AMA record. For example, Bellcore has its own AMA format, called BAF, which defines a set of AMA data modules. In some cases it is required by downstream applications to take a single AMA record with multiple BAF modules and create multiple AMA records, one for each BAF module.

In some situations, the opposite is true: some data elements in AMA records need to be removed before the AMA records are delivered to the applications. One situation is where the downstream application simply does not need the data. Another situation is where some elements of data are prohibited from being passed to an application. For example, AMA data records may be exchanged between service providers. Certain proprietary information in AMA data records cannot be transferred to another service provider.

AMA record aggregation

In some cases, multiple AMA records need to be combined into a single AMA record through the summarization of the data contained in those records. One motivation for the record aggregation is to conserve the network bandwidth required for transferring the AMA data records. Also some downstream applications in the billing and charging process may require aggregated data. One extreme example is that an application requires the call duration, instead of the start and ending time of a call. Thus, two AMA records, one for call start and one for call completion, can be combined into one that contains the desired call duration.

A more realistic example is an aggregated count of an event such as arrival of packets or ATM cells over a specified period of time. If the aggregated count required by an application is over a period of time longer than the one generated by the usage metering system, then shorter-period counts are summarized to arrive at the count of desired period of time. Also it may be required to summarize the service usage on a per-customer and per-service basis, based on the service provider's criteria.

Usage surveillance

This functional component monitors all AMA data to detect usage patterns deviating from an established or service provider–defined norm. By doing so, a service provider can detect errors in translation, general software, or network routing. An additional purpose is to detect potential fraudulent activity.

Charging and Billing

This is the final stage of the accounting management process. This process starts with the AMA data passed in by the AMA data-processing system in the format required by the charging and billing systems. From the AMA records, the charging system aggregates the information related to one service that may spread across AMA records into a service transaction record and applies the price information to it. The billing system applies tax, discount, and related tariff information and generates bills for individual customers. Charging and billing as described in this section are mainly for the usage of the network service by individual subscribers rather than for network resource usage by the service provider (e.g., network access charge).

A general functional model for charging and billing is shown in Fig. 11.9. A similar model for ATM billing can be found in van Hoorn and Kuiper (1996). The functional components included in the charging and billing systems are presentation, rating, message processing, and an interface to the AMA metering data-processing system. Charging and billing also interact with other support components such as customer administration and invoice management.

Customer service usage processing

The primary function of the functional component customer service usage processing is to build service usage records out of AMA data records and associate service records with a particular customer. There are no standard formats for customer service records, for the obvious reason that there is a great diversity of services. In general, three types of data can be found in customer service records:

- Customer-identifying information
- Service-identifying information
- Service usage information

The four steps of this process are shown in Fig. 11.10.

AMA data are retrieved from local data storage. The AMA records should be searchable by customer name, service types, or other criteria. A query to the customer management subsystem shall yield customer identifier, DN associated with the customer, services subscribed by the customer, and other identifying information (e.g., address and social security number). The information serves as a reference point for selecting the AMA records associated with this customer.

A service may be characterized by many attributes. For example, the ATM-based data service can have a transmit quality-of-service class, transmit peak cell rate, transmit sustainable cell rate, and burstiness. A contract between a customer and a service provider may include only a subset of the service

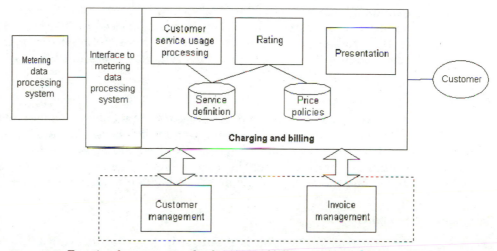

Figure 11.9 Functional components of a charging and billing system and its context.

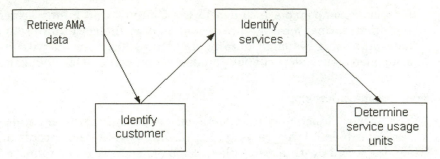

Figure 11.10   The process of generating customer service usage data.

attributes. This step maps the contracted service attributes to those defined by a service provider and associates the AMA data records that recorded the service attributes for the customer with a billable service. Then based on the identified services and service attributes, the last step of the process calculates the chargeable units of service usage.

Rating

The central goal of this functional component is to calculate the actual charge for the units of service usage calculated above and provide financial components of a billing record. The calculation of the basic charge is straightforward: charge = total units × unit price. However, some other financial components may need to be factored into the calculation. Some examples of the financial components include internal cost, initial nondiscount rate, discount rates, rates of special offering, wholesale rate, or retail rate. Usage charge aggregation means totaling the rated usage charges for a billing period and applies any appropriate discounts, rebates, and surcharges.

The service level agreement (SLA) also needs to be taken into account. For example, if the services offered are below what the SLA stipulates in terms of quality of service, the penalties against the service provider should be reflected in the final service charge. In addition, this functional component provides the capability to compute tax tables related to different types of usage data.

Presentation

This functional component determines the final charge on a bill and the format and layout of the bill. To determine the final charge, various tariffs should be applied. The layout and format of a bill should reflect customer's needs and should be easily presentable to customers. The layout and format may also depend on the preagreed payment method: automatic payment or alternative.

Information repositories

A variety of information is required in the charging and billing processing, among which are the two primary ones, service definition and price and tariff information.

Service definition repository. The service definition repository categorizes services and defines common characteristics of a class of services. The common classifications include PSTN service, IN services, enhanced PSTN services, operator services, and wireless services. An earlier section of this chapter listed examples of the services. Within each class of services, finer subcategories of services are often defined to capture the unique features of those services.

The characteristics of a service can be categorized for the convenience of managing the service definitions. One way of categorization divides them into internal static characteristics and external characteristics. The internal static category features the technical characteristics of a service. For example, for ATM services, the defining characteristics may include bit rate type (e.g., variable bit rate, constant bit rate, or other) and connection type (permanent virtual connection or switched virtual connection). The external characteristics are those service parameters that customers can negotiate. For example, some parameters of a call forwarding on no answer are settable by the customer: calls can be forwarded to one number for a specified period and to another number for a different time period. For ATM services, quality of service and bandwidth are customer negotiable.

Price and tariff repository. A variety of information related to price and tariff can be in the repository. This discussion only serves up some examples. The price plan for unit-based services shall be defined and available in the price and tariff repository for use in the charging and billing process. To increase the flexibility in service charging, more than one price plan can be defined for a service. A price plan basically consists of component costs of the service. For example, a price plan for an ATM-based service may have.

- Cost of call setup or connection establishment
- Usage cost during the connection

A rate table is established for each service that takes into account the time service is rendered and other factors such as unit-price plan. For example, for a long-distance call the rate table may look like that in Table 11.6.

Besides the rate table and price plans, the other information may include, but is not limited to,

- Tariff rates that may include tax levied by both local and nonlocal governments
- Charging discount
- Charging methods
- Accounting formats

Support functions

There are other functions that are not part of the charging and billing process but provide support for them. Two such critical function components are customer administration and invoice management.

TABLE 11.6   A Sample Rate Table for a Unit of 1 Minute

Time	cost per minute		
	Weekdays	Weekend	Holiday
Peak hours (7 a.m.–8 p.m.)	0.25	0.10	0.05
Nonpeak hours (8 p.m.–12 a.m.)	0.15	0.10	0.05
Night (12 a.m.–7 a.m.)	0.10	0.10	0.03

Customer administration.   Customer administration is not part of the charging and billing processes but provides support for them. Its responsibilities include administering customer accounts and profiles and managing customer inquiries.

One task of the customer account administration is handling delinquent accounts by providing access to those accounts, arranging collection actions, etc. The account administration is also responsible for computing account balances, computing one-time and recurring charges, and adjusting account balances according to credit and debit information from invoice assembly and receipt of payment functional units. Customer profile administration receives information from configuration management and keeps track of individual customer information (e.g., subscribed services, customer address, ID).

The customer administration provides two kinds of customer support. One is support for in-call service requests to accommodate on-demand, real-time requests for usage information while a call is in progress or at its completion. The second type of support is for off-line inquiry after service is rendered. A customer may inquire about the current balance, review the service contract, and contest service charges.

Invoice management.   Like the customer administration, invoice management is part of the customer care system, instead of part of the charging and billing process. The responsibilities of invoice management include receiving payments, sending invoices, and storing invoices.

Summary

Accounting management in general, and billing in particular, is considered one of the core business processes of a telecommunications service provider and one of the most proprietary ones. The primary goal of accounting management is to measure, charge, and bill the service and network resource usage by customers. Accounting management is becoming increasingly complicated largely due to the fact that telecommunications services are rapidly increasing in complexity and variety.

This chapter discusses three processes of accounting management: usage metering, AMA data processing, and charging and billing. A usage metering system interacts with a network node such as a switching system or SCP to collect formatted usage data, performs very basic processing on the data, and formats the data into predefined AMA records before shipping the records to an AMA record-processing center.

An AMA record-processing system receives AMA data records from multiple usage metering systems and may perform any number of the following tasks. It may correlate the data records from multiple usage metering systems to produce a consistent view of AMA records on a per-customer or per-service basis. It may aggregate AMA records to either reduce the number of AMA records to be transferred or meet the requirements of downstream applications. AMA records may be converted from one format to another, one record expanded into multiple ones, or multiple records condensed into a single one, again depending on the requirements of the downstream applications.

The primary downstream applications are charging and billing. Charging and billing generate billable usage units on a per-service basis for a customer for a given billing period and apply a unit price to arrive at the total charge. Then a whole range of rating factors needs to be applied to the service charge: tariffs, rate plan, discount rate, special promotion offering, and others, in order to produce the final bill.

Looking forward, we can say that accounting management is facing enormous challenges, maybe more so than other areas of telecommunications network and service management, as a result of rapid changes in the regulatory environment and marketplace, new customer requirements, and technological advances.

Some of the newer customer requirements will help illustrate the point. One is real-time billing that provides the real-time charging and billing information while a call is in progress. For example, a wireless service user may want to know the charge of a call up to the moment. For another, the integrated billing combines charges of different types of services into one bill. As telecommunications services increasingly become indispensable in our daily life, an increasing number of customers will use more than one type of service. Soon customers will not only desire but demand an integrated bill that combines the charges for services such as the Internet service, regular PSTN and wireless services, TV cable, and multimedia services, even though the services are provided by different service providers.

Exercises

11.1 What is the main difference between the AMA metering method and the multi-metering method?

11.2 What are the three major processes in accounting management, as discussed in this chapter?

11.3 In the ITU usage metering model, what is the relationship between the usageMeteringControl object and usageMeteringData object?

11.4 In the ITU usage metering model, which managed object contains the AMA data?

11.5 Explain why the AMA data records from usage metering need to go through intermediary processing at the AMA data-processing system before being used in the charging and billing process.

11.6 Give an example where AMA records need to be correlated before being sent to charging and billing applications.

11.7 Give an example where AMA records need to be converted from one format to another before being sent to charging and billing applications.

11.8 Describe the input and output of the charging functional components.

12

Security Management

Outline

- The "Introduction" provides a high-level view of a security management process model that consists of three subprocesses: fraud prevention, fraud detection, and fraud recovery.

- The first section, "Fraud Prevention," discusses proactive measures to prevent frauds.

- The second section "Fraud Detection," discusses ways to detect a fraud promptly and efficiently once it occurs.

- The third section, "Fraud Containment and Recovery," discusses how to contain and limit the harm brought about by frauds.

- The fourth section, "Security Services," introduces five types of security services, i.e., authentication, integrity, confidentiality, access control, and nonrepudiation, that can be used to prevent or detect frauds.

- The last section, "Security Mechanisms," introduces security mechanisms such as encryption, digital signature, secure hash functions, and notarization that are used to implement security services.

Introduction

The primary goal of security management is to ensure the security of telecommunications networks, the management system that manages the telecommunications networks, and management transactions. Specifically, security management is intended to prevent frauds if at all possible, detect frauds promptly, and recover from and limit the consequences of frauds as efficiently as possible. In the current telecommunications environment, with the fast expansion of telecommunications services, fraud is becoming a serious problem. Examples of fraud include

- Fraudulent use of a service (e.g., service theft)
- Fraudulent use of a network resource (e.g., malicious circular forwarding of calls)
- Unauthorized management operations (e.g., assuming the identity of a manager and issuing management directives)
- Modification of information (e.g., intercepting a management message and changing it)
- Disclosure (e.g., breaking the encryption) of confidential information

The security management process, shown in Fig. 12.1, consists of three subprocesses. The first goal is to prevent fraud from taking place and thus prevent any harm a fraud may bring about. Since no system is fraudproof, the next best thing that can be hoped for is to detect frauds as soon as possible and to minimize the damage caused by the frauds. All three processes in the security management model use security services to achieve their goals. Security services include authentication, data integrity, confidentiality, access control, and nonrepudiation services. A security service is implemented through one or more security mechanisms, and examples of the security mechanisms include encryption, digital signature, data padding, and notary.

Fraud Prevention

The primary goal of fraud prevention is to keep network resources out of reach of potentially fraudulent users by denying them access to the resource. While other preventive measures are in use, such as security screening that verifies that a customer has a good credit record to indicate an acceptable risk for providing services, by far the most important measure for fraud prevention is access control. Details of access control are presented in the "Security Services" section later.

Fraud Detection

Perhaps there are as many ways to detect frauds as there are types of frauds themselves. This section first provides a high-level description of several commonly used approaches and concludes with an introduction of a security alarm report model and a security audit trail model. The two models are not exclusively defined for the purpose of fraud detection. Instead, fraud detection is a good application of the security alarm reporting and security audit trail.

Common fraud detection approaches

One approach is customer profiling. This approach creates profiles of customers and analyzes customer information such as credit history, payment records, and other related information in order to assess the risk of providing services to such customers. High-risk customers may be flagged for special attention.

Another approach is usage pattern analysis. This approach, similar to customer profiling, analyzes audit trail information and service usage data of a customer or a group of customers to detect substantial deviation from established patterns. Two kinds of audit trails are in common use: those of customers and those of networks. An audit trail of a customer is a chronological record of all events and activities associated with the customer. Examples include the call history of the customer, which may include data such as directory number of terminating party, time of day, and length of each call. An audit trail of a network is a chronological record of all events and activities associated with the network. Examples include records of equipment maintenance personnel accessing the routing table.

Analysis of billing records may reveal irregularities and usage abnormalities. An example is a sudden surge of overseas wireless phone calls that may be worth further investigation to determine whether the originating terminal is reported stolen. Another example is monitoring the relationship between the revenue and network resource usage. Deviation from an established pattern may indicate fraudulent use of network resources.

Both customer profiling and usage pattern analysis require the capability to process large amounts of data, identify the underlying patterns, and unearth the hidden knowledge. Many of the data mining techniques such as tree induction, association rules, and cluster identification can find a good fit in fraud detection applications.

Another approach is traffic and activity pattern analysis that can also indicate a security breach or fraud. One example is looking for a spike of resource

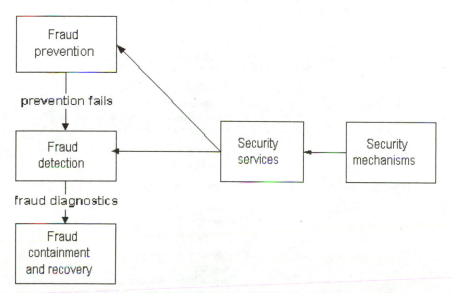

Figure 12.1   Security management model.

usage such as trunk usage that cannot be explained by known factors. Analysis of security alarms can also indicate a potential security breach. For example, abnormally frequent connection and disconnection may indicate some unauthorized party's attempt to gain access to network resources.

Security alarm reporting. An OSI-model–based security alarm reporting function, as defined in CCITT Recommendation X.736 (1992), provides a mechanism to alert a managing system of an event indicating an attack or potential attack on a system's security. Security alarm reporting is based on the OSI event reporting model as described in Chap. 10. Security-related events are forwarded by an event forwarding discriminator object to a managing system through the CMIS service M-EVENT-REPORT. The following five types of security alarms are defined, and the possible causes of each alarm type are listed in Table 12.1.

- *Integrity violation.* Indicates that a potential interruption of information flow has occurred such that information may have been modified, inserted, deleted, or duplicated by an unauthorized entity

- *Operational violation.* Indicates that the provision of the requested service was not possible because of the unavailability, malfunction, or incorrect invocation of the service

- *Physical violation.* Indicates that a break of the physical resource has been detected

- *Security service or mechanism violation.* Indicates that a security attack has been detected by a security service

- *Time domain violation.* Indicates that an event has occurred outside the permitted time period

TABLE 12.1   Security Alarm Types and Causes

Event type	Security alarm causes
Integrity violation	Duplicate information, information missing, information modification detected, information out of sequence, unexpected information
Operational violation	Denial of service, out of service, procedural error, unspecified reason
Physical violation	Cable tamper, intrusion detection, unspecified reason
Security service or mechanism violation	Authentication failure, breach of confidentiality, nonrepudiation failure, unauthorized access attempt, unspecified reason
Time domain violation	Delayed information, key expired, out-of-hours activity

Five classes of security alarm severity are specified as follows.

- *Critical.*   A breach of security has occurred that has compromised the system.

- *Major.*   A breach of security has been detected, and significant information or mechanisms have been compromised.

- *Minor.*   A breach of security has been detected and less significant information or mechanisms have been compromised.

- *Warning.*   A security attack has been detected. The security of the system is not believed to have been compromised.

- *Indeterminate.*   A security attack has been detected, and the integrity of the system is unknown.

Note that the types of security alarms as described in this subsection are mainly for OSI-based systems and many more will need to be defined for telecommunications applications.

Security audit trail

ITU-T Recommendation X.740 (1992) defines the kinds of events that should be contained in a log for the purpose of a security audit. The security-related log, also called a security audit trail, can be used to detect security attacks that are difficult to detect as they occur. For example, modification of usage metering data by a person that has the privilege to access the billing data is of this type. In contrast to the security alarm reporting function discussed in the preceding subsection that alerts a managing system of an attack as it happens, a security audit trail provides a history of events that may indicate a fraud that already took place.

A security audit trail is based on the OSI log control function as described in Chap. 9. Figure 12.2 presents an OSI-model–based security audit trail process model. In this model, the security audit trail service uses the CMIS service M-EVENT-REPORT to report security-related events from a managed system to a managing system.

There are three sources that may provide security log events: the managed object's notification; an incoming CMIP PDU that is from the manager's CMIP message related to security of communication; a received event report from another event report function. Security log preprocessing is responsible for filtering information and putting the information into the correct format. A security audit trail log, once created from one of the sources, is sent to one or more managing systems that have registered to receive the security log.

The type of security-related events that may be subject to security auditing include

- Connections
- Disconnections
- Security mechanism utilization

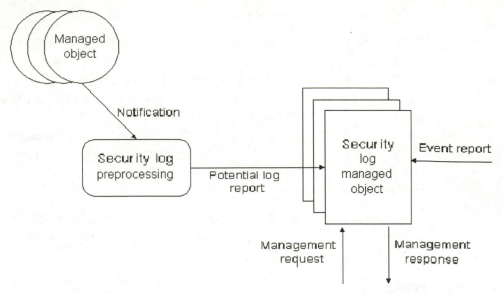

Figure 12.2   A security audit trail process model.

- Management operations
- Usage accounting

The security audit trail reports are divided into the following two categories.

- *Service report.*   Those related to the provisioning, denial, or recovery of a service
- *Usage report.*   Those related to resource or service usage of a statistical nature

In an OSI model, a security audit trail is generated as an event that is sent from the managed system to a managing system via M-EVENT-REPORT. The following lists the causes for generating a service-related event report:

- *Request for service.*   The notification has been originated because of a request for the provision of service.
- *Denial of service.*   The notification has been generated because a request for a service has been denied.
- *Response from service.*   The notification has been generated because the request for a service has been met.
- *Service failure.*   The notification has been generated because an abnormal condition that has caused service failure has been detected.
- *Service recovery.*   The notification has been generated because a service has been restored.
- *Other reason.*   The notification has been generated for reasons other than those listed above.

Fraud Containment and Recovery

Fraud containment seeks to minimize the harm caused by a fraud to a network through means such as isolating viruses that cause data corruption or revoking a customer's privileges to perform certain operations that may compromise the security of the network. Recovery means to restore services and network integrity by reinstalling services and data that have been corrupted.

Customer and network revocation

Revocation administration maintains a list of all customer public keys and access control certificates for the current time period that are known or suspected to be invalid for a security violation. The purpose of the list is preventing if at all possible and at least minimizing the damage that may be caused by the compromised component of a security system. For example, a known stolen secret key or a key belonging to a customer that has moved elsewhere would be on a revocation list. The list will be used to guard against any attempt by a fraudulent user to use network resources or services using the compromised keys or certificates.

Revocation administration also maintains a list of all network public keys and access control certificates for the current time period that are known or suspected of being invalid because of a security violation. The known stolen secret key or access control certificate for a network element that is already removed will be on the revocation list. Again this is for the purpose of guarding against any unauthorized attempt to access the network resource.

Severing connections

As a result of a detected security violation that threatens to spread over to other parts of the network or if the impact of the fraud is uncertain, one common measure is to sever connections with outside customers to contain data and system corruption. One example is physically severing connections such as disconnecting a phone line jumper or setting a customer profile to inactive to prevent the customer from accessing the network. Another example is severing connections with an internal user to contain data and system corruption if the internal user is suspected of being the source of a security breach. Revoking access privileges is one way of severing connections with an internal user.

Intrusion recovery

Intrusion recovery has two aspects: service and network intrusion recovery. The service intrusion recovery allows a customer with a network management responsibility to access backup files in order to restore service after detection of a security breach. For example, all service data can be reloaded when the copy in memory is suspected of being tampered with. Network intrusion recovery includes allowing the internal user to access backup files to restore the network configuration after the detection of a security violation. One example is reloading the whole routing table to reconfigure the network.

Protected storage of data

The capability to have backup and protected storage data is essential for fraud recovery. The data include customer data, service data, system software, network configuration data, and network element configuration data.

Security Services

This section introduces five security services that are defined in ITU-T Recommendations X.810 (1993) through X.815 (1995): authentication, access control, data integrity, confidentiality, and nonrepudiation service. For each service, a set of key concepts is introduced, the service operation process is described, and the information elements required for the service are discussed. The security services can be used to prevent or detect frauds. For example, access control and authentication can be used to prevent potential frauds, and data integrity and confidentiality can be used for detecting frauds.

Authentication service (ITU-T Recommendation X.811, 1995)

Authentication is a process to ensure the real identity of the sender of a digital message i.e., to make sure he or she is who he or she claims to be and to ensure the integrity of the message. The goal of authentication is to provide assurance of an identity of a principal. Authentication is meaningful only in the context of a relationship between a principal and verifier, or a party whose claimed identity needs to be authenticated and a party who requires an authenticated identity.

Threats and attacks to authentication.   What types of threats and issues does authentication service intend to address? Basically two kinds of threats, or security problems: relay and replay. Let's assume the following scenario to help illustrate the issues under discussion. A wireless caller, in initiating a call, has the mobile station number transmitted to the nearest base station as the unique identifier of the mobile station. An intruder intercepts and clones the mobile identification number and starts making calls pretending to be the intercepted mobile station. This is called a *replay attack* on authentication where an intruder intercepts the information passed between two entities (i.e., a base station and mobile station) and replays the information (i.e., a mobile station number in this case).

Another scenario is that an intruder pretends to be a base station and receives the mobile station number and then relays the number to the real intended base station. In doing so, the intruder can send his or her own message along with the relayed message. This is called a *relay attack*.

Basic concepts.   We first clarify several basic concepts that are important for understanding the authentication process. Summary definitions for the concepts can be found in Table 12.2.

TABLE 12.2   Authentication-Related Concepts and Definitions

Term	Definition
Authenticated identity	A distinguished identifier of a principal that has been assured through authentication
Authentication	The process of assuring the claimed identity of an entity
Authentication certificate	A security certificate that is guaranteed by an authentication authority and used to assure the identity of an entity
Authentication exchange	A sequence of one or more transfers of exchange authentication information for the purpose of performing an authentication
Authentication information	Information used for authentication purposes such as a private key or an authentication certificate
Authentication initiator	The entity that starts an authentication exchange
Challenge	A time variant parameter generated by a verifier
Claimant	An entity that is a principal for the purpose of authentication
Distinguished identifier	Data that unambiguously distinguish an entity in the authentication process
Masquerade attack	A security attack initiated by an entity that pretends to be a party other than itself
Off-line authentication certificate	An authentication certificate binding a distinguished identifier to verification authentication information, which may be available to all entities
On-line authentication certificate	An authentication certificate obtained by a claimant directly from the authority who guarantees it
Principal	An entity whose identity is to be authenticated
Unique number	A time variant parameter generated by a claimant
Verifier	An entity that requires an authenticated identity. A verifier includes the functions necessary for engaging in authentication exchange.

A principal.   A *principal* is an entity whose claimed identity needs to be authenticated. A principal must have one or more distinguished identifiers to identify it. Once authenticated, a principal's entity is an authenticated identity that can be trusted within a security domain. Who can be a principal? Practically any entity that interacts with another entity whose resources require protection. Examples include human users, processes, a software system, protocol layer entities, or enterprises. Assume a scenario where a user attempts to gain access to the network server; then the user at the workstation is a principal.

A distinguished identifier distinguishes a principal from others in a security domain. Typical distinguished identifiers include directory name, network addresses, object identifiers, name of a person, and a unique number such as a passport or social security number. When authentication takes place between domains, a distinguishing identifier alone may not be sufficient to unambiguously identify an entity, because the same identifier can be used in different security domains. In this case, a distinguishing identifier must be qualified with an identifier of a security domain.

A claimant.   The term *claimant* is used to represent the combination of a principal and the functions necessary for the principal to engage in an authentication process. In the scenario of a user at a workstation, the workstation is a claimant that acts on behalf of the principal (user) to send the authentication information to other entities involved in an authentication instance.

A verifier.   The term *verifier* is used to represent the combination of the entity that requires an authenticated identity and the functions necessary for the entity to engage in the authentication exchanges. An entity involved in mutual authentication assumes both roles of claimant and verifier. For example, two parties exchanging messages and authenticating each other serve both as claimant and verifier. A security authority or its delegate is a third party trusted by both verifier and claimant.

Trusted third party.   A trusted third party is an entity other than the claimant and verifier engaging in an authentication process that is neutral to and can be trusted by both parties. For example, a dedicated authentication center is a typical example of a trusted third party.

Authentication information.   The authentication information (AI) refers to the information exchanged between a verifier and a claimant. Three types of authentication information are identified: exchange authentication information, claimant authentication information, and verifier authentication information.

Exchange AI is the information exchanged between a claimant and a verifier during the process of authenticating a principal. Examples of exchange AI include

- Claimed distinguished identifier
- Password
- Challenge
- Response to challenge
- Unique number
- Verifier distinguishing identifier
- On-line certificate
- Off-line certificate
- The result of a transformation function applied to or using claim AI and other data such as time stamp, random number, counter, and digital fingerprints

Claim AI is the information used to generate exchange AI required to authenticate a principal. Examples of claim AI include a password, secret key, and private key. Verification AI is the information used to verify an identity claimed through exchange AI. Examples of verification AI include a password, secret key that is related to the identity of a principal or authority, and public key that is related to the identity of a principal or authority. The relationships between claimant, verifier, trusted third party, and AI are shown in Fig. 12.3.

An authentication process model. We use an authentication scenario to illustrate the authentication concepts and general process.

   An authentication scenario. It is assumed in this scenario that a user (principal) at a workstation (claimant) attempts to gain access to the network resource through a network server (verifier) and that an authentication center is responsible for verifying the user-supplied data. What are the major steps the user has to go through to be authenticated, including the initial steps of setting up an authentication session? The major steps include installation, changing authentication information, and distribution of exchange authentication information. Note that the sequence of actions described below is illustrative and just one possible sequence out of many. In this scenario, the following terms are used interchangeably: server and verifier, user and principal, claimant and workstation.

1. First, a management system defines the claim AI and the verification AI. For example, the claim AI, the authentication information used by the

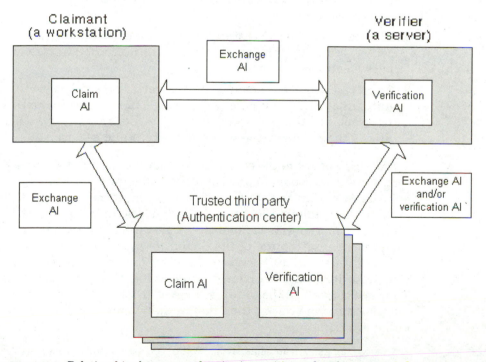

Figure 12.3 Relationships between authentication entities and authentication information.

workstation, can be the user password. The verification AI can be the work-station private key encrypted according to an algorithm known to the verifier and the authentication center.

2. If so desired, the security system changes both the claim AI and the verifi-er AI. For example, a new private key can be generated at the beginning of each session. The frequency of change and the content of the change are determined by the management system.

3. The claimant (workstation) sends a message to the authentication center to request an authentication certificate. The exchange AI the claimant sent consists of a distinguished identifier of the user on the workstation, the server identifier, and the password of the user. Then the management sys-tem distributes a verification AI to the claimant, the verifier, or both for use in verifying exchange AI.

4. The authentication center checks its database to determine whether the user has supplied the correct password for this user and whether this user is per-mitted to access the server. If both tests pass, the authentication center accepts the user as authentic and must convey the information to the server. The authentication center issues an authentication certificate to the work-station that consists of the user's identifier, network address, and the server's identifier. The authentication certificate is encrypted using a secret key shared between the authentication center and the server. The workstation cannot alter the authentication certification because it does not know the encryption key.

5. The claimant (i.e., workstation) sends an exchange AI to the verifier (i.e., server) containing the encrypted authentication certificate and the princi-pal's identifier.

6. Upon receiving the exchange AI, the verifier (i.e., the server) decrypts the authentication certificate and verifies that the principal's identifier is the same as the decrypted user identifier. If they match, the verifier grants the claimant the requested access and service.

Each component in the exchange AI from the claimant serves a purpose. The certificate is encrypted to prevent alteration or forgery. The verifier's identifier is included so that the verifier can verify that it has decrypted the certificate properly. The claimant identifier is included to indicate that the certificate is issued on behalf of this claimant. The network address is included to counter the possible interception of the certificate by an unauthorized entity from a different workstation. For example, if an unauthorized party intercepts the certificate, it can use the named claimant identifier and send AI to the server. The verifier can use the network address to verify the certificate is transmit-ted from the same workstation that initially requested the certificate from the authentication center.

Trusted third party in authentication. A trusted third party can play different roles in the authentication process. In the simplest case, a claimant and a verifier directly exchange the exchange AI without involvement of a third party. This requires that a claimant and verifier be able to generate and verify the exchange AI themselves and the verification AI be installed beforehand. Without a trusted third party, an entity is restricted to a limited number of communication partners, because in the worst case, each verifier is required to have verification AIs for all principals in a security domain, with the total information requirement growing as the square of the number of entities involved.

In the cases where a trusted third party is involved in the authentication process, the involvement can be classified into in-line, on-line, and off-line categories, according to the manner in which the trusted third party is involved.

In in-line authentication, a trusted third party such as an authentication center sits between a claimant and a verifier and serves as an intermediary. The trusted intermediary vouches for the identity of a principal, and the verifier trusts the intermediary to have properly authenticated the principal. In in-line authentication, a claimant does not directly exchange any exchange AI with a verifier during the authentication process, as shown in Fig. 12.4.

On-line authentication, like in-line authentication, has one or more trusted third parties actively involved in every instance of an authentication exchange in real time. However, unlike in-line authentication, the on-line trusted third party does not function as an intermediary between a claimant and a verifier. Instead, it issues authentication certificates upon a request from a principal and verifies a principal on behalf of a verifier. It serves as an assistant, and a claimant directly communicates with a verifier to be verified as authentic. The scenario depicted in Fig. 12.3 (the authentication process model figure) is an on-line authentication model.

Off-line authentication is characterized by non-real-time involvement of one or more trusted third parties in the authentication process. A third party is not directly involved in each instance of authentication exchange. Instead, it issues authentication certificates, a certified list of revoked certificates, certificate time-outs, and other authentication-related information in an off-line manner, in advance of an actual exchange of exchange AI's. One motivation for this approach is that it reduces the real-time messages required for authenticating a principal, because the trusted third party does not directly interact with either a claimant or a verifier during the exchange of exchange AIs.

Authentication certificates. A brief definition of an authentication certificate was introduced a little while ago, and now a more detailed discussion will be presented after a discussion of the different roles of a trusted third party in the authentication process.

Figure 12.4 Trusted third party in in-line authentication.

Examples of components found within an authentication certificate include

- Identification of the method and/or key used to encrypt information components.

- The identity of the authentication authority and the identity of the agent that issues the authentication certificate.

- The creation time of the authentication certificate that may be used for audit purposes or may be used when the validity period of the authentication certificate is not present.

- The validity period of the certificate that specifies the time window within which the certificate is valid.

- The security policy applicable for the authentication certificate such as checking the validity period or verifying the origin of messages.

- Type of certificate such as encrypted or plaintext.

- Identity or attributes of the verifier for which the authentication is intended. For example, in the workstation scenario, the server's identity is included in the certificate.

Mutual versus unilateral authentication. Two general types of authentication are distinguished according to the number of entities being authenticated: mutual and unilateral authentication. Unilateral authentication, also called one-way authentication, authenticates only one principal. One example of unilateral authentication is an e-mail message authentication where only the sender of a message is authenticated to verify that the message indeed came from the alleged sender. The above example of a user at a workstation requesting service from a remote server is also a one-way authentication. An authentication that verifies identities of both principals is called mutual authentication. It is not required to use the same authentication mechanism in both directions.

Authentication mechanisms. Authentication mechanisms are the facilities that provide authentication services. The mechanisms can be classified in many ways. Discussed here are the characteristics and classifications of the mechanisms.

Symmetric versus asymmetric authentication mechanism. Authentication of a principal is often based on secrets shared between two entities involved in authentication. The shared secrets can take the form of a password, key, or others. Authentication involves demonstrating the knowledge of the secret. There are two general categories of methods for key-based authentication: symmetric and asymmetric.

For a symmetric method, both entities share common authentication information. For example, a password or authorization code for accessing a network service (e.g., long-distance call) must be known to both the owner of the password or code and the service-providing system. Another example is a challenge encrypted using a symmetric key that is shared only between a pair of communicating parties.

For an asymmetric scheme, not all the authentication information is shared by both parties. One example of an asymmetric method is the asymmetric key technique that will be discussed shortly.

Encryption-based and non-encryption-based authentication mechanisms. At the core of authentication is how to protect the authentication information. One type of authentication mechanism is characterized by the use of encryption algorithms to protect authentication information. Authentication based on symmetric, asymmetric, or a hybrid of encryption techniques belongs to this category.

The second category of authentication mechanisms is based on nonencryption techniques such as passwords or a challenge-and-response table. Password-based authentication by nature is symmetric.

Access control service

This subsection presents an access control model that provides access control service, as defined in ITU-T Recommendations X.812 (1995) and X.741 (1995). The goal of the model is controlling access to data, processes, services, and other network resources. In this model it is assumed that access is made by a process that may be acting on behalf of humans or other processes. The control of access can be in the same security domain or across security domains.

The introduction of the access control model consists of a high-level description of the mode components, the steps for setting up an access control system based on the model, and how the model operates in general.

Access control model description. Figure 12.5 presents an access control model that conveys a logical view of a potential access control system. Included in the models are all major entities involved in the access control process.

Access initiator is the party attempting to gain access to the network resource or service that is represented as the target of access. It can be a user, a process, or an application within a process.

Access control enforcement function is a logical entity that sits between an access initiator and the target of access to administrate the access control. It ensures that an access decision is made by the access control decision function. For example, in the event an access is granted, this function provides to the access initiator the access to the requested resource, and in the event of access denial, this function is responsible for logging the denial and related information.

Access control decision function is a logical entity that is responsible for making decisions on either granting or denying an access request from an access initiator, based on the security policy of this security domain and the information provided by the access initiator.

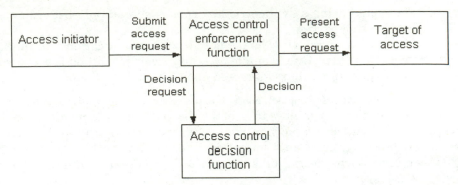

Figure 12.5   An access control process model. (*From ITU-T Recommendation X.812, 1995.*)

Target of access represents the network resources and services that are the target of an access attempt by outside entities. Examples of target of access include the network server, files, telecommunications services, and the access control mechanism itself.

Commonly used terminology related to access control is listed in Table 12.3.

Initiation of access control information and an access control system.   Prior to any access control operation, a set of activities is required to set up an access control system.

Establish access policy and representation of the policy.   Two categories of security policy, rule-based and identity-based, are distinguished based on the application scope of a security policy. Rule-based access control policies are intended to be applicable to all access requests by any initiator on any target in a security domain. The identity information of individual group members is not factored into the access control decision. The membership of a particular group is considered an access right. Of course, the ability to create or modify groups must be subject to access control. The identity-based security policies are specific to each individual initiator, and identity information of each initiator is part of the access control information to be presented to an access control enforcement function. The two types of security policies are not mutually exclusive and can be used together in various combinations.

The general access control policies stated in the form of security requirements must be converted to engineering representation. Security labels are one type of representation for rule-based access control policies. A security label is a unit of information or marking that is bound to a resource and that names and specifies the security attributes of the label. Initiators and targets are separately associated with named security labels. Access decisions are based on a comparison of the initiator and target security labels.

Establish access control information.   The access control information (ACI) is exchanged between two communicating systems as part of the access control operation, and the parties involved must have an understanding of the syntax and semantics of the access control information. The information required in an access control operation can be classified into three categories: initiator and target ACI, access ACI, and contextual information.

The initiator and target ACI concerns the access initiator and target. Examples of the information contents include

- Access control identity of an individual initiator or target
- Identifier of the access group of which the initiator is a member
- Identifier of the roles that the initiator may take
- Identifier of the entity that contains the target
- Sensitivity markings and integrity markings (see Table 12.3 for the definitions)

Access request ACI concerns an access request from an initiator. Examples of this type of ACI include

- Allowed class of operations such as read and write
- Integrity level required for use of operation
- Data type of the operation

Examples of contextual information used in access control include

- *Time period of a granted access.*   Such as day, week, or month
- *Route used for an access.*   An access is required to have certain characteristics before the access request is granted
- *Location.*   An access may be granted to initiators at a specific location such as a system, workstation, or terminal

TABLE 12.3   Concepts Related to Access Control

Concept	Definition
Access control information (ACI)	Information that is used for access control purposes such as a password, identity of an entity initiating an access request, and a security label
Initiator	An entity that initiates an access request
Integrity marking	Marking in numeric value or descriptive terms of a resource's integrity
Security domain	A set of elements that are subject to the same security policy for a set of specified security-related activities
Security domain authority	An entity that is responsible for the definition, implementation, or enforcement of security policy
Sensitivity marking	Marking in numeric value or descriptive terms of the characteristic of a resource that indicates the value, importance, and maybe vulnerability of a resource
Target	An entity to which accesses may be attempted

- *System status.*   An access request may be granted only if the system has a particular status
- *Strength of authentication.*   An access request may only be granted when authentication mechanisms of at least a given level are used
- Other accesses currently active for this or other initiators.

In an access control operation, a combination of the above categories of access control information may be used.

Distribute and bind ACI to initiator and targets.   Once created by the appropriate management entities, the ACI needs to be distributed to associated initiators and targets. The main issue is determining what information is distributed and where. In general, the ACI distributed to an entity includes the ACI about the identity, the ACI about the relationship between this entity and associated entities, and the contextual ACI required by the entity. For example, the information distributed to an initiator may include the initiator ACI, the target ACI, and the contextual ACI. The distribution is dictated by the security policy of each specific security domain.

Bind ACI to entities.   Binding an ACI to an entity such as an access target creates a secure linkage between an entity and the ACI allocated to the entity. The binding provides assurance to access control functions such as the access control enforcement function and the access control decision function that an ACI has indeed been assigned to a designated entity and that the integrity of the ACI is protected since the binding was made. The integrity of binding can be achieved through such integrity mechanisms as digital signature. The integrity of the ACI shall be protected at both the initiator and target and during the exchange of ACI as well. The time when an initiator or a target has ACI bound to it depends on each entity involved. For example, all targets will have ACI bound to them by a security domain authority or its agents by the time targets become accessible.

Operational procedure of access control model.   Again we use the scenario of a user attempting to gain access to a network resource to illustrate the general behavior of the access control model just described.

1. First, the initiator (the user) provides initiator ACI that includes initiator location (e.g., at the workstation) and the initiator access control identity (e.g., the password) in the access request message. The access control enforcement function performs one or more of the following verifications depending on the specification of security policy.
   - Validate the integrity of the information passed in from the initiator.
   - Verify the validity of the information by checking that the origin of the information is a recognized security domain authority.
   - Verify the content of the information by ensuring the values in the information are within the permitted range.

2. Once the initiator ACI is verified and validated, the access control enforcement function identifies the request targets to ensure that, first, the requested access targets are available in this security domain and are in the state that allows access. The target identification can be achieved in various ways. For example, the security management may maintain a management information base that contains all access targets in the domain along with other attributes such as security label, state, and status.

3. The access control enforcement function generates an initiator access control decision information (ADI) by selecting those information items relevant to this instance of access control operation from the initiator ACI.

4. The access control enforcement function creates an access request ADI from the original access request by including only the information required by the access control decision function for making a decision on the access request. The access control enforcement function sends the initiator ADI and the request ADI to the access control decision function.

5. The access control decision function, as shown in Fig. 12.6, first retrieves required information such as access control policy rules, target ADI (the portion of target ACI that is relevant for this instance of access control operation), and contextual information. No assumption is made on the location of the retrieved information, which could be local or remote.

6. The access control decision function then performs a set of tests to reach a decision on whether or not to grant access to the initiator. If the test is successful, the requested access is granted. Otherwise, access is denied. The input to the test includes the initiator ADI and a rule identifying operations permitted on requested targets. The test may compare the security label, the identity, or contextual information of the request initiator against a set of allowed identities, security labels, or contextual information, all depending on the security policy rules.

7. The access control decision function sends the access control decision to the access control enforcement function to carry out the decision. In the case where the requested access is granted, the access control enforcement function allows access through and logs the access information if required by the security policy. In the case where the access request is denied, the access control enforcement function provides the initiator with a denial response along with the reason for the denial. Other denial enforcement actions may be defined as specified by the security policy of the security domain.

Maintenance operation.   The access control system is also responsible for the maintenance operations. When new entities are added in the security domain, new ACI and ADI are created and allocation and binding of ACI is performed. The ACI allocated and bound to an entity may be modified to reflect changing security attributes and requirements. The change may result in revocation of the ADI. The access control decision function may retain certain commonly

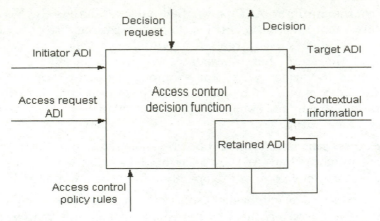

ADI -- Access control decision information

Figure 12.6 Illustration of an access control decision function.

used ADI at its local storage space for efficiency purposes. The retained ADI may require modification in the event of any change in the binding between ACI and an entity, contextual information, role an entity takes, and others.

Confidentiality service (ITU-T Recommendation X.814)

The confidentiality service ensures that information is available only to those authorized to access the information. The confidentiality service along with the confidentiality mechanisms mainly addresses the following attempts to compromise the confidentiality of a data item.

- Attempt to understand the semantics of the data
- Attempt to understand the associated attributes of the data
- Attempt to consider the context of the data, e.g., the other data object associated with the data
- Attempt to observe the dynamic variation of the data representation

There are two general categories of mechanisms to protect data confidentiality. The first one ensures that access to the data is limited to those authorized. The second category represents the data in such a way that the data semantics remain accessible only to those who possess critical information such as encryption keys.

Threats to confidentiality. The threats to confidentiality-protected information, namely, disclosure of the protected information, can appear in various ways, and in general they fall into two categories: threats to access prevention mechanisms and threats to information-hiding mechanisms. The threats to the mechanisms preventing unauthorized access include

- Penetrating the access prevention mechanism such as exploiting weakness in physically protected channels, and masquerading certificates, or implementation of the prevention mechanism.

- Penetrating the services the prevention mechanism depends on. Examples include masquerading when access is based on identity authentication and penetrating the integrity mechanism used to protect certificates.

- Penetrating system utilities that may disclose information about the system.

- Using covert channels to gain access to data.

The threats to information-hiding mechanisms include

- Penetrating cryptographic mechanisms using cryptanalysis, purloined keys, plaintext attack, or other means to directly access the confidentiality-protected data

- Analyzing traffic patterns that may indicate the content of traffic

- Analyzing PDU headers to gain a peek at the content carried in the PDUs

It is possible to distinguish between active and passive attacks to data confidentiality, depending on whether or not an attack results in a system change. Passive attacks normally don't result in any change to the system under attack. Examples include eavesdropping and wiretapping, traffic analysis, analysis of PDU headers for the purpose of disclosing the PDU contents, copying PDU data to systems other than the intended destination, and analysis of encryption algorithms for the purpose of breaking them.

In contrast, active attacks do result in changes to the system under attack. Examples include Trojan horses (i.e., code with covert features that are intended to breach system security); penetration of the mechanisms that support confidentiality; spurious invocations of the cryptographic mechanisms, such as chosen plaintext attack; and covert channels.

Confidentiality mechanisms.   This subsection provides a high-level introduction to three types of confidentiality mechanisms:

- Mechanisms that prevent unauthorized access to data

- Mechanisms that hide data but leave it accessible

- Contextual mechanisms that make data partially accessible, so that data cannot be completely re-created from the limited amount of accessible data

Confidentiality through access prevention.   A simple and straightforward scheme is strictly controlling access to the physical medium that stores data. Confidentiality is ensured by never granting access to unauthorized entities. A second method is achieving confidentiality protection through routing control. Simply put, confidentiality is ensured by using only trusted and secure facilities to route data.

Confidentiality through enciphering. The purpose of this type of confidentiality mechanism is to hide the semantics of data, either in transit or in storage, from any unauthorized entities. One method is data padding that prevents a potential attacker from knowing the real size of a confidentiality-protected data item. It increases the size of a data item so that as a padded data item its size bears little relation to its original size. For example, random data can be added to the beginning or end of the data item. This must be done in such a way that the padding is recognizable as such by authorized entities but is indistinguishable from the original data by unauthorized entities. This can be achieved by using data padding in conjunction with cryptographic transformation.

Another method uses dummy events to prevent any potential attacker from inferring information based on the rate that a given event occurs. For example, this method is used in network layer security protocols that seek to hide the volume of traffic exchanged over untrusted links by producing pseudo-events that only authorized parties can identify as such.

Yet another confidentiality mechanism uses time-varying fields in conjunction with enciphering to keep a potential attacker from inferring based on dynamic variation of data items. The time-varying fields are added to a data item, and the resulting data item is enciphered in such a way that a potential attacker cannot determine whether changes in the representation are caused by changes in data or by changes in the time-varying fields. For example, in PDU transmission, a time-varying field is placed in front of the protected part of each PDU and the combined data are then enciphered using a cryptographic algorithm.

Confidentiality through contextual location. This type of confidentiality mechanism provides such a large number of contexts for a data item that it is impractical, if not impossible, for a potential attacker to examine all possible contexts before the context is changed. For example, a large number of physical or virtual channels can be provisioned for transmitting information. Or, the transmission of information is through hidden secondary communications channels that are concealed within a primary channel. It is essential for this type of confidentiality that an unauthorized recipient cannot obtain the information necessary to identify the currently correct context. The context information itself must be protected by a confidentiality service.

Data integrity service (ITU-T Recommendation X.815)

The integrity service is used to protect the integrity of data and to counter a number of ways in which data integrity can be compromised. Listed below are major threats to data integrity.

- A potential attacker can compromise data integrity by attacking the data transfer channel. For example, an unauthorized entity can tap into a communication channel and create, modify, delete, insert, or replay data.

- A potential attacker can compromise data integrity by attacking the data origin. For example, assume that the communicating entities A and B are using an encoding scheme to protect the integrity of data sent from A to B

but the origin of the data is left unprotected. Then the data that were sent from A to B can be later submitted to A as if originating from B, and this is called *reflection attack*.

- A potential attacker can compromise the medium where data are stored or in which data are transmitted.

Integrity mechanisms.   Various integrity mechanisms are introduced to counter the integrity threats mentioned above. In order to provide integrity service, a mechanism shall provide three basic functions: shield, validate, and unshield.

The *shield* function, also called data shielding, applies integrity protection to data using shield integrity information (SII) such as private keys, secret keys, algorithm and associated cryptographic parameters, and time-variant parameters (e.g., time stamps). The *validate* function, also called data validating, checks the integrity-protected data for unauthorized alteration using information such as public keys and secret keys. The *unshield* function, or data unshielding, converts integrity-protected data into the original form prior to being shielded using information such as public keys or secret keys. Different ways of providing the three functions result in different integrity mechanisms, which are introduced next.

Integrity through encryption.   Two classes of encryption-based integrity mechanisms can be distinguished. One is based on symmetric encryption techniques in which the validation of integrity-protected data is possible through knowledge of the same secret key. One is based on asymmetric encryption techniques in which the validation of integrity-protected data is through knowledge of the public key corresponding to the private key used to shield the data.

Integrity through sealing.   Sealing provides integrity by appending an encryption check value to the data to be protected. The same secret key is used to protect and to validate the integrity of the data. This mechanism supports detection of data modification through three steps as follows:

- *Shield* is achieved by attaching an encryption check value on the data to be protected. One example is computing a one-way function over the data to be protected.

- *Validate* is implemented by using the data, the encryption check value, and the secret key to determine if the data match the seal.

- *Unshield* removes the encryption check value once the data have been validated.

Integrity through digital signature.   Digital signatures, the electronic equivalent of signing a message, are computed using a private key and an asymmetric algorithm. The shielded data can be validated using the corresponding public key.

Data shielding is achieved by attaching a cryptographic check value on the data to be integrity-protected. One example is generating a digital signature using a digital fingerprint of the data to be protected combined with a private key and possibly other parameters (e.g., time stamps). Data validation is achieved by verifying the received data using a digital fingerprint of the data received, the digital signature, and the public key along with an encryption algorithm. Failure of the digital signature verification indicates the received data have been altered. Data unshielding is achieved by removing the cryptographic check value, after the data have been validated.

Integrity through context. This type of integrity mechanism provides integrity protection by storing or transmitting data in one or more preagreed contexts.

Data replication. Integrity is provided by replication of the data over space or time. For example, multiple copies of the same data are created at different times or stored at different locations. This integrity mechanism assumes that a potential attacker cannot compromise all replicas simultaneously and, whenever attacked, the data can be restructured from an original copy.

Data shielding is achieved by replicating the same data multiple times either successively in time or at different locations. Data validation is achieved by gathering and comparing different copies of data. If they are not all identical, an integrity violation is considered to have occurred. Data unshielding is achieved when some preestablished criteria are met. For example, if 95 percent or more of the values of the data agree, the data are considered integrity-protected.

Preagreed context. This type of integrity mechanism provides deletion detection to integrity-protected data and is often used in conjunction with other integrity mechanisms. Data shielding is achieved by providing data at a predetermined time or location or both to the recipient of the data. Data validation is achieved by expecting the data at the given time and location. It is like a rendezvous between two communicating parties, and failure by either party to be present at a specified time and location is considered an integrity violation.

Integrity through detection and acknowledgment. This type of integrity mechanism uses idempotent operations to detect integrity violation. An operation is deemed *idempotent* if multiple, successive executions of the operation yield the same result. The data shielding is achieved by repeating the same action until either a positive acknowledgment is received from the other party or the integrity policy dictates otherwise. Data validation is performed on each instance of shielded data, and successful validation results in sending a positive acknowledgment to the party performing data shielding.

Nonrepudiation framework (ITU-T Recommendation X.813, 1995)

The nonrepudiation service is used to resolve disputes concerning the occurrence or lack of occurrence of an event or action in an instance of communication (e.g., a data transfer and a bank transaction). For example, a dispute over a bank transaction may take the form of the alleged transaction originator denying having initiated the transaction. Or the alleged recipient denies the occurrence of an alleged transaction. A system that provides nonrepudiation services resolves disputes like above by collecting, maintaining, making available, and validating irrefutable evidence concerning a claimed event or action.

Basic concepts. The nonrepudiation service as discussed here involves the generation, verification, and recording of evidence, and the subsequent retrieval and reverification of the evidence in order to resolve disputes. The nonrepudiation service may be requested by entities other than those involved in the event or action. Examples of actions include sending a message across a network, inserting a record in a database, and invoking a remote operation. In case of data transfer, the nonrepudiation service provides evidence or proof of the origin, identities of the originator and recipient, and the integrity of the data.

Thus nonrepudiation services will provide facilities for the following functions:

- Generation of evidence
- Recording of evidence
- Verification of generated evidence
- Retrieval and reverification of evidence

Disputes may be settled between parties through inspection of the mutually agreed upon evidence. However, a dispute may have to be resolved by an adjudicator, an entity that is entrusted with the authority by both parties to evaluate the evidence and determine whether or not the disputed action or event occurred. The basic terms related to nonrepudiation services are defined in Table 12.4.

TABLE 12.4   Common Terms Related to Nonrepudiation Services

Term	Definition
Adjudicator	An entity that is neutral to both parties involved in a dispute and is charged with the responsibility of resolving the dispute
Evidence	Information that is used to resolve a dispute regarding a data communication. Examples include a distinguishing identifier of the message originator or recipient, a digital signature, and a time stamp.
Evidence subject	An entity whose involvement in an event or action is to be verified. For example, the originator or recipient of a message is an evidence subject.
Evidence user	An entity that uses the nonrepudiation evidence to resolve a dispute or for another purpose such as evidence bookkeeping. An evidence verifier can be an evidence user.
Evidence verifier	An entity that verifies nonrepudiation evidence to resolve a dispute
Nonrepudiation service	The service provided by a security management system to resolve disputes concerning occurrence or nonoccurrence of an event or action in an instance of communication or data transaction
Notary	An entity that can provide assurance about the properties of data communicated between two entities such as origin, recipient, and time of data creation or delivery

Nonrepudiation operation. Figure 12.7 presents a basic process model for providing the nonrepudiation service. The model intends to be a high-level description of the nonrepudiation process short of any design level details. We use a scenario to illustrate the behavior of the model. Assume that a large fund is transferred into a bank account through a teletransfer over a network. Various potential disputes may arise over the transaction: the alleged originator of the transaction may be denied the fund for lack of identity information; the receiving side of the transaction may deny having received the transferred fund; the alleged amount of the fund may not match the amount claimed to have been received. The process of generating evidence and resolving the above disputes is shown in the following steps.

1. A party involved in the dispute, through an interface, sends a request for evidence to the evidence generation requestor to generate evidence for the dispute over the transaction.

2. The evidence generation requestor sends an internal message to the evidence generator specifying the types of requested evidences such as time stamp, identifier, and password.

3. The evidence generator determines the evidence subject involved in the transaction and sends an observation to the evidence subject. An evidence subject is an entity whose involvement in an event or action is to be verified. For example, the originator or recipient of the transaction is an evidence subject. The observation sent to the evidence subject specifies the observed evidence and the types of evidence requested.

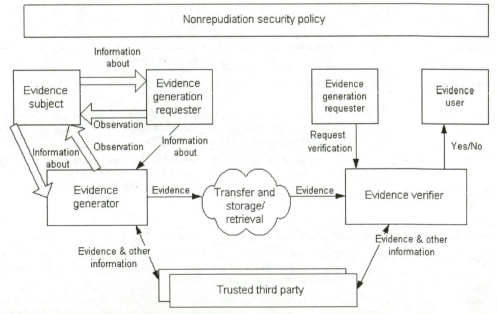

Figure 12.7 Entities involved in the generation, transfer, storage, retrieval, and verification of the nonrepudiation process.

4. The evidence subject collects the evidence according to the specification in the observation and sends the information about a category of evidence to either the evidence generator or the evidence requestor.

5. The evidence generator combines the observation and other related information into evidence and sends the evidence across the network to the evidence verifier. The evidence generator may also choose to store the evidence into a database that is known to the evidence verifier and send a message to the verifier asking that the stored evidence be verified.

6. The evidence verifier, an entity trusted by the evidence user, verifies the supplied evidence to ensure the evidence is sufficient and adequate in the event of a dispute. One or more trusted third parties might be involved in the verification at the request of the evidence verifier.

7. An adjudicator is requested to resolve the dispute. It collects evidence from disputing parties and from one or more other trusted third parties if needed. How an adjudicator resolves disputes depends on predefined nonrepudiation policies and falls outside the scope of this discussion. The involvement of an adjudicator is optional, and the disputing parties might settle on the occurrence or lack of occurrence of an event themselves.

The components of the nonrepudiation process model described above are logical entities, which may or may not map to a system entity. For example, an evidence subject and an evidence generator may be the same entity, or the evidence subject may be combined with the evidence generation requester. A variety of combinations is possible: an evidence generator and a trusted third party; an evidence generator, an evidence generation requester, and a trusted third party; and an evidence user and an evidence verifier.

Information of nonrepudiation services. There are many forms of disputes. Among them is the repudiation associated with data transfer. The potential disputes include repudiation of the originator's involvement in an event and of the recipient's involvement in an event of data transfer. For example, an alleged originator can claim that the message was either forged by the recipient or forged by a masquerading attacker. An alleged recipient can claim the message was either not sent, lost in transit, or only received by a masquerading attacker.

A variety of nonrepudiation evidence is required to resolve various types of disputes. Examples include

- An identifier of the nonrepudiation security policy
- The distinguishing identifier of the originator
- The distinguishing identifier of the recipient
- A digital signature or secure envelope
- The distinguishing identifier of the evidence generator
- The distinguishing identifier of the evidence generation requestor
- The message or a digital fingerprint of the message

- The message identifier
- An indication of the secret key needed to validate the security token
- An identification of a particular public key needed to validate the digital signature
- The distinguishing identifier of the notary, time stamping trusted third party, or in-line trusted third party
- A unique identifier for the evidence
- The date and time that the evidence was deposited or recorded
- The date and time the digital signature or security token was generated

Nonrepudiation mechanisms. Nonrepudiation services can be implemented by a variety of nonrepudiation mechanisms, and five are surveyed next: security token, tamper-resistant modules, digital signature, time stamping, and notary.

Nonrepudiation using a trusted-third-party security token. In this scheme, nonrepudiation evidence consists of a security token, sealed with a secret key known only to a trusted third party. The trusted third party generates a security token at the request of the evidence generation requestor and also verifies the security token for the evidence user or adjudicator. The candidate inputs to the trusted third party for generating a security token include

- Specification of the method or encryption algorithm used to ensure the integrity of the security token
- Specification of the method or encryption algorithm for ensuring the confidentiality of the security token
- The distinguishing identifier of the evidence subject such as originator of a message or a transaction
- The distinguishing identifier of the evidence generation requestor
- The date and time of the event or action

Imagine the scenario where the originator of a message sends the message along with a security token to a recipient of the message. The originator cannot deny the event of transmitting the message because the security token was originated in response to a request from the originator and it contains the time and date the security token was generated.

Nonrepudiation using security tokens and tamper-resistant modules. In this scheme, nonrepudiation evidence consists of a security token, sealed with a secret key, which is stored within tamper-resistant encryption modules. All involved parties, the evidence generator, the evidence verifier, and the adjudicator, have access to the modules but with different privileges. The evidence generator can use a secret key inside the modules to create a sealed token, while the evidence verifier and the adjudicator can only use the same secret key to perform token verification. All parties involved must trust that the secret keys have been installed correctly.

If a dispute arises in which the message originator denies having ever sent the message, the evidence user can present the sealed token to the adjudicator and show that the message originator is the only one capable of generating the security token using the secret key.

Nonrepudiation using a digital signature. A digital signature is a digitally signed data structure, which has two associated keys. A signing key is used to generate the signature, and a verification key is used to verify the signature. In this scheme, the digital signature is generated either by the evidence subject or a trusted third party in a signature generation role. The evidence user or the adjudicator must obtain the corresponding verification key to verify the digital signature. A digital signature that is generated by the evidence subject is called a *direct digital signature,* and a digital signature generated by a trusted third party is called a *mediated signature.*

In case of a dispute where a message originator denies ever sending a message or performing the transaction, the recipient of the message can present the digital signature generated by the message originator to the adjudicator. If the adjudicator can successfully verify the signature, the origin of the message is proven.

Nonrepudiation using time stamping. Time stamping adds the time, date, and a seal or digital signature to the data. Often a trusted time reference is needed to determine the original order of actions or events in order to determine whether a message is authentic or forged. When the clock provided by an entity that produces the digital signature or security token cannot be trusted, a trusted third party is called to provide time stamping. For example, time stamping provided by a trusted third party can be used to establish that a message was signed before the signature key was created and thus cannot be a forgery. Time stamping often constitutes an integral part of a digital signature instead of serving alone as a digital signature.

Time stamping may be requested by a variety of entities. It may be an evidence generator, a nonrepudiation service requester, an evidence user, or an evidence verifier. Time stamping itself does not require the authentication of the entity that requests a time stamp. It is the responsibility of an evidence verifier to authenticate the identity of the requesting entity.

Nonrepudiation using a notary. A notarization mechanism provides assurance about the properties of data exchanged between two entities, such as its origin, destination, or time of creation or delivery. A notary, like a trusted third party, must be trusted by both entities involved to hold the necessary information for providing assurance in a verifiable manner. The notary will be used in combination with a digital signature, encryption, and integrity mechanism.

Security Mechanisms

Security mechanisms provide means and functions for providing the security services discussed above. There is no single security mechanism that can provide all the security services. This section provides a high-level overview of the commonly used security mechanisms with emphasis on encryption, by far the most frequently used. Other security mechanisms include digital signature, secure hash functions, notarization, and access control mechanism.

Encryption

Two forms of encryption are in frequent use: conventional (or symmetric) encryption and public-key (or asymmetric) encryption. It is used in many security services such as authentication, integrity, confidentiality, and nonrepudiation services. The basic concepts related to encryption are listed in Table 12.5.

Encryption is a process that transforms original, human readable text, or plaintext, into random, nonreadable text, also called ciphertext. All encryption schemes have two basic components in common: one or more encryption keys and an encryption algorithm. Input to an encryption is a plaintext document or message, which is converted into ciphertext using a key and an encryption algorithm. The values of the output ciphertext of encryption are a function of the encryption key. Decryption is an inverse of the encryption process: convert the ciphertext back to the plaintext using the same key used to encrypt the original text or a different key.

Conventional encryption.   Conventional encryption is characterized by the fact that the same algorithm with the same key is used for both encryption and decryption. In fact, the sender and receiver must share the algorithm and the key. As shown in Fig. 12.8, the encryption process operates as follows. User A sends the plaintext into the encryption module that consists of an encryption algorithm and an encryption key. The encryption algorithm and key are applied to the readable plaintext to produce ciphertext. The secret key shared between user A and user B is passed from A to B. At the receiving end, the secret key along with its algorithm is used to convert the ciphertext back to it is original, human readable format.

The Data Encryption Standard (DES) is the most widely used conventional encryption algorithm among many algorithms of this type that have been developed. It is the official encryption algorithm of its type for the U.S. government since the late 1970s. Put in simple terms, the DES takes a 64-bit block of plaintext and a 56-bit key as input and produces a 64-bit block of ciphertext as output. The decryption algorithm takes a 64-bit block of ciphertext and the same 56-bit key as input and produces the original 64-bit block of plaintext. This standard symmetric encryption algorithm is often implemented in hardware and thus can execute very quickly and efficiently. It is also proven to be very secure in applications.

The way the secret key is distributed directly impacts the security of DES. Some well-known methods for distribution of encryption keys are use of the Diffie-Hellman algorithm by two communicating entities to establish a secret key; use of IEEE standards for key management over a LAN/MAN; and use of a third party such as a Kerberos server to securely distribute a secrete key to two communicating parties.

TABLE 12.5   Basic Concepts Related to Encryption

Terms	Definition
Asymmetric cryptosystem	A cryptosystem in which the encryption key is different from the decryption key
Ciphertext	The transformed unreadable data
Cryptography	A discipline that studies encryption algorithms and process
Cryptosystem	A system that specifies the encryption algorithm and decryption key and performs the whole task of encryption and decryption
Decryption	Transformation of data back into original readable form using the secret key
Encryption	Transformation of data into a form unreadable by anyone without a secret decryption key, in order to achieve privacy
Plaintext	The original, human readable data
Symmetric cryptosystem	A single key serving as both the encryption and decryption key

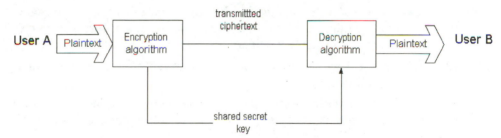

Figure 12.8 Model of conventional encryption.

The conventional encryption model is symmetric because of the fact that the same key is used for both encryption and decryption by both the sender and recipient. The symmetric encryption schemes protect the two parties exchanging a message from attack by a third party but do not prevent the two parties from attacking each other. For example, a sender can deny ever sending a message or claim the message was forged by a third party. Or the recipient may deny ever receiving a message. This issue is addressed by the public-key-based encryption.

Public-key-based encryption. The development of public-key cryptography is viewed as perhaps the most important milestone of modern cryptography. Until recently, virtually all cryptographic systems have been based on the simple notion of a secret key, permutation, and substitution. Public-key cryptography represents a radical departure from all conventional schemes. First, public-key algorithms are based on mathematical functions rather than on substitution and permutation. Second, more importantly, public-key cryptography is asymmetric, using two separate keys, instead of one key as in conventional cryptography. The use of two keys has a significant impact on areas of key distribution, confidentiality, and authentication.

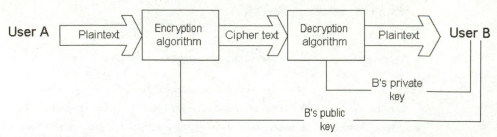

Figure 12.9   Public-key encryption model.

Only a brief overview of how public-key encryption works is provided and for a detailed understanding see Stallings (1999). As shown in Fig. 12.9, the encryption model operates as follows.

- Each end system at user *A* and user *B* generates a pair of keys to be used for encryption and decryption of messages it will receive.

- Each system publishes one key by placing it in a public registry or a public folder, and a public key is accessible to any interested party. Each system keeps the other key private and secret.

- If user *A* wishes to send a message to user *B,* it encrypts the message using *B*'s public key.

- When *B* receives the message, it decrypts it with its matching private key. No other recipient can decrypt the message since only *B* has the private key.

Public-key cryptography addressed two problems associated with conventional encryption. The first problem is key distribution. The key distribution under conventional encryption requires that either of the two parties already shares a secret key to begin with or uses a key distribution center. Either case can create problems of its own. In public-key cryptography, the only key that requires distribution is the public key, which does not have secrecy requirements.

The second problem is that of digital signatures, the electronic equivalent of a signature used in paper documents. The requirement is to devise a method to prove, to the satisfaction of all parties involved, that a message has indeed originated from the alleged sender. The private key of a message sender, which is known to no one but its owner and is generated locally, satisfies the requirements. Table 12.6 provides a summary of some of the important aspects of conventional and public-key encryption.

River-Shamir-Adelman (RSA), developed by Ron River, Adi Shamir, and Len Adleman at the Massachusetts Institute of Technology (MIT) in 1977, is a widely accepted and implemented public-key encryption algorithm. In fact, it has achieved standard status in many aspects. For example, ITU-T Recommendation X.509 lists RSA as a security standard. It has been built into current or planned operating systems by vendors such as Microsoft, Apple, Sun, and Novell. For a detailed introduction to RSA, see Stallings (1999).

Digital signature

A digital signature is an important security mechanism widely used in security services such as nonrepudiation, authentication, and access control. In essence, a digital signature is an electronic equivalent of signing a document so that the recipient knows that the document must have come from the party that signed the document. Digital signatures consist of two parts:

- A method of signing a document such that forgery is infeasible
- A method of verifying that a signature originated from the person who signed the document

For a digital signature to be practical, it must be relatively easy to generate and must contain some information unique to the sender, to prevent both forgery and denial by the sender. On the recipient side, a digital signature should be easy to recognize and verify. A recipient must be able to verify the author, the date and time of the signature, and the contents of a message at the time of the signature.

A digital signature should be computationally infeasible to forge either by constructing a new message for an existing digital signature or by constructing a fraudulent digital signature for a given message.

Various approaches to implementing digital signatures fall into two categories: direct and arbitrated. The direct signature involves only the two communicating parties (i.e., sender and recipient). This type of digital signature assumes that the private key of the sender is encrypted into the digital signature in one form or another and the recipient knows the public key of the sender. A common weakness shared by digital signature functions of this type is that the validity of the scheme depends on the security of the sender's private key. The private key can be lost or claimed to be lost.

The second category of digital signatures, the arbitrated digital signature, addresses the weakness of the direct digital signature. There is a variety of

TABLE 12.6   Conventional and Public-Key Encryption

Conventional encryption	Public-key encryption
Only one key, called a secret key, is shared by communicating parties.	A pair of keys, one called a pubic key and one a private key, is used at each side.
The same algorithm with the same key is used for both encryption and decryption.	For encryption and decryption, one algorithm is used with a pair of different keys, one for encryption and one for decryption.
The sender and receiver must share the algorithm and the key and must find a way to pass the key and algorithm from one party to the other.	The sender and receiver must each have one of the matched pair of keys for the encryption and decryption to work.
The key must be kept secret.	Each end must keep one key secret.
To provide security, it must be impossible or at least impractical to decipher a message if no other information is available.	To provide security, it must be impossible or at least impractical to decipher a message if no other information is available.

arbitrated signature schemes. In general, they all behave as follows: every signed message from a sender to a recipient goes first to an arbiter, who verifies and authenticates the message and its signature. The message is then dated and sent to the intended recipient. This solves the problem faced by the direct signature schemes: the sender might deny ever having sent the message. This type of digital signature scheme requires that both the sender and recipient trust the arbiter.

The National Institute of Standards and Technology has published a Digital Signature Standard (DSS), which presents the Digital Signature Algorithm (DSA) that uses a secure hash function to compute the signature. The DSS was first published in 1991 and has gone through several revisions since then.

Secure hash functions

The secure hash function is a security mechanism widely used in combination with other security mechanisms to implement security services such as data confidentiality, data integrity, and authentication. A hash function, as shown below, takes as input a message and applies a hash algorithm to compute a hash value

$$v = H(M)$$

where M is a variable-length message and $H(M)$ is the fixed-length hash value. The hash function can be used to ensure the integrity of a message. For example, a hash value may be appended to the message at the source at a time when the message is created. The receiver can check the integrity of the message by recomputing the message. A hash function is characterized by the following properties as summarized in Stallings (1999):

- The hash function H can be applied to a block of data of any size.
- The hash function H shall produce a fixed-length output.
- $H(x)$ shall be relatively easy to compute for any given x, making both hardware and software implementation practical.
- For any given code h, it is computationally infeasible to find x such that $H(x) = h$. This is also referred to as a one-way property.
- For any given data block x, it shall be computationally infeasible to find $y = x$, using $H(y) = H(x)$. This is also referred to as a weak collision resistance.
- It shall be computationally infeasible to find any pair (x, y) such that $H(x) = H(y)$. This is sometimes referred to as a strong collision resistance.

All hash function algorithms strive to eliminate if at all possible and at least minimize the likelihood that two messages chosen at random will produce the same hash code.

The MD5 message-digest algorithm is one of the most widely used secure hash functions. This algorithm takes a message of arbitrary length as input and produces a 128-bit message digest as output. The MD5 algorithm has the property of every bit of the hash code being a function of every bit of the input. The complex algorithm of MD5 makes the likelihood very low that two messages chosen at random, even with similar characteristics, will have the same hash code. MD4, a predecessor to MD5, published in 1990, is slightly simpler than MD5.

There are other widely used secure hash functions. The secure hash algorithm (SHA), developed by the National Institute of Standards and Technology, and published as a federal information processing standard in 1993, is based on the MD4 algorithm. The RIPEMD-160 message-digest algorithm was developed under the European RACE Integrity Primitive Evaluation (RIPE) project and published in 1996. It takes as input a message of arbitrary length and produces as output a 160-bit message digest.

Access control mechanisms

Access-control-related mechanisms either authenticate the identity of an entity attempting to access a network resource or enforce the control of capabilities granted to an entity. The mechanism may include

- An access control information base, where the access rights of peer entities are maintained. The information may be maintained by a trusted third party or the network entity that is being accessed. The information may take the form of an access control list or a matrix of hierarchical or distributed structures.

- Authentication information such as passwords and authorization certificate.

- Capabilities, access rights, and restrictions.

- Access-related records such as time of attempted access, route of attempted access, and duration of access.

Access control mechanisms can be applied at either end of a communication association or at the intermediate point. When control is applied at the originating side, it determines whether the sender is authorized to communicate with the recipient.

Other security mechanisms

Data integrity mechanisms.   The commonly used integrity mechanisms combine data padding with encryption to either detect or prevent modification, deletion, re-creation, duplication, insertion, and replay of data items being transmitted between two entities. Data padding appends supplementary information such as a block check code or a cryptographic check value that is a function of data itself and may itself be enciphered. A secure hash function is often used to compute the check value or a block check code.

Notarization mechanism.   A notarization mechanism provides assurance on the properties of the data communicated between two or more entities, such as its integrity, origin, destination, and time. Usually a trusted third party serves as a notary, using other security mechanisms such as encryption, digital signature, and integrity mechanisms.

Summary

Security management is a very broad and complicated area. This chapter first discusses a security management process model that consists of three parts: fraud prevention, fraud detection, and fraud containment and recovery. Fraud prevention means taking proactive measures and keeping unauthorized users from ever accessing the network resources and service. The most effective fraud prevention scheme is access control that prevents fraudulent users from ever accessing network resources. Discussed next is the fraud detection that aims to detect and thus stop frauds promptly and efficiently. A brief description of several approaches to fraud detection (e.g., usage pattern analysis and traffic pattern analysis) points to some directions the reader might pursue further. Fraud containment and recovery are used to minimize the harm done by frauds. Measures that can be taken to this end include customer revocation, severing connections, and service and network reconfiguration data reloading.

Five security services, i.e., authentication, integrity, confidentiality, access control, and nonrepudiation are introduced. The security services are provided through a set of security mechanisms. A high-level description is provided of a set of security mechanisms, including encryption, digital signatures, secure hash functions, and a notarization mechanism, with emphasis on encryption.

Exercises

12.1   Following is a realistic fraud scenario. A fraudulent user dials for an operator-assisted collect call from a coin-operated phone (i.e., a pay phone). The operator connects the calling party to the called party as indicated by the phone number given by the calling party. It turns out that the called party is a coin-operated phone. Discuss in general terms the potential methods to counter this type of fraud from the perspective of fraud prevention and fraud detection.

12.2   Here is another realistic fraud scenario. A group of hackers conspire to attack a PSTN network by forming a call-forwarding loop. For example, A calls B, B forwards the call to C, and C forwards it back to A. A forwards the call originated from A to B, and an endless forwarding loop just starts. Each forwarded call is treated as a new call and a new circuit is allocated. This can quickly use up trunks and potentially bring an entire network to a halt. Discuss the possible remedies from the perspective of fraud detection and prevention.

12.3   Describe some scenarios where nonrepudiation service is required.

12.4   Discuss the goals and major operations of fraud prevention and detection, using examples if possible.

12.5   Find some scenarios where authentication services are required.

12.6   Describe at a high level how the public-key encryption model works and how it addresses the problem of key distribution associated with the conventional encryption.

12.7 Describe at a high level how conventional encryption works and how the secret key is distributed.

12.8 What are the main attacks on data integrity and confidentiality?

12.9 Explain the difference between secret-key-based encryption and public-key-based encryption. Use examples to explain how encryption can be used in all the security services, i.e., authentication, confidentiality, integrity, access control, and nonrepudiation.

12.10 Explain how public-key encryption addresses the key distribution issue.

Applications of Standards

Outline

- The first section, "An Application Scenario—Design a Network Management System," presents a step-by-step process for designing a sample network management system. The summary part at the end reveals the point of the application scenario.

- The second section, "Overview of NMF Solution Sets and Component Sets," provides a summary overview of each of the NMF solution sets and component sets.

- The last section, "API Architectures for Network Management Applications," introduces the efforts in standardizing the implementation technologies for telecommunications network management applications.

Introduction

To best understand this chapter, put yourself in the position of a designer or developer of a network management system. Where are you going to start and how can you best apply the management standards you have learned so far? As a designer facing a large body of standards that can easily prove overwhelming, on the one hand you don't want to reinvent the wheel and on the other hand you don't want to spend as much effort in locating the relevant established models as you would in developing them.

The first part of this chapter presents a scenario of applying telecommunications network management standards, largely based on one of the NMF solution sets. The goal is twofold. First, the application scenario will familiarize the reader with how to apply the established network management standards to solve a network management problem. Second, the scenario is designed to walk the reader through the basic steps of designing and developing a network management system. After this application scenario, the reader is introduced to the NMF solution sets, each of which presents one complete solution to a specific network management problem. Then an overview of the application programming interface (API) architectures that have been developed for telecommunications network management applications outlines the issues addressed by the API architectures and general approaches to the APIs.

An Application Scenario—Design a Network Management System

Assume the following scenario. You are given the task of designing a TMN-compliant network management system to manage the interconnections as shown in Fig. 13.1. Two switching nodes are shown in the network although the number of switching nodes the designed system is capable of managing may be much larger. Each node is connected to a group of end users, another switching node in the same network, and a node from a different network. The connections management system is used to manage the connections between switching nodes. Among the management tasks are

- Maintaining a representation of the network switching nodes and their interconnections
- Representing and changing the management state of the network resources
- Creating and deleting circuit subgroups and circuits between switching nodes
- Adding circuits to circuit subgroups and removing circuits from circuit subgroups

The interconnections between switching nodes carry network traffic. The traffic patterns on the connections are assumed to change under temporary fault conditions, in response to predictable growth or introduction of new services into the network. To ensure quality of service for customers and to reduce operation costs, it is essential to design an interconnection management system that automates as many management operations as possible and provides the capabilities to handle the dynamic traffic variation. The switch interconnection management (SIM) system thus shall improve the control of the network connections; allow service providers to manage circuit subgroups and circuits within a subgroup; and maintain the network interconnections in response to growth, fault conditions, or service additions.

The design of a management system starts with a set of requirements. In this scenario, it is assumed that the above high-level description of the responsibilities serves as the set of requirements. The design process consists of the following five steps, and each is described in detail in this section. The intention is to demonstrate the general process that makes use of the defined network management standards.

1. Determine the management context.

2. Model the network resources. A designer first analyzes and lists the managed resources to define the boundaries of the management system. Then the designer builds an information model by defining managed object classes to represent the network resources that fall in the domain of the management system to be designed and defining the operations each managed object shall perform.

3. Perform a functional analysis to define support objects that mainly provide support management operations without directly representing any network resource and ensuring all requirements are met. This step will complete the basic design of a management system.

4. Define the information models.

5. Check the underlying implementation technologies such as development environment and management protocol to be used to see if any constraints imposed by the environment are violated in the design. If yes, go back to step 1 or 2.

Determine the management context

The contents of this step can vary from application to application. In general, it lays out the overall boundary for the system to be designed and defines the

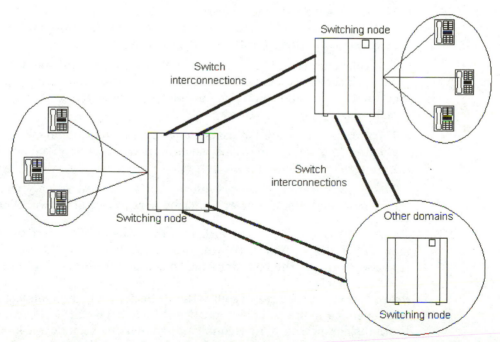

Figure 13.1   A sample switching network and interconnections.

dependencies the system has on the outside systems and environment. The following list gives some examples of the issues to be addressed by this step.

- Define the scope of the management system.

- Decide the management information exchange framework, e.g., manager-agent paradigm and the role each party plays in general terms.

- Define where the system belongs on the TMN map (the five logical layers by the five functional areas). This is important because a clear understanding of the position on the TMN map will help locate the relevant existing management standards.

- Decide on the communication environment for the management information exchange and the management protocol to be used (i.e., CMIP, CORBA, or another alternative).

- Define the system layout: where is the management system, who does it talk to, and what does it talk about?

Model the network resources

First, the requirements of the network resource modeling should be clarified because they define the criteria for checking whether the resulting information models are complete and meet the management requirements. The general requirements of the information modeling can be specified as follows.

- The representation of the network resources, i.e., switching node, interconnections, and other related resources, shall be vendor-independent and shall be consistent and applicable to all vendor equipment.

- The representation shall be consistent for and independent of underlying network technologies, whether it is voice, data, analog, or digital network equipment; in addition, it should also be independent of the transmission technology.

- The representation shall be consistent for and independent of the services provided on the interconnections, such as plain old telephone service (POTS), Intelligent Network services, or virtual private network (VPN) services.

- The representation shall provide an abstract view of only those network resources that are of concern from the perspective of network management.

This step defines all the network resources that are subject to the management of the SIM system and of which the SIM system must have a logical view. It serves to clarify the boundaries of the resource representation.

Network. A network represents the collection of interconnected network components and management entities, which can be either logical or physical. From the perspective of interconnection management, a network is a collection of interconnected circuits and the switching nodes they connect to.

Network element. A network element represents network equipment or TMN entities within a network and has one or more *Q* (*Q3* or *Qx*) interfaces. From the perspective of interconnection management, a network element includes the switching nodes and the associated network element management system.

Equipment. Equipment represents the physical components of a network element, including replaceable components. Equipment can represent the physical terminations of a trunk subsystem supporting a number of termination points.

Circuit. A circuit is a combination of two transmission channels that permits bidirectional transmission of signals between two points in a network. In the context of the switching network to be managed by the SIM system, a circuit may be a time slot or low-order path within a physical bearer, or link. A circuit may be analog, digital, or mixed and is terminated by two circuit termination points within two network elements that the circuit connects to.

Circuit termination point. A circuit termination point is a point within a network element where a circuit terminates. Within a network element, there is only a finite number of termination points. A termination point has to be allocated for a circuit when the circuit is in use and deallocated when the associated circuit is taken down. Managing the relationships between a circuit and a circuit termination point is part of the management tasks.

Circuit subgroup. A circuit subgroup is a logical grouping of a set of circuits between two network elements that share similar physical and traffic characteristics. A circuit group is terminated by two circuit subgroup termination points. Again, the management system is responsible for keeping track of the relationship between circuit groups and termination points. There may be more than one circuit group between network elements.

Bearer termination point. A bearer termination point represents the physical termination of a circuit bearer system in a network element. For example, the termination point of a T1 or E1 within a switching node is a bearer termination point. A bearer termination point contains a number of circuit termination points, just as a bearer such as T1 or E1 contains a set of circuits (24 and 32 circuits, respectively). Note that a bearer termination point does not represent the physical characteristics of the bearer system and that it is independent of the transmission technology associated with the bearer system. The relationship between a bearer termination point and a set of circuit termination points is shown in Fig. 13.2.

Define management operations

This step is to define the functional boundaries of the switch interconnection management system, i.e., what the system does and what it does not do.

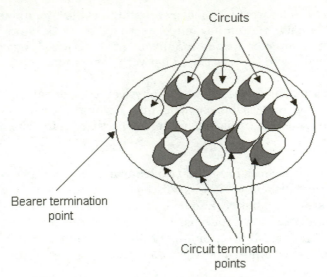

Figure 13.2 Relationship between a bearer termination point and circuit termination points.

Provisioning. When the network first starts up, the managed system is responsible for creating logical views of the managed resources. The logical views shall include what we just identified: a network, managed element, equipment, circuit termination point, circuit subgroup termination point, and bearer termination point. The managed system reports the views to the SIM system, and the SIM system in turn builds an overall view of the network resources. Note that the provisioning as used here means allocating the resources necessary for a network to operate, as opposed to the service provisioning that readies the network to provide the subscribed service to the customer.

Configuration of circuit and circuit subgroup. Once the circuit termination points have been provisioned, the SIM system establishes a logical view of all the circuits and may associate each circuit with a pair of the provisioned circuit termination points. A circuit has exactly two termination points, and a circuit termination point may be associated with only one circuit. The managing system may request a managed system to delete or modify a circuit, and once the requested operation is completed, the managing system must receive an event report from the managed system to indicate whether or not the requested operation is successful. The managing system should be able to know from a circuit termination point which circuit it is associated with and likewise, from a circuit, which two circuit termination points it is associated with.

In similar fashion, once the necessary circuit termination points have been provisioned, the SIM system establishes a logical view of all the circuit subgroups and may associate a circuit subgroup with a pair of the provisioned circuit subgroup termination points. A circuit subgroup has exactly two circuit subgroup termination points, and a circuit subgroup termination point may be associated with only one circuit subgroup. The managing system may request the managed system to delete a circuit subgroup and modify the circuit subgroup configuration once the operation is completed. The managing system shall keep and maintain the knowledge of the association between circuit subgroups and circuit subgroup termination points.

Creation of a circuit subgroup.   Based on the predefined rules for forming circuit subgroups, which in turn are dependent on the predicted traffic pattern and available resources, the managing system shall be able to add or remove a circuit from a circuit subgroup. Adding a circuit into a circuit subgroup will result in the creation of an association between a circuit and a circuit subgroup and between a pair of circuit termination points and a pair of circuit subgroup termination points. Removing a circuit from a circuit subgroup will result in deletion of the associations.

Fault recovery.   Another function that the managing system requires is to recover from a system failure and to maintain an overall view of the networkwide interconnections that is consistent with the states of network resources. Once the communication between the managing and managed systems is reestablished, the managing system needs to obtain all updates on the changes to the network resources to realign its view of the network resources with that of the managed system.

State management.   Once the switching interconnections have been provisioned and configured, an important remaining task is managing the states of the interconnection. We may find that the four types of states defined in CCITT Recommendation X.731 (1992) can adequately meet the needs of managing switch interconnections. The defined states only indicate the types of information a state captures. Still to be defined are the detailed semantics of each state, a set of rules for state transitions, and rules for state interactions.

The usage state of a resource indicates whether a resource is usable and whether it is being used at the moment. Normally only the natural operation of resources can cause a usage state transition to occur between three possible state values, i.e., *idle, active,* and *busy.* The managing system only needs to know the value of this state but does not change it. Sometimes in place of the usage state, the operational state is used, which can be set to either *enabled* or *disabled.*

A third state, the availability state, is used to qualify the administrative and operational state attributes. It provides additional information on the state of a network resource, for example, by indicating a resource has been disabled because of a change in state of another resource on which it depends.

The administrative state provides the managing system with a means to control the managed resource by setting the state value to *locked, unlocked, shutting down,* or *user quit.* A designer needs to define the detailed semantics of this state and state transition rules for each managed resource that requires this state as an attribute. We can define the administrative state transitions as follows.

Locked. The transition to this state occurs when the managing system intends to unconditionally lock the resource such as a circuit, a circuit subgroup, a circuit termination point, a circuit subgroup termination, or a bearer termination point. No new call will be allowed, and existing calls will be discarded.

Unlocked. The transition to this state occurs when the managing system intends to make a resource administratively available for use. Note that when the administrative state of a resource is in the unlocked state, it does not necessarily mean that the resource is available for use. Its operational state may be disabled due to a fault within the resource, or its usage state may be busy.

Shutting down. This is a graceful way of administratively locking a resource, allowing the existing calls to proceed to completion before the resource is locked.

The next task is to define the state interaction rules, i.e., how a state of one resource affects the states of all other associated resources. Following are some examples of the interaction rules.

1. When the administrative state at an endpoint resource (e.g., a circuit termination point) becomes locked, then the operational states of the associated connectivity resource (e.g., a circuit) and the other associated endpoint resource (a circuit subgroup termination point) transition to the disabled state.

2. When the administrative state of a connectivity resource (e.g., a circuit) becomes locked, then the administrative states of the associated endpoint resources (a pair of circuit termination points) transition to the locked state and the operational states of the associated endpoint resource (e.g., a circuit subgroup termination point) transition to the disabled state.

3. When a connectivity resource is unlocked, then the administrative states of the associated endpoint resources transition to unlocked and the operational states of the associated endpoint and connectivity resources transition to enabled.

4. When the operational state of a resource changes, then the same state value is set in the operational states of all contained resources. In turn, the same transition occurs in the operational states of any connectivity resources associated with the contained resources and of any endpoint resources associated with the connectivity resources.

5. When the administrative state of a resource is locked, then the operational state of any related resource, as defined in Table 13.1, transitions to disabled.

A dependency table such as the one shown in Table 13.1 describing the dependency relationships between the resources is a useful means for defining the interaction rules.

The sequence of management operations. A designer need define not only the management operation but also the sequence of the management operations. The high-level definition of a management operation sequence provides a system-level view that facilitates system design. For example, the operation sequence can be used for system-level object identification and later for the check of design completeness. One example of the management operation sequence to create a circuit subgroup is shown in Fig. 13.3.

Define the information model

Information modeling decisions need to be made with regard to how to model or represent each network resource and the relationships between the network resources. For example, a resource can be modeled as an independent object or an attribute of an object. In this application, the decision is to model each resource as an individual managed object class. The next step is to check the defined management standards to avoid reinventing the wheel. Many identified managed object classes may have already been defined in the standards. For this application, more than half of the classes can directly come from existing standards, as listed in Table 13.2.

It is decided that the following four managed object classes cannot be found in the defined management standards and thus need to be defined specifically

TABLE 13.1   Related Resources

Resource	Related resource
Circuit subgroup termination point	Circuit termination point
Circuit subgroup	Circuit
Equipment	Bearer termination point

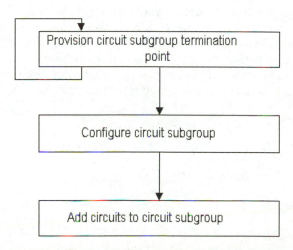

Figure 13.3 Sequence of management operation for creating a circuit subgroup.

TABLE 13.2 **Managed Object Classes for the SIM System**

Network resource	Managed object class	Source	Definitions
Network	network	ITU-T Rec. M.3100 (1995)	See App. D and Chap. 7.
Network element	networkElement	ITU-T Rec. M.3100 (1995)	See App. D and Chap. 7.
Equipment	equipment	ITU-T Rec. M.3100 (1995)	See App. D and Chap. 7.
Bearer termination point	trailTerminationPoint-Bidirectional	ITU-T Rec. M.3100 (1995)	See App. D. and Chap. 7.
Circuit	interExchangeCircuit	Application specific	An instance of this class represents a single circuit on a switch.
Circuit subgroup	circuitSubgroup	Application specific	An instance of this class represents a set of circuits sharing the same characteristics grouped for routing.
Circuit subgroup termination point	circuitSubGroupXtp	Application specific	An instance of this class represents a circuit subgroup termination point.
Circuit termination	circuitXtp	Application specific	An instance of this class represents a circuit termination point.

for this application. It occurs often in designing management applications that you don't find all the managed object classes in the existing standards. The rule of thumb is using as many defined classes from standards as can be found and then defining the rest of them. Even if you cannot find a complete managed object class for reuse, it is more than likely that you can find components of classes, i.e., attributes, actions, and notifications, as building blocks for the managed object class you have to define yourself, as shown in this case.

For brevity and convenient reading, simple text notations rather than GDMO are used in definitions of the following application-specific managed object classes. No distinction is made between mandatory and conditional packages partly because the implementer of the design may choose some attributes over others for a particular application context anyway.

Managed object class: circuitSubGroup

Superclass: CCITT Recommendation X.721 (1992), top

Attributes

    circuitSubgroupId
    aTPSGInstance
    zTPSGInstance
    circuitsInCircuitSubGroupList
    ITU Recommendation X.721: availabilityStatus
    ITU Recommendation X.721: administrativeState
    ITU Recommendation X.721: operationalState

Actions

  addCircuitToCircuitGroup
  removeCircuitFromCircuitGroup

Notifications

  ITU Recommendation 721: objectCreation
  ITU Recommendation 721: objectDeletion
  ITU Recommendation 721: attributeValueChange
  ITU Recommendation 721: stateChange

The managed object class circuitSubGroup represents the circuit subgroup network resource that is identified in a previous subsection. The three state attributes and four notifications are all from the defined standards, and their definitions can be found in the Managed Object Class Dictionary (App. D). A circuit subgroup is terminated by two group termination points, represented by the attributes aTPSGInstance and zTPSGInstance, respectively. The other attribute, circuitsInCircuitSubGroupList, is a list of circuits in this circuit subgroup. The two actions, addCircuitToCircuitSubGroup and removeCircuitFromCircuitSubgroup, implement the management operations of adding a circuit to and removing a circuit from a circuit subgroup.

Managed object class: circuitSubGroupXtp

Attributes

  directionality
  modificationOfClass
  OffNetConnection
  Recommendation X.721: availabilityStatus
  Recommendation X.721: usageState

Notifications

  Recommendation X.721: equipmentAlarm
  Recommendation X.721: qualityofServiceAlarm

The managed object class circuitSubGroupXtp represents the exchange circuit subgroup termination points that are identified in a previous subsection. For those attributes and notifications that are from the defined standards, see App. D for their definitions. The usageState attribute supports the active and busy states. When one or more, but not all, of the circuits associated with the circuit subgroup termination point carry user traffic, the circuit subgroup termination points are said to be in the active state. They are said to be in the busy state if all the associated circuits are carrying traffic. The attribute, offNetConnection, indicates whether this circuitSubGroupXtp is connected to an off-network connection and the default value can be set to FALSE. The circuit subgroup exchange termination point directionality attribute is used to specify the direction of the associated circuits, and it can take one of three values: incoming, outgoing, or bidirectional. Both the incoming and outgoing directionalities indicate that the associated circuits are reserved for one-way traffic, either incoming or outgoing. It is evident that the associated circuits with a bidirectional directionality can carry traffic in both directions. The modificationOfClass attribute supports the ability to modify or set a default calling party category for incoming calls by differentiating the following categories of calling parties: ordinary calling subscriber, calling subscriber with priority, operator, data call, test call, and pay phone. The uAconvRequired attribute indicates whether the associated resource supports the U-law or A-law conversion.

Managed object class: circuitXtp

Attributes

CDS (circuit dual seizure)
huntingNumber
offNetConnection
Recommendation X.721: availabilityStatus
Recommendation X.721: administrativeState

Notifications
Recommendation X.721: equipmentAlarm
Recommendation X.721: qualityOfServiceAlarm

The managed object class circuitXtp represents the exchange circuit termination points that are identified in a previous subsection. See App. D for definitions of those attributes and notifications that come from the defined standards. The attribute offNetConnection, as in the circuitSubGroupXtp managed object class, indicates whether the termination point is connected to an off-network connection. The attribute administrativeState allows the managing system to control the resource. The attribute huntingNumber specifies the order in which circuit termination points within a subgroup of circuit termination points are checked to find an available circuit. The circuit termination points are seized in a sequential order according to their respective hunting number. The attribute CDS indicates if the circuit termination point has a priority on a dual seizure attempt of a circuit.

Managed object class: interExchangeCircuit

Superclass: Recommendation M.3100 (1995): connectivity

Attributes

circuitId
circuitType
circuitBandwidth
associatedOwningCircuitSubGroup
Recommendation X.721: availabilityStatus
Recommendation X.721: usageState

Notifications

Recommendation X.721: qualityOfServiceAlarm

The managed object class interExchangeCircuit represents the connection between two switching nodes, and a circuit is always terminated by two circuit termination points. The associatedOwningCircuitSubGroup attribute identifies the circuit subgroup object instance to which this circuit belongs. It indicates the circuit has not been assigned to any circuit subgroup if this attribute value is null. The attribute circuitBandwidth represents the bandwidth capacity of the circuit that is measured in thousands of bits per second. The administrative state of an interexchange circuit must reflect the states of the associated circuit termination points. For example, when the administrative state of either one of the interexchange circuit's termination points transitions to the locked state or the operational state of a termination point transitions to the disabled state, the operational state of the circuits must transition to the disabled state.

Next to be defined are the two types of important relationships between managed object classes already defined: inheritance and containment relationships. One of the main goals of using inheritance relationships is to take advantage of and to reuse already defined managed object classes. The first place to look when one is modeling network resources is the TMN generic network information models, as discussed in Chap. 7. A set of information models representing a generic telecommunications network devoid of network technology details is meant for reuse. For example, the managed object class circuitXtp apparently has characteristics and behavior in common with the managed object classes in the termination point fragment as discussed in Chap. 7. It is left as an exercise to compare the managed object classes defined for this application with the generic information models defined in ITU Recommendation M.3100 (1995) to determine the inheritance relationships.

When deciding on the containment relationships between managed object classes, one should pay attention to the following two issues. First, the containment relationships are used to model the "has a" or "is part of" relationships in the real world. If the network resources that are being modeled do have this kind of relationship, it is natural to use it. For example, an equipment holder object has a card object as one of its components. The second consideration for using this type of relationship is for the purpose of organizing the managed objects. For example, if the OSI MIB structure is used, a containment relationship can be defined between two object classes, which may or may not have a "has a" or "is part of" relationship.

Define management information exchange

As defined in the management context, an OSI CMISE (CMIP and CMIS) provides a management protocol and the protocol interface for applications. This step defines precisely what CMIS service shall be used to accomplish the management operations required of the SIM system.

Provisioning of network resources. Provisioning a network resource means that a managed system creates an instance of an object representing the resource and reports the event of object creation to the managing system so a management view of the resource is kept at the managing system. It is the responsibility of the managing system to see that required event forwarding discriminator (EFD) objects are created and filtering parameters are set at the managed system. The CMIS service message exchanges to accomplish provisioning are shown in Fig. 13.4.

Configuration of a circuit and circuit subgroup. Configuration of a circuit or circuit subgroup is to create an association between a circuit and a pair of circuit termination points and between a circuit subgroup and a pair of circuit subgroup termination points. This is accomplished by sending a request to the managed system to create such an association at the managed system and then reflecting the created association in the managing system, as shown in Fig. 13.5. A circuit is represented by an instance of the interExchangeCircuit managed object class.

Circuit subgroup management. Managing a circuit subgroup, including adding a member circuit to and removing a member circuit from the circuit group, is achieved by issuing an M-ACTION request from a managing system to a managed system, as shown in Fig. 13.6.

State management. For the managing system, the main task of state management is to set administrative states of network interconnections and enforce the state interaction rules as described above. This is achieved by

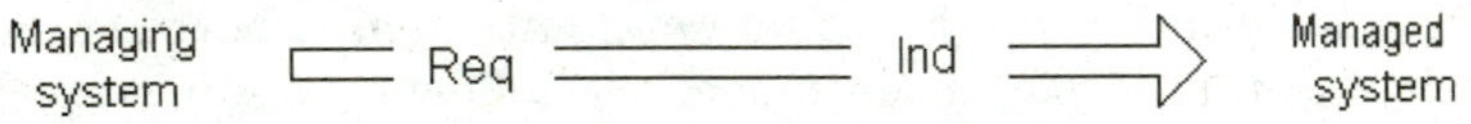

```
Managing         Req                Ind          Managed
system                                            system

M_CREATE invoke ( ....
        managedObjectClass = eventForwardingDiscriminator
        superiorObjectInstance = managedElement or network instance distinguished name
        .......
        attributeList = { ......}
)
```

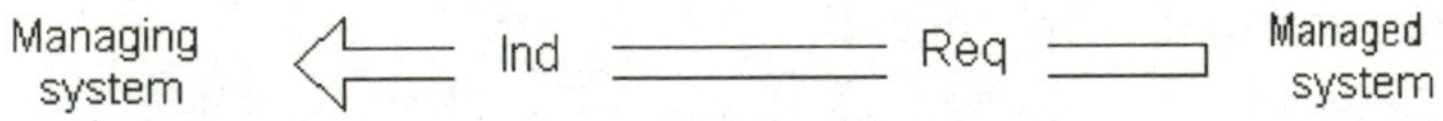

```
Managing         Ind                Req          Managed
system                                            system

M_EVENT_REPORT ( ....
        managedObjectClass = network / managedElement / equipment / circuit / circuit
                        SubGroup / circuitTerminationPoint / circuitSubGroupTerminationPoint
        superiorObjectInstance =  instance distinguished name
        eventTime = ...,
        eventType = objectCreation,
        eventInfo = = { ......}
)
```

Figure 13.4 Management information exchanges for resource provisioning.

sending an M-SET request to a managed system to set the administrative state of a resource, as shown in Fig. 13.7.

Summary

The two goals we wanted to achieve by going through this application scenario were (1) introduce the general steps in designing a telecommunications network management system and (2) introduce the work that has been done by

```
Managing          ┌──┐ Req ══════════ Ind ═══════▷   Managed
system            └──┘                               system


M_CREATE invoke ( ....
        managedObjectClass =  enterExchangeCircuit
        superiorObjectInstance =  network instance distinguished name
        ........,
        attributeList = {
                {   attributed = nameBinding,
                    attributeValue = interExchangeCircuit-network,
                }
                {   attributed  = associatedOwningCircuitSubgroup,
                    attributeValue = circuit / circuit subgroup distinguished name,
                }
                {   attributed = a-TPInstance,
                    attributeValue = circuitXtp / circuitSubGroupXtp instance distinguished
name;
                }
        }
)
```

Figure 13.5 Management information exchange for configuration.

```
Managing          ┌──┐ Req ══════════ Ind ═══════▷   Managed
system            └──┘                               system


   M_ACTION invoke ( ....
            managedObjectClass =  circuitSubGroup,
            superiorObjectInstance =  circuitSubGroup distinguished name
            ........,
            actionInfo  = {
                actionType = addCircuitToCircuitSubGroup /
removeCircuitFromCIrcuitSubGroup,
                actionInfoArg = id of circuit to be added or removed
                ......
            }
            ......
   )
```

Figure 13.6 Management information exchange for circuit subgroup management.

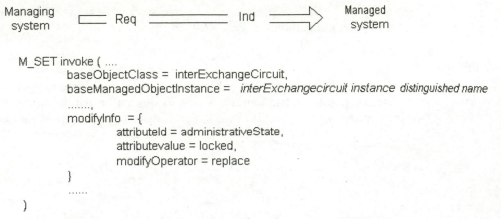

Figure 13.7 Management information exchange for state management.

the NMF in hopes that developers can leverage the large amount of existing work and avoid reinventing the wheel.

This section describes a process of designing a switching interconnection management system. One purpose of the exercise is to illustrate the work by the NMF. The design described above is derived from one of the NMF solution sets entitled "Switch Interconnection Management Configuration Management Ensemble" (NMF 035). The NMF, to be discussed further in the following section, has produced a set of solutions for network management applications based on TMN standards. An NMF solution set provides one or more technical specifications for one management problem, such as SIM, leased circuit management, or switched service feature administration. In general, a technical specification may consist of a detailed information model, management function specifications, standards references, and conformance requirements.

To many telecommunications network management application designers and developers, the formal standards can become overwhelming or even intimidating because of their sheer volume and complexity. An NMF solution set selects those applicable portions of standards, defines those that cannot be found from the standards, and provides a complete management solution for one problem. A cardinal principle of management application development, or for that matter, all software development, is to reuse existing designs and components whenever possible.

Overview of NMF Solution Sets and Component Sets

The NMF is a worldwide consortium consisting of major telecommunications service providers, equipment vendors, and software vendors related to telecommunications network and service management. The declared mission of the NMF is to provide a forum that brings together a wide spectrum of players in the telecommunications industry to produce "practical, cost-effective solutions for improving the management of public networks and services." The NMF publishes a range of business and technical agreements, reference guides, and standards-based specifications. In recent years, industrial consortiums such as the NMF are playing an increasingly important role in defining industry standards by feeding their work to official standards bodies and by producing business and technical agreement-type documents that have won wide industry acceptance.

Among the many endeavors of the NMF, the following two aspects are most relevant to the discussion of this chapter:

■ The Open Management Interoperability Point (OMNI*Point*) consists of solution sets and component sets. An OMNI*Point* solution set is a set of implementation agreements designed to solve a single specific telecommunications network management problem. An OMNI*Point* component set defines the underlying technologies such as management protocols that are used to implement a solution set.

■ A management API suite is used to facilitate the development of standards-based management applications and to reach industrywide agreement on the implementations of the management standards in order to achieve real management interoperability.

The remainder of this chapter will discuss each of these aspects.

OMNIPoint solution set

This subsection reviews the solution sets of OMNI*Point*. The reader may come across the terms OMNI*Point* 1 and OMNI*Point* 2, which refer to the two phases of the OMNI*Point* program, but this review does not make the distinction.

A solution set provides a complete solution for a single management problem and consists of two parts: a product descriptor and a technical solution. The product descriptor provides a brief summary of the solution set. At the core of the technical solution is an ensemble, a document that includes a management information model and a set of management functions that together implement the solution to the management problem. The application scenario just described is based on one of the solution sets, the SIM. A brief overview provides an indication of the breadth of the solution sets.

Generic alarm monitoring. This goal of alarm monitoring is to automate the process of monitoring, analyzing, and correlating across several component alarms in order to identify a problem promptly. This solution set provides a general-purpose facility that can be used across different types of networking equipment to monitor, log, filter, and route alarms in a consistent way. This is accomplished by grouping several components as a single subsystem object for each element type and then applying a generic set of interface capabilities to provide a common alarm interface. This results in a single, consistent way for any network management system to collect alarms from a wide variety of network and element subsystems.

Leased circuit management. Leased lines, also called *private lines,* cross two management domains: the domain of a service provider that owns the leased line and the domain of a service user, normally a business or an institution. Because leased lines operate over a fixed set of facilities, reliability is of critical importance. It is difficult for either service provider or service user to maintain a desired level of service quality without some type of comprehensive end-to-end view of the leased lines. This requires a management interface between the two management domains. This solution set specifies selected alarm, performance, and configuration capabilities at the service management level. It provides the managed objects and application functions that define a TMN-compliant X interface between the two management domains.

Switched service feature administration.   Provisioning services to customers has been a tedious and labor-intensive process. Complicating the issue is the proprietary nature of the process that is specific to a vendor's switching equipment. Based on ITU Q.824, this solution set provides the managed object classes that are sufficient to model the complete provisioning process which begins with a service order from the customer and ends with allocation of necessary resources to implement the subscribed services. In this solution set, a customer is represented in a customer service profile that defines the service available to the customer and the resource necessary to deliver the services. The services include bearer services that define how information is accessed and teleservices that describe the type of services (fax, freephone, etc.). The solution set provides a set of managed object classes for the standard $Q3$ interface between a provisioning system and a switching node.

Multinetwork bandwidth management.   Bandwidth provisioning across networks is a critical step in responding to customer demands in a timely manner. It is difficult to allocate bandwidth end-to-end across networks because it requires the exchange of information between different networks and network management systems. This solution set defines a set of standard management commands to allow a manager to automatically reconfigure bandwidth at network elements, according to the requirements of a customer service. In addition, it allows a service provider to correlate alarms from different vendor's equipment, create a unified view of network bandwidth, present network topologies, and configure service parameters.

LAN alarm interface.   An enterprise LAN consists of many network elements that in the majority of cases are managed using SNMP, while many management systems are based on OSI Common Management Information Protocol (CMIP). This solution set enables a network manager to use a single management system to monitor alarms from LAN network elements, including those that use SNMP to communicate with the underlying network resources. This is achieved through the use of proxy agents that convert the SNMP commands into the CMIP counterparts.

Trouble ticket—electronic bonding. This solution set provides an interface for the exchange of trouble ticket information between two service providers. Information exchange of this sort is becoming important because providing end-to-end services to customers increasingly involves more than a single service provider. The focus of this solution set is on the specification of standard information models that are exchanged across administration boundaries; the information models are also called a common "profile" of trouble ticket information. The electronic exchange of management information between service providers (e.g., between a local carrier and an interexchange carrier) is also termed *electronic bonding*.

Trouble ticket—customer/provider. This solution set provides an interface for the exchange of trouble ticket information between a service provider and its customers. This interface enables the customer to report service trouble and check the trouble status electronically. Again the focus of this solution set is on the specification of standard information models so that a customer can access the trouble ticket information of multiple service providers using the same interface.

Performance reporting. This solution set aims to provide a standard set of terminology and service performance parameters and a generic process for reporting service performance to customers. One major issue with customer service reporting is that there are few standards available with regard to the service performance and there are no common definitions of terms to be used at service level. Consequently this causes confusion in communication between customers and a service provider. For example, the term *service availability* may mean that the application is running without a problem to customers, while the service provider uses the term to mean that the service is working even if impaired. The mismatch of expectations will lead to misunderstanding and dissatisfaction on the customer's part. This solution set provides the precise definitions of a set of service performance parameters that address customer's concerns and the methods for computing them. In addition, the solution set also defines a generic process for reporting customer service performance.

Order exchange. This solution set aims to automate and standardize a generic service order process that will permit a quick exchange of information between a customer and a service provider and between two service providers. The service ordering process has traditionally been labor-intensive and error-prone. The automation of this process is a key step for a service provider to become competitive in the telecommunications market. The focus of this solution set is on the *process* of information exchange as well as the *information* that is exchanged between parties. This solution set defines a set of information models to standardize the generic aspects of the ordering process that are applicable to all service providers.

API Architectures for Network Management Applications

To achieve true management interoperability intended by the network management standards, a standard API technology plays a crucial role. To some degree, it can be viewed as standards of implementing standards, or implementation standards. The fact that multiple vendors implement the same standards does not guarantee interoperability between their systems. The standardization at the implementation level can ensure the interoperability intended by the standards. This section introduces the API suite developed by the NMF over the years for network management applications in hopes that the reader will

1. Develop an awareness of the industry-accepted management APIs
2. Gain a general understanding of how a management API works and how to apply and even develop an API

There are two fundamental requirements for a management API, and for that matter, any API.

- The API must be simple, easy to use, and a natural part of the programming language used in the application program.
- The application must provide application source level portability.

API for ASN.1

ASN.1 is an OSI standard specification language for representing data in a way that is platform-independent and programming language–neutral. ASN.1 is widely used in telecommunications network management applications.

Even with ample knowledge of ASN.1, a programmer would rather not deal with ASN.1 directly in a management application program. For one reason, it is extremely tedious to convert the ASN.1 types into the types of a programming language. For another, either this application may not work once it is moved to another platform, or it may not be integrated with a system that uses a different conversion scheme. These are precisely the issues the ASN.1 API is designed to address (NMF 040-1 and 040-2, 1997). Stated formally, the ASN.1 C++ API is designed to

- Provide portability of applications that use ASN.1 so that those applications can operate in any environment that implements the standardized API
- Shield a programmer from any details of ASN.1 data structure, some of which are uncommon in programming languages

When considering the fact that the expertise in ASN.1 is scarce as opposed to the relative abundance of expertise in common programming languages like C++, the ASN.1 C++ is even more important. The ASN.1 API is introduced from the perspective of a programmer, with emphasis on the user's view rather than on the internal workings of the API.

As shown in Fig. 13.8, a network management application deals only with C++ objects that represent an ASN.1 abstract data instance and with C++ classes that represent ASN.1 abstract data types. An ASN.1 data type is represented as a C++ class. For example, under the superclass abstractData, there are two groups of subclasses, representing ASN.1 primitive types and complex types, respectively. The C++ classes for primitive ASN.1 types include BitString, Boolean, Choice, Enumerated, and Integer. The C++ classes for complex ASN.1 types include set, setOf, sequenceOf, and sequence. Some of those classes have subclasses of their own.

Besides the abstractData superclass and its subclasses, C++ encoding and decoding functions, iterators for traversing input or output data in ordered sets, and error objects for exception handling are also available to an application. The API also provides some auxiliary objects for maintenance purposes to help the application manage the interface objects.

In summary, the NMF ASN.1 C++ API greatly reduces the difficulty for an application programmer to use ASN.1 and limits to a minimum the knowledge of ASN.1 a programmer must possess. Even more importantly, if the API is widely adopted in the industry and becomes a de facto standard, it will be one step closer to achieving application level system integration and interoperability.

CMIS C++ API

As shown in Fig. 5.1 and discussed in Chap. 5, the CMISE consists of two parts: the CMIS and the CMIP. As shown in Fig. 13.9, the CMIS provides the interface to the application program and passes through the association service from the ACSE to the application. The CMIP is the protocol to implement the CMIS services using its protocol messages. CMIS services can be categorized into three groups: association services, management event notification services, and management operation services, as shown in Fig. 5.1. The CMIS provides an interface

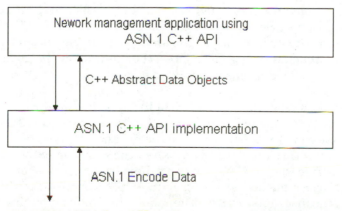

Figure 13.8   ASN.1 C++ API context.

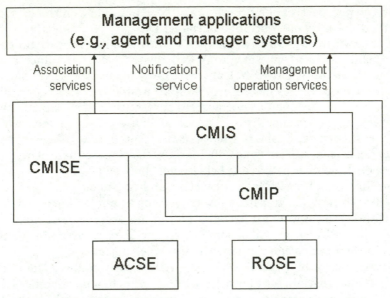

Figure 13.9   The CMIS and CMISE.

to the outside world, specifying the types of management operations represented by a management message and parameters required for the message.

The API standardizes implementation of CMISE, CMIP, and ACSE services. If not standardized, one vendor's implementation can be different from another vendor's. An application coupled with one vendor's implementation may not work once it has been moved to another vendor's management environment. The application has to be rewritten to custom fit into the new vendor's implementation. Once a standard API is in place, each CMISE tool vendor is free to optimize and modify its implementation without concern of being noncompatible with existing applications.

As shown in Fig. 13.10, an API must interface two sides: an external interface to the management application and an internal interface to the implementation of the CMISE and ACSE. This introduction, based on NMF 041 (1997) focuses on the external interface for an application programmer. It is necessary for a tool vendor to understand both interfaces, the internal one in particular, which, however, is not covered in this overview.

API for association services.   As described in Chap. 5, the ACSE provides three basic services for a management application to establish an association with another application (e.g., between a manager and an agent) or to take down the association either prematurely or when the communication is completed. The three services are A-Associate, A-Release, and A-Abort. The ASCE services are accomplished by using four ACSE service primitives: request, indication, response, and confirm. The requesting application first sends an A-Associate.request primitive to request an association, and the request arrives

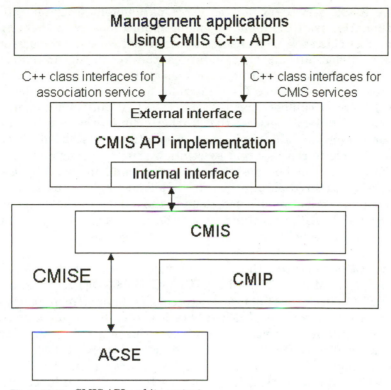

Figure 13.10 CMIS API and its context.

at the responding application as an A-Associate.indication. The responding application returns an A-Associate.response to indicate agreement on the association, and the response arrives at the requesting application as an A-Associate.confirmation.

The API provides C++ classes to model those service primitives, which include the following C++ classes:

- A-ASSOCIATE-Req

- A-ASSOCIATE-Ind

- A-ASSOCIATE-Rsp

- A-ASSOCIATE-Cnf

- A-RELEASE-Req

- A-RELEASE-Ind

- A-RELEASE-Rsp

- A-RELEASE-Cnf

- A-ABORT-Req

- A-ABORT-Ind

The ACSE service primitives carry parameters from one application to another. Also available to management applications are a set of ACSE service parameter classes, such as AssociationInfo that contains the association information, ReleaseResult that indicates the result of an association, and ReleaseResultReason that provides reasons for release of an association.

If using a standardized C++ interface for a managing application association still proves cumbersome, the CMIS C++ API offers automatic association management (AAM) services that further simplify the interface for establishing and managing the application associations. Instead of directly dealing with the objects representing association primitives such as A-ASSOCIATE-Req, an application obtains an association handle and specifies the characteristics of the desired association. All details of establishing and managing an application association are done behind the scenes by the API. Unfortunately, this part of the API, as it stands now, is optional, and it is likely that some vendors may choose not to implement it.

API for CMIS services.   The CMIS provides seven services (M-GET, M-SET, M-CREATE, M-ACTION, M-CANCEL-GET, M-DELETE, and M-EVENT-NOTIFICATION) that fall into two categories: management operation services and notification services. Each service is accomplished using four primitives: request, indication, response, and confirm, mirroring the ACSE service primitives. The CMIS C++ API, as described in NMF 041 (1997), provides corresponding C++ classes for the management and notification services.

The C++ classes for CMIS services are broken down into four categories, all abstract classes inheriting from a superclass primitive: Req (Request), Ind (Indication), Rsp (Response), and Cnf (Confirm). From each of the four abstract classes are derived the CMIS service primitive classes that are concrete classes an application program can use directly. As shown in Fig. 13.11, M-GET-req is derived from the abstract class Req. Those are the interface classes that may be directly used in an application program.

The parameters the CMIS service primitive classes require are modeled as C++ classes too. For example, the parameter classes include such familiar CMIS service primitive parameters as attributeList, actionInfo, actionReply, eventInfo, and eventReply.

In summary, the CMIS C++ API aims to standardize the implementation of the CMIS interface. Chief among the benefits of the standardized interface are the following three aspects. First, a management application becomes truly portable when it moves to another management environment as long as the standard API is implemented. Secondly, management application development tool vendors (those who develop CMIS and CMIP tool kits) are free to optimize their implementation as long as the external interface is provided. Third, application developers are insulated from much of the CMISE details, and this results in faster development of management applications. Thus an application programmer only needs to understand the external interface part of the API.

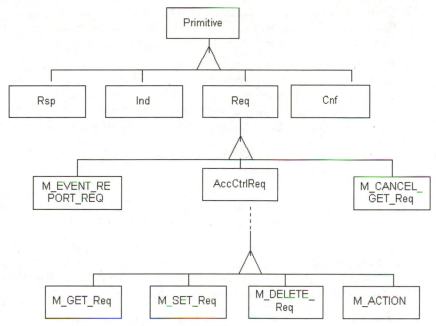

Figure 13.11   Inheritance hierarchy of CMISE request classes.

API for GDMO

GDMO C++ API actually offers more than what the name suggests, i.e., translating GDMO notation into C++ so an application programmer need only deal with C++ classes and objects. It provides a whole package for building a TMN management system, whether it is a manager system or an agent system. Building a management system involves a host of tasks other than establishing and managing an application association and sending, receiving, and processing CMIS service primitives over a management protocol. The items that a programmer must take care of in a management system include

- *Association management.*   As discussed in Chap. 5, a manager system and an agent system exchange management information in the form of requests and responses. Among the management tasks associated with the exchange are keeping track of the outstanding requests and associating a request with a right response.

- *System configuration management.*   A request a manager sends to an agent eventually will go to a managed object that supports the requested operations. The configuration information must be available to a manager on what managed objects are supported by which agent. An agent must have the configuration information of managed objects in order to route a request to an appropriate managed object.

- *System management.*   A management system must interact with the underlying operating system to start and shut down a management application and obtain such information as time and system events.

- *Managed object management.*   The network resources are represented as managed objects in an agent system. This logical view of managed resources must be reflected in the manager side in order for a manager to perform management operations. It is a nontrivial task to have an efficient, yet sufficient representation of the managed resource available to a manager system.

Support for building a TMN manager application.   As shown in Fig. 13.12, the API, as defined in NMF 042-1, 042-2, and 042-3 (1997), provides a manager application developer with a manager framework, managed object handles, managed object handle management facilities, association management facilities, and configuration management capabilities.

Manager framework.   The manager framework is a general facility that provides for the interactions between an application and the API implementation. For example, it provides methods for an application to access other parts of the API, such as managed object handle collections.

Managed object handles (MOHs).   Relieving a manager application programmer of the burden of having to maintain a view of managed objects and all associated management chores, the managed object handles provide a simple, yet efficient way to manage the managed objects in a manager's domain. Those managed object handles present a logical view of managed objects to a manager application by enabling a manager application to access all necessary information about a managed object such as name, state, and class-level metadata. In addition, they allow the application to invoke synchronous and asynchronous CMIS operations on the corresponding managed object in the agent domain and provide default logic for processing indication and confirmation associated with the managed objects. The logic is in the form of virtual methods that can be specialized to fit the need of a particular application.

Managed object handle factory registry.   The managed object handle factory registry provides an application with the capability to construct object handles either with default parameters or with application-specific characteristics.

Managed object handle collections.   These collections are containers of managed object handles and provide a manager application with the ability to organize and manage the managed object handles. Two types of collections are distinguished: enumerated collections and rule-based collections. The enumerated collections allow an application to add and remove an object handle using methods from the C++ Standard Template Library (STL). The rule-based collections contain those object handles that allow an application to invoke scoped and filtered operations on agents in the network.

Manager send control objects.   This type of management object allows an application to control the communication parameter of the CMIS requests the manager sends out.

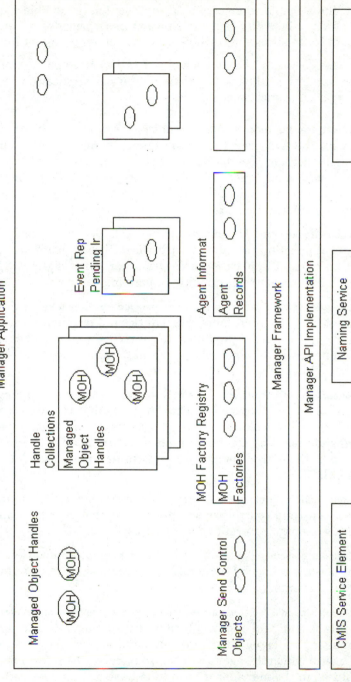

Figure 13.12 GDMO C++ API support for a manager application. (*From NMF 042-1, 1997.*)

Agent handle registry and agent information base.   The agent handle registry allows an application to represent a known agent in the network as an agent handle, just like a managed object being represented as an object handle in the manager's domain. The agent information base provides an application with the capability to store and retrieve information on an agent handle.

Callback objects.   The callback objects allow an application to register callback functions. The callback functions enable an application to invoke CMIS services on an asynchronous basis.

Support for building a TMN agent application.   As shown in Fig. 13.13, the support the API provides for an agent application developer closely mirrors the support for the manager application. The support includes an agent framework, a managed object factory registry, management information tree, agent send control objects, callback objects, a manager handle registry, and a manager information base.

Agent framework.   The agent framework provides general facilities for an agent application to interact with the API. For example, through the framework, an agent application accesses other parts of the API, such as the manager handle registry or queuing of pending requests.

Managed object factory registry and managed object factory.   In an agent domain, a managed object represents a network resource being managed. The factory registry provides the capability to construct a managed object either based on the generic GDMO object class default parameters or the characteristics of the agent application.

Managed information tree.   The MIT provides an agent application with the ability to organize managed objects in a GDMO naming tree, perform naming resolution, and look up a requested managed object. Remember that OSI recommendations specify the naming structure of an MIB but not the implementation of the MIB. The MIT intends to provide a standard for implementing the OSI MIB.

Agent send control objects.   Like their counterparts for manager applications, the agent send control objects provide control over the communication parameters of outgoing event report requests.

Callback objects.   These objects allow an agent application to register application-specific callback functions for the requests sent from managers as well as for local events.

Manager handle registry, manager handles, and manager information base.   A manager handle is a logical representation of a manager in the agent's domain. The manager handle registry allows an agent application to register a known manager with which the agent will communicate. The information about managers is stored in the manager information base as manager records.

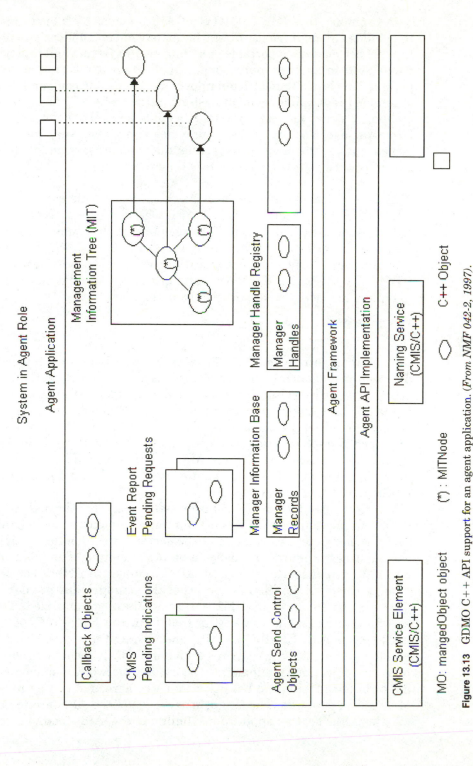

Figure 13.13 GDMO C++ API support for an agent application. (*From NMF 042-2, 1997*).

In summary, the NMF GDMO C++ API provides a set of standard services to achieve two main goals. First, the management application built on top of the API becomes truly portable. A management system shall function the same when used in another management application if the standard API is supported. This lays a foundation for developing reusable components that can be assembled into a management system. Second, the API facilitates the development of a management system. It shields most of the details of the underlying management protocol and provides simple yet very efficient interface objects such as the managed object handle and manager handle to manipulate the associated objects and resources they represent.

TMN C++ API suite.   The set of APIs described in this section, ASN.1 C++, CMIS C++ API, and GDMO C++ API, together form a TMN C++ API suite for implementing TMN-based management applications. The relationships between the APIs are shown in Figure 13.14. The CMIS API, in addition to providing an interface to an application system, also provides services to the GDMO C++ API. The TMN API suite provides a foundation that is sufficient to build a management application from the start.

Summary

This chapter introduces the efforts in standardizing implementation of the network management standards discussed in earlier chapters. A primary goal of the large volume of network management standards is to make a management system interoperable with management systems of different vendors on different computing platforms. Without a uniform or standard approach to implementing management standards, the standards will not be useful in a practical sense. The NMF and other organizations bring the standards into a practical management context and define a set of standards for implementing the standards.

We first introduce NMF solution sets by going through a scenario of designing an application system for management of switch interconnections. Step by step, by the end of the section, a complete solution, based on the existing standards, is ready to be implemented. The point of this exercise is to demonstrate what is in a solution set. An NMF solution set provides a complete solution for one specific management problem. Following the scenario, each of the NMF solution sets is summarized. The following section focuses on the three APIs that form the NMF TMN API suite, i.e., ASN.1 C++ API, CMIS C++ API, and GDMO C++ API. The APIs are the result of efforts by expert NMF members, which include service providers, telecommunications equipment vendors, and network management tool vendors. Since the APIs are based on industry agreement, they have the effect of de facto standards. The two primary goals of the APIs are to (1) standardize the APIs so the applications built on the APIs become portable and

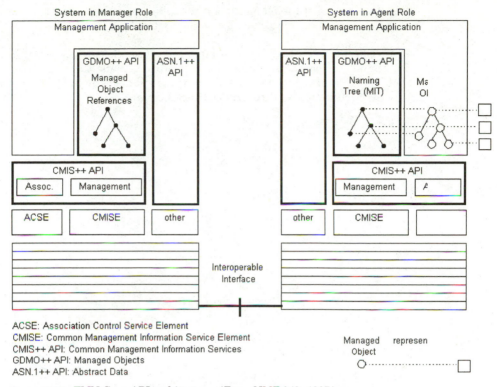

ACSE: Association Control Service Element
CMISE: Common Management Information Service Element
CMIS++ API: Common Management Information Services
GDMO++ API: Managed Objects
ASN.1++ API: Abstract Data

Managed   represen
Object

Figure 13.14 TMN C++ API architecture. (*From NMF 043, 1997.*)

(2) reduce the efforts required for development of a management system, since the standardized components can be reused.

Exercises

13.1 List the major steps in designing a network management system in the context of a TMN management framework.

13.2 NMF solution sets cover only a small set of network management applications. When you design a system not covered by the NMF solution set, what design steps will you take?

13.3 What are the relationships and differences between an NMF solution set and a component set?

13.4 In order for two applications based on different management protocols to interwork with each other, what are the two basic types of conversions that must take place?

13.5 Briefly describe the relationships between the three APIs of the TMN API suite in terms of services one provides to others, if any.

13.6   Discuss what the managed object handle provides to a manager system developer, as opposed to the managed objects themselves.

13.7   Check the six fragments of managed object classes discussed in Chap. 7, and determine the appropriate superclasses for the following three managed object classes that are identified for the switch connection management system: interExchange-Circuit, circuitSubGroupXtp, and circuitSubGroup.

Telecommunications Service Management

Outline

- The "Introduction" first provides an overview of telecommunications services, such as regular phone services, Intelligent Network services, and wireless services, and then surveys three areas of standards to arrive at the functional areas to be discussed in this chapter.

- The first section, "Service Configuration Management," covers two topics: service deployment and service maintenance.

- The second section, "Service Ordering," first describes a service order interface that allows customers to order telecommunications services and then discusses the functions required to process service orders.

- The third section, "Service Provisioning," discusses the service provisioning functions of the service management layer and the element management layer.

- The last section, "Service Performance Management," discusses service performance reporting, service performance measures calculation, and service performance reporting to customers.

Introduction

This chapter discusses service management, an area that is becoming increasingly important and yet for which there are few formal, systematic standards. For example, the word *service* can invoke vastly different perceptions of what it might mean. There is just as little consensus on what a service management system is supposed to do.

Thus the first order of business is establishing a basic understanding of the services that are being discussed here and what are the functional areas involved in service management. These are the two issues the remainder of this introduction section will address: it will seek to classify the telecommunications services based on common industry practices and to define the functional areas of service management based on available standards and literature. The remainder of this chapter will discuss each of the identified areas of service management.

Telecommunications services

Classifying telecommunications services in a dynamic telecommunications environment has proven to be a tricky business. Nonetheless, to facilitate the discussion, we borrow the service categories commonly used by service providers for provisioning services to customers. The service categories include plain old telephone service (POTS), Intelligent Network (IN) services, supplementary services, data services, and wireless services. A note of caution is that these are not mutually exclusive categories of telecommunications services and the criteria used to define the service may vary from service to service. By definition, a telecommunications service is a capability implemented on a telecommunications network and offered to customers.

Basic PSTN or POTS services.   POTS services mainly refer to direct line-to-line calls, the service for which the telephone was invented. POTS services are switch-based because they are implemented solely on a switch. From the perspective of service management, POTS services are very simple. It is the other services that require sophisticated management operations.

Supplementary services.   This is a hodgepodge category that includes a wide range of services that may not necessarily best fit under one category. Since these are the earlier additions to the conventional PSTN services, they are also referred to as the *enhanced PSTN* services. Supplementary services may be implemented on one single switch or on multiple network nodes, depending on the network infrastructure. A few samples out of the large number of services in this category include

- Call forwarding, which may have service features such as call forwarding on busy, call forwarding on no answer, and call forwarding unconditional
- Call waiting
- Calling line identification (i.e., caller ID)
- Call transfer
- Directory inquiry service

This category of service falls somewhere between the POTS and the IN services to be described next.

Intelligent Network services.   The IN services are network-based because the implementation of these services involves multiple network nodes [i.e., service control point (SCP), service switching point (SSP), and intelligent peripheral (IP)]. There are two variations of the IN services, and one seems to be a much broader category than the other one. The Intelligent Network in North America is called the Advanced Intelligent Network (AIN), and the conventional AIN services seem to be restricted to those services that require a database lookup in the call processing. Examples of the AIN services include

- *Free phone (800-number calls).* A service that charges the called party for the call.

- *Credit card calls.*   A service that allows a caller to make a call and charge it to a credit card.

- *Prepaid card calls.*   A service that allows a caller to prepay for a block of call time at a fixed rate.

On the other hand, the IN services, as defined by the official international standards organization ITU-T, encompass a much broader range of services. Three sets of the IN services have been defined, also called capability sets (CSs). The three sets, i.e., capability set 1 (CS1), capability set 2 (CS2), and capability set 3 (CS3), are defined by ITU-T over a period of time on a continuous basis. CS1 and CS2 are well defined, whereas the CS3 service definitions are in progress.

CS1, as defined in ITU-T Recommendation Q.1211 (1993), includes over 50 services. Examples of the services include

- Credit card calls that allow a caller to make a call and charge it to a credit card.

- Customized recorded announcements that allow a customer to record a message to be played to calling parties. (Remember your last call to a doctor's office?)

- Conference calls that allow the connection of multiple parties in a single conversation.

- Free phone that allows a call to be charged to the called party instead of the calling party.

- Call transfer that allows a subscriber to place an active call on hold, establish another connection, and then transfer the call on hold to the other connection.

- Call waiting that allows a subscriber to receive a notification that another party is trying to reach this number while he or she is in conversation with a different calling party.

- Selective call forwarding that allows a subscriber to have calls from a selected list of calling parties diverted to another number.

- Split charging that allows a service to charge both calling and called parties for a call.

CS2 has over 60 customer services that include many wireless services and also overlap some of the CS1 services. Examples of CS2 services as defined in ITU-T Recommendation Q.1221 (1995) include

- Completion of calls to a busy subscriber that allows a subscriber encountering a busy destination to be informed when the destination becomes free and automatically connect to the destination without making another call attempt
- Conference calling (same as CS1)
- Service negotiation that allows parties involved in a call to negotiate network capabilities such as bandwidth
- Private numbering plan that allows a subscriber to maintain a numbering plan within a subscriber's private network that is different from the public numbering plan
- Radio paging that allows a user to send a message to a selected pager terminal or a group of terminals

Broadband services. This is a broad category of services that normally require a broad data bandwidth. The underlying technologies of these services are changing rapidly; thus we only describe the services in generic terms.

- *Video conferencing.* Also referred to as visual telephone service, a two-way telecommunications service that can carry live or static pictures and associated speech
- *Televoting.* Enables subscribers to survey public opinion using the telephone
- *Video on demand.* Involves several connections that provide for the transfer of digitally compressed and encoded video data from a server to a client on a demand basis where it is reassembled, uncompressed, and converted to an analog signal for presentation on a monitor
- *Telemedicine.* A special application of video on demand service combined with teleconferencing to allow transfer of x-ray images and two-way communications
- *telefax.* A public facsimile service between subscribers through a public switched network
- *Virtual private network service.* Uses public network resources to provide private network capabilities without necessarily using dedicated network resources

Wireless services. A wireless service can be any of the above PSTN services, the supplementary services, the IN services, or the broadband data services. A wireless service in essence is no different from its wireline counterpart except that the service is provided to a mobile station as opposed to a stationary phone.

This brief survey of telecommunications services should be sufficient to provide a background for the service management to be discussed in the rest of the chapter. Now we turn to the functional areas of service management.

Functional areas of service management

Three areas of telecommunications standards are surveyed to arrive at the functional areas of service management for this chapter: service management in TMN logical layered architecture, the IN service management, and an NMF service management process model.

Service management in TMN logical layered architecture. As discussed in Chaps. 6 and 7, a TMN architectural principle is the logical layered architecture that consists of five layers: network element layer (NEL), element management layer (EML), network management layer (NML), service management layer (SML), and business management layer (BML). The intention is to create different levels of abstraction of the complicated management information so that the difficult management tasks can be handled at progressively abstract levels, as shown in Fig. 14-1, with increasing levels of abstraction.

It is the SML, as shown in Fig. 14.1, that is the focus of this chapter. The contents of the SML have only been sparsely defined in formal standards. The TMN management functions as specified in ITU-T Recommendation M.3400 (1992) mainly cover the NEL, leaving the functionality of the NML and SML largely unspecified. Some regional or quasi-standards bodies such as European Telecommunications Standards Institute (ETSI) and Bellcore have defined functional areas for higher TMN layers. The following areas are identified for the service management layer in Bellcore GR-2869 (1995).

- *Service planning and negotiation.* Covers customer need identification, customer feature definition, customer service planning, etc.

- *Provisioning.* Covers access route and directory number determination, service profile administration, etc.

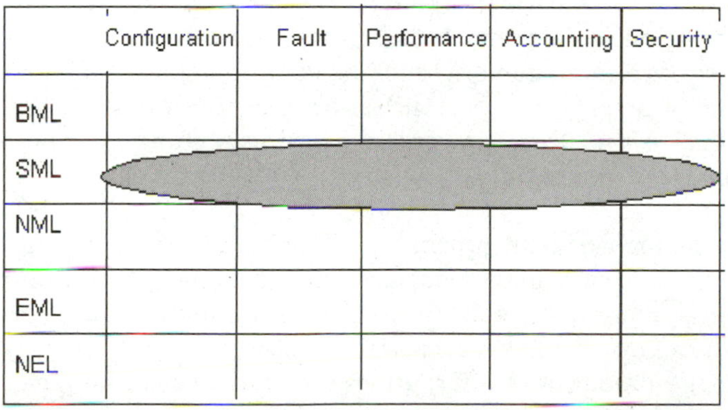

	Configuration	Fault	Performance	Accounting	Security
BML					
SML					
NML					
EML					
NEL					

Figure 14.1 Service management in the context of TMN layered architecture.

- *Service performance monitoring and analysis.* Covers customer service performance summary and reporting of quality of service, performance assessment, etc.

- *Service testing.*   Tests a new service

- *Service trouble administration.*   Resolves the service troubles such as quality of service degradation

- *Service usage measurement and tariffing.*   Define how to measure the service usage and how to charge for the usage

IN service management. The IN service management has some special requirements. The service management, as defined in ITU-T Recommendation Q.1224 (1996) and implemented via a service management function (SMF), is intended for the management of IN CS2 services as discussed above. Specifically, the SMF is divided into five functional areas.

- *Service deployment functions.*   Include loading service logic, service data, signaling routing data, and trigger data into appropriate network elements and service testing

- *Service provisioning functions.*   Include provisioning and maintaining subscriber-specific data in a network element

- *Service operation control functions.*   Include service activation, deactivation, configuration and reconfiguration, service maintenance such as updating a service logic or service data to a newer version, and service dismantlement from a network

- *Billing functions.*   Include generating and maintaining service charging records on an individual customer basis

- *Service monitoring functions.*   Include collecting service performance measurement data, analyzing the data, and reporting the service usage and performance

Since CS2 services cover a broad set of services, the IN CS2 service management function is also very broad in its scope. Its emphasis on the control aspects of service management, such as service deployment and activation or deactivation of a service, is a manifestation of the intention of the IN designers to automate the whole process of creating, managing, and executing services in an IN network.

NMF's service management processes. One main reason that international standards on service management are hard to come by is that many aspects of service management are specific to a telecommunications service provider. This is where an industry consortium like the NMF steps in. The NMF invested large amounts of effort in creating a service management business process model that represents the industry consensus on the issue. This industry-owned process model comes from 2 years of effort on research and study of the common portions of service management identified by service providers surveyed by the NMF (Adams 1997).

Shown in the model are two sets of key processes involved in service management from the perspective of service providers, i.e., the customer care processes, and the service delivery and maintenance process. The interface to customers is included in the customer care processes. The third process, the network management process, falls in the domain of the network management layer and out of the scope of the discussion.

The customer care processes consist of

- *Sales.* This process is responsible for learning about the needs of each customer and communicating the telecommunications service information to customers.

- *Order handling.* This process is responsible for accepting a customer's service order, tracking the progress of the order, and notifying the customer when the order is completed.

- *Problem handling.* This process is responsible for receiving service complaints from customers and resolving them to the customer's satisfaction.

- *Performance reporting.* This process is responsible for providing whatever service performance reports that were requested by the customer and available from the service provider.

- *Invoice collection.* This process is responsible for sending invoices to customers and processing payments.

The service delivery and maintenance processes, the second category of service management processes, focus on service delivery and maintenance as opposed to the management of the underlying network or day-to-day interactions with customers.

- *Service planning and development.* This process is responsible for designing technical capabilities to implement services, to ensure that the service is properly installed, and to ensure that it correctly executes.

- *Service configuration.* This process is responsible for the installation and configuration of services for specific customers, including the installation and configuration of customer premises equipment.

- *Service problem handling.* This process is responsible for isolating the root cause of service failures and interacting with other systems to resolve them.

- *Service quality management.* This process is responsible for monitoring service quality on a class-of-service basis.

- *Rating and discounting.* This process is responsible for applying the correct rating rules to usage data on an individual customer basis.

Although each service provider may structure the processes differently, the identification of these processes provides a conceptual framework for establishing common ground for standardizing information models used in the processes. The processes in Fig. 14.2 that are marked with small squares are identified by the NMF as having a high potential of involving external organizations and thus a high probability of having agreements on the information content and format exchanged between the organizations in the processes. Examples of the processes include service problem resolution and service configuration.

The focus of this chapter is on the service management layer of the TMN logical layered architecture. On this layer, we combine the service management areas for IN services and several processes in the NMF service process model to arrive at the following four areas of the service management to be discussed in the remainder of this chapter.

- *Service configuration management.* Consists of two areas: service deployment and service maintenance. This is a combination of service configuration in the NMF process model and the IN service deployment and service maintenance.

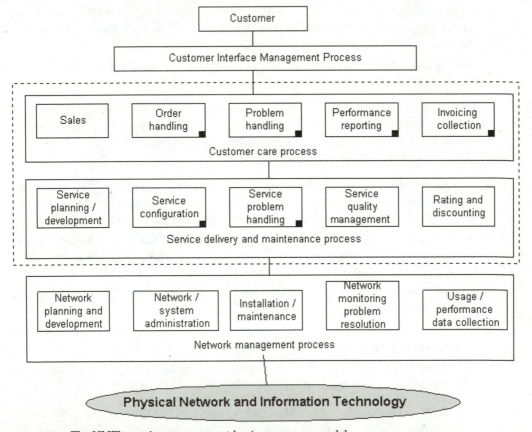

Figure 14.2 The NMF's service management business process model.

- *Service ordering.*   Covers the order processing in the NMF service process model and service negotiation and planning.

- *Service provisioning.*   Discusses the IN service provisioning process.

- *Service performance management.*   Covers service performance monitoring, service performance reporting, and service performance measure calculations.

Several processes of the NMF service management process model are covered in other chapters: service trouble handling is discussed in the trouble administration section of Chap. 10 and rating, tariffing, and invoice collection are briefly discussed in Chap. 11.

Service Configuration Management

Service configuration has two primary responsibilities: first, once a service is created, it is responsible for deploying the service into the network elements. Second, once the service is in a network, it is responsible for maintaining the service in working order; called service state management, it will suspend, resume, activate, and deactivate a service.

Service life cycle and service state management

There are few formal standards specifically on service states, although the usage states, administrative states, and operational states as defined in ITU-T Recommendation X.731 (1992) have been augmented to accommodate service-specific aspects of state management. Figure 14.3 shows one state transition diagram for service management.

The service state diagram presents the perspective of the service provider, indicating the availability and operability of the service. The service operation control (SOC) restricts operations that can be performed to a service based on the state of the service (e. g., not allowing deactivation of a service that has already been deactivated). The states are described as follows.

- *created.*   The service has been created in a service creation environment automatically or manually.

- *available.*   The service is retrieved from the service creation environment to the service management system, ready to be deployed; this is an operation of the service management layer, as shown in Fig. 14.3.

- *deployed.*   The service is deployed from the service management system to an appropriate network element; at this point, the service is on a local disk and is not in execution. Deploying a service into a network element is an operation on the service management layer, as shown in Fig. 14.3.

- *disabled.*   The service is loaded into memory, ready to be activated but not in service yet. This is an operation of the EML.

- *enabled.*   The service is activated by the management application and is up and running. Activating a service is an operation of the EML.

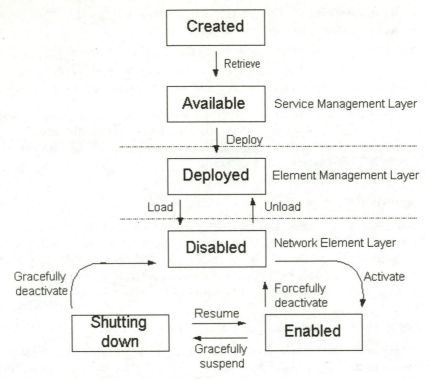

Figure 14.3   Service state transition diagram.

- *shuttingDown.*   The service is suspended. In this state, the service is still in memory and will continue to process the in-progress calls to completion, but it will reject any new call attempt. From this state a service can either resume or move into the enabled state.

Service deployment

The way a service is deployed is closely associated with the type of service. Traditionally, the service logic is tightly woven into the switching logic inside a switch. Service deployment merely means loading the switch software. The scenario became different and more complicated when many different kinds of new services came onto the scene.

There are few standards on service deployment. The discussion of service deployment is largely based on the deployment of IN services. Nonetheless, service deployment is applicable to many other types of services as well because new generations of telecommunications network architectures, either wireless or wireline, are based on the same principles IN is based on: separating networking logic from the service logic. The old practice of interweaving service logic with network switching control logic is a thing of the past.

Service-related data.   From the perspective of a network, a service is a resource that consists of service logic software and service data. When the service logic executes, it uses network resources, service data, and customer-specific data to offer utilities to subscribers. The service logic and service data can be very simple or extremely complicated, all depending on a specific service. For example, the task of deploying a POTS service is very simple. IN services probably are on the other end of the spectrum and can be quite complicated. In general, for those relatively complicated services, the following categories of data need to be deployed into appropriate network elements.

- *Service logic programs.*   Are the core logic of a service that can be in one of various forms: executable, textual instruction to be fed into an execution environment, or others.

- *Signaling routing data.*   Specify signaling routing requirements specific for this service. For example, for a free phone service, the 800 numbers may need to be routed to a designated SCP for translation.

- *Trigger data.*   Specify events in call processing that will cause transfer of control and require special processing. Trigger data are often stored in a table called a trigger table; examples of trigger data include the service key, version number, trigger type, event detection point, detection point trigger criteria, destination for event routing, and fault handling.

- *Specialized resource data.*   Examples of this type of data are an announcement that you often hear over the phone or tone sounds.

- *Data schema.*   Defines the format and fields of service data in a database.

- *Service and service feature interaction rules.*   May contain multiple service features. For example, call forwarding is a service and call forwarding on busy, call forwarding on no answer, and call forwarding unconditional are three features of the call-forwarding service. When the number of different services offered by a network gets up into the hundreds, some of the services or some feature of a service are bound to interfere with other service or service features in an unexpected, adverse way. This is called *service interaction.* Four types of service interactions can be distinguished, and thus four types of rules are required to either prevent service interactions or to provide instructions on what to do in case of an interaction.

  - *Interservice interaction rules.*   They identify the service pairs that cannot be subscribed to by a customer at the same time. One example of such a pair is call-forwarding and follow-me services. The follow-me service allows a subscriber to receive a call terminating to his or her number on another line by entering special codes so that the calls terminating to his or her number can be forwarded to the predesignated number. When the subscriber's line or the line to be served has been designated "terminating call barring," "call forwarding," "absentee service," or "do not disturb," the follow-me service cannot be used. The rules are meant for the service order-entry system to prevent a subscriber from having two conflicting services.

■ *Interfeature interaction rules.* They identify the feature pairs within a service that cannot be subscribed to at the same time. Either some of these features of a service are mutually exclusive or one is included in another. For example, call forwarding unconditional includes call forwarding on busy, and a customer does not need to have both of them. This interaction is also meant for the service order-entry system to prevent a subscriber from having either two conflicting service features or two features with one completely subtending other.

■ *Interfeature-across-services interaction rules.* They identify the pairs of features of different services that cannot be subscribed to at the same time. The rules are also meant for the service order-entry system to prevent a customer from having two conflicting service features.

■ *Interstate interaction rules.* They identify the pairs of call-processing states of two different services that conflict. A subscriber may be allowed to have two individual services that don't conflict. But a service in a particular state may prevent another service in a given state from functioning properly. This type of rule specifies the action to be taken in case of a conflict situation.

In order to perform service deployment, a service management system must possess either all or part of the following three types of knowledge:

■ Knowledge of a network configuration that allows a service to be deployed to appropriate network elements.

■ Knowledge of a service configuration that provides information on where a service component is deployed, which version it is, and what are the service environment requirements.

■ Knowledge of network and specialized resource configurations. The resources required for a service can be located in various network elements, and it is the responsibility of a service management system to ensure the resources required by a service are in place for a service to execute.

Service deployment functions. The functions required and the procedure for service deployment vary from service to service. In general, the following functions are necessary for the deployment of a sophisticated service such as an IN service.

Validating service. Before deploying components of a service, a service management system must check that all necessary components are present, versions are consistent, and the required special resources are available to ensure that the service is complete. A knowledge-based system is often a logical choice for dealing with large amounts of knowledge about services.

Loading service. This function loads the service logic program and service data into appropriate network elements and sets the state of the service. Sometimes existing service data are used for a new service, and in this case, the version of the existing data shall match what is required by the new service. Once all components of a service are loaded and tested, the interested parties should be notified of the availability of the service.

Testing service. Either before or right after a service is loaded, the service management system performs service testing to ensure the service executes in the expected manner in the target environment. When the service is tested is a decision local to each service provider. Service testing may be performed by a testing system that is independent of the service management system that deploys the service. But the service management system at least shall have an interface to the testing system to request service testing be performed and to receive the testing result.

Managing service configuration. Managing service configuration involves storing; deleting; and updating information on services, i.e., its version, its location of deployment, and its components, execution environment requirements, and resource requirements. An interface to network configuration management is required to obtain information on network configuration and resource configuration.

A service deployment scenario. We use a scenario of deploying a new service to illustrate the process of service deployment. Assume that a service is created at a service creation environment that is outside of the service management system.

1. The service management system receives a notification from the service creation environment that a newly created service is ready for deployment. Included in the notification are the new service name, address of the service repository where it can be retrieved, and the package ID for the service.

2. The service management system downloads the service package via a file transfer protocol such as file transfer protocol (FTP) or file transfer access management (FTAM), unpacks it, and groups the service-related data according to the deployment destinations.

3. The service management system validates the service by ensuring that the special resource data that are required by the service are available at the appropriate network elements, all necessary components of the service are present, the service version is consistent, and so on.

4. The service management system creates the portion of the service data that can only be created by the service management system and processes the received service data. For example, the service key, a unique key to identify a service, may be assigned by the service management system. Then the service management system converts the received service logic program in a form that meets the requirements of the destination network element. Not all services require this step.

5. The service management system deploys service logic and/or service data to appropriate network elements. For example, in case of an IN service, the service logic program goes to an SCP, the trigger data goes to an SSP, and the first three categories of the service interaction rules go to a service order-entry system that takes service orders from customers.

6. The service management system keeps a record on the deployed service (e.g., service key, service version, deployment destinations, and service characteristics) and may optionally keep a copy of the service data itself.

7. The service management system interacts with a service test system to test the deployed service before the service is activated. The deployment process terminates upon successful completion of the service test.

Service maintenance

Once a service is deployed, routine maintenance and control operations of the service shall follow. This includes updating an older version of a service, maintaining service states, suspending and resuming a service, and activating and deactivating a service.

Service activation and deactivation.   Activating a service means loading all the components of a service into memory and setting the service state as enabled. At this point, the service is ready to execute. There are several steps involved before a service is activated. Service deployment is responsible for placing all components of a service into appropriate network elements. Once deployed onto a network element, service components are put on a disk. In order to activate a service, coordination of activation of all required resources and other components the service depends on must take place.

Service deactivation has the opposite effect of service activation. A deactivated service is no longer operational. When a service is deactivated, all components of the service are removed from memory and put into local storage space.

Deactivation comes in two different flavors, namely, forceful deactivation and graceful deactivation, and the difference is in how the in-progress calls are handled. When a service is forcefully deactivated, all in-progress calls are prematurely terminated. The service transitions from the enabled to the disabled state. The graceful deactivation allows the service to first transition to an intermediate state, the shuttingDown state in which all in-progress calls proceed to completion before all components of the service are removed from the memory.

Service suspending and resuming.   In contrast, a service is suspended when the service components are not removed from the memory but its state changes to shuttingDown. In the shuttingDown state all new service requests are rejected.

Resuming a service means setting the state of the service back to enabled so that new service requests are accepted. The cases where a service needs to be suspended or resumed include updating a service to a new version. The old version must be suspended before a new version of the service can be activated.

Service updating.   Service updating means to replace the existing version of a service with a new version. It uses service deployment to load a service into disk storage of appropriate network elements and then uses the service suspending functions to put the existing version on hold. After the new version is activated, it is possible that reverting to the old version is needed when the new version does not behave as expected. The suspended version of the service is resumed after the new version is deactivated or suspended, depending on the specific situation.

Service Ordering

Once deployed into a network and activated, a telecommunications service is ready to serve customers. A customer needs to go through two steps before he or she can use the service. First, the customer has to subscribe to the service through a service ordering process, and second, the subscription must be processed and put into the appropriate network element before the network can provide the service to the customer. This and the following sections discuss the two processes, namely, service ordering and service provisioning, respectively. This section first describes a service ordering process model, and then discusses each of the major components in the model.

A service ordering process model.   At a conceptual level, a functional model for service ordering, as shown in Fig. 14.4, consists of several parts: an interface to the customer, an order-processing component, an interface to the service provisioning system, and an interface to order-entry systems of other service providers.

The goal of the service ordering system is to support a wide range of activities in the process of customer ordering services. The activities range from a simple quote on a service price to configuring a complicated set of services. The following scenario illustrates the general behavior of the service order model shown in Fig. 14.4.

1. Via the customer interface, a customer initiates an ordering process either by issuing an enquiry with a set of telecommunications needs in mind or by simply placing a service order.

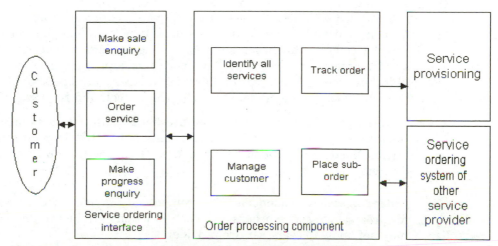

Figure 14.4   General service ordering process model. (*From NMF 504.*)

2. The order-processing component provides the intelligence for retrieving and generating information needed to answer a customer query and for creating a service order from the data the customer provides.

3. The order-processing component determines whether the involvement of another service provider is required for this service order. The cases where more than one service provider is needed include that a secondary service provider (e.g., telecommunications service reseller) communicates with a primary service provider to complete a service order. The information exchange is through the interface to the service ordering systems of other service providers.

4. Via the interface to the service provisioning system, the order-processing component sends the service order to the service provisioning system to implement the customer services.

5. Via the interface to the customer, the order-processing component informs the customer of the status of the service order.

Next discussed is each of the components in the model.

Service order interface.   First, note that the customer as referred to here is not limited to residential subscribers but may be an organization such as a business enterprise. Depending on the type of customers the order-entry system is designed for, the functions of the service order interface will vary. The three general components are identified and described below.

Make a sale enquiry.   The interface shall allow a customer to make an enquiry on the available services and request a service proposal that meets the customer's communication needs. This process may consist of a series of interactions between the customer and service provider, such as back-and-forth queries and clarifications. The result of the enquiry shall yield sufficient details for the customer to make a decision on whether to proceed to the phase of placing an order.

Order service.   The interface shall allow a customer to enter service order data in a friendly and flexible fashion. It is possible that a customer can directly come to this process without going through the preorder enquiry. A service ordering interface shall support a variety of means for a customer to place an order: phone, fax, or the Internet. Mechanisms should be provided to make it easy for a customer to change, cancel, and resubmit an order.

Make a progress enquiry.   The interface should allow a customer to make a progress enquiry on an order that has been submitted for processing. Flexible means should be provided for a customer to query the progress on an order, e.g., via e-mail, fax, or phone.

Order-processing component.   This component provides a set of core functions for processing service orders. Six such functions are identified, which may or may not apply to every service provider.

Manage customer. This function obtains the customer service order data, enquiry, and other input from the customer via the service order interface and determines where to send the information. In the other direction, it also retrieves the information requested by a customer and presents it to the customer via the same service order interface.

Identify services. This function helps carry out a negotiating session with a customer. Assuming that a customer expresses the telecommunications needs in the customer's terminology, this function analyzes customer needs and comes up with a solution proposal that would satisfy the needs. In order to reach a solution, all possible service options, both internal and external, are gathered and searched. The best services are selected and checked to see whether each can fit the requirements. Once a solution is formed, a description of the offer is stored for future use.

Create a service order. When a service offer is accepted by the customer, a credit check is performed if necessary and then a service order is created based on the predefined form. A service order can contain several categories of information, such as service order information, customer information, and service information. An example of service orders is described below.

Track order. This function is responsible for monitoring and tracking the progress of a service order. This function can be triggered either by a query from a customer or by a schedule. It may need to query the service provisioning system to get an update on the status of service implementation.

Sample service order. Because of the diversity of services and complexity of each service provider's operation environment, it is not realistic to standardize the service order format and attributes. However, the information in a service order can be broken down into several general categories, such as order attributes, customer attributes, and service attributes that are common to most service orders. A sample of service orders with the three categories of information is shown in Table 14.1. There are some other types of information that may or may not be used, depending on the local context. For example, service orders may include the location to be served, installation contact, and service design contact.

Interface to the service provisioning system. Once created, a service order goes to the service provisioning system that implements the ordered service for the customer. This interface is responsible for sending service orders to the provisioning system and providing any additional information requested by the provisioning system. This interface has gained more attention recently in part because the service provisioning system and the order-entry system are likely to come from two different vendors. A standard interface will certainly make the life of both vendors easier and make either system more portable and interoperable.

TABLE 14.1   Sample Service Order Data Attributes

Attribute name	Description
Order Attributes	
Order ID	A unique identifier for the order
Contract ID	A unique identifier for the contract specifying the terms and conditions of the order
Customer ID	A unique identifier for the customer for whom the order is placed
Charge date	The date at which charges will be billed
Charge code	The customer charge code for the order
Project code	The project code used to associate related orders
Requester ID	A unique identifier for the person placing the order
Engineer ID	A unique identifier for the person to be contacted with technical issues regarding service implementation
Installer ID	A unique identifier for the person to be contacted with issues regarding the installation
Order description	A summary description of the order
Order priority	A numeric value indicating the priority
Order status	A numeric value indicating the service order status
Order type	An indicator for either an initial or subcontracted order
Requested date	The date order fulfillment was requested
Committed date	The date order fulfillment was committed to by the provider
Fulfillment date	The date order fulfillment actually occurred
Ready for service date	The date all requested services are operational
Facility test date	The date cooperative testing of the service is scheduled
Cut over date	The date all requested services are cut over
Acceptance date	The date the order fulfillment is accepted by the customer
Service item list	The count of the entries in the service item list
Billing account number	The service provider's account number for charges associated with this order
Billing address	The address to which invoices are to be sent for charges associated with this order
Billing contact ID	The person to be contacted with queries regarding payment of the invoice associated with this order
Customer Data	
Customer ID	A unique identifier for the customer
Customer type	Business, government, or residential
Customer address	The address of the customer
Customer status	New, approved, active, or discontinued

TABLE 14.1   Sample Service Order Data Attributes   *(Continued)*

Attribute name	Description
	Service Attributes
Service ID	A unique identifier for a service ordered
Request type	New service, disconnect, move, change
Service type	PSTN, X.25, wireless, etc.
Service order status	The current status of the service order
Service order priority	The order priority that is predefined
Number of locations	The number of locations to be served
Location ID list	A list of location identifications each defined for a location
Number of end users	The number of end users to be served
End-user list	A list of end users' unique identifiers
Service-specific attributes	A list of attributes required to define a specific service

Service Provisioning

The next stop for the service order is the service provisioning system. The term *service provisioning* as used in this chapter means to provision services to a customer. The process begins with receipt of a service order from a service order-entry system and ends with the subscription data being loaded in the appropriate format into a network element. The subscription data include information about the subscriber or the subscribed service. In general, the whole process of provisioning services to a subscriber involves the management functions of both the service management layer and the element management layer. This section provides a high-level description of these functions.

Service provisioning functions of the service management layer

Conventionally service provisioning has mostly been done at a network element. In part this is because most of the services were simpler and implemented at the network element (e.g., a switch node). As telecommunications services become more complex and the principle of the Intelligent Network (i.e., separation of service logic from the networking logic) is widely accepted in the newer generations of telecommunications networks, as discussed in the service deployment section, the provisioning functions of the service management layer become indispensable. Two such functions are described briefly.

One such provisioning function creates or obtains the customer-specific service data necessary for the implementation of the service. For example, when a customer subscribes to a free phone service (e.g. 800-number service), a free phone number must be assigned to the customer. The available 800 numbers are administrated by a third-party administrator, and the service provisioning system must be able to interface the 800-number administrator to obtain an available number for the customer.

Another provisioning function distributes components of the subscription data to appropriate network elements. A subscription consists of multiple data components, and the components may go to different network elements. This is because multiple network elements are involved in implementing a customer service, and thus a networkwide view is required. An example in point is the free phone service. For a customer with free phone service, the free phone number and other service parameters need to go to an SCP and the customer data need to go to an SSP.

Service provisioning functions of the element management layer

Once processed by the service provisioning function at the service management layer, the subscription data arrive at a network element. In order to implement the subscription, the required logical and physical resources need to be allocated, and the service and subscriber data need to be at the right place and in the right state. The following two general provisioning functions accomplish the implementation of a customer service.

Allocate resources. The resources that are discussed here are meant to be representative rather than exhaustive. They illustrate the resources required for provisioning POTS services, ISDN services, and many supplementary services. As new types of services are introduced into the market, new types of resources are required.

The type of resources required for a customer service varies with the type of the service. For example, the resource required for a POTS service may be different from that for an ISDN service or an IN service. Discussed here are general resources that are common to most of the services.

Directory number. Network providers allocate a unique directory number, which is normally called a telephone number, to uniquely identify an individual customer in its network. The number allocation is based on a worldwide standard numbering plan, called the ISDN era numbering plan, as defined in CCITT Recommendation E.164 (1991). For more details, see Chap. 16.

A directory number serves several purposes. First, it identifies a customer in a network. Before the ISDN era, there was a one-to-one mapping between a directory number and a customer. Second, it uniquely identifies an access port and the equipment at a customer's premises (e.g., a telephone at your home). For analog switching systems, a direct number has a one-to-one mapping to an access port. In the ISDN era, the relationships between directory number, subscriber, and access port have changed. Multiple directory numbers may be associated with a single customer (e.g., multiple lines into a home), and multiple access ports may be associated with a directory number. Allocating directory numbers and managing the complicated relationships is the responsibility of the service provisioning system.

Access port and channel.   An access port represents the point of access to the switching node. In a conventional telephone switch, an access port is the point at which a telephone line from the customer equipment connects to the switch.

The concepts of access port and access channel are described first. An access port can be categorized in several different ways, primarily depending on the type of the customer equipment the customer wants to connect to the line. Two broad categories are ISDN and non-ISDN. The ISDN type of access ports represents digital communication and is further divided into two types: basic and primary. The basic type, also called basic rate interface (BRI), provides a nonmultiplexed service for individual customers, and the primary ISDN type, also called primary rate interface (PRI), provides multiplexed services to groups of customers. The non-ISDN types of access ports can be further divided into two types: digital and analog. One example of the non-ISDN digital access port is leased-line services at a variety of data rates. Examples of nondigital analog access ports include those for PSTN services, analog leased-line, and fax. An access channel is a single bearer or signaling channel. A number of access channels may be contained in a single access port. Figure 14.5 shows a logical view of the resources. It is the responsibility of a service provisioning system to allocate access ports and manage the access port–related information, including type, associated service type, and status.

Manage subscriber and service data.   When a new customer subscription first arrives at a network element, generally two sets of data are created:

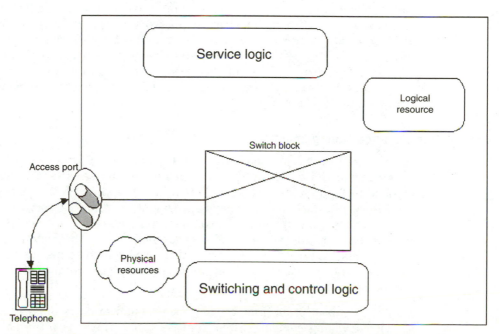

Figure 14.5   Resources for service provisioning at a network element.

customer data and service data. The services subscribed to by a customer, the resources provisioned for the services subscribed to, and the association between them come together into one place called the customer profile. A *customer profile* is a network view of a customer that contains all the information or links to the information necessary for the network to serve the customer. The type of customer data contained in the customer profile really depends on the type of network and the services subscribed to. Examples of the subscriber profile data include

- A list of services subscribed to by this customer.
- The directory number assigned to the customer.
- A pointer to the access port and access channel associated with the directory number.
- The status of the subscriber, e.g., active or suspended. A customer can be suspended for administrative reasons such as nonpayment.
- Any restrictions put on the subscriber (e.g., no international calls, no 900-number calls).
- A mobile station identification number if this is a wireless subscriber.

The contents of the second set of data, the service data, depend on the type of services subscribed to by the customer. In general, this is the service feature data specific to a customer. For example, if the customer has a free phone service, a unique free phone is assigned to the customer. Another example is the call-forwarding service. A customer may have a time-dependent routing feature associated with the service: during work hours, all calls are forwarded to a given number; during after-work hours, all calls are forwarded to a different number; and during the weekend, all calls are forwarded to yet another number. It is the responsibility of the service provisioning system to manage all these service data. How the service data are represented depends on the network technologies in use.

Service Performance Management

In essence, the main responsibility of the service performance management is reporting to the customer the performance of the services the customer subscribed to. To do so, service management must go through the following steps or processes: establish a service level agreement, monitor the service performance, calculate service performance measures, generate a service report, and finally report the calculated service performance measures to the customer. These steps are summarized in the following scenario where a customer has an agreement for the service provider to periodically report the performance of the services he or she has subscribed to.

1. At the subscription time, the service provider and customer establish a service level agreement (SLA) that specifies the commitment the service provider agrees on to the customer in aspects such as quality of service and pricing policy.

2. Once the services are activated for the customer, the performance data are collected on a continuous basis. The service performance data can come from a number of sources: the management systems at the network management layer such as the fault and performance management systems or the service fault management of the service management layer.

3. At the end of the performance report period, the service performance measures are calculated.

4. A service performance report is generated according to the predefined format.

5. The generated service report is sent to the customer through an external interface.

The remainder of this section discusses each of these steps or processes in detail.

Manage service level agreement

As the competition in the telecommunications market intensifies, the SLA becomes a widely used means for a service provider to attract and hold on to customers. An SLA is a formal negotiated agreement between a service provider and a customer that specifies many aspects of the relationships between the customer and the service provider. An SLA may cover performance of service, customer care, billing, provisioning, and others. The contents of an SLA can vary from one service provider to another and are not subject to standardization. A service provider may choose certain items out of a typical SLA list and not use others.

A set of service-related measurable parameters is a core part of an SLA, and it serves as a yardstick to determine whether a service provider has lived up to its commitments. A typical list of measurable quality-of-service parameters may include, but is not limited to,

- Service availability, which will be discussed next
- Time to identify the cause of a customer-reported malfunction
- Time to repair a customer-reported malfunction
- Time to provision a new service
- Performance measures specific to a service

The SLA quality-of-service parameters can be broken down into two general categories of criteria: operational criteria and service performance criteria. The operational criteria measure the performance of the service provider organization and the criteria include mean time between failure, mean time to repair, and mean provisioning time.

The service-specific performance criteria depend on each individual service. The following examples illustrate some of the possible service-specific parameters:

- Network cell delay or errored cell ratio for a digital nonswitched service such as ATM permanent virtual circuit (PVC)

- Call establishment time, network delay, or end-to-end distortion for a switched analog service such as basic PSTN voice services

- Bit error rate, unavailable seconds, errored seconds, or severely errored seconds for private line or leased-line service

- Cyclic redundancy error check, committed information rate quality, discarded frames, or network delay for frame relay PVC service

But one of the important measures of quality of service in an SLA that is common to all services is service availability, which will be discussed shortly.

Monitor service performance

This is performance monitoring at the service management layer, as opposed to at the network or element management layer, where performance data are directly collected from the objects representing network resources, as described in Chap. 9. The service performance data are collected from other management systems as shown in Fig. 14.6.

Data from performance management of the network management layer.   From the performance management of the network management layer, the following four types of performance data are collected:

- *Error performance data.*   Indicates the type and occurrence frequency of error that adversely affect network performance and services. Examples of ATM error performance data include
  - Errored cell ratio, which is the percentage of cells transmitted or received that are in error
  - Bit error rate (BER), which is the percentage of bits transmitted or received that are in error
  - Unavailable seconds (UAS)
  - Severely errored second ratio (SES)
  - Errored second ratio (%ES)
  - Background block error ratio (BBER)

- *Quality of service.*   Measures quality of service for a particular type of network. Examples of ATM quality-of-service data include
  - Average end-to-end delay measured in milliseconds as experienced by customers
  - Probability of average delay above a threshold value
  - Throughput, the amount of data transmitted from one end to the other within a given amount of time
  - Cell delay deviation, the change in the interarrival times between cells at the peak cell rate

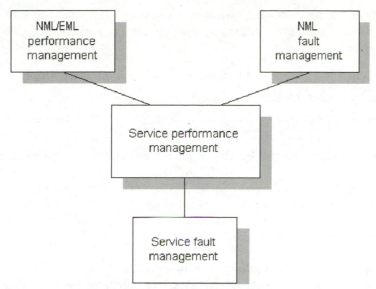

Figure 14.6 Data collection for service performance monitoring.

- *Traffic data.* Provides an overall picture of traffic conditions on the network. Examples of ATM traffic data include
  - Cells transmitted in either direction
  - Cells transmitted with cell loss percentage = 0 in either direction
  - Cells received in either direction
  - Cells received with cell loss rate = 0 in either direction
  - Cells discarded in either direction
  - Cells discarded with cell loss percentage = 0 in either direction
- *Resource utilization data.* Measures how much the network resource of a particular type of network is being used for a customer service. Examples of the ATM network resource utilization data include

  - *Average utilization.* The utilization of the PVC based upon the traffic flowing for a period of time divided by the sustainable information rate (SIR). Traffic is defined as the number of cells multiplied by 53 bytes per cell. This is an average for the time period determined by the service provider in the granularity of the report. This is available for each direction of a PVC.

$$\text{Utilization} = \frac{\text{bytes transferred}}{\text{SIR}}$$

  - *Peak utilization.* The peak average utilization of the PVC and the time it occurred for a fine granularity determined by the service provider during the reporting period. This peak may be an hourly peak over 1 month and is available for each direction of a PVC.

Data from the network fault management system. This is the data on the fault conditions that have affected the customer services. Examples of the fault conditions include

- System down and the duration for which customer services are affected
- Alarms caused by the threshold crossing for traffic data
- Alarms caused by service quality degradation such as network traffic congestion
- Service outage caused by any other reason and the outage duration

Data from service fault management. The service fault management interfaces customers to receive service trouble reports. The data of interest to service performance monitoring include the service trouble report that includes service outage and the perceived severity of the trouble.

Interfaces for service performance monitoring. The interfaces between the service management system and other management systems as shown in Fig. 14.6 for the performance monitoring data collection are of the TMN Q class, since they are the interfaces between operating systems of different TMN layers. However, very few information models have been defined for the service performance monitoring application.

Service performance measures computation

Some of the performance data collected above by service performance monitoring can be directly reported to the customer. However, some of the data that represent a network and technical view of a service are of no interest to the customer. One challenge of service performance management is to convert some of the data from a network perspective to a customer perspective and come up with customer service performance measures that reflect the customer's perspectives and interests. Computing service performance measures at the service management layer may involve aggregating and analyzing the collected performance data from the management systems of the network management layer and other sources. This subsection first discusses the basic requirements of service performance computation and then describes in detail the computation of a key customer performance measure, the service availability. The computation described below is based on NMF 701.

One requirement is to be able to calculate the end-to-end service availability measurement. This is not a simple summarization of performance measures of individual network elements. Another requirement is to determine the effect of a service fault on quality of service in terms of the service outage the fault causes:

- No service outage, i.e., service fully available
- Partial service outage, i.e., service degraded but still available
- Complete service outage, i.e., service unavailable

Also the computation algorithms should be able to handle boundary conditions such as service outages crossing reporting period boundaries and be able to reflect service degradation and performance degradation.

The service availability (SA) is defined in terms of service unavailability (UA).

$$SA\% = 100\% - UA\%$$

Thus the whole approach centers on the calculation of the service unavailability. There are three ways of calculating the service unavailability, each taking different factors into consideration. The simplest way is as follows:

$$UA\% = \frac{\Sigma \text{ outage interval}}{\text{activity time}} \times 100\%$$

An event affecting the service at the SAP, which will be described shortly, is defined as an outage. The duration of this specific event is called the *outage interval*. A specified period of time is called the *activity time*.

It is hardly a black-and-white relationship between an event and a service outage: an event either causes or does not cause a service outage. Rather the issue is to what degree an event is causing a service outage. In order to address this issue, the concept of the service degradation factor (SDF) is defined to represent the degree of a service outage an event is causing. In other words, the SDF represents how much an event adversely affects a service. When the SDF is factored into the calculation of service unavailability, we have the following:

$$UA\% = \frac{\Sigma(\text{outage interval} \times \text{SDF})}{\text{activity time}}$$

where $0 \leq SDF \leq 1$.

In most cases the SDF can be quantitatively defined. For example, x minutes of peak hour downtime may result in the SDF $= y$. The definitions of SDF are completely based on mutual agreement between a service provider and a customer. A set of SDF values with the corresponding event type can be defined in the SLA as discussed above. A list of SDFs is provided in Table 14.2 as an example.

Now we turn to the concept of the service access point (SAP), which is the point where a service is delivered and becomes visible to the customer. The SAP concept is designed to distinguish the boundary between the customer domain and the service provider domain and enable a customer to report a service fault in terms of a faulty SAP. Each service contains one or more SAPs, and an SAP can only be associated with one service. For example, a service consisting of a server site connected to multiple client sites may have multiple SAPs: one for the server and one for each client. The SAP-related concepts are listed below.

- *SAP.* A logical element located on the interface between the customer domain and the service provider's domain representing the point to which a service is delivered. An SAP can be weighted according to a business critical factor as defined in an SLA.

TABLE 14.2   SDF Examples

Agreed SDF value	Event type	Duration source (time stamps information)
1	Service fully unavailable	System 1 or 2
0.8	Outage type *A*	System 3
0.6	Outage type *B*	System 4
0.5	Outage type *C*	System 5
⋮	⋮	⋮
0	Service considered available	Customer happy with it

SOURCE: NMF 701, 1998.

- *SAP activity interval.*   The time interval of a specific active period when a customer requires service from the SAP within a defined reporting period.

- *SAP activity time.*   The total time of all SAP activity intervals of a specific SAP within a defined reporting period.

- *SAP outage interval.*   The time interval of a specific outage period within the defined reporting period. An outage period is a period of service unavailability which occurs during an SAP activity interval for a given SAP.

- *SAP outage time.*   The total time of all SAP outage intervals of a specific SAP within a defined reporting period.

- *SAP weight.*   A value reflecting the relative priority and perceived importance attached to an SAP.

- *Service degradation factor (SDF).* A set of preagreed values between 0 and 1 assigned to the different degrees of service degradation. It is similar to the membership function of fuzzy logic, if the reader happens to be familiar with the topic.

- *Service outage.*   An event affecting the quality of service.

- *Service outage duration.*   The duration of the event.

The next important concept is SAP weight. Each SAP may have a weight attached to it indicating the perceived importance of the SAP to the customer. A service may consist of many SAPs which may or may not have weight factors; if all have associated weight factors, they may not all be weighted equally. For example, a service involving a server (host) site connected to multiple client sites can have several SAPs: one for the server and one for each client. In this example the weight of the server site may be set higher than the one assigned to the client sites, indicating that a problem with the server will impact the service more severely from the customer's perspective.

When the SAP weight is factored into the calculation of service unavailability, we have the following:

$$UA\% = \frac{\Sigma[\text{ SAP weight} \times \Sigma\,(\text{SAP outage interval} \times \text{SDF})]}{\Sigma\,(\text{SAP weight} \times \text{SAP activity time})}$$

One of the main issues with the service availability calculation is how to select the appropriate information sources for all the required formula elements. Performance management requires information from other management applications to calculate the service unavailability, as shown below.

- *Outage interval.* The duration of a service-affecting event that can be obtained from an element level fault management system
- *Activity time.* The total amount of time for service availability that can be obtained from the performance monitoring
- *SAP weight.* The perceived importance of an SAP from a customer's perspective that is assigned based on the customer's business needs and mutual agreement between a customer and a service provider

Generate service performance report

A variety of information contained in a service performance report can be classified into five categories:

- *Customer information.* Customer name, ID, contact information, etc.
- *Service provider information.* Service provider name, contact information, etc.
- *Service information.* Service type, service profile description (i.e., description of configuration parameters and corresponding values), SAP information (SAP address and weight), etc.
- *SLA information.* Guaranteed values of service availability, bit error rate, cell loss rate, etc.
- *Service performance information.* The core of a performance report that may contain information such as
  - Report type, i.e., the quality-of-service (QoS) report or traffic data report
  - Reporting period, i.e., the start and end times of the reported interval
  - Indicator for boundary conditions such as outages crossing reporting interval boundaries
  - Suspect flag to indicate the report is suspected to contain unreliable or incomplete data

The QoS portion of the report consists of two parts: service-independent measurements and service-specific measurements. As discussed above, the key service-independent measurement is service availability. Other measurements of this category may include

- Total number of SAP outage intervals
- Time to restore service for a specific SAP

- Percentage of outages where service was not restored within the committed time as specified in the SLA
- Mean restoration time for a specific SAP
- Mean time between failures for a specific SAP

The second part of the QoS report is service-specific measurements which are specific to each service. In general, they fall into the four categories of performance data collected during the service performance monitoring, i.e., traffic data, service-specific QoS data, resource utilization data, and error performance data. For example, the error performance data for leased-line service can include

- Bit error rate
- Unavailable seconds
- Severely errored seconds
- Errored second ratio
- Severely errored second ratio
- Background block error ratio

Service-specific service quality data for analog services may include

- Call establishment time
- Network delay
- End-to-end distortion

Again, it is up to a service provider to decide categories of service performance data to report, based on the condition local to each service provider.

Report service performance to customer

The interface to the customer mainly supports distribution of performance reports. The methods of report distribution shall be flexible, including fax, e-mail, CMIP-based interface, and even postal mail. The interface shall accommodate various modes of distribution—for example, reporting with and without acknowledgment, or reporting on demand. The interface shall support different reporting frequencies, such as synchronized with the billing cycle, on demand, weekly, or monthly. A customer should have the option of changing reporting mode, disabling or enabling a reporting route (i.e., where a report goes), and even specifying the desired performance parameters.

A performance report can be initiated under various circumstances. For example, a customer may request a performance report on a subscribed service. Or a periodic service performance report may be part of the service agreement between a service provider and a customer when the service is subscribed. Note that the definition of "customer" here might be different from what is commonly known. It refers to more than just residential end users; a customer can also be another service provider that subcontracts from the primary service provider or a corporate customer that subscribes to a whole range of services and has its own management system in the customer premises equipment.

Summary and Discussion

Summary

This chapter discusses four areas of service management: service configuration, service ordering, service provisioning, and service performance management. However, the discussion begins with the introduction section which provides an overview of telecommunications services to provide a background for service management. Then three areas of the standards on service management are surveyed, that is, the TMN logical layered architecture and the functional specification of the service management layer, the IN service management, and the NMF's service management process model.

The service configuration section covers three topics: the service life cycle and state management, the service deployment, and the service maintenance. The service provisioning section first looks at the two service provisioning functions of the service management layer, i.e., creating subscription data and distributing the subscription data to appropriate network elements. For the element management layer, two general service provisioning functions are described: allocating resources required for a customer service and managing subscriber and service data. The service performance management section discusses the service performance monitoring, the calculation of the customer service performance measures, and service performance reporting to customers.

Discussion

It is typical in today's networks that each category of services is managed by a service-category-specific management system. For example, POTS services, which conventionally have been implemented solely on a switch, are managed by the management functionality that is bundled in switch-level OAM&P. POTS service provisioning and deployment are normally performed at a switch.

Existing service management systems are often service-content-dependent. For example, data services, in contrast to voice services, are often managed by systems based on different standards, principles, and practices.

Today's service management systems are often technology-specific. Telecommunications technological innovations have resulted in the emergence of new types of services at unprecedented rates, e.g., wireless, commercial Internet, and multimedia services, to name just a few. In the past, each new technology resulted in a separate service management system.

Existing service management systems are also often vendor-specific. This is rooted in the fact that telecommunications network architectures historically are exclusively proprietary and service management functions are deeply embedded into the proprietary network architectures. One vendor's service management system typically manages services supplied by the equipment of the same vendor. Any interoperability between different vendors' equipment requires heavy participation of all vendors involved.

The new realities of the telecommunications industry and marketplace are making it very difficult to continue the conventional practices of service management. The following new realities must be taken into consideration in designing the new generation of service management systems.

- Service providers are likely to operate in a multivendor equipment environment and support multiple telecommunications network technologies (e.g., wireless, wireline, broadband, narrowband, Internet, cable) to provide diverse services such as voice, data, multimedia, and virtual private networks.

- Customers desire to bundle all subscribed services including local, long distance, data, wireless, and Internet services into one service-offering package.

- Service providers are forced to consolidate management functions in order to cut down on operation costs and be more competitive in today's market.

The limitations of the current service management practices combined with new realities of the telecommunications industry and marketplace give rise to the following requirements for the new generation of service management systems.

The new generation of service management systems should be network-technology-independent. The goal is for a service management system to support services that are based on such diverse network technologies as wireless, conventional PSTN, narrow-band ISDN (N-ISDN), broadband ISDN (B-ISDN), ATM, and asynchronous digital service line (ADSL). With the deregulation of the telecommunications market, service providers will soon be able to bundle such a wide range of services as local, long distance, internet, personal communications services (PCS), paging, and cable services into a single service-offering package. An open, technology-independent service management system that can manage the services of such diverse technologies is crucial.

The service management system should be service-independent. The goal is for the service management system to support rapid service deployment and provisioning. No service-specific features should be built into the service management system architecture and design. It should be able to adopt a new service without a large-scale redesign and time-consuming reengineering. Although it is necessary to have service-specific code to support a service feature, there should be no dependency at the architectural and design level. One example of a design level service dependency is that the billing data structure, designed for POTS services, cannot accommodate wireless services without a major data structure change.

The service management system should be vendor- and equipment-independent. The goal is for the service management system to manage services provided by the equipment of a different vendor without major changes being made to the service management system and without heavy participation of the vendor.

The service management system should have an open interface. The goal is for the service management system to integrate with an overall network management system smoothly and efficiently. Other component systems of a network management system may be developed by a different team of the same company or even by a different vendor.

The service management system should support a customer-centric service management philosophy. The goal is for the service management system to provide the flexibility for an end user to tailor a subscribed service to meet customer-specific requirements. This requires that end users be given control over the management of services.

The service management system should support distributed computing environments efficiently. For reasons such as performance concerns, the components of the service management system may be distributed across different platforms at different locations. The architecture and design of the service management system should be highly modular in order to adapt to new requirements.

Exercises

14.1 Give a few examples of IN service and discuss what distinguishes an IN service from a conventional PSTN service.

14.2 Explain why formal standards for service management on the SML are sparse.

14.3 Describe the differences between two types of service processes in the NMF's service process model, i.e., between the customer care processes and service delivery and maintenance processes.

14.4 In terms of the TMN five-layer model, which layer does this chapter focus on?

14.5 What does service deployment mean as discussed in this chapter (which is based on ITU IN architecture)?

14.6 What is service provisioning? What are the major steps in the service provisioning process?

14.7 Use an example to explain what a service access point (SAP) is.

14.8 What are the service degradation factor and SAP weight?

14.9 What information does the service unavailability measurement with service degradation factor and SAP weight provide?

Appendix: CS1 Services

The ITU-T has published its view of the IN architecture and services in the Q.1200 series of recommendations. Organized (per chronological development) by capability sets, the ITU-T has defined CS1 and CS2, and now CS3 is under development. All the capability sets are described in terms of a service plane (not included for CS1), a global functional plane, a distributed functional plane, and a physical plane.

Examples of the CS1 services include

- Abbreviated dialing
- Account card calling
- Automatic alternative billing
- Call distribution
- Call forwarding
- Call rerouting distribution
- Completion of call to busy subscriber
- Conference calling
- Credit card calling
- Destination call routing
- Follow-me diversion
- Free phone
- Malicious call identification
- Mass calling
- Originating call screening
- Premium rate
- Security screening
- Selective call forwarding on busy or don't answer
- Split charging
- Televoting
- Terminating call screening
- Universal access number
- Universal personal telecommunications
- User-defined routing
- Virtual private network

Examples of the CS2 services include

- Internetwork free phone
- Call transfer
- Internetwork premium rate

- Call waiting
- Internetwork mass calling
- Hot line
- Internetwork televoting
- Multimedia
- Global virtual network service
- Terminating key code screening
- Completion of call to busy subscriber
- Message store and forward
- Conference calling
- International telecommunications charge card
- Call hold
- Mobility services

Chapter 15

Distributed Telecommunications Network Management

Outline

- The first section, "Introduction to ITV Open Distributed Processing (ODP)," provides an overview of the ITU's distributed processing architectures and the architectural components and principles.

- The second section, "Introduction to CORBA," introduces the CORBA architectural components, interface definition language, services, and facilities.

- The last section, "Applications of CORBA to Telecommunications Network Management," discusses issues related to development of CORBA-based network management systems, CORBA applied to TMN interfaces (Q, X, and F interfaces), and CORBA interworking with TMN.

Introduction

We start the chapter with a discussion of the motivations for adopting distributed processing technologies in the field of telecommunications network management. First, telecommunications networks are becoming increasingly distributed with the rise of IN and AIN that distribute network intelligence and allow distributed realization of a network system. Second, multivendor environments for both network hardware equipment and software are inherently distributed and heterogeneous, and there is a strong need for a distributed framework to tie together the diverse, distributed network elements and to have them interoperate with one another. Third, telecommunications network and service management systems in essence are specializations of information systems. The general computing industry's distributed, client-server architectures can be borrowed off the shelf with lower costs in development and maintenance. An increasingly competitive marketplace exerts enormous economic forces on the move toward a distributed computing environment for telecommunications network management. Another noted factor is that the new distributed technologies tend to level the playing field for newcomers to the telecommunications industry. Established players must start at the same place as the newcomers. In addition, the established players are handicapped by the need to maintain their legacy systems.

Introduction to ITU Open Distributed Processing (ODP)

ODP is a response of the telecommunications industry to the advance and wide acceptance of distributed computing technology. It is a distributed processing framework defined by ITU-T to standardize the distributed computing environments in the telecommunications environments. An ODP system, as used throughout this chapter, provides such a distributed environment for user applications. ITU-T recommendations specify a set of principles and constructs for building such an ODP system rather than the system specification itself (see ITU-T Recommendations X.902 through X.904).

The word *distributed* could mean different things to different people. ITU-T defines the term distributed by precisely specifying what a distributed system should provide. The services provided by an ODP system are termed *ODP distribution transparencies*. Assume that you have an application running in a distributed communication environment provided by an ODP system; the following seven distributed transparencies are available to your application. Though it is not necessary for a distributed application to be developed in an object-oriented language, the recent distributed computing technologies are closely coupled with the object-oriented concepts.

Access transparency. An object of your application should not know the specific representation of a piece of data it needs or a method specific to data for retrieving it. It only needs to know a standard way of data representation, i.e., the object and the interface of the object. The application object is oblivious to the fact that the data it is interested in could be in a relational database and represented in a proprietary format.

Location transparency. The application object should not know the location of another object of another application with which it is communicating. It could be right next to your application object on the same machine and in the same process. Or it could be in an application running on a machine that is located across the continent. The application object treats a remote object the same way it treats a local object.

Failure transparency. The application object should not know that an object it depends on has failed. The service performed by the failed object should be resumed by another object, and your application object should not be aware that the service provider has changed.

Transaction transparency.   When your application object operates on data, it should not be burdened with the responsibility of ensuring the data consistency and integrity, e.g., checking for access privilege and consistency between different levels of access privileges. Your object should be oblivious to the underlying activities to ensure transaction consistency.

Migration transparency.   When an object your application is communicating to is moved from one location to another, for reasons such as load balancing and sharing, it should not impact the object of your application and, in fact, should be completely transparent to the object of your application.

Relocation transparency.   The interface of your object may be bound to the interface of another object. When the interface your object interface is bound to has relocated, it should not impact your object interface and, as a matter of fact, should be transparent to your object.

Persistence transparency.   When your application object communicates to another object, it should not be burdened with any knowledge of the state the other object is in. For example, it should be oblivious to whether the other object is activated or not.

Replication transparency.   An object of your application should not know whether a service it receives is provided by one object or a group of objects. Often a group of objects can enhance performance, availability, and reliability.

ODP architecture overview

The ODP framework specifies five viewpoints that, in a simplistic way, can be viewed as five sets of system and architectural requirements. Together they provide an overall view of the behavior of an ODP system, as well as a set of principles and guidelines for developing such a system. The five viewpoints are enterprise, information, computational, engineering, and technology. Each viewpoint has a language to express the view. Thus there are five viewpoint languages, i.e., enterprise viewpoint language, information viewpoint language, computational viewpoint language, engineering viewpoint language, and technology viewpoint language. Each viewpoint language consists of a set of concepts, structuring rules, and conformance and reference points. The concepts provide a vocabulary set, and in some cases, constructs for defining an ODP system from the perspective of a particular viewpoint. In addition to viewpoint-specific concepts, a set of concepts common to all viewpoints is defined so that they are used across viewpoints. Many of the common concepts are directly adopted from object-oriented vocabulary. The rules specify the behavior of an ODP system from the perspective of the viewpoint. The conformance and reference points specify the requirements for a system to be ODP-compliant and the interface (reference) points that will be tested against the standard specification.

Enterprise viewpoint and language.   The enterprise viewpoint should define the purpose, scope, and policies of an ODP system in terms of roles played by the system, activities undertaken by the system, and policy statements about the system such as the computing environment restriction. Two important concepts in the enterprise viewpoint language are community and federation. A *community* is defined as a configuration of objects formed to meet an objective. A configuration of objects is a set of objects with well-defined relationships between the objects in the set. A *federation* is a community of a specific domain.

The structuring rules define how the concepts of enterprise language shall be used to define the roles, purposes, and policies. For example, the role an ODP system plays should be defined in terms of the roles the objects in a community play. Then the role an object plays is defined in terms of the permissions, obligations, prohibitions, and behavior of the object.

Information viewpoint and language.   The information viewpoint defines the semantics of information and information processing in terms of static and dynamic state schemata. For example, the relationships between objects can be modeled as part of the state of the involved objects, using the static schema. Object behavior, including object creation and deletion, can be modeled as allowable state changes.

The concepts specific to this viewpoint include invariant, static, and dynamic schemata. An *invariant schema* is a set of predicates on one or more objects that must always be true. The invariant schema is used to constrain possible states of an object. A *static schema* specifies the state of an object at a particular point in time. A *dynamic schema* specifies the allowable state changes of one or more objects.

Computational viewpoint and language.   The computational viewpoint specifies the functional decomposition of an ODP system into objects and interactions between objects. Thus an ODP function can be viewed as consisting of configurations of interacting objects.

The concepts that are defined for the computational viewpoint language to describe the interactions between objects include signal, signal interface, signal interface signature, operation, operation interface, and operation interface signature. A *signal* is defined as an atomic action that results in one-way communication from an initiating object to a responding object. A *signal interface* is a logical point where two objects interact with each other by sending signals. A *signal interface signature* consists of a set of action templates, one for each signal type in the interface. An *action template* comprises the name of the signal and the number, names, and types of its parameters.

An *operation* is an interaction between a client and a server. An interaction can be one of two kinds: either a client sends a request to a server, which is called an *announcement* in ODP terminology, or a client and a server exchange a message, which is called *interrogation*. An *operation interface* is a point where two objects interact with each other. Similarly, an *operation interface signa-*

ture consists of a set of announcement signatures and interrogation signatures. An *announcement signature* is an action template containing the name of the invocation and the names and types of its parameters. An *interrogation signature* comprises one action template for the client and one action template for the server.

Several sets of structuring rules specify behavior and behavior constraints of object interactions. Some of the important rules are summarized here. The naming rules specify that all names defined in computational language must have an associated context. This includes signal, parameter, invocation, and operation names. For example, a signal name in a signal interface is an identifier of the context of that signature.

The interaction rules specify the constraints on the interactions between objects for each type of interaction, e.g., signal-based interaction and operation-based interaction. For example, a client object using an operation interface can only invoke the operations named in the interface signature.

The parameter rules specify the format of signature parameters. For example, a formal parameter that is an identifier of an interface shall be qualified by an interface signature type.

The binding rules specify how an object, say, a client object, can be bound to another object for distributed communication. There are two types of bindings: *explicit* binding and *implicit* binding. An explicit binding uses one of the two kinds of binding action: primitive binding action or compound binding action. A primitive binding action directly binds two objects. A compound binding action binds two objects via a third object, a binding object. It is an implicit binding if in an interaction there is no place to express binding action. For example, if an invocation by a client object references a server object operation interface to which the client object does not have a way to explicitly bind, an implicit binding needs to be set up. Normally the procedure for setting up an implicit binding comprises several steps:

- Create a client operation interface of complementary signature type to the server interface.

- Bind the client operation interface to the server operation interface.

- Invoke the server object using the client operation interface.

- Delete the client interface when the operation completes.

Engineering viewpoint and language. The engineering viewpoint language defines the mechanisms and functions that provide the core of a distributed computing environment. A detailed discussion of these mechanisms and functions is the focus of the following subsection.

Technology viewpoint language. The technology viewpoint language focuses on the choice of technologies for implementation of an ODP system and how the identified technology can be used to implement the specification of an ODP system.

Summary. An ODP system is intended to provide a distributed computing environment for applications in areas such as telecommunications. Table 15.1 summarizes the core ideas of the five ODP viewpoints.

Distributed computing concepts of ODP

The engineering viewpoint adopts the distributed computing concepts developed in the general computing industry, particularly those that have been standardized by the Object Management Group (OMG). These distributed computing concepts are the focus of the rest of this section, with an emphasis on their applications in the telecommunications field.

Concepts. The concepts of the engineering viewpoint language provide the specification of the infrastructures for a distributed computing environment. These concepts can be broken into two categories: the support concepts that define the infrastructure supporting the distributed environment and the concepts that define the infrastructure of a distributed environment.

The first category of concepts includes the following:

Node. A configuration of engineering objects that has a set of storage, processing, and communication functions.

Nucleus. Provides services such as coordinating processing, storage, and communication functions to other parts of a distributed environment.

Engineering object. An object that either provides or requires support of the infrastructure of a distributed environment.

Capsule. A configuration (i.e., a set) of engineering objects forming a single unit for the purpose of encapsulation of processing and storage functions. A capsule is contained in a node.

Capsule manager. An engineering object that manages the objects in a capsule.

TABLE 15.1   Summary of the Five Viewpoint Languages in the ODP System

Viewpoint	Summary ideas
Enterprise	Defines the scope, goals, and policies of an ODP system in terms of roles identified objects play
Information	Defines the behavior of an ODP system in terms of static and dynamic states
Computational	Defines how a client and server object shall interact with each other in a distributed environment
Engineering	Defines the core of the infrastructure of a distributed environment
Technology	Specifies the technology required to implement an ODP system

Cluster. A configuration of basic engineering objects for the purpose of deactivation, checkpointing, reactivation, recovery, and migration. That is, it is a set of objects that can be migrated, reactivated, or deactivated as a single unit. A cluster is contained in a capsule.

Cluster manager. Manages the objects in the cluster. Each cluster has one.

The relationships between these concepts are shown in Fig. 15.1.

The support concepts just described are an adaptation of similar concepts from general computing fields for telecommunications applications. The correspondences of these concepts to their counterparts in the general computing world are shown in Table 15.2.

The second category of concepts provides the core of the distributed environment.

Channel. A logical communication link connecting two objects that may be remotely located from each other. It supports the interactions between the two objects, which include

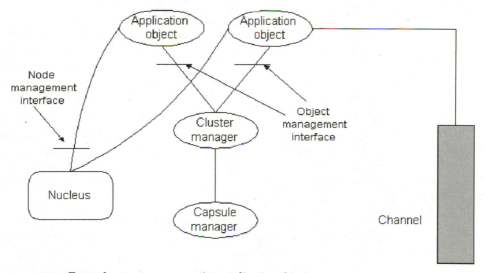

Figure 15.1 Example structure supporting application objects.

TABLE 15.2   ODP Support Concepts and Their Counterparts in the Computing Industry

ODP concepts	Counterpart in general computing
Capsule	A process
Cluster	A set of objects
Engineering object	An object in the distributed environment
Nucleus	An operating system
Node	A computer with supported software

- Operation execution between a client and a server object
- A group of objects multicasting to another group of objects

Through a channel, two objects exchange interface references, data, and cluster templates.

Stub. The point where the application objects are locally bound. A stub can perform the following:

- Convert the data passed through a channel to a form locally understood
- Apply control and keep records for security or accounting purposes

Normally a stub interacts only with objects that are communicating through a channel. It could also interact with objects outside the channel when it is necessary. One such case is where the stub talks to an outside object for authenticating a client object.

Binder. Manages the end-to-end integrity of the channel. As shown in Fig. 15.2, a binder has at least one interface to a stub and one or more interfaces to protocol objects.

Protocol objects. Responsible for furnishing communication functions to the application objects. As shown in Fig. 15.2, a protocol object has an interface for interaction with a binder and at least one communication interface to other protocol objects. If protocol objects in a channel are of different types, an interceptor is required to provide protocol conversion.

Interceptor. Stands at the boundary between two domains and provides checks and transformations on the interactions that cross domain boundaries. Depending on the boundary crossed, an interceptor may require different types of information to convert a message from one domain to another. For example, it will need to know the signature of the object interface bound to the channel.

Client object. An object that receives service from other objects. For example, in a banking system, a client object can represent a client that requests banking services.

Server object. An object providing service to other objects. For example, in a banking system, a server object can be an account manager that can create a counter, check account balances, and answer queries from a client.

Behavior of the ODP engineering model. The following application scenario provides an illustrative description of the behavior of an ODP system, in which a client object communicates with a server object via the distributed environment provided by an ODP system.

1. Both client-side and server-side applications form clusters from the application objects based on the application requirements and criteria such as degree

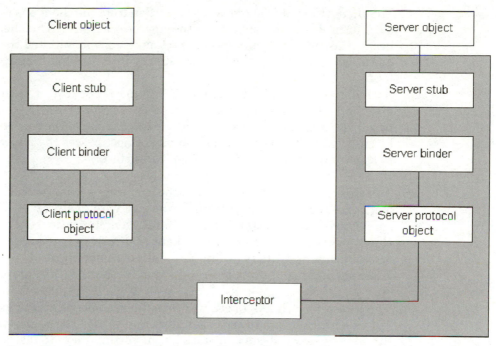

Figure 15.2   An example of a client-server channel. (*From ITU-T Recommendation X.903, 1997.*)

of coupling between the objects. A cluster is a group of application objects that can be migrated, relocated, activated, or deactivated as a single unit.

2. A client object that intends to communicate with a server object requests that a nucleus create a channel in order to communicate with a remote server object. The following major activities take place as a result.

   ■ The nucleus creates appropriate stubs, binders, protocol objects, and interceptors as part of channel establishment. The nucleus for each object to be bound creates a configuration of stubs, binders, and protocol objects at its node to support the interface of the client object.
   ■ The created protocol objects are connected to their respective communication interfaces, which enable them to use the native communication environment such as a local communication protocol.
   ■ The client object that is bound to the channel is assigned a binding endpoint identifier for each interface it has to the channel. This identifier is used to identify an interaction to be carried on the channel.

3. A client object initiates an interaction session with a remote server object by performing the following:

   ■ Binds itself to a client stub
   ■ Provides the information on the service it requests to the client stub

4. The client stub may convert the information from the client object into a standard form if it is not already in the appropriate form. The client stub also keeps a record of the initiated interaction (e.g., who initiated the interaction and when). The client stub then passes the interaction initiation to a binder.

5. A binder object has in its domain information about all the server objects and binds the requested interaction to the server object. Note that at this point the client object does not bind the interaction to the server object itself but instead to its proxy. Then the binder object passes the interaction information to the protocol object via their interface.

6. The client-side protocol object uses the underlying communication functions (e.g., the TCP/IP protocol) to communicate with the protocol object at the server side and to establish a physical communication channel between the two sides. Assuming that the client and server objects reside in two different domains with two different native communication protocols (e.g., Ethernet and token ring), the interaction is passed to the interceptor.

7. The interceptor checks to see whether the requested interaction is permitted in both client and server domains and masks the underlying protocol differences between the two sides by converting them to a standard representation. The interceptor also resolves any naming conflict between a client object and server protocol object.

The server side responds as follows:

1. The server-side protocol object receives an interaction (e.g.), request for service in its native form after the protocol conversion and information versioning performed by the interceptor. The interaction is then passed on to the server binder.

2. The server binder performs a security check as required and finds the location of the server object through the object locator.

3. The server stub, which has the server objects bound to it, converts the data (e.g., parameters) requested in the interaction to a form known to the server object and invokes the server object as needed.

4. The server object performs the requested services and may return the results back to the requesting client object.

In summary, the ODP specifications provide a set of principles and constructs necessary to develop a distributed communication environment. However, the ODP does not mandate a particular implementation. The specifications must be realizable through a particular implementation technology. OMG's CORBA architecture, which has influenced the ODP specifications, is a logical choice (if not the only choice) for implementing an ODP system and is the focus of the following section.

Introduction to CORBA

CORBA is an object-based distributed architecture that has achieved the status of the industry standard and has begun to be widely accepted in telecommunications network management applications. This section introduces the basic concepts of CORBA to lay a foundation for the ensuing discussion on the application of CORBA to telecommunications network management.

Introduction

This introduction consists of background information on CORBA, motivations for using CORBA, and an overview of the CORBA architecture.

Background information

Object Management Group (OMG).   Established in 1989, the OMG is one of the largest industry consortiums. Its declared mission is to "promote the theory and practice of object technology for the development of distributed computing systems" and to provide a common architectural framework for object-oriented applications based on widely available interface specifications.

Object Management Architecture (OMA).   OMA is the overall umbrella architecture of OMG on which all OMG standardization activities are centered. Specifically, OMA consists of the following components:

- *Object request broker.*   Commonly known as CORBA, the object request broker (ORB) is the communication cornerstone of the distributed architecture. It provides an infrastructure that enables objects to communicate with each other, independent of specific computing platforms, the underlying network technologies, and programming languages used to implement the objects.

- *Common facilities / CORBA facilities.*   A set of general-purpose application capabilities that can be applicable to multiple domains. Examples include generic printing, document management, and database facilities that can be used in word processing, bank transactions, and domain-specific applications.

- *Object services / CORBA services.*   A set of components that standardize the life cycle of management objects. The services provide common interfaces for an application to create objects, control access to objects, keep track of the location of an object, and help maintain the relationships between objects. Object services provide a standard way of managing objects and facilitate the development tasks of programmers.

- *Domain interfaces.*   A set of capabilities of direct interest to end users in a particular application domain. Domain interfaces may use common facilities and object services combined with domain-specific functions.

- *Facilities.*   A set of application level services either applicable across domains (horizontal) or exclusively to a particular domain (vertical).

Common Object Request Broker Architecture (CORBA).   CORBA is a specification of a standard object-oriented architecture for distributed applications. The basic CORBA concepts were first published in the *Object Management Architecture Guide* by the OMG in 1990. Since then, there have been several major enhancements (see Object Management Group 1998).

Motivations for CORBA.   CORBA in essence is a middleware that masks the differences in the underlying computing environments and allows an object to talk to another object on a totally different computing platform just as it would to an object in its native environment. The need for such a middleware is evident.

It is quite common that an enterprise's computing environment consists of a wide range of computing technologies including mainframes, workstations, desktop computers, and network servers. Those computers at different times run on equally diverse ranges of operating systems such as Windows, IBM OS/2, or Apple Macintosh. Networks connecting the computers are highly heterogeneous from network protocols (FDDI, Ethernet, etc.) to the underlying network technologies (LAN, WAN, etc.). A number of information systems may coexist in the same enterprise, relational database, object-oriented database, and proprietary information system. In an environment where both information and applications are distributed, imagine the multitude of efforts a developer has to make in dealing with the computing network, computing platforms, information access method, and programming environment!

All these point to the need for a middleware that sits between the application layer and underneath network technology-specific layers to shield the diversity of the networks, protocols, and operating systems. There are many such middlewares, and the problem is that they are proprietary. Applications built on one vendor's middleware cannot talk to the applications built on the middleware of a different vendor. Thus, it is utterly important to develop a middleware standard, and CORBA is such a standard middleware.

An architectural overview.   CORBA provides an architecture that formally separates a client from a server in a distributed computing environment. A client, i.e., a system functional entity like an object that requests another functional entity, a server, to perform some operation on its behalf, does not concern itself with knowledge about the server that will provide the service. It does not know where the server is located, in what programming language it is written, on what computing platform it is running, or for that matter whether it is running at all. All a client needs to know is how to make a service request. On the other end of the spectrum, a server is equally ignorant of or isolated from a client. It does not know who the client is, where it is, or what it is. All it knows is to provide the requested service and return the result to the party that has passed in the service request.

The whole process of a client sending a request and a server responding to the request is termed an *interaction*. The process of sending a request is called an *invocation,* and for an invocation to be successful, a request must have the following components:

- The name of the operation that a client requests a server to perform
- A reference to the object on which the operation is performed
- Zero or more arguments specific to the requested operation
- A mechanism to return exception information about success or failure of the operation

How can a client afford to ignore such important details as the computing platform and programming language of the server that will perform the requested service? Because of a platform-independent and programming language–neutral specification language called the Interface Definition Language (IDL). The IDL is an intermediary to and from which a request is converted.

The remainder of this section is devoted to the introduction of the client-side components, server-side components, and IDL.

Client-side components

As shown in Fig. 15.3, the client side consists of a client application, a client stub, a dynamic invocation interface, an interface repository, and the client-side ORB.

Client application.   The client-side application is a user application written in a language that can be the same or different from the language used to write the server-side object implementation. The client-side application performs three functions. First, it makes a request for operations on objects using one of two ways: static or dynamic invocation. Second, it receives and processes the operation results passed back from the server side. Third, it handles exceptions in case an error occurs.

Client-side stub.   Logically, the client stub is an object sitting between a client application and ORB. On the one hand, it provides an application programming interface (API) to the client so that the client application can send a request through the API. This is one of two ways for a client to send a request.

A client-side stub is generated by an IDL compiler and not written by a programmer. The client stub code maps OMG IDL operation definitions for an object type (defined by an interface definition) into a programming language–specific procedural routine that the client application will call to invoke a request. The client stub code is programming language–specific: the IDL compiler, according to OMG standard language mapping, converts the IDL definition into the code of a programming language that is chosen by the client-side developer. Of course, the programming language must be supported by the OMG.

In addition, a client stub works with the client ORB to marshal the request. That is, the stub helps to convert the request from its representation in the programming language to one suitable for transmission to the target object. In contrast to the IDL-defined, standard client-to-stub interface, the stub-to-ORB interface is proprietary. Since this part is hidden from user applications,

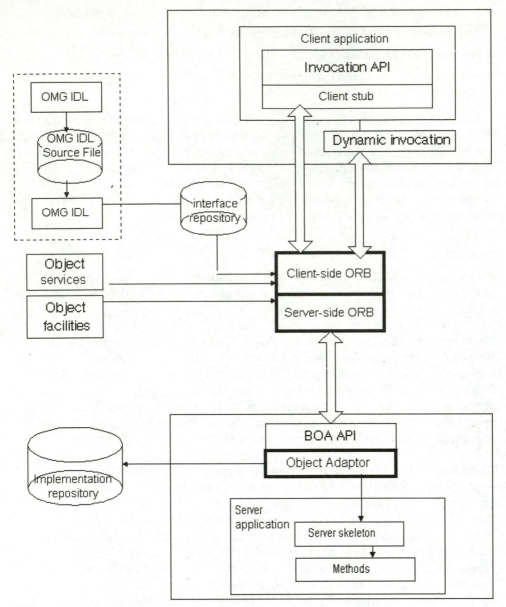

Figure 15.3 An architectural view of CORBA.

there is no need to standardize this interface and it is left for each vendor to optimize for performance and reliability.

Invocations by a client application via a static invocation interface (SII) are normally synchronous in the sense that the client application must receive either the result of a request or an exception before it can proceed further. A client application has the choice of invoking an operation in a deferred mode that defers the binding to a server until the last moment.

Dynamic invocation and dispatch. The CORBA dynamic invocation interface (DII) provides a client application with the capability of invoking any operation on any object that it may access over the network. Remember that via the SII, a client application can only invoke the operations for which it has stubs. In contrast, via the DII, a client application can invoke the operations on objects it has no compile-time knowledge of, objects newly added to the network, or objects discovered through a naming or trading service. The DII is very useful for interactive programs such as browsers.

The DII supports all three modes of invocation: synchronous, deferred synchronous, and one-way invocation. In the synchronous mode, a client application invokes the request and then blocks waiting for the response. From the client's perspective, it is equivalent to a remote procedure call (RPC). In the deferred synchronous mode, the client invokes an operation, continues processing while the request is dispatched, and later collects the result. In the one-way invocation mode, the client invokes an operation and then continues processing; there is no result to be returned. Of course, there is a tradeoff for the freedom a client application has in choosing a target object for invocation at run time. The DII is more complicated than the SII to write. A client application needs to go through the following four steps for a dynamic invocation, compared to only going through the last step for the SII:

- Identify the target object for invocation.

- Retrieve its interface.

- Construct the invocation.

- Invoke the request and receive the result or exception.

One common way to identify and locate a target object is through the CORBA standard trader service. The trader service is like a combination of the yellow pages and a product order catalog. It lists all the objects available on a network with additional information on each object. There are two ways of using the trader service: as a browser for a designer to manually browse through available objects or as a section of code in the client application to do the browsing. The trader service can provide the object reference that is necessary for invoking an operation on the chosen object.

The next step is retrieving the target interface. The target interface comes from several sources. One source is the interface repository (IR) that provides the syntax of the interface. A client application can use the object reference obtained from the trader or naming service to query the ORB and get an object reference that returns the top-level components of the interface. Using the top-level components to query the IR, a client application can obtain the target interface's operation and its parameters and types.

Constructing the invocation is straightforward: the DII provides a standard interface for building a request. All the client application needs to do is supply the operation name and parameter list obtained from the previous step and call the standard interface.

The final steps of invoking a DII request are similar to those for invoking an SII request except that a client application is required to explicitly specify an invocation mode, i.e., synchronous, deferred synchronous, or one-way invocation.

Interface repository.   The IR is a crucial component of CORBA. The IR stores the definitions of all objects known to an ORB. Thus each ORB is required by the CORBA standards to implement the standard IR interface, allowing the IDL definitions of the objects to be stored, modified, and retrieved. The CORBA specification lists the following ways an ORB can use the IR and the object definitions stored in the IR:

- To provide interoperability between different ORB implementations
- To provide type-checking of the signature of a request that can be issued through the SII or the DII
- To check the correctness of inheritance graphs
- To manage installation and distribution of interface definitions around a network
- To allow an application designer to browse and modify interface definitions
- To allow the language compiler to compile stubs and skeletons directly from the IR instead of from the IDL files

There is not a one-to-one correspondence between IRs and ORBs; an ORB may be shared by more than one ORB; an ORB may access more than one IR. The requirement is one ORB must be able to access at least one IR. Each IR has a repository ID, and an ORB keeps IDs of all the IRs it has access to.

IR components include modules, interfaces, operations, attributes, parameters, type definitions, exceptions, and contexts. OMG specifications do not require an IR to store interfaces in IDL form, although many do. The IR components are stored in a hierarchical structure to allow the original IDL file to be re-created. An IR interface provides access to one or more ORBs, not to a client application. For this reason, the implementation of IR is not standardized and is left for each vendor to implement in a way that is best for the target platform and operating system.

Client-side ORB interface and ORB core.   The ORB interface has a part for a client application as shown in Fig. 15.3. The most important part of the interface is the initialization component. When a client application first starts up, it must have an object reference to the ORB it is associated with, a naming service, and the IR. The ORB interface to the client side standardizes the initialization of all the three components and allows the application to specify which ORB, naming service, and IR it desires to connect to. Once the client-side ORB is initialized using the ORB interface, the ORB core can provide the following functions to a client application.

An ORB delivers requests to objects and returns any responses to the clients making the requests. The ORB does both marshaling and unmarshaling and shields the client from the underlying communication mechanism and protocols (e.g., TCP/IP, shared memory, local method call).

We briefly discuss how a client finds an object reference before we move on to server-side components. To make a request, the client specifies a target object by object reference. An object reference is created for each object at the object creation time. There are several ways a client can obtain an object reference:

- *Object creation.* A client can create a new object to get the object reference.

- *Directory service.* A client can use a lookup service to obtain the object reference.

- *Convert to string and back.* An application can ask the ORB to turn an object reference into a string, and this string can be stored into a file or a database. This string can be converted back into an object reference.

Server-side components

As shown in Figs. 15.3 and 15.4, the server side consists of the server-side ORB, static IDL skeleton, dynamic skeleton interface, object adapter, and server-side implementation.

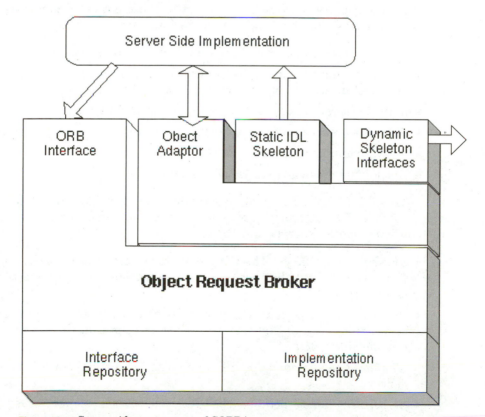

Figure 15.4 Server-side components of CORBA.

Server-side ORB. An ORB is one single component but has some behavior specific to the client side and other behavior specific to the server side. It is the server-side behavior that is discussed in this section, which we refer to as server-side ORB. The server-side ORB is responsible for the following.

- Receive requests from a client stub.
- Determine the best implementation to satisfy the request.
- Prepare the implementation to receive the request.
- Communicate the data in the request.
- Marshal and unmarshal requests and results.

Basic object adapter. The OMG standard version of the object adapter, the basic object adapter (BOA), is mainly responsible for masking the differences in object implementations to achieve portability. Objects can be implemented in a variety of ways and have various requirements for invocation. For example, as explained in Siegel (1996), some objects reside in a remote process and require activation before invocation; others may co-reside on the same platform as the client application; still others have different invocation requirements. It is the responsibility of the BOA to convert the different requirements into a set of standard ones. Specifically, the BOA provides capabilities to

- Register implementations in the implementation repository
- Generate and interpret object references (the interpretation means binding an object interface to a server implementation)
- Map object references to their corresponding implementations
- Start up a server process in which an object can be activated, if needed
- Activate and deactivate object implementations
- Invoke methods via a static IDL skeleton

When does the BOA decide to activate a server process for an object invocation? The BOA interface supports four activation policies, covering different configurations of servers and objects.

- *Shared server policy.*   The BOA activates a server only for the object that is first invoked, and the rest of the activated objects will have to share the same server.
- *Persistent server policy.*   Similar to the shared server policy in terms of one server shared by multiple running objects except that the server is activated outside of the BOA.
- *Unshared server policy.*   Only one object at a time can be activated on a server.
- *Server-per-method policy.*   The BOA activates a server for each method invocation and deactivates it once the request is served.

Shared-server policy is very common because of the efficient use of computing resources and the reduced overhead in process initiation and management. A disadvantage is that the operating system is not available to keep one object from violating the security policy of another object. Persistent server policy is good for the database transaction-oriented applications. Unshared servers are suitable for the case where a server is needed for an exclusive resource such as a file server. One server per method allows isolation of servers from each other for security or other reasons.

Though the BOA is the only standard object adapter that has been widely implemented, two other object adapters are also specified by the OMG. A library object adapter, for one, is like a lightweight adapter that is used in the case where client and server objects are running in the same process. No complicated activation and authentication are required for library objects. The other is an object-oriented database adapter that uses the capabilities of an object-oriented database system to provide a method and persistent storage.

Static IDL skeleton.   The static IDL skeletons on the server side play a role similar to that played by the client stub on the client side. The skeletons are generated by a compiler from OMG IDL definitions and are specific to a particular interface and object adapter. Specifically, the responsibilities of the skeletons include

- Mapping IDL types in a request to a type in a specific programming language on the server side

- Linking the BOA to the appropriate implementation and method for the request, after the request is received by the ORB and the BOA has selected the appropriate implementation

Dynamic skeleton interface (DSI).   The DSI allows an ORB to bind a request to any object implementation or proxy it can contact on the network, not just the ones it connects to via static stubs. It can serve as a bridge between two ORBs to support interoperability. Through the DSI, one ORB can communicate an invocation of an object at a remote ORB and get a response back. Without the DSI, this would be possible only if the invoking ORB had linked a skeleton for the target object, an unrealistic scenario in a dynamic network.

Implementation repository.   The implementation repository contains information that allows the ORB to locate and activate an object implementation. Using the information, an ORB can bind object references to the corresponding object implementations the same way an ORB associates object references with their interfaces using the interface repository. Since the object implementations are specific to a programming language and operating environment, most of the information in an implementation repository is specific to an operating environment or vendor's ORB. The interface to the implementation repository is not standardized. Note that the implementation repository contains the information describing implementations, not the implementations themselves.

Server-side implementation. The implementation provides the actual behavior and state of an object that provides the service requested by the client-side application. The objects may be implemented in a programming language that is different from the one used to implement the client application. With the exception of a few often-used objects, most of the other objects are not running at the time a request is made. When a requested object is not active and its service is requested, the ORB, CORBA persistence service, and the server-side management facility work together to activate the instance of the requested object and restore its state. Of course, this must happen before the invocation of any operation on the object can be processed. This relieves the client application of the burden of knowing anything about the server objects other than the operation interface. This behavior is mandated by the OMG, and one requirement from the behavior is that the state information of an object must be saved when the object is deactivated.

The client application can also request the creation of an object on the server side through a factory object, a special-purpose object that creates instances of other objects. The factory object creates a key, also called an ID, along with the created object instance for the future persistent storage. The factory object passes the object instance along to the implementation repository, and the new object instance becomes part of the server-side implementation.

A server application contains one or more implementations that are part of the server that can accomplish a client's request for operation on a specific object. It contains code to do the following:

- Initialize the server application
- Shut down the server application
- Handle service requests

OMG Interface Definition Language

The IDL is at the core of making CORBA a distributed computing environment that is independent of a specific programming language and computing platforms and that achieves full separation of a client from a server. Both client applications and server implementations use the IDL to specify the operation an object can perform, the input and output parameters that are required, and exceptions that may be generated in the operations. A client application uses the same interface definition to build and dispatch invocations as the object implementation at the server side uses to receive and respond. This common, intermediary language makes it possible for the client applications and server-side object implementations to be implemented in different programming languages, on different computing platforms, yet be able to interact with each other.

Definition. The OMG IDL is a programming-language-independent, declarative language for defining object interfaces. Note that the IDL itself is not a programming language and does not have the control structures normally seen in a programming language. It is a strongly typed interface language that requires every declared variable to be typed. Yet it also bears a resemblance to a programming language, especially C++, because its grammar is a subset of C++ grammar and it adopts many of the C++ syntactic rules.

As shown in Fig. 15.5, the top-level construct of an IDL specification is a module. An IDL module is like a program that may comprise one or more interfaces, defined types, and structures that are defined with *struct* or *union*.

An interface may consist of a set of operations and defined types. An IDL interface, like a class in C++, is the basic unit of inheritance. An operation, very much like a function or method declaration, must have a return type, an operation name, and a parameter list. Each parameter declaration in the list must have a type declaration, which can be a built-in type or defined type; a parameter attribute, which is *in, out,* or *inout*; and a parameter name.

An operation declaration may optionally have an exception. An exception declaration consists of an exception name and a parameter list. A list of one or more exceptions may be associated with an operation using the *raises* keyword.

The second optional parameter of the operation declaration is the one-way declaration, as shown in the example IDL module structure. A one-way declaration must have a void return type; only input parameters are allowed and no exceptions are allowed. This declaration indicates that a client application, after sending an operation invocation, continues with its normal process and does not expect a response result back. One potential issue is that the ORB does not guarantee the delivery of the invocation, only its best effort.

```
module exampleModule {
    typedef string Astring;
    struct myStruct { string name;
                  int          Id;
                  };

    interface sampleInterface1 {
       typedef string Bstring;
       oneway void sendMsg (in string myMsg);
       exception outOfRange {short dummy};
       int operation1(in short arg1) raises (outOfRange);
    };


    interface sampleInterface2 {
       attribute short objState;
    };


    . . . . . .
    };
```

Figure 15.5 Example structure of IDL specification.

The third optional parameter of the operation declaration is the context object, not shown in the example. A context object sets environment variables for an operation.

The key word *attribute* in a variable declaration requires that the variable retain its assigned value until the value is explicitly altered. It has two implicit functions to allow a client to retrieve and set the variable value. This is one way to implement object state persistence, as required by the OMG specification.

All IDL constructs are scoped. That is, a variable declared in a module is visible to the whole module while a variable declared in an interface is visible only inside the interface. Operations, structs, unions, and exceptions are scoped as well. The OMG IDL supports single as well as multiple inheritances, and the syntax is similar to that of C++.

IDL types.   The IDL types consist of three categories: built-in, constructed, and template. The important types of each category are briefly described. Examples of built-in types include most of the primitive types you would normally see in programming languages like C or C++ such as long, short, float, char, boolean, and enum. CORBA specifications precisely define the sizes of all the built-in types to ensure interoperability. For example, the built-in type *short* represents the range of 1^{15} to $2^{15}-1$. The constructed types are built using struct and union. The type *struct* represents data aggregation as in C and C++. The type *union* represents a discriminator that keeps track of which alternative out of a set is currently valid. The template types include *string* and *wstring* that represent bounded and unbounded strings and *sequence* that defines one-dimensional arrays. In addition, the *any* type allows the specification of values that can express any OMG IDL type. Table 15.3 provides some examples of mapping from IDL types to C++ types.

TABLE 15.3   Mapping between IDL Types and C++ Types.

IDL type	C++ mapping type
long, short	long, short
float, double	float, double
enum	enum
boolean	bool
octet	unsigned char
any	any class
struct	struct
union	class
string	char*
wstring	wchar_t*
sequence	class
object reference	pointer or object
interface	class

Language mapping. An IDL specification has to be mapped to the constructs of a programming language at the client side and at the server side. Up to now, the mappings from IDL to the following programming languages have been standardized.

- C
- C++
- Smalltalk
- Ada
- Java

CORBA services

This section summarizes the CORBA services, the core component of the OMA. Those services that are deemed more relevant to the telecommunications network management applications are described in more detail than others. Note that the CORBA services are being defined on a continuing basis, and services other than those discussed here will be available in the future.

Naming and trader services

Naming service. This is a vital service of CORBA because objects in one domain have to resolve the name of objects in another domain in order to invoke operations on them. As you may recall, an object reference, one for each object in the network and assigned by an ORB at the object creation time, is the key to an object. A client application can invoke an operation on a remote object only by its object reference. The naming service provides object references for a client application. Given an object name, the naming service returns an object reference for the named object. An object name is formed according to a syntax-independent, in-memory hierarchical naming structure. The naming service can be best compared to the white pages of a phone book: given a name, it supplies a phone number and address that enable you to locate the person.

Trader service. The purpose of the trader service is also for a client to obtain object references. If a good analogy for the naming service is the white pages of a telephone book that allow a client application to search by name, an equally good analogy for the trading service is the yellow pages of a phone book. In the yellow pages, you can locate an auto repair shop by searching the auto repair section. In a similar fashion, the trader service allows you to locate an object reference by searching objects by categories and groups. The trader service allows a developer to browse objects manually or to code a search routine with a set of built-in search criteria. When such a search routine is combined with a DII client, the search and invocation process becomes automatic.

Life cycle and relationship services

Life cycle service.   Life cycle service defines services and conventions for creating, deleting, copying, and moving objects. The creation of new object instances is achieved through a specialized object called a *factory object*. Aside from creating a new object instance, a factory object can determine the location of a new object, check the resources required for creating a new object (e.g., memory and storage space), obtain an object reference for the newly created object, and notify the concerned parties of the object creation (e.g., a BOA).

Copying, moving, and deleting objects, just like creating objects, are basic capabilities that can be provided by object-oriented programming. But the CORBA life cycle service provides additional functions to do so in the ORB environment. For example, after deleting an object, the resources associated with the object are distributed across a network and need to be released in a distributed fashion.

Relationship service.   The relationship service provides information on the relationships between objects in a network. Relationships between objects can be defined from multiple dimensions. The CORBA relationship service defines the following dimensions of relationships:

- *Type.*   A relationship can be typed just as an object can. For example, teaching is a relationship between a teacher and a group of students.

- *Roles.*   The role an object involved in a relationship can play. In the teaching relationship, for example, Mr. Diamond assumes the role of teacher.

- *Degree.*   The number of required roles. For example, the teaching relationship has two roles: teacher and student.

- *Cardinality.*   The number of relationships that may involve a particular role.

- *Semantics.*   The attributes that may describe the semantic aspect of a relationship. For example, grade level and subject of teaching can further describe the teaching relationship.

The relationship service enables an application to traverse from one object to another through the relationship between the two objects or through the role an object plays. This service becomes very useful for searching for an object of interest.

Event service.   The CORBA event service provides the capabilities for an object to send, register for, and receive an event of interest in the CORBA environment. Figure 15.6 shows a generic event service model defined by OMG.

The object sending an event is called the *supplier,* and the object receiving an event is called the *consumer.* A supplier and a consumer are connected by an event channel, which itself is an object. The event channel object decouples the communication between the supplier and the consumer. An event channel can be shared by one or many suppliers and one or many consumers. A supplier can have as many event channels as it desires and so can a consumer.

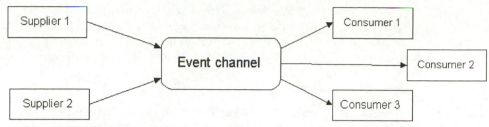

Figure 15.6 A CORBA event service model.

The event service supports both the push event model and pull event model, also commonly known as the notification and polling models, respectively. In a push model a supplier sends an event to the event channel object whenever it wants. The channel must accept the event and send event notification to all objects that have registered for this event. In a pull model the channel plays the role of client and polls the suppliers for an event at the channel's discretion either when it receives a pull from the consumer or at timed intervals. The CORBA event service provides standard interfaces for an object to send and receive an event.

An application object can register to receive or send events through a set of standard interfaces provided by the CORBA event service. The object that intends to send events creates an event channel and decides who can register as an event consumer and a supplier. This will avoid any conflict at the supplier end and ensure that only one object can send events while making the consumer end public so that any number of objects can receive events.

Transaction and concurrency services

Transaction services. The OMG object transaction service (OTS) provides the capability to support transaction processing in the CORBA environment. Aside from satisfying the basic requirement of transaction processing, i.e., guaranteeing atomicity, consistency, isolation, and durability (ACID), the OMG OTS provides a mechanism to make an object-oriented system, CORBA, and a transaction process work together. Among the supports provided by the OTS are

- Support for various configurations of ORB and transaction services such as one ORB per transaction service, multiple ORBs per transaction service, multiple transaction services per ORB, and multiple transaction services with multiple ORBs

- Support for both flat and nested transactions

- Support for integration of an existing or legacy transaction processing system

- Support for executing multiple transactions concurrently and executing client, server, and transaction services in separate processes

Concurrency control service. The CORBA concurrency service is designed specifically to provide locking management to the object transaction service discussed above. It provides five lock levels to enable an application to choose from very high level to fine-grained locking to suit different needs.

Other OMA services

Persistent object service (POS). Every object's state must remain unchanged from one invocation to the next unless the state is intentionally changed or the object is deleted. The POS is used to provide a capability for a client or a server application to maintain the persistent state of an object. There are many ways to achieve this, and the POS provides a set of standard interfaces and standardizes the process.

Externalization service. Similar to the POS, the externalization service deals with the persistence of object states. It defines interfaces, protocols, and methods to externalize object states into a standardized stream of data that can be saved on disk, played back, or sent over the network. Like the POS, it records the state of objects in a persistent fashion. Unlike the POS, it does not make persistent changes to a variable of an object immediately, but only at a fixed time and location.

Property service. Object property service allows a client to assign properties to an object dynamically (versus statically). The natural way to add an attribute to an object is changing the IDL file. But sometime changing a vendor package or files developed by someone other than you can become complicated and cumbersome. The property service allows a developer to hang additional properties onto an object from an existing package. The service provides standard interfaces for adding, listing, and deleting properties for an object.

Query service. The object query service allows a user to perform OODB-type queries on the CORBA object in a system. The necessary components for the query service to work include a query language, a standard interface for the application to invoke a query on an object and to execute the query invocation, and finally, a standard way to organize a collection of CORBA objects so that they can be queried. Two query languages are chosen, SQL and object-oriented query language (OQL) (defined by the Object Database Management Group), because one is most widely used and one directly supports object level queries. The OMG has specified standard interfaces to query an object or a group of objects and to manage queries. Other services, such as relationship and naming services can be used to help construct a collection of CORBA objects for query.

CORBA facilities

CORBA facilities are another part of the OMG OMA. While the CORBA services provide object level services, the CORBA facilities are intended to provide application level services. The CORBA facility architecture has two components: a horizontal one and a vertical one. The horizontal component intends to establish application facilities that can be used across industries and application domains. One such example is compound document services that can be used by all businesses. The vertical component seeks to establish and develop standard status application facilities for a particular domain or industry. So large and ambitious is the scope of the CORBA facilities that the CORBA facilities threaten to dwarf the CORBA and CORBA services eventually.

The horizontal CORBA facilities, meant to be useful across domains, currently consist of four basic categories: user interface, information management, system management, and task management. Examples of the user interface facilities include rendering management that deals with the general-purpose presentation of objects on screen, paper, or other media; user support facilities including help, spell and grammar checking; and automatic scripting. The information management facilities are those for information modeling, storage, retrieval, encoding and decoding, and exchange. The system management facilities are for management of complex, multivendor information systems. The task management facilities cover work flow, also called work process; rule management; automation facilities; and others.

While CORBA, CORBA services, and CORBA horizontal facilities provide a foundation, it is the vertical CORBA facilities that will bear the fruit of utilizing the distributed architecture in each specific industry. The participation in this effort can only come from each industry segment, and a wide array of industries (ranging from finance, healthcare, to telecommunications) are active participants. The OMG Telecom Task Force, established in 1996, is responsible for defining the vertical CORBA facilities for the telecommunications industry.

The efforts of the OMG Telecom Domain Task Force so far fall into two general categories: general architecture framework and a reference model of telecommunications network management based on CORBA architecture; the mapping of the telecommunications-specific information modeling language GDMO/ASN.1 to CORBA IDL.

Application of CORBA to Telecommunications Network Management

This section surveys efforts of the OMG Telecom Task Force and discusses various scenarios of applying CORBA architecture to telecommunications network management. At the center of the discussion is the scenario of using CORBA as an implementation technology for TMN.

Introduction

As described earlier, the ITU-T ODP specifications express the requirements of a distributed computing environment in terms of the seven distributed transparencies. A quick comparison indicates that the available CORBA, the CORBA services, and the CORBA facilities, can meet most, if not all, of the requirements of the seven transparencies. The access, location, replication, and migration transparencies are already supported by ORB in combination with naming and trading services; the transaction and distribution transparencies can be achieved using a combination of ORB, object transaction service, and naming and trading services. The failure transparency will become achievable as the CORBA fault-tolerant capabilities are enhanced. It has been widely acknowledged that CORBA is a viable implementation technology of the ITU-T ODP.

This section discusses different aspects of applying CORBA to telecommunications network management. First, we look at the issues related to developing a fully CORBA-based TMN, where CORBA is used as an alternative TMN implementation technology to the CMIP. Then, the focus of discussion turns to CORBA being applied to the TMN interfaces, namely, Q, X, and F interfaces, to take advantage of the interoperability CORBA offers. Finally, we look at how CORBA interworks with existing OSI-based TMNs by summarizing the efforts of the OMG Telecom Task Force in this direction. The efforts fall into two general categories. One area is mapping TMN's managed object classes to CORBA's IDL interfaces and mapping the SNMP information model to the IDL-based information model. The other direction is mapping the run-time service such as CMIS services to CORBA object services.

CORBA-based TMN

As CORBA is being accepted as a standard object-oriented distributed technology in the computing industry, more applications are emerging to demonstrate not only the feasibility but also the advantages of applying CORBA to telecommunications network management. Strong interest has been expressed in not only piecemeal application of CORBA to telecommunications management, but also in the development of fully CORBA-based TMNs.

We first have a very brief review of the key elements of a TMN to put the discussion of a CORBA-based TMN into context.

TMN architectures. The ITU-T TMN recommendations define a set of three TMN architectures: functional, information, and physical. Together the architectures specify the boundaries of a TMN, the internal building blocks, and the external interfaces.

Standard interfaces. The TMN specifications are interface-centric in the sense that a primary goal of TMN standards is to achieve management interoperability among diverse network elements through standard interfaces. An interface is the point where two disjointed systems exchange management information and is characterized by two elements: an information model and a management protocol. Three classes of standard interfaces have been defined, namely, Q, X, and F interfaces, to connect a TMN managing system to a managed system, an external TMN system, and a workstation that interfaces an end user, respectively.

TMN information models. An information model consists of managed objects that represent the managed network resources. The standard information model is the key to achieving interoperability because it serves as a common language that enables diverse network elements and management systems to communicate with each other. All existing TMN information models have been defined in GDMO, an OSI-defined, management protocol–dependent information modeling language (see Chap. 4 for details on GDMO).

TMN management protocol. A management protocol determines how the management information is transported from one end to the other. Although the TMN specifications do not preclude other alternative management protocols, the CMIP arguably is the only viable management protocol that has been implemented in real-world TMN-based systems. (This has led to the erroneous notion that equates TMN to CMIP!)

Manager-agent paradigm. The TMN recommendations adopt the widely used manager-agent communication pattern within the boundary of a TMN. A management system conveys management instructions to and receives management data from a set of agents. The interface between a manager and its agents is standardized (i.e., $Q3$ interface), and much of the TMN specification is based on the manager-agent paradigm.

Although this brief review does not cover all the important elements of the TMN standards, it suffices to give a background for the discussion of a fully CORBA-based TMN. It has been pointed out, perhaps correctly, that TMN recommendations specify one implementation technology, namely, the OSI framework, but they do not preclude other alternative or complementary implementation technologies. If CORBA is chosen as a TMN implementation technology, will CORBA measure up to the TMN requirements and what are the major issues and implications? We look at a CORBA-based TMN from three perspectives, i.e., the managing system, managed system, and interface between them.

CORBA-based managing system. A managing system that performs the manager role requires three parts at a minimum: a set of internal function blocks, the capabilities to process information models, and a communication scheme. The building blocks implement management tasks, as specified in ITU-T Recommendation M.3400 (1992), that are divided into five areas (i.e., fault, configuration, accounting, performance, and security management). The capabilities to process managed objects that represent managed network resources require the understanding of both semantics and syntax of the information model. The communication scheme covers both the communications between the internal components and between the managing system and the outside world.

A CORBA-based TMN reference model, as shown in Fig. 15.7, provides a general architecture of a CORBA-based managing system. For more details, see OMG Telecom Domain Task Force (1996).

First, the TMN building blocks, as identified in ITU-T Recommendation M.3400 (1992), and also called support objects, will be implemented by the

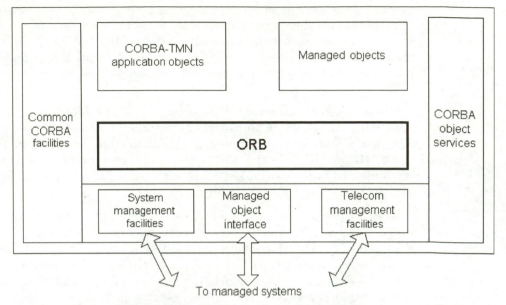

Figure 15.7 A reference model for the CORBA-based managing system.

application objects using CORBA object services. These CORBA-based application objects have the potential to become standard plug-and-play objects, and in fact efforts are under way within the de facto standards organizations such as the NMF to turn the generic building blocks into off-the-shelf Javabean-based components for developing a CORBA-based TMN.

A managing system must have a view of managed objects that represents managed network resources. In a fully CORBA-based TMN, network resources whose representation will go across network element boundaries will be represented as managed objects in the IDL. Within a managing system, the IDL managed objects are treated in a way similar to the way other CORBA objects are treated; this certainly facilitates the integration of views of diverse network elements into the managing system, because of the interoperability the IDL provides.

The communication between the internal building blocks will be based on the ORB and client-server framework. It is all objects communicating with each other without knowledge of the location of the other objects. Communication with managed systems will also be based on ORB and the CORBA-based client-server framework. A CORBA-based management reference model does not distinguish an internal system component from a managed system. The goal is to turn the portion of a managing system that interfaces with outside systems into standard CORBA telecommunications facilities that are built on top of the already defined CORBA common facilities and CORBA object services, as shown in Fig. 15.7.

What are the implications of a CORBA-based TMN to the manager-agent paradigm? The reference model of a CORBA-based TMN managing system will maintain the classification of a system role as either a managing system or a managed system but with some new wrinkles. The role of a managing system will likely be distributed rather than concentrated in one system. Multiple configurations of a role are possible. First, a role can be collectively played by several systems. For example, a manager role may be distributed with one distributed component playing the monitoring part of the role and another distributed component playing the controlling part of the role. Second, a system may play both manager and agent roles. Combinations of the two configurations and anything in between are also possibilities in a CORBA-based TMN.

CORBA-based managed system. An agent system, also referred to as a managed system, on the one hand, directly interfaces managed network resources via managed objects and, on the other hand, interfaces a managing system to receive management instruction and to report on the conditions of the managed network resource, either solicited or unsolicited. At a minimum, a CORBA-based managed system requires a systematic way to represent the managed network resources, a scheme to organize the representations, and a component responsible for communicating with a managing system.

Managed object and managed object classes. In a CORBA-based managed system, network resources are represented as IDL interfaces. This object-oriented approach to system management closely parallels the ITU-T approach of using GDMO to define managed objects. We take a quick look at whether the IDL interface is sufficient to meet the TMN managed object requirements, as specified in the TMN architectural document ITU-T Recommendation M.3010 (1996).

- Has the attributes visible at its boundary
- Capable of management operation that may be applied to the object
- Capable of emitting notification
- Exhibits well-defined behavior in response to management operations or in reaction to other types of stimuli
- Provides flexible representation of the underlying resources: one object per resource or multiple objects per resource
- Represents physical resources as well as logical resources, including the resource internal to a management system
- Provides an abstract view of resources that are represented by other managed objects
- Represents a large resource that contains other resources, themselves modeled as subentities of the larger object (in other words, capable of containment relationships)

These are basic requirements of object-oriented information modeling. In principle, the IDL interface is capable of meeting all the requirements, though some tweaking of the generic IDL definition may be needed. For example, in order to make some attributes visible at the object boundary, a special arrangement may be needed to make them public. Other requirements, such as inheritance, containment relationships between managed objects, and flexible mapping between managed objects and network resources, are readily supported by the IDL interface.

A critical issue will be whether a set of base managed object classes defined in IDL can achieve the status of standard information models as those existing TMN information models defined in GDMO. These base managed object classes can be specialized for a specific technology and, for that matter, for each individual application. Efforts in this direction are already under way, but it will take time and a large number of deployed CORBA-based TMN systems for IDL information models to achieve standard status.

MIB. A management information base (MIB) defines the way managed objects are organized, stored, and retrieved. In the TMN information architecture, the ITU-T recommendations do not mandate the standardization of an MIB structure and its implementation. However, in practice, most of the deployed TMN systems employ the OSI-defined hierarchically structured MIB, simply because it is consistent with the structures of GDMO information models, naming conventions, and other infrastructure such as directory service. Listed below are some of the basic requirements for an MIB, either implicitly or explicitly stated in TMN recommendations:

- Support for object-oriented representation of network resources

- Support for a naming convention for consistent naming of managed objects

- Support for a systematic object identifier structure that is easy to process, can scale up for a large number of managed objects, and supports the addressing requirement for TMN interworking

- Support for both network managed objects and support objects (or application objects)

- Support for persistent data storage

The X.500 directory service is one implementation option mentioned in the TMN recommendations to support the TMN naming and addressing requirements. It seems necessary that the telecommunications-specific CORBA facilities be required to meet the TMN MIB requirements. These facilities may incorporate transaction services, query services, and other CORBA services and facilities into a widely deployed MIB structure.

CORBA-based interfaces. As stated earlier, an interface consists of two key components: an information model and a management protocol. Included here are only those aspects of an interface that have not been discussed already. From the perspective of TMN interfaces, the basic requirements include

- Support for transaction-oriented information exchange between a managing system and an agent system

- Support for the interoperability that requires common management knowledge shared between a managed system and managing system that includes

  - Supported protocol capabilities so that a managing system knows what can and cannot be sent over the network to a managed system

  - Supported managed function, i.e., the knowledge of what other parties can and cannot perform

  - Authorized capabilities, i.e., knowledge of not only what other parties can perform but also what they are allowed to perform

  - Supported managed object instances, i.e., the knowledge of what other parties know about the managed network resources

  - Containment relationships between objects, i.e., the knowledge of which object contains which because this knowledge determines how to access an object of interest in an OSI-based MIB

The transaction-oriented exchange is readily supported in CORBA. It is apparent that some of the shared management knowledge requirements, derived from the CMIP-based management environment, may prove a challenge for a fully CORBA-based TMN system. The knowledge of supported functions and protocol capabilities of other parties really requires a standard base set of management capabilities and protocol capabilities that are known to all application developers in order to achieve a wide range of interoperability. The CORBA DII proves useful in certain cases, but a developer cannot count on it to develop management applications. This leads us to the topic to be discussed next, the management protocol and protocol services.

TMN management protocols can be broken down into two groups: lower-layer protocols and upper-layer protocols. The lower-layer protocols cover OSI layers 1 through 4 and the upper-layer protocols include layers 5 through 7. An array of alternative protocols is listed for both groups for TMN $Q3$ and X interfaces [see ITU-T Recommendations Q.811 (1997) and Q.812 (1993) for details]. None of the protocols, with the exception of the ones at the application layer and in some cases the presentation layer, directly concern management applications. The application layer protocols, for network management purposes, fall into two categories: interactive and file-oriented. In most cases, the management applications are shielded from directly dealing with the management protocol. Instead, what a management application sees is the front end of the management protocol or a set of standardized protocol services. For example, TMN management applications rarely directly deal with the CMIP, if at all. Instead, the CMIS, the front end, provides a set of standard services to be used by any two communicating management systems: M-GET, M-SET, M-DELETE, M-EVENT-REPORT, M-ACTION, M-CREATE, and M-GET-NEXT. These services with their well-defined parameters collectively serve as a standard interface for the user to access a protocol while hiding the protocol details.

CORBA can achieve the same level of protocol detail encapsulation with the ORB on a per-application basis. However, in order to achieve a large-scale application level interoperability, a set of standard management APIs with well-defined syntax and semantics must be defined. Efforts are under way within the OMG Telecom Task Force to map the CMIS services into a set of CORBA facilities or CORBA object services. The ultimate goal is to elevate the defined set of CORBA IDL-based management interfaces to the standard status.

The other components of shared management knowledge such as the authorized capabilities, object instances, and object containment relationships should be easy to come by once a set of standard CORBA IDL-based management interfaces is established. In addition, the CORBA environment provides support for the following complementary services required for a management interface to support interoperability:

- Multithread and concurrency control services
- Both synchronous and asynchronous messaging services
- Object filtering services
- Event notification for alarm, fault, and performance data

CORBA applied to TMN interfaces

While the preceding subsection discusses the TMN interface requirements in the context of a CORBA-based TMN system, this subsection takes a look at the issues related to applying CORBA to each individual TMN interface, i.e., the Q, X, and F interfaces. Since the $Q3$ interface is well understood, the discussion will dwell more on the other two types of interfaces, X and F.

Q **interface.** The Q class of interfaces includes Q3 and Qx. Since Qx is an underdefined subset of Q3, we only discuss Q3. The Q3 interface is indisputably the most widely used TMN interface. As discussed in Chap. 6, a Q3 interface connects a managing system to a network element, a mediation device, or a Q adapter.

Recently CORBA has begun to be widely applied to the development of the $Q3$ interface, as shown in Fig. 15.8. It is used as the underlying communication mechanism between a managing system and an agent, between a managing system and a mediation device, or between a mediation device and a non-TMN agent system. Among the attractions of the CORBA-based $Q3$ interface are the ready availability of CORBA, the expressive power of CORBA IDL in representing managed network resources, and the relative ease of CORBA-based object-oriented programming, accompanied by the abundance of expertise in CORBA. However, as discussed earlier, among the biggest obstacles to the development of the CORBA-based $Q3$ interface is the lack of standard information models and management protocol services.

X **interface.** The TMN X interface supports exchange of management information between two TMNs that often cross jurisdiction boundaries, as shown in Fig. 15.9. Examples of the information flowing through this interface include

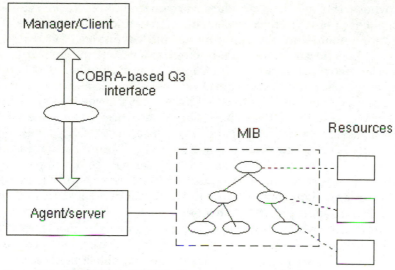

Figure 15.8 CORBA-based $Q3$ interface.

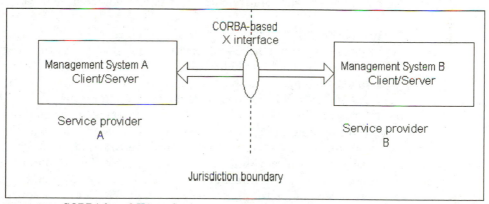

Figure 15.9 CORBA-based X interface.

- Billing data for services involving multiple service providers
- Service order entry and provisioning data that involve more than one administration domain
- Service performance reports for those services crossing multiple networks
- Trouble tickets for network or service faults that involve more than one network

There is much less specification of both information models and protocol services for the X interface, compared to the $Q3$ interface. This is a double-edged blessing. Without detailed specification, there are few guidelines to follow and few legacy burdens and constraints to applying CORBA to this interface. This is where the de facto standards organizations such as the NMF step in. Practical pilot projects have been developed under the auspices of the NMF in part to demonstrate the applicability and advantage of CORBA at this interface. It is believed there is great potential for CORBA application at this interface for two reasons. First, the information exchanged at this interface (e.g., customer data and service data) has very few dependencies on the network technology and should be handled by generic information systems which have a large installed CORBA base. This is an incentive to use CORBA at this interface simply for interoperability reasons. Second, it will be difficult to standardize the information models for the X interface because the data are largely proprietary. The information model will be developed on a per-application, per-service-provider, or per-group-of-service-providers basis. CORBA is better suited for this case because of its relative low cost in developing such individualized information models.

But there is a hitch: security. The information exchanged at this interface can be confidential and proprietary. For example, strong mutual authentication, access control at management association establishment, access to managed objects, confidentiality, data integrity, and data origin authentication are among the security requirements of the X interface. Until the CORBA security mechanisms are sufficiently mature, security will remain a concern.

CORBA-based *F* interface. A TMN F interface connects a workstation to a management system where the workstation is a system that translates TMN information to a displayable format or vice versa, as shown in Fig. 15.10.

As few formal specifications as there are on the X interface, there are even less on the TMN F interface. At least the CMIP is specified as one management protocol for the X interface. There is reported development of X interfaces (Weese et al. 1996), and active standardization activities are under way in standards organizations such as ETSI. In contrast, little has been done on the F interface and not much is in progress. Neither a management protocol nor a single formal information model has been formally specified.

There has been a strong interest in using CORBA to transport management information from a managing system to the workstation and using Java or web-based tools as the front end to display the information to the end user. The advantages include the interoperability provided by CORBA in masking the difference of programming languages and the computer platforms of workstations and the relatively low development cost of CORBA applications.

CORBA is also being used to support conventional human-machine interactions. Transaction Language 1 (TL1) is a standard human-machine language in North America for the Regional Bell Operating Companies (RBOCs), which provides a command line interface for controlling network elements. In some applications, the CORBA ORB is used to transport management information from a network element to a workstation, while a CORBA server at the workstation wraps TL1 commands as objects to support end-to-end CORBA operations.

CORBA interworking with TMN

Both CMIP and SNMP have a large installed base, and a critical issue any CORBA-based TMN must address is interworking with the TMNs that are based on other implementation technologies such as CMIP. Extensive efforts are now under way to define CORBA facilities to accomplish the interworking. The efforts are comprised of two distinctive parts: information model mapping and interaction mapping. The former maps static information models in GDMO and ASN.1 or SNMP MIB in SMI into an IDL equivalent. The second step converts dynamically between an operation in one domain and another operation in the other domain, without either party being aware of the conversion. One example of such a conversion is when a manager issues a CMIS command M-CREATE. The agent side converts it to a CORBA request and returns the result back in an M-CREATE reply without the CMIP-based manager knowing about the conversion.

Interworking scenarios.   Four interworking scenarios that require a gateway function are listed in JIDM (1998) and OMG Telecom Task Force (1997). The first two scenarios are shown in Fig. 15.11, where a CORBA manager is in the managing systems and either a CMIS agent or SNMP agent is in the managed system. Either a CMIP interface (i.e., the CMIP management protocol and GDMO information model) or SNMP interface (i.e., the SNMP management protocol and SMI information model) connects the two sides. In either case, a management instruction issued by a CORBA manager will be translated at a CORBA-to-CMIP or CORBA-to-SNMP gateway into a CMIP PDU, which in turn is translated into a CMIS or SNMP operation. A referenced CORBA managed object in the operation will be translated into a GDMO managed object or SMI object by the gateway too. In the other direction, a report sent from a CMIP agent or SNMP agent to a CORBA manager will be translated from a CMIS or SNMP operation into an operation on the CORBA

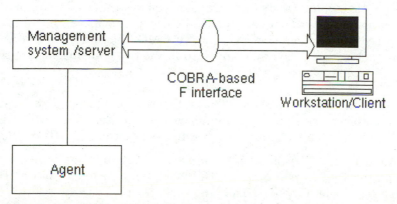

Figure 15.10   CORBA-based *F* interface.

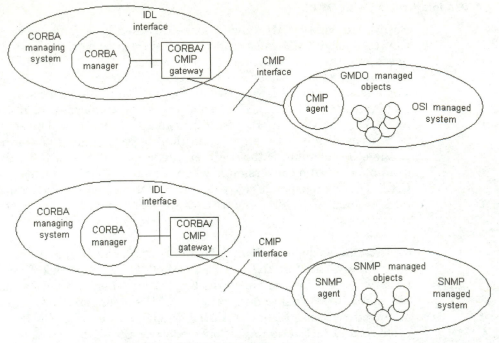

Figure 15.11   CORBA-to-CMIP and CORBA-to-SNMP gateways.

manager and a referenced managed object in SMI or GDMO will be translated into an IDL interface.

Two more interworking scenarios are shown in Fig. 15.12, where a CORBA agent connects to either a CMIP manager or a SNMP manager via an IDL interface. A management instruction issued by a CMIP or a SNMP manager will be translated at a CMIP-to-CORBA or SNMP-to-CORBA gateway before being sent over to the CORBA agent. Any referenced managed objects will have to be translated from either GDMO or SMI form into IDL form.

It is clear that two distinct translations are required: one for the management operation and one for managed objects. The operation translation is also referred to as interaction mapping in OMG Telecom Task Force (1997). The approach and generic issues involved in the two translations are the topics for the following two subsections.

Information model translation.   This subsection provides a high-level view of the translation rules that map an information model in GDMO and ASN.1 or SNMP SMI to an IDL interface module, as described in JIDM (1998).

GDMO to IDL.   The following sample translation rules indicate the general steps involved in mapping a GDMO information model to an IDL interface module. For details, see JIDM (1994).

- Map each GDMO document file to an IDL file and the corresponding GDMO document to an IDL module in the corresponding file.

- Map the GDMO managed object class template into an IDL interface.

- Treat the conditional and mandatory packages in the same way by the translation compiler; the developer makes this distinction during the implementation.

- Map each GDMO managed object class template to an IDL interface as a subclass of the generic interface CoreMOCTemplate.

- Map each GDMO package template to an IDL interface that is a subclass of the generic interface PackageObj.

- Map each GDMO attribute to an IDL interface as a subclass of the interface AttributeObj.

- Map each GDMO notification template to an IDL interface as a subclass of the generic interface NotificationObj.

- Map each GDMO action template to an interface as a subclass of the generic interface ActionObj.

- Map each GDMO attribute group template to an IDL interface as a subclass of the generic interface AttributeGroupObj.

- Map each GDMO parameter template to an IDL interface as a subclass of the interface ParameterObj.

- Map each GDMO name binding template to an interface as a subclass of the interface NameBindingObj.

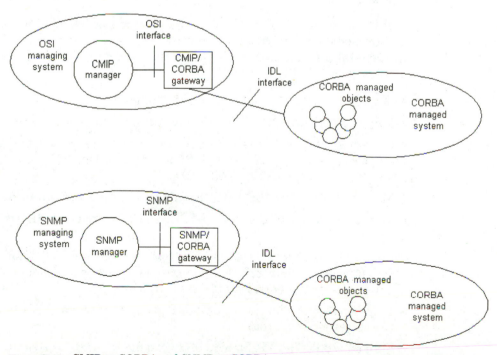

Figure 15.12   CMIP-to-CORBA and SNMP-to-CORBA gateways.

- Map each GDMO behavior template to an IDL interface as a subclass of the generic interface BehaviorObj.

In summary, an IDL interface is created for each of the GDMO templates such as managed object class, package, action, parameter, notification, and behavior; each instance of a GDMO template in a specific GDMO specification is mapped to a subclass of the corresponding IDL interface.

ASN.1 to IDL.   The mapping from ASN.1 to the CORBA IDL is straightforward since both are strongly typed specification languages. The following translation rules are intended to provide an indication of the general process rather than prescriptive steps. For more details, see JIDM (1994).

- Map each ASN.1 module in a GDMO MIB document to an IDL module within the scope of the document module.

- Map only TypeAssignment and ValueAssignment and ignore other types.

- Resolve the conflict in ASN.1 items due to the case-insensitive nature of IDL identifiers by generating an intermediary ASN.1 description that records additional information necessary to resolve any naming conflict caused by case sensitiveness of ASN.1.

- Classify every ASN.1 type into a built-in type or a constructed type to align with IDL types for mapping.

- Rearrange the order of the ASN.1 types such that there are no forward references.

SNMP SMI to IDL.   The following examples of the translation rules map an SNMP module in SMI to an IDL interface module. The intention is to indicate the general translation process rather than the detailed conversion steps. For more details, see JIDM (1998).

- Map an SNMP SMI module to an IDL module.

- Map the SMI IMPORT/FROM clause to the include statement.

- Generate a separate file to record the information about mapping an IDL scoped name for interfaces and attributes, and Object Identifier (OID) constants to corresponding SNMP OIDs.

- Map the SMI MODULE DESCRIPTION to formatted comments.

- Map an SNMPv2-SMI OBJECT-IDENTITY macro to the IDL constant literal of string type.

- Map a group or a table entry in SNMPv2 OBJECT-TYPE macro to a base IDL interface; one interface is generated for each group, and one interface is generated for the entries of the table in a group.

- Map nontabular objects of an SNMP group to attributes of the interface for the corresponding group.

- Map an OBJECT-TYPE macro of a variable to an IDL attribute.

- Map an SNMPv2-SMI NOTIFICATION-TYPE macro to a notification IDL interface and optionally some other interfaces derived from the notification interface depending on the involved event type.

Interaction translations. The second part of the interworking between CORBA and other management environments dynamically converts protocols and behaviors used in the two domains. The conversion is also referred to as an *interaction translation*. The efforts in this direction, undertaken by the Joint Inter Domain Management Task Force (JIDM), under the joint auspices of the NMF and other organizations, are still ongoing. Only a brief review is provided here, and for more details, the reader is referred to OMG Telecom Task Force (1997) and JIDM (1998).

The conversion is to be accomplished by a set of telecommunications CORBA facilities with the goal of enabling a CORBA-based management environment to interwork with a non-CORBA management environment. Each management environment has a reference model, such as the OSI management reference model and the Internet management reference model, that defines the basic principles and structure of the management environment. There are some aspects that are generic to all management models, and some other aspects specific to a given management model.

The proposed conversion principles divide the proposed CORBA facilities into three categories. The first category comprises generic interfaces that are network-management-model independent. These facilities provide a generic framework to access a managed domain, independent of the management reference model being used. The second category of facilities consists of generic interfaces but for a particular management model. Two management reference models, i.e., OSI and Internet management models, are considered. The third category of facilities consists of interfaces that provide functionality specific to a given information model and the associated management reference model.

Comparison between CORBA and OSI CMISE

The CMISE (CMIP and CMIS; see Chap. 5 for details) has traditionally been the designated implementation technology of TMN and thus equated to the TMN itself, correctly or not. Table 15.4 provides a high-level comparison between the OSI management reference model and CORBA-based management systems.

Discussion

The telecommunications industry is at a crossroads where traditional telecommunications and data communications are converging and multiple network technologies are converging. New technologies from the general computing industry will play an increasingly more important role in telecommunications network management. CORBA can be used to provide an implementation alternative to the OSI upper-layer protocols.

TABLE 15.4   Comparison between OSI Management Reference Model and CORBA

	OSI management reference model	CORBA
Application domain	Telecommunications specific	General computing industry
Communication protocol	CMIP	ORB
Object naming convention and reference	OSI naming tree and distinguished name (DN) and relative distinguished name (RDN)	CORBA naming and trader services and object OID
Association services	ACSE	ORB
Fault tolerant	Confirm and nonconfirm options for a management operation	No built-in facility available
Security services	No built-in application layer security mechanism	Newly defined security services
Information encoding and decoding	ASN.1 encoding and decoding	Marshaling and unmarshaling
Information representation	GDMO	IDL
Management information base	Containment hierarchy	Generic DBMS

There are several factors working in favor of CORBA as the distributed technology of choice for the telecommunications industry in general and telecommunications network management in particular. The first factor is the potential of CORBA to support interoperability between standards-based and non-standards-based systems, and between systems based on different standards. It is important that telecommunications management systems offer the flexibility to support network elements with different types of interfaces (i.e., proprietary interface, SNMP, CMIP). This may be effectively done via the introduction of a middleware such as CORBA.

Second, the economic force, which is the ultimate driver behind any technological innovation, is in favor of CORBA for the following reasons.

- CORBA, as a technology from the general computing industry, enjoys broader support than any telecommunications-exclusive software technology. The adoption of such technology from the general computing industry will result in much lower platform procurement costs and more off-the-shelf reusability.

- CORBA expertise is more abundant than that in the field of telecommunications network management such as the CMIP.

- CORBA is fast becoming the standard of distributed computing technology and thus has a large installed base, which in turn makes it easier to achieve interoperability.

Finally, CORBA provides a level playing field for telecommunications new-comers. The entrenched equipment vendors and service providers alike have accumulated very extensive expertise in telecommunications-exclusive tech-nology over a very long period of time, which makes it difficult for a newcom-er to compete against the established player. Any new technology from the general computing industry will certainly help tilt the imbalance of expertise in favor of the newcomer.

Equally true are the challenges facing CORBA. As discussed before, the lack of standard information models and the lack of standardized manage-ment operations are among the biggest obstacles. The lack of mature security mechanisms certainly limits its application in some areas. It is a fact of life that multiple standards will coexist for a long, long time, if not forever, and the CORBA purist's vision of fully CORBA-based TMN systems may take an equally long time to materialize. Instead, the success of CORBA will, to a large extent, depend on its ability to interwork with the existing network management standards and installed systems. It will take time for large-scale CORBA-based network management applications to emerge to demon-strate its scalability. The phrase "seeing is believing" still rings true for many practitioners.

Summary

This chapter starts with a discussion on the motivations for applying distrib-uted computing technology to telecommunications network management. Chief among them is the trend of telecommunications switching technologies moving away from a centralized, monolithic environment to one that is both logically and physically distributed, with intelligence spreading to the edge of the network, a trend that closely parallels what already happened in the gen-eral computing industry over a decade ago.

ODP, the ITU's response to the trend, is the specification of a distributed computing environment for the telecommunications industry. It defines the functional requirements (seven distributed transparencies), architectural requirements (computational and engineering viewpoints), technology require-ments (technology viewpoint), and a set of concepts and constructs necessary for defining such an environment. ODP is a specification of requirements of a distributed system instead of the system itself. ODP requires an implementa-tion technology to turn it into a distributed system.

CORBA is such an implementation technology. CORBA may not be the only choice of ODP implementation technologies, but it is a logical one. For one rea-son, CORBA has achieved standard status in the computing industry. Another reason is that the ODP specification itself is influenced by CORBA.

CORBA basic concepts and architecture are introduced along the line of server and client largely because CORBA fully realizes the separation of a client from a server. The client side has a client application, a client stub, an interface repository, and client-side ORB functions. The server side consists of server-side ORB functions, a server skeleton, an implementation repository, and a server implementation object. The components of both the server and client sides are introduced in detail.

CORBA IDL is a platform-neutral and programming-language-independent specification language that provides cross-platform and cross-programming-language interoperability.

The chapter concludes with a discussion on the application of CORBA to telecommunications network management. First discussed are the issues related to developing a fully CORBA-based TMN, followed by a closer look at the CORBA-based TMN Q, X, and F interfaces. Finally a brief review of the efforts in the interworking between CORBA and some existing TMN implementation technologies such as OSI CMIP and SNMP.

Exercises

15.1   Briefly describe two motivations of adopting a distributed computing environment in telecommunications network management.

15.2   What do the ODP access and failure transparencies mean?

15.3   What is the main difference between relocation transparency and migration transparency?

15.4   Describe the functionality of a client stub, using an analogy if possible.

15.5   When the result of an interaction is sent back from the server to the client, what are the client-side ORB functions? What are the server-side ORB functions? You should be able to infer these from the process of sending a request from a client to a server.

15.6   Describe the CORBA event service and compare it with the CMIS M-EVENT-REPORT.

15.7   Describe the difference between the CORBA trader service and naming service.

15.8   Discuss the issues involved if OSI naming conventions and identifier structure, as described in Chap. 4, are directly adopted in the CORBA IDL?

15.9   Find a scenario where the CORBA property service is useful.

15.10   What is a CORBA facility? How is it different from a CORBA service?

15.11   A GDMO-to-IDL translator will convert a GDMO managed object definition into an IDL definition. Can an exact inverse of the translator do the translation in the opposite direction, i.e., from the IDL definition to the GDMO definition? If yes, why? If not, what kind of issues may occur? Use examples to explain.

15.12   Explain the advantages and disadvantages of using CORBA at the TMN X interface.

Numbering Plans and Local Number Portability

Outline

- The first section, "Numbering Plans for Data Networks," introduces the ITU numbering plan for public data networks, including the number structure and guidelines for the private data network identification code.

- The second section, "Numbering Plans for Telephone Networks," introduces the ITU numbering plan for telephone networks and the North American numbering plan. The structure of each plan is discussed in detail.

- The third section, "Number Portability Architectures," first introduces the issues of local number portability in the context of the current telecommunications market. Then two architectural solutions to the problem, IN-based and non-IN-based, are introduced. The impact of the solutions on the network and on service management is discussed.

- The last section, "Number Pooling," introduces the concept of pooling directory numbers and a few proposed solutions.

Introduction

Number portability is a network capability that allows a subscriber to change service provider, location, and service without changing his or her public directory number (i.e., phone number). More specifically, there are three types of number portability services: location portability, service portability, and local number portability.

Service portability is a capability that allows a subscriber to retain the original directory number after changing services. For example, when a subscriber trades in regular telephone services for ISDN, he or she should not have to be assigned a new phone number. Another example is allowing a subscriber to replace existing wireline services with wireless services while keeping the original phone number. Location portability is a capability that allows a subscriber to retain the original directory number after moving from one area to another. Service provider portability is a capability that allows a subscriber to retain the original phone number after switching the service provider. In the United States and most European countries, network service providers will be obliged to introduce number portability, as specified in the Telecommunications Act of 1996 by the U.S. Congress and A Numbering Policy for Telecommunication Services in Europe by Commission of the European Communities, 1996. The primary driving force behind number portability is the opening of the telecommunications market and the introduction of competition into highly regulated local telephone services. Another factor is the advent of personal communications, such as, smart card, which is used in the Global System for Mobile (GSM) wireless networks, and Universal Personal Telecommunications (UPT), which allows a directory number to be assigned to a person rather than to a network access, such as a phone line.

The goal of this chapter is twofold. First, we provide background information on local number portability by introducing the ITU-T-defined numbering plans for data networks and PSTNs. Then we present the architectures to address the local number portability issues that are currently being used or are being considered for use. This includes both IN-based and non-IN-based solutions and number pooling.

Numbering Plans for Data Networks

The international numbering plan was developed by ITU-T to facilitate the operation of public data networks and to provide for their interworking on a global basis. The goals of the numbering plan can be seen from three perspectives. That is, the plans are to allow multiple data networks to exist within a country without address conflict, to provide for unique identification of a country as well as a specific public data network in that country, and to provide for a mechanism to interwork with other numbering plans. A guideline was also developed on how to efficiently create data network identification codes to conserve the scarce numbering resource. This section presents the general structure of the numbering plans for public data networks and introduces the identification codes for private data networks.

Structure of the numbering plan for data network identification

Though both the public and private data network numbering plans are discussed, the emphasis is on the former. Included in a numbering plan are the following: a generic data network identification code, the code for a particular country and particular type of network (e.g., satellite network), and supplementary information such as a suffix and an escape code.

Data network identification code. A data network identification code (DNIC) can be assigned to a variety of networks. It can be a public data network or a group of public data networks within a country. It can be a global service such as the public mobile satellite system, a global public data network, a public switched telephone network (PSTN), or an ISDN that can accommodate data calls from data terminal equipment connected to the PSTN or to the ISDN. It can also be a group of private data networks within a country connected to a public data network.

All DNICs shall consist of four digits, as shown in Fig. 16.1, and belong to one of the two categories. If the first digit is any digit 2 through 7, the DNIC identifies a public data network in a specific country or geographic area, as shown in case (a). If the first digit of a DNIC is the digit 1, the DNIC identifies a public mobile satellite system or a global data network, as shown in case (b).

Data network identification codes for a country. In the case of a DNIC identifying a data network within a country, the first three digits are called the data country code (DCC), which identifies the country. The fourth digit is the network digit, which identifies a data network within the country.

Each country should be assigned at least one 3-digit DCC. A DCC in conjunction with the network digit can identify up to 10 public data networks (i.e., 0 through 9). When a country requires more than 10 DNICs, additional DCCs can be allocated. For example, through October 1996, seven DCCs, ranging from 311 to 316, have been assigned to the United States. The first digit of a DCC is restricted to 2 through 7, inclusively. The digit 1 is reserved

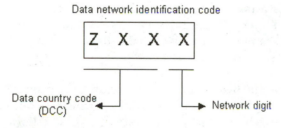

Data network identification code

$$Z \quad X \quad X \quad X$$

Data country code
(DCC) → Network digit

X -- any digit from 0 through 9
Z -- any digit from 2 through 7

(a) Structure of country-specific data network identification code

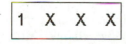

Data network identification code

$$1 \quad X \quad X \quad X$$

X -- any digit from 0 through 9

(b) Structure of global data network identification code

Figure 16.1 Structures of data network identification code.

for identifying the global data networks or public mobile satellite systems. Digits 8, 9, and 0 are used as escape codes and are not part of a DNIC. The first digit of a DDC is also used to identify a world zone or a geographic delimitation. For example, the United States, Canada, and other North American countries form zone 3. Theoretically, there are 600 DCCs, and thus a maximum of 6000 DNICs are available for allocation.

The assignment of network digits to form a DNIC is decided by each country's appropriate authority, rather than by the ITU-T. However, each country is required to notify the ITU-T of any new assignment, reallocation, or removal of network digits.

DNICs for global data networks and mobile satellite systems.   The DNICs currently allocated to public mobile systems are 111X, where X can be any digit from 0 through 9 and denotes ocean areas. Only four digits, 1 though 4, i.e., 1111, 1112, 1113, and 1114, are currently assigned to denote Atlantic Ocean east, Pacific Ocean, Indian Ocean, and Atlantic Ocean west, respectively.

Two categories of data networks qualify as global public data networks. The first category is those data networks that are operated either by an international treaty organization such as the United Nations or its various charter organizations. The second category is the networks that provide public data transmission services and are operated by one or a group of network operators. The networks must provide open and nondiscriminatory access to all users who may wish to subscribe to the service and span several countries or geographic regions. For those global data networks, the DNICs 1000 through 1109 and 1120 though 1999 are reserved. With only a total of 990 codes available for future global data networks, stringent eligibility criteria and procedures for the allocation of such DNICs have been set up and administrated by ITU-T through its Telecommunications Standardization Bureau. For more details, see ITU-T Recommendation X.121 (1996).

International data number.   The international data number assigned to data circuit-terminating equipment or data terminal equipment (DCE/DTE) consists of the DNIC or the public data network, followed by the network terminal number (NTN), or of a DCC followed by a national number (NN) where an integrated numbering system exists within a country. That is,

$$\text{International data number} = \text{DNIC} + \text{NTN} = \text{DCC} + \text{NN}$$

An NTN should consist of the full address used to call the data terminal from within its serving public data network. An NN should consist of the full address used to call the data terminal from another terminal within the national integrated numbering scheme. International data numbers have a maximum of 14 digits and a minimum of 5 digits. With the DNIC fixed at 4 digits and the DCC fixed at 3 digits, an NTN can have up to 10 digits and the NN up to 11 digits. The choice of NTN or NN length will depend on the specific network and is not mandated by the ITU-T.

Prefixes and escape codes.   A prefix is an indicator made up of one or more digits, for the purpose of selecting different types of address formats (e.g., national data number or international data number formats). Prefixes are not part of the ITU-T Recommendation X.121 international data number format as discussed in this chapter and are not signaled over internetwork or international boundaries. The use and format of the prefix are a matter local to each country.

An escape code is a one-digit indicator indicating which numbering plan the following digit string belongs to. When required, it can be carried across internetwork and international boundaries. Digits used as escape codes are the digits 8, 9, and 10. The escape codes are not part of the international data number but are part of the international X.121 format. The definitions of the escape codes are as follows.

8.   Indicates the following digits are from the F.69 numbering plan, a plan for telex destination codes and telex network identification codes.

9.   Indicates the following digits are from the E.164 numbering plan that will be discussed shortly, and an analog interface on the destination network (PSTN or integrated ISDN/PSTN) is required.

10.   Indicates the following digits are from the E.164 numbering plan and a digital interface on the destination network (ISDN or integrated ISDN/PSTN) is required.

Private data network identification codes.   Though not mandated, a numbering plan for private data network identification is recommended by the ITU-T to maintain the numbering consistency with public data network identification codes. This is needed especially when a private data network is to be connected to a public data network.

According to the recommended numbering scheme, a private data network identification Code (PNIC) consists of up to 6 digits. The international data number of a terminal on a private data network is of the following form:

International data number = DNIC + PNIC + private network terminal number

PNICs can be of either fixed or variable length. The PNIC administration and allocation are a matter local to each country. However, ITU-T has defined some general criteria and guidelines for efficient use of the scarce resource. Note that the assignment of a PNIC matters only when a private network intends to connect to a public switched data network.

If a variable length of PNICs is used, the unstructured allocation of two- to six-digit PNICs will incur considerable overhead in digit analysis for the purpose of routing a data call. Thus it is suggested that some rules be used in allocating variable-length PNICs. For example, four-digit PNICs could be allocated starting from the lowest to the highest value within a range 1000 to 3999, and three-digit PNICs could be assigned starting from the highest to the lowest value within a specified range, say, 699 to 400. In this way, the number analysis becomes less complicated because of the restricted number space.

Numbering Plans for Telephone Networks

A basic understanding of the PSTN numbering plan, the scheme that defines the directory number structure and assigns the numbers that will distinguish one country from another, is important for understanding the local number portability issues. This section introduces the ITU-T numbering plan and the North American numbering plan for public switched telephone networks.

The ITU-T numbering plans

The rapid advances in a wide range of telecommunications network technologies coupled with increased diversification of customer service needs (telephone, fax, data, wireless, etc.) have created a need for uniform access and network structure, or the ISDN. A consistent, forward-looking numbering plan for the ISDN era is a basic step toward a worldwide integration of telecommunications services. Introduction of such a plan defined by the ITU-T lays a foundation for the understanding of the local number portability-related issues. Some key concepts on the ITU-T numbering plan are listed in Table 16.1.

Structure of the ITU-T numbering plan.   A standardized international telephone number, or an E.164 number as defined in ITU-T Recommendation E.164 (1991) and with the structure shown in Fig. 16.2, intends to allow a network to determine the following of a called number:

- The appropriate country or area out of a group of countries or areas
- The appropriate routing network that reflects economic and other network factors
- The appropriate charging information such as charging rate

An international ISDN number is composed of a variable length of decimal digits, organized in two code fields. One code field is the country code, and the other is the national (significant) number. A country code (CC) in most cases is used to select a destination country and in some cases also a geographic area. A country code ranges from one to three digits in length. For example, the United States has the country code of 1, China 86, and Ireland 353.

The national significant number (NSN) is used to determine the destination subscriber. Specifically, the NSN is used to select a destination network. An NSN is comprised of two portions: a national destination code followed by the subscriber number. The national destination number (NDN) is of variable length depending on each individual destination country. The subscriber number (SN) varies in length from one destination country to another. The three portions of an international telephone number are introduced individually below.

TABLE 16.1   Terminology for the ITU-T PSTN Numbering Plan

Term	Definition
Country code	One, two, or three digits that uniquely identify a called country
Destination network code	An optional code filed in an E.164 number that identifies the destination network serving the destination subscriber
E.164 numbering plan	A scheme that defines the structure of international numbers for PSTN networks in the ISDN era. A number based on the E.164 numbering plan is also called an E.164 number.
Escape code	An indicator that consists of one or more digits and is defined in a given numbering plan which indicates the format of the digits that follow the escape code
International prefix	A combination of digits to be dialed by a subscriber in a call to another subscriber in a different country to obtain access to the automatic outgoing international equipment of a service provider in the home country of the calling subscriber
National destination code	A code field of the E.164 numbering plan, when combined with the subscriber number, that will constitute the national number of the international ISDN number
National prefix	A digit or combination of digits to be dialed by a subscriber in a call to another subscriber in his or her own country but outside his or her own numbering area
National significant number	The number to be dialed following the national or trunk prefix to obtain a subscriber in the same country but outside the same numbering area
Subscriber number	The number to be dialed to reach a subscriber in the same local network or numbering area
Trunk code	A digit or combination of digits that uniquely identifies a called numbering area within a country

Country code (CC). The purpose of a country code is to route a call to the required country in an automatic operation. The ITU-T is responsible for assigning a country code to all the countries in the world. The number of digits in a country code varies from one to three digits and is determined according to the foreseeable telecommunications and demographic development needs of the country concerned. Table 16.2 is a sample of the country codes assigned.

Note in Table 16.2 that the country codes for the United States and Canada are both 1. This is because a country code is assigned to an integrated numbering area instead of a single country and all the countries within the integrated area are distinguished by the national significant number. In fact, all countries and areas in the North American numbering area, including the United States, Canada, Jamaica, British Virgin Islands, and Bermuda, have the country code of 1.

The world is divided into nine numbering zones represented by 1 through 9. The first digit of each country code indicates the numbering zone of the country

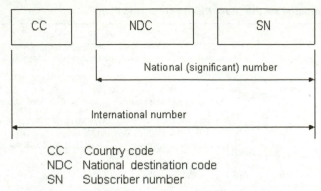

CC   Country code
NDC   National destination code
SN   Subscriber number

Figure 16.2   ITU-T number structure.

TABLE 16.2   A Sample of ITU-T Assigned Country Codes and Corresponding World Numbering Zones

Country name	Country code	World numbering zone
United States	1	1
Canada	1	1
Egypt	20	2
Algeria	21	2
France	33	3
Great Britain and Northern Ireland	44	4
Brazil	55	5
Guam	671	6
Hong Kong	852	8
Jordan	962	9

or area. For example, the country code starting with the digit 1 belongs to the numbering zone 1, the country code starting with the digit 2 belongs to the numbering zone 2, and so forth, as shown in Table 16.2.

There is a pool of designated spare country codes for each numbering zone to accommodate future extension. For example, the entire Soviet Union had the country code of 7, and when later political developments required additional country codes, the spare codes of the numbering zone were used. The ITU-T defines the following rules governing the use of spare country codes.

- All new country codes will be assigned on a three-digit basis.

- In case all the spare country codes in a world numbering zone have been exhausted and a new country code is needed, a spare country code from an adjacent numbering zone is assigned.

- In case no spare country code from an adjacent zone or zones is available, a spare code from a numbering zone with the most spare codes will be used.

- The following rules are followed in assigning a spare country code within a numbering zone:
  - Single, isolated three-digit codes should be assigned before any of a three-digit consecutive code series is used.
  - Assignment of a country code within that zone should start with the lowest-numbered three-digit code in ascending order.
  - Assignment of a country code to another world numbering zone should start with the highest-numbered three-digit code in descending order.

National destination code (NDC). An NDC, as mentioned earlier, is used to identify a destination network within a designated country and ultimately a subscriber. It is the responsibility of each individual country's numbering administration to design the format and structure of the national NDC. According to ITU-T-recommendations, each national numbering plan should be such that

- The appropriate network routing and charging information are indicated.

- A typical subscriber is always called by the same number in the trunk service and the number should be applicable to all incoming international calls.

- The digit analysis shall not exceed established limits.

The number of digits in the national destination code is variable depending on the requirements of the destination countries and may have one of the following structures.

- A destination network code that is of one or more digits and can be used to select a destination network serving a destination subscriber.

- A trunk code that is a set of one or more digits that designates a called party within a country. The trunk code has to precede the subscriber's number in the numbering plan and needs to be dialed where the calling and called subscribers are in different numbering areas.

- Any combination of destination network and trunk code.

Subscriber number. A subscriber number is the number to be dialed to reach a subscriber in the same local network or numbering area. The subscriber number varies in length depending on the requirements of the destination country.

North American numbering plan (NANP)

The North American numbering plan (NANP), the national numbering plan for the North American countries as a whole, is defined by the North America Numbering Council (NANC), the governing body in charge of national numbering plans for North American countries. The NANP is a specific implementation of the national destination code (NDC) of the E.164 numbering plan. It consists of a 10-digit dialing plan that in turn consists of two basic parts. The first three digits refer to the numbering plan area (NPA), or *area code* as commonly referred to. The remaining seven digits are yet again divided into two parts. The first three numbers represent the central office code. The remaining four digits represent a subscriber number under the central office. Each of the three parts of the North American numbering plan is described below.

Numbering plan area (NPA) codes. NPA codes (area codes) are provided in the form of *NXX*. That is, an area code consists of three digits where the first and second digits *N* are a value of 2 through 9 (the second digit was restricted to either a 1 or a 0 until recently) and the third digit is a value of 0 through 9. When the second and third digits are both 1, they together denote a special code. For example,

- 411 = directory assistance
- 611 = repair service (where implemented)
- 811 = business office (where implemented)
- 911 = emergency number
- 211, 311, 511, and 711 numbers are reserved

Additionally, the NPA codes also support service access codes (SAC). These codes support 700, 800, and 900 services.

Central office codes. The assignment of central office codes within an NPA is performed by the serving Bell Operating Company (BOC). Reserved for special use are

- 555 = toll directory assistance
- 844 = time service (where implemented)
- 936 = weather service (where implemented)
- 950 = access to interexchange carriers under feature group B access
- 958 = plant test
- 959 = plant test
- 976 = information delivery service

Also reserved for use are some *NN*0 (last digit "0") codes.

Access codes. Aside from 10-digit NANP numbers, the access code is a special digit series of variable length, which must be dialed in order to gain access to certain services. For example, a 1 is transmitted as the first digit to indicate a long-distance toll call. The examples of two-digit access codes include

- 00 = interexchange operator assistance.
- 01 = used for international direct distance dialing (IDDD).
- 10 = used as part of the 10*XXX* sequence. *XXX* specifies the equal-access interexchange carrier (IC).
- 11 = access code for custom calling services. This is the same function as achieved by the DTMF * key.

The 10*XXX* sequence signifies a carrier access code (CAC), a code for a long-distance carrier to access its customers via the facility of a local phone company. The *XXX* is a three-digit number assigned to the long-distance carriers through Bellcore, such as

222 = MCI	223 = Cable and wireless
288 = AT&T	432 = Litel (LCI Intl.)
333 = U.S. Sprint	234 = ACC Long Distance

Number provisioning process. Figure 16.3 shows the number provisioning process. NPA and *NXX* are administrated and allocated by the North America Numbering Plan Administration (NANPA). There is an industrywide inventory pool containing all available NPAs and *NXX*s within each NPA. Once an *NXX* is provisioned to a service provider, a 10,000-number block under the *NXX* (i.e., from 0000 to 9999), is made available to the service provider in the service provider's inventory pool to be provisioned to the customers within the region by the service provider's provisioning system.

Geographic versus nongeographic numbers. The 10-digit NANP numbers are geographic in nature because a called party phone number contains sufficient information about the geographic location of the subscriber that the call can be routed to the correct area (NPA), central office switch (*NXX*), and customer home location (*XXXX,* the subscriber number). It is this coupling between a phone number and a specific location that prevents a customer from changing service provider, service, or home location while still retaining the same phone number.

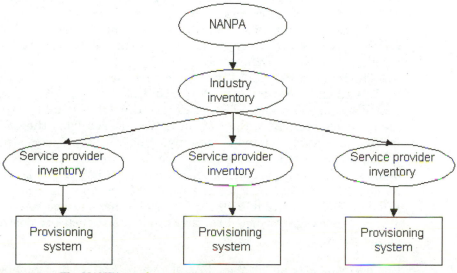

Figure 16.3 The NANPA number provisioning process.

In contrast, a nongeographic number does not contain any information about the home location of the called party. Currently the widely used nongeographic numbers are the toll-free services, (e.g., 800- and 888-number services) or special service phone numbers, such as 900 services.

Number Portability Architectures

As mentioned earlier, the number portability is a network capability that allows a subscriber to change service provider, location, and service without changing his or her public directory number (i.e., phone number). As it stands now, the implementation of number portability has been mostly, if not exclusively, focused on the service provider portability. That is, the number portability enables a subscriber to change service provider without changing the original directory number.

This section introduces two types of architectures, IN-based and non-IN-based, to accomplish the service provider portability. A simple scenario will help illustrate the number portability issues to be addressed, and the same scenario will be used throughout the section to facilitate the discussion. As shown in Fig. 16.4, a subscriber originates a call and the called party number, 972-555-2222, belongs to the donor switch. However, the called subscriber DN has already moved to the recipient switch, a switch of a different service provider. The number portability architectures center on how, when, and where to determine whether a number is a ported one and what is the intended directory number. The basic terminology on the local number portability is listed in Table 16.3.

Intelligent Network–based methods

Intelligent network. The Intelligent Network–based solution uses the Intelligent Network (IN) that in turn uses the SS7 signaling network. Figure 16.5 shows a simplistic view of an Intelligent Network. Related terminology can be found in Table 16.4. The LNP architecture is based on the local routing number architecture which in turn is built on the existing SS7 signaling and Intelligent Network architectures.

As introduced in Chap. 1, one major distinction between an IN call and a POTS call is the involvement of an SCP in the IN call. When an IN call first comes to the SSP, the SSP recognizes the IN call and a trigger will be invoked. Subsequently, the control is transferred from the SSP to the SCP for further service-related processing, which is often a database lookup. One typical example is a free phone service. Once an 800 call is received and recognized by the call-processing software at an SSP, the call is triggered to the associated SCP. The SCP translates the 800 number into an actual routing number and returns it to the SSP after other service-related processing. The SSP uses the returned routing number to reach the destination of the called party.

Number portability architecture. As shown in Fig. 16.6, a number portability architecture has the following components added to an IN architecture: a

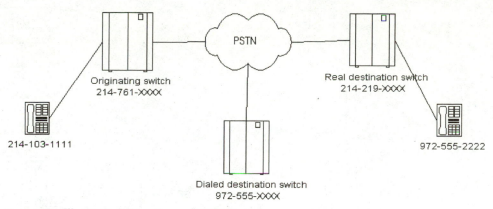

Figure 16.4 Illustration of number portability issue.

TABLE 16.3 Terminology Related to LNP

Term	Definition
Central office code	The *NXX* part of the North American numbering plan, i.e., NPA-*NXX-XXXX*
Donor switch	The switch the DN was initially ported from, sometimes also called the original destination switch
Geographic numbers	The directory numbers that indicate the geographic location
Location portability	The network capability to allow a subscriber to retain the original DN after changing the physical location
Location routing number	A digit string sufficient to uniquely identify a switch that has ported numbers. In the U.S., a 10-digit number is used.
Nongeographic numbers	A directory number that does not reflect the geographic location
Non-NP-capable switch	A switch that does not support number portability
NP query	An IN number portability query to a ported number database
NP trigger	An IN trigger specifically for number portability that has a trigger detection point and trigger detection criteria that must be met before an NP query is launched
Originating switch	The switch where a call originates
Ported number	A DN that has been moved from one switch to another that may or may not belong to the same service provider
Rate center	An area defined within the coverage area of a switch for the purpose of billing
Recipient switch	The switch the DN is ported to, sometimes also called the current serving switch
Service portability	The network capability that allows a subscriber to retain the original DN after changing service
Service provider portability	The network capability to allow a subscriber to retain the original directory number after changing service providers
Serving switch (or current serving switch)	Another term for the recipient switch

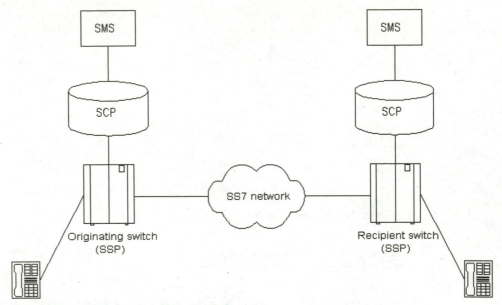

Figure 16.5 A simple Intelligent Network example.

TABLE 16.4   Intelligent Network and SS7 Architecture-Related Terminology

Term	Definition
ISUP (ISDN user part)	The call control part of the SS7 signaling protocol, responsible for setting up, coordinating, and taking down a call. The call control is accomplished via a set of ISUP messages, including initial address message (IAM), address complete message (ACM), answer message (ANM), release message (REL), and release complete message (RLC).
Service control point (SCP)	A major component of the Intelligent Network (IN) for providing IN services such as 800-number calls and credit card calls. The SCP provides a database and the translation and routing capabilities. It translates an 800 number to the required routing number. The SCP is separated from the switch, making it easier to introduce new services.
SMS (service management system)	SMS is responsible for provisioning and updating information on subscribers and services in near real time for billing and administrative purposes.
Service switching point (SSP)	A switch that is capable of recognizing IN calls and can route and connect the IN calls as directed by the SCP
Trigger	An event combined with a set of criteria. The occurrence of the specified event and satisfaction of the criteria will cause certain call-related processing at the SSP. For example, the occurrence of an off-hook event (i.e., a subscriber picks up the phone) accompanied by the satisfaction of certain criteria will trigger the transfer of call control from SSP to SCP.

number portability administration center (NPAC), a routing database and management in SCP, a local service management system (LSMS), and a number portability trigger function in SSP.

The *number portability administration center,* normally administrated by a third neutral party, is the central control point for the provisioning and administration of networkwide number portability services with the following main responsibilities:

- Resolve the conflict between service providers. The copies in a service provider's routing database may be out of synchronization or erroneous and the copy in the administration center is the reference point the other copies are aligned to.
- Populate all the routing databases under its jurisdiction.
- Control the access to and administration of porting subscribers.
- Provide the ported number downloads to the SMS.

The *LRN routing database* is where a ported called party number is translated to a location routing number at the call setup time. This database can be owned, shared, or leased. In the United States, it is owned by the dominant local service providers of the region.

The *local service management system* is responsible for downloading the ported subscriber numbers into the routing database and ensuring the ported subscriber numbers are synchronized with the copy in the NPAC.

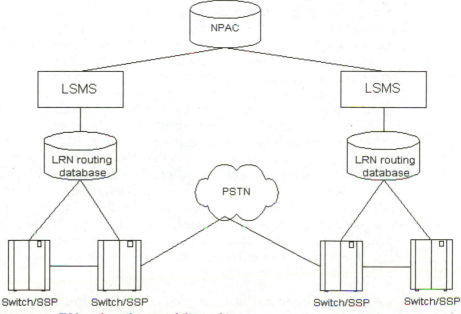

Figure 16.6   IN-based number portability architecture.

The *NP trigger in SSP* is a new trigger added to the existing call-processing software. Included are the corresponding trigger detection point, trigger detection criteria, and the trigger processing functions.

The provisioning process. A scenario of a subscriber switching service providers and the process of provisioning a ported number will help illustrate the IN-based number portability architecture described above. For more details, see Litcher (1997). It is assumed that before any porting subscriber can be provisioned, a number portability trigger has already been provisioned by adding the trigger into the trigger database and setting the trigger status to activated. Then,

1. The subscriber puts in a switch request to the new service provider and the new service provider notifies the old provider of the change.

2. After both service providers notify the NPAC of the change, the NPAC provisions a routing number, or location routing number, for the ported directory number.

3. The new service provider notifies the NPAC to activate the routing number.

4. The NPAC downloads to all service providers the record history file and logs any failure.

5. Each and every LSMS updates its routing database to include the translation from the ported directory number to the provisioned routing number.

6. The new service provider tests the service, and the old service provider deletes the directory number and the translation from its database.

Donor-switch triggered method. There are two different flavors of this IN solution, called donor-switch triggered and originating-switch triggered, for lack of a better term. IN-based solutions to the number portability problem are characterized by the use of a number portability trigger, or NP trigger. The location routing number (LRN) is the number returned as a result of a database query from the SCP routing database. There is an LRN for each ported phone number, and the translation from the dialed number to the LRN is done by the routing service management, similar to SCP.

The donor-switch triggered solution (see Fig. 16.7) is to have all calls routed to their respective donor switches, the destination switches using the dialed called party number, regardless of whether it is ported or not. It is the responsibility of a donor switch to determine that the dialed DN is ported and release the call back up to a point in the network where the routing service can be triggered. The point can be some switch between the originating switch and the donor switches or the originating switch itself. The steps for a ported number call using the donor-switch-originated trigger method are as follows:

1. A call to a ported number is originated and received at the originating SSP with the called party number 972-555-2222.

2. The SSP routes the call to the donor switch based on the called party number.

3. The donor switch fails to find the called number in its database and sends an ISUP release message back.

4. The NP trigger is invoked upon receiving the release message at the originating SSP. The call is triggered up to the SCP with a routing database for a lookup for the called party number. A location routing number for the recipient switch is returned.

5. The originating switch routes the call to the recipient switch using the LRN.

6. The recipient switch connects the call to the called party and the call setup is completed.

Originating switch triggered solution.   It is quite simple how this solution works: all calls are triggered as IN calls, resulting in a query to the routing database and return of a routing number. An example is illustrated in Fig. 16.8 and the steps are described as follows:

1. A call to a ported number is originated and received at the originating SSP with the called party number 972-555-2222.

2. The NP trigger is invoked and the call is triggered up to the SCP with an LRN routing database for a lookup for the called party number. The LRN for the called number is returned.

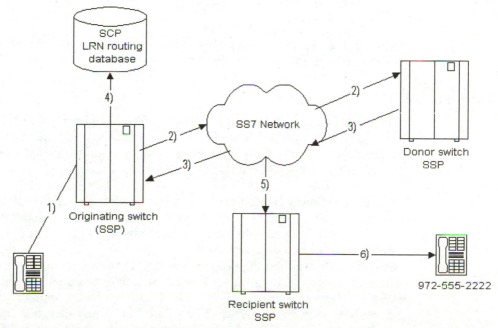

Figure 16.7 Donor-switch triggered solution.

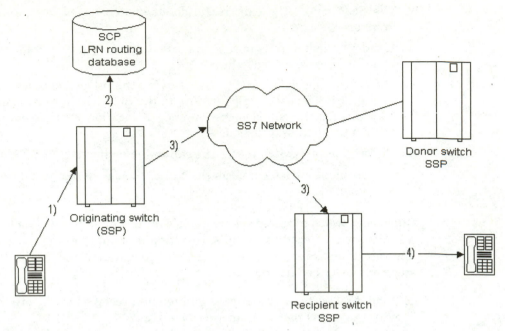

Figure 16.8 Originating-switch triggered solution.

3. The originating switch routes the call to the recipient switch using the LRN.

4. The recipient switch connects the call to the called party and the call setup is completed.

Among the advantages of this solution are the support of the service provider, as well as location and service portability.

Alternative methods

The alternative methods of implementing number portability that do not use IN triggers are needed because in some areas the IN infrastructures are unavailable. In some cases, it is mandated by regulation that both IN- and non-IN-based solutions be implemented.

Forwarding method. In this method, the originating switch routes the call to the donor switch, using the donor switch identity in the called party number. The donor switch routes the call to the recipient switch, using a database it maintains on all exported numbers.

One obvious disadvantage of this method is the extra burden on the donor switch: It has to get involved in setting up a call to a ported number that no longer exits on its switch. The tariff and cost implications make this method difficult to implement in certain cases. Another disadvantage is that the extra call setup step required may not be very desirable from both traffic and real-time performance perspectives.

Call dropback. With this method, the originating switch treats a call to a ported number as any other regular call, routing it to the donor switch based on the switch address in the called party number. The call setup fails because the directory number is no longer on this switch and a call rejection message is sent back (dropback) to the originating switch.

There are two variants of this method, mainly with respect to where the routing data are kept. With one version, the donor switch maintains all the exported numbers and their current serving switches, as in the case of forwarding method. The address of the current serving switch for this ported number is in the rejection message sent back to the originating switch. This enables the originating switch to reroute the call to the recipient switch.

In the other version of this method, it is up to the originating switch to determine and reroute the call to the recipient switch. Once a rejection message is received indicating that the message was rejected because the called party is a ported number, the originating switch performs a query to the database. Obviously this method requires that all switches maintain or have access to the most up-to-date ported number and routing database.

Impact on network and service management

The network infrastructure changes required to provide local number portability will not be discussed here; rather the impacts of implementing local number portability on billing systems and future numbering schemes, the two aspects closely associated with network and service management, will be discussed. For information on the former, see ATIS initial report.

Impact on billing systems. Many billing systems use the directory number to identify the location, switch, service provider, and rate center for a subscriber. With the introduction of number portability, the relationships between a DN and other pieces of information may no longer exist. Number portability requires that the structure of billing records be changed to take into consideration the following data.

- Location routing number
- Actual location of both called and calling parties
- Identity of those network elements that are involved in the number portability and processing, such as donor, originating, and recipient switches
- Number of database queries required to obtain the location routing number.

Local number portability also requires the modification of the process of generating billing records to accommodate

- Local number portability-related queries
- Calls entering a service provider's network, which is not the serving network, purely due to number portability
- Calls received at the recipient switch

Many regulatory issues with regard to billing local portability-related costs are still being discussed. The resolution of the billing issues may have the potential to tilt the balance between different types of service providers (e.g., incumbent service providers versus competitive local exchange carriers, local carriers versus long-distance carriers).

Impact on numbering plans and addressing scheme. The implications of number portability on future numbering plans and thus on customer services in general are enormous. Imagine that area codes become obsolete. The relationship between area code and location is decoupled, and the geographic numbering scheme is phased out. Then the tariff will become location-independent. The directory number allocation will be much more flexible than the current practice where a number administrator (e.g., NANPA) allocates blocks to each service provider and the provider allocates to its customer a number within the block.

With the flexibility, the digits and letter dialing become more available and a subscriber may be addressing by name instead of an abstract number. Alphanumeric name addressing and support of multiple addressing schemes are important steps for introducing multimedia services.

Number Pooling

With the explosion of telecommunications networks and services, directory numbers have become scarce resources. The increasing demand for telephone numbers fueled by the demand for second lines for both residential and business applications, plus the use of fax, modems, and wireless services, has resulted in the shortage of telephone numbers. This situation is compounded by the current scheme of assigning telephone numbers in North America and in many other areas in the world. The inefficient use of the available numbers creates the incentive for number pooling. It is the number portability technology that makes number pooling technically plausible.

The issue

In order to understand the need for number pooling, a basic understanding of how the number assignment scheme works is necessary. In North America, telephone numbers are assigned to a service provider based on a central office code, i.e., *NXX* of a 10-digit phone number. A whole block of a central office code (i.e., an *NXX* and 10,000 numbers with it, from 0000 to 9999) are assigned at one time to one service provider. There are many cases where a large quantity of numbers from a full block of central office codes remain unassigned, while a new central office code is requested and assigned. Figure 16.9 illustrates the problems.

An observing reader may have noticed the analogy between the problem that number pooling is addressing and that of computer memory coalescing: make full use of allocated blocks. The only difference is that the number assignment problem is much grander in scale and complicated in nature due to historical, political, and legal implications and causes of the problem. Note

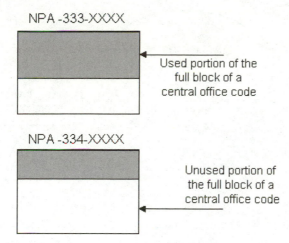

NPA -333-XXXX

Used portion of the
full block of a
central office code

NPA -334-XXXX

Unused portion of
the full block of a
central office code

Figure 16.9   Inefficiency created by the current telephone number assignment scheme.

that number pooling is intended to rectify the current number assignment scheme and relieve the shortage of central office codes by making full use of all available numbers. The total of available telephone numbers is finite such that it cannot be changed by any scheme.

With the implementation of number portability comes the possibility for number pooling. Number pooling can only be implemented in a location where local number portability has been implemented. Specifically, number pooling will only be implemented where an LRN-based LNP is available.

Principles of number pooling

Number pooling by definition means to pool all unused geographic numbers into a number reservoir to be shared by multiple service providers and to allocate to each service provider a block of numbers in finer granularity than the current block of 10,000 numbers. The numbers to be pooled can be either blocks of new, unassigned central office codes or the unused numbers in a block of an assigned office code. A generic number pooling model is depicted in Fig. 16.10, where the pooled numbers are in the industry inventory. Each participating service provider's number inventory both contributes and receives the pooled number to and from the industry number inventory.

Based on the granularity of number assignment, the number pooling comes in two categories: one is allocation by smaller number blocks, say, 100 or 1000, and one is allocation by individual telephone number, also referred to as block-based pooling and individual telephone number–based pooling, respectively. It is normally assumed that the numbers in a block are in sequential order for a block-based pooling.

In order to implement number pooling, the following basic components are required, as shown in Fig. 16.10. While an industry inventory administrator is

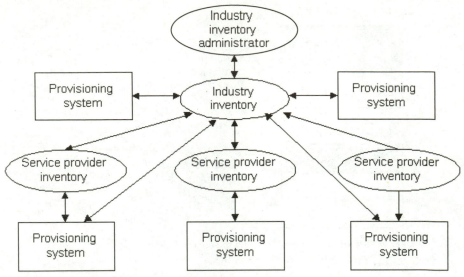

Figure 16.10 Administrative structure for number pooling.

responsible for the industrywide number inventory, each service provider's inventory system is responsible for administrating the portion of the shared numbering resource allocated to it.

A high-level process flow of number provisioning in block-based pooling is shown in Fig. 16.11. For more detail, see the ATIS initial report.

1. The industry inventory pool administrator creates the industry number inventory by pooling either unassigned new *NXX-X* telephone number blocks, unused telephone numbers of assigned *NXX*s, or both, based on forecast demand.

2. Each service provider that has contributed blocks of numbers reconfigures its systems to prevent the pooled number blocks from being allocated again.

3. When a service provider requests a new block of numbers from the industry inventory pool administrator, the administrator forwards a number block to the requesting service provider. It then notifies other service providers and the number administration authority, like the number portability administration center (NPAC), involved in the transaction.

4. The requesting service provider loads the received number blocks into its service number inventory system but cannot use the numbers until the following step is completed.

5. The requesting service provider sends LNP service requests for the received number blocks to the appropriate number portability administration center. It begins the assignment and allocation of the received numbers only after receiving confirmation and an activation message from the NPAC.

Summary

This chapter introduces numbering plans and discusses local number portability-related issues. The numbering plans provide a context and background for the discussion on the local number portability-related issues. First described is the public data network numbering plan, a topic sparsely discussed in existing literature. An understanding of the public data network numbering plan will become as important as the PSTN, and conventional PSTNs are fast converging.

The E.164 numbering plan is the most widely used numbering plan in the world for the public switched telephone networks. The plan consists of three parts: country code, national destination code, and subscriber number. While the assignment and allocation of the country code is controlled by the ITU-T, the numbering plan only specifies a guideline for the other two parts, leaving details for the numbering plan administrator of each country. The North American numbering plan defined by North American Numbering Council is a specific implementation of the latter two parts of the E.164 numbering plan.

Though number portability is discussed in general, the focus is on local number portability (LNP) that allows a subscriber to change the service provider without changing the original directory number. Two types of LNP solutions are discussed with an emphasis on the IN-based solutions. An IN-based LNP architecture adds a number portability administration center (NPAC) and LNP-specific SMS to the existing IN network to accomplish the number portability.

The number pooling, made possible by the IN-based LNP architecture, is intended to allow service providers to share the allocated but not used directory numbers among themselves so that the scarce number resources can be better utilized.

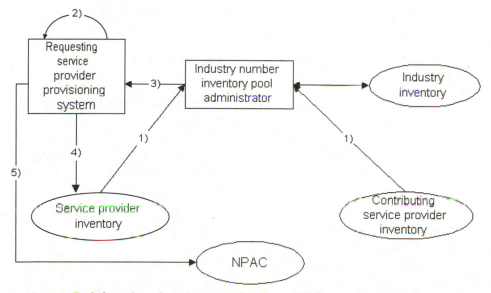

Figure 16.11 Pooled number administration, allocation, and assignment process.

Exercises

16.1 What is the structure of a data network identification code (DNIC)? To which of the two types of DNIC does the digit combination 1009 belong?

16.2 What is the destination country code (DCC)? What is the relationship between a DCC and DNIC?

16.3 Describe the two parts of an E.164 number.

16.4 Describe the two parts of a national significant number (NSN).

16.5 What does the first digit of China's country code 86 stand for?

16.6 What is the relationship between the E.164 numbering plan and the North American numbering plan?

16.7 Describe the three components of the North American numbering plan, i.e., NPA-*NXX-XXXX*.

16.8 Compare the structure of the public data network numbering plans with that of the PSTN numbering plans in terms of commonality and differences between the two. Discuss the general issues involved in converting from one structure to the other.

16.9 What are the geographic numbers and nongeographic numbers? Give an example for each category.

16.10 What are the main components of the IN-based LNP architecture? Describe the responsibilities of a Number Portability Administration Center.

16.11 Explain the concept of donor switch and recipient switch and their respective roles in the LNP architecture.

16.12 Explain what number pooling is about and why there is a need for number pooling.

16.13 Explain why the LNP makes number pooling plausible.

Introduction to ASN.1

Introduction

This appendix consists of three parts: an introduction, ASN.1 syntax, and encoding rules. ASN.1, an abstract syntax definition language developed by ITU-T (formerly CCITT) in the late 1980s, provides a common abstract syntax to represent application data to enable applications on different systems to communicate with each other. ASN.1 and the basic encoding rules (BER) are defined in ITU-T Recommendations X.208 (1998) and X.209 (1998), respectively.

First of all, a solid understanding of what ASN.1 is used for is essential for understanding the details of ASN.1. At a conceptual level, a network can be viewed as consisting of two components. The data transfer component consists of the session layer and all the layers below it in the OSI network reference model and Transmission Control Protocol (TCP) or UTP layer and all the layers below it in the Internet network architecture. The application component uses the data transfer component and is concerned with the application. In the OSI network reference model, it consists of the application layer. In the TCP/IP model, the application component consists of the applica-tion/process layer that has application layer protocols such as the Simple Network Management Protocol (SNMP), FTP SMTP, and TELNET.

One application component exchanges management information with another application component on a different system using a data transfer component. The data transfer component deals only with a binary stream. The application component deals with application data, which is concerned with data semantics and data value. For two application components to understand and communicate with each other, they must share a common structure of the application data, or the application data will be a stream of meaningless characters to the receiving party, even if both parties use the same data transfer component and every single character is transferred and translated correctly. A simple example will help illustrate the point. Application *A*, as shown in Fig. A.1, sends a piece of information to application *B*

"call-drop-rate 10"

"alarm-severity 1"

to mean that the call drop rate at the sending end is 10 percent and the alarm severity level is of a major class represented as 1. Without a common structure, there is nothing stopping the receiving end from interpreting the received data any way it pleases other than the intended way.

This is precisely what ASN.1 provides, an abstract syntax, more commonly known as a set of types, to structure application data so two communicating parties can understand each other.

In summary, from the perspective of network management, ASN.1 is used to

- Define the abstract syntax of application data

- Define management protocol data units (PDUs) such as CMIS's M-GET and M-SET

- Define a management information base (MIB) for both the SNMP and OSI CMIP

ASN.1 is widely used not only in ISO- and TCP/IP-based network management applications, but also in other types of network applications, such as transfer of signaling messages in the Signaling System No. 7 (SS7) network.

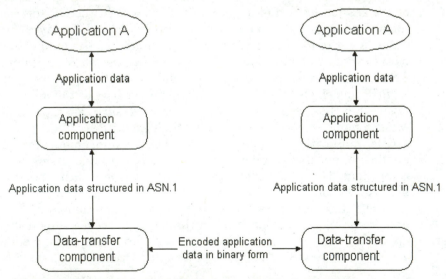

Figure A.1   Illustration of using ASN.1 by two communicating applications.

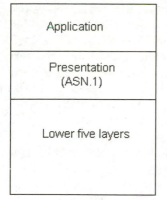

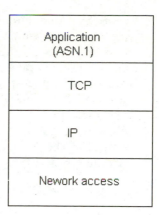

Figure A.2   ASN.1 in OSI and TCP/IP network models.

Figure A.2 shows where ASN.1 is used in the OSI network reference model and the TCP/IP-based Internet network model, respectively.

ASN.1 Types—Abstract Syntax

This section introduces the abstract syntax, mainly the ASN.1 types. The ASN.1 types can be classified into the following categories: simple types, structured types, tagged types, subtypes, and ASN.1 macros. But the ASN.1 module definition and syntactic conventions are introduced first.

- *Simple types.*   A simple type is a basic type that cannot be further decomposed into component types. Examples include INTEGER, REAL, and BOOLEAN.

- *Structured types.*   A type that is made up of component types that can be either of simple or constructed type or both.

- *Tagged types.*   A tagged type is derived from other types with a tag.

- *Subtypes.*   A type that is derived from an existing type and can take on a subset of the values of the source type.

- *ASN.1 macro.*   A mechanism that allows one to extend the established ASN.1 syntax to accommodate the need of an application.

ASN.1 module definition

All ASN.1 definitions for a particular application are collected in an ASN.1 module that is very much like a programming module consisting of a set of programming statements with the following form.

```
ModuleIdentifier DEFINITION TagDefault ::=
        BEGIN
        module body
        END
```

where the ModuleIdentifer is a user-defined identifier, like a program identifier, that distinguishes this module from other ASN.1 modules. TagDefault takes one of two values IMPLICIT TAG or EXPLICIT TAG. IMPLICIT forces all user-defined tags to be implicitly encoded unless they are overwritten by an explicit tag inside the module body. If no TagDefault is present, the default value is EXPLICIT. The concept of tag will be discussed in detail in the tagged type subsection. The ASN.1 definitions of a user application reside in the module body, enclosed by BEGIN and END.

Most of the ASN.1 definitions are type assignments defining application-specific types. A type assignment has the following general form.

```
typeReference ::= TYPE
```

where typeReference is a new type that has just been defined and TYPE is one or more existing types, either built-in, user defined, or both. The following syntactic conventions are followed in type assignments.

- Layout is not significant; the blank spaces and empty lines only serve as separators between identifiers.
- The ASN.1 character set that can form an ASN.1 identifier consists of both capital and lowercase alphabetic characters, digits 0 through 9, and the following special characters: : = , { } < . ( ) [ ] - ' and ".
- A new type can be recursively defined.
- ASN.1 reserved words cannot be used as user-defined identifiers.
- All ASN.1 built-in types are in capital letters, e.g., INTEGER, REAL, or BOOLEAN.
- The last character of an identifier cannot be a hyphen.
- Comments start with a pair of hyphens (--) and end with either a pair of hyphens or the end of the line.
- A type or module name begins with an uppercase letter.
- A user-defined identifier begins with a lowercase letter.

Simple types

A simple type is a built-in type that is atomic in the sense that it cannot be decomposed further into other types and it can be used to build other more complex types. For the convenience of description, the ASN.1 simple types can be categorized into several subgroups: basic types, object types, character string types, and miscellaneous types.

Basic types

- *BOOLEAN.*   A set of two values, TRUE or FALSE.
- *INTEGER.*   The set of positive and negative integers, including 0. An integer can be assigned a name to indicate a specific meaning.

- *OCTET STRING.*   An ordered set of 0 or more bits in increments of 8 which can be defined as a string of either binary or hexadecimal digits.

- *BIT STRING.*   An ordered set of 0 or more bits. Individual bits can be assigned names to indicate specific meanings and the actual value of a BIT STRING can be a string of either binary or hexadecimal digits. The number of bits in a BIT STRING does not have to be an increment of 8.

- *REAL.*   Numbers expressed in scientific notation of mantissa, base, and exponent; that is, $M \times B^E$.

The ENUMERATED type consists of an explicitly enumerated list of integers and an associated name for each integer. The ENUMERATED type can be viewed as the INTEGER type with all the values explicitly named. One restriction that applies to the ENUMERATED type is that although its values are integers, arithmetic operations cannot be applied on enumerated values.

Object types.   The second subgroup of the simple types is object types that are used to name and describe information objects such as standards documents and managed object classes. An OBJECT IDENTIFIER is a unique identifier for an information object. An OBJECT DESCRIPTOR provides a plain textual description of an information object.

Character string types.   The subgroup of the simple types is a group of character string types as shown below.

- *NumericString.*   A string of digits 0 through 9.

- *PrintableString.*   A string consisting of uppercase and lowercase letters, digits, punctuation marks, and spaces.

- *IA5String.*   It stands for the International Alphabet Number 5, the international standards body's equivalent of ASCII.

- *GraphicString.*   A character set defined by ISO.

- *GeneralString.*   A general character string.

- *TelexString.*   A character set defined by ITU-T Recommendation T60.

- *VideotexString.*   A set of alphabetic and graphic characters defined by ITU-T Recommendations T.100.

- *VisibleString.*   A character set defined by ISO (equivalent to ASCII).

Miscellaneous types.   Finally, there are some miscellaneous types as well. They include NULL, EXTERNAL, UTCTime, and GeneralizedTime. The NULL type is used as a placeholder where a value may or may not be present. The EXTERNAL type is another placeholder where the values are not specified in the ASN.1 module; they are specified in some other documents such as a different standards document. UTCTime and GeneralizedTime are two types that use different formats to specify local and global time.

Now we have examples to illustrate the simple types just described.

```
EventSent ::= BOOLEAN
FileSize ::= INTEGER
BestColors ::= INTEGER{red(0), gold(1), green(2)}
FunctionalUnits ::= BIT STRING
SunnyDaysOfTheWeek ::= BIT STRING {
        Sunday(0), Monday(1), Tuesday(2), Wednesday(3),
        Thursday(4), Friday(5), Saturday(6) }
FileName ::= OCTET STRING (SIZE(1..256))
Record ::= OCTET STRING (SIZE(128))
```

Structured types

A structured type is defined in terms of other ASN.1 types, and the component types may be built-in primitive types or user-defined types. We describe six structured types of ASN.1.

- SEQUENCE
- SEQUENCE OF
- SET
- SET OF
- CHOICE
- ANY

SEQUENCE and SEQUENCE OF. The SEQUENCE type is used to define an ordered list that consists of values of one or more other ASN.1 types. It is analogous to the record structure found in many programming languages such as C. The syntax of the SEQUENCE type consists of an ordered list of elements, each specifying a type and, optionally, a name. As shown in the following example, three elements are in the SEQUENCE type PersonRecord, and they are of the type OCTET STRING, INTEGER, and BOOLEAN and have the element names lastName, age, and single, respectively. Note that each element name begins with a lowercase letter and the order of the elements is fixed. This makes it possible that element names are not transmitted from one end to the other because of the fixed order of the elements.

An example of the SEQUENCE type.

```
PersonRecord ::= SEQUENCE {
    lastName    OCTET STRING,
    age     INTEGER,
    single BOOLEAN }
```

The SEQUENCE OF type is the same as the SEQUENCE type with the one exception that the element types must be the same. For example,

```
errorCount ::= SEQUENCE OF {
    numberOfMajorError INTEGER,
    numberOfMinorError INTEGER,
    numberOfNewError   INTEGER
}
```

SET and SET OF. A SET is similar to a SEQUENCE except that the order of the list is not fixed; the elements may be in any order when they are encoded into a particular representation. That is, a SET consists of elements that can be of different types, and the order of the elements within a SET is not significant. For example, as shown below, when a Connect-pdu is transferred over, the first element may be dest-id instead of source-id. The sequence numbers in brackets are of one type of tag and are used to uniquely identify the elements in a SET and will be explained shortly.

```
Station-Id ::= INTEGER(0|1|2|3|4|5|6|7)
Connect-pdu ::= SET {
source_id       [0] Station_Id,
dest_id         [1] Station_Id,
path      OCTET STRING }
```

A SET OF is the same as a SET with the one exception that the elements of a SET OF can be of only one type.

CHOICE. The CHOICE type is used to select one of the alternative elements from a collection. A CHOICE type consists of a list of alternative known types out of which only one can be assigned a value at any given time. It is very similar to the union structure found in C or C++ programming language. For example, in the example shown below, an EntityId can assume either an oid or code, but not both. Note that no tag is needed for a CHOICE type, because when a particular value is assigned to a CHOICE type, the type of the value is also assigned.

An example of the CHOICE type is

```
EntityId ::= CHOICE {
        oid  [0] IMPLICIT OBJECT IDENTIFIER,
        code [1] IMPLICIT INTEGER }
```

ANY. The type ANY is used to model variables that can be one of a variety of types but where the exact type is not known until the run time. An example of the ANY type is

```
MessageContents ::= ANY
```

Tagged types

All data types in ASN.1 have an associated tag. Tags are used to eliminate ambiguity of variable types within a structured type. For example, when the data of the user-defined type ObjectName are transmitted, the receiving end must have a way of knowing whether it is a GeneralString or OBJECT IDENTIFIER. A unique tag associated with a type will remove any ambiguity.

There are two categories of tags: EXPLICIT and IMPLICIT. An IMPLICIT tag does not need to be transferred along with data to the other end while an

EXPLICIT tag does. All tags are EXPLICIT by default unless specified as IMPLICIT. That is, all types without a tag have an EXPLICIT tag. If an ASN.1 module is specified as IMPLICIT, then all types in the module have an IMPLICIT tag. The general tag syntax is shown here.

```
Tag syntax:
TaggedType ::=
    Tag type |
    Tag IMPLICIT Type |
    Tag EXPLICIT Type
Tag ::= [Class ClassNumber]
ClassNumber ::=
    number |
    DefinedValue
Class ::=
    UNIVERSAL |
    APPLICATION |
    PRIVATE |
    context-specific
```

A user-defined tag consists of a class and a number within the square brackets. There are four classes of user-defined tags: UNIVERSAL, APPLICATION, PRIVATE, and context-specific.

UNIVERSAL tag.   The UNIVERSAL tag is used for built-in types only, and the assigned value for a built-in type is universally used, as specified in CCITT Recommendation X.208 (1988). An application can only use the UNIVERSAL tags and is not allowed to reassign the tag value. The built-in and assigned UNIVERSAL tag values are shown in Table A.1.

APPLICATION tag.   An APPLICATION tag is used to uniquely identify a type within an ASN.1 module and thus is significant only within the module. For the example shown below, the type Address, Age, and Sex each have an APPLICATION tag with assigned tag values 1, 2, and 3, respectively. Obviously each APPLICATION tag value has to be unique within the module.

PRIVATE tag.   The PRIVATE tag can be used to uniquely identify data types within an organization. In the following example, the type Name with a PRIVATE tag can be used by other ASN.1 modules within the same organization. This class of tag is not used as often as the other three are.

Context-specific tag.   The context-specific tag is used to distinguish component types within a SET, SEQUENCE, CHOICE, or other structured type. For example, for a type of CHOICE, if the elements are not specifically mentioned, it is difficult for the receiving end to know which element is sent. The context-specific tag will remove the ambiguity as shown in the following example. Four context-specific tags are defined for the elements within the PersonnelRecord SET, i.e., [0], [1], [2], and [3]. A context-specific tag value needs to be unique only within the structured type where it is defined.

Here is an ASN.1 module to illustrate the four tag classes:

TABLE A.1   UNIVERSAL Class Tag Assignment

Tag	Type
UNIVERSAL 1	BOOLEAN
UNIVERSAL 2	INTEGER
UNIVERSAL 3	BIT STRING
UNIVERSAL 4	OCTET STRING
UNIVERSAL 5	NULL
UNIVERSAL 6	OBJECT IDENTIFIER
UNIVERSAL 7	Object descriptor
UNIVERSAL 8	EXTERNAL
UNIVERSAL 9	REAL
UNIVERSAL 10	ENUMERATED
UNIVERSAL 11–15	Reserved
UNIVERSAL 16	SEQUENCE and SEQUENCE OF
UNIVERSAL 17	SET and SET OF
UNIVERSAL 18	NumericString
UNIVERSAL 19	PrintableString
UNIVERSAL 20	TelexString
UNIVERSAL 21	VideotexString
UNIVERSAL 22	IA5String
UNIVERSAL 23–24	Time
UNIVERSAL 25	GraphicString
UNIVERSAL 26	VisibleString
UNIVERSAL 27	GeneralString
UNIVERSAL 28+	Reserved for addenda of ISO 8824

```
TagModule DEFINITION IMPLICIT TAGS ::=
BEGIN
   ADDRESS ::= [APPLICATION 1] OCTET STRING
       Age ::= [APPLICATION 2] OCTET STRING
       Sex ::= [APPLICATION 3] EXPLICIT INTEGER
       Name ::= [PRIVATE 1] IMPLICIT SEQUENCE {
          firstName        PrintableString,
          initial          PrintableString,
          surname          PrintableString
     }
PersonnelRecord ::= [APPLICATION 0] IMPLICIT SET {
          name             [0] PrintableString,
          dateOfHire       [1] Date,
          nameOfSpouse     [2] PrintableString,
          children         [3] IMPLICIT SEQUENCE OF
                               Name
}
END
```

Subtypes

A subtype is derived from an existing type by assigning to it a subset of values of the existing type. The existing type is also called a *parent type*. The subtyping can be nested, meaning that a subtype itself can be a parent type from which a subtype is derived. Subtyping can be very useful in certain situations. For example, if two or more types have characteristics in common, and the common portion is frequently used, a subtype can be created to extract the common characteristics. In some cases, it is convenient for an application to limit the size of an existing type and create a subtype to hold the desired range of values. The six different forms of subtypes are explained next.

Single-valued subtype.   A single-valued subtype is an explicit listing of all the values the subtype may take on. For example,

```
Week ::= ENUMERATED { Monday(1), Tuesday(2), Wednesday(3), Thursday(4),
Friday (5), Saturday (6), Sunday (7) }
Weekdays ::= Week(Monday, Tuesday, Wednesday, Thursday, Friday)
Weekend :: = Week (Saturday,Sunday)
```

Contained subtype.   A contained subtype is formed by including all the values of the subtypes that it contains. For example,

```
LongWeekend := Week (INCLUDE Weekend ¦ Monday)
```

where the contained subtype LongWeekend is formed by including the existing subtype Weekend and a value of the subtype Week, Monday.

Value-ranged subtype.   This subtype applies only to INTEGER and REAL types. It is formed by specifying a specific range of the values with a starting point and an endpoint. The endpoint is either closed or open. The first example shown below has a closed endpoint while the second example has an open endpoint. For example,

```
ASCIIValue ::= INTEGER (0..127)
PostiveInteger ::= INTEGER (1..PLUS-INFINITY)
```

Permitted alphabet subtype.   This subtype may be applied only to character string types. A permitted alphabet subtype consists of all its possible strings that are formed using a subset of the parent type. For example,

```
DigitString ::= IA5String (FROM ("0" ¦ "1"¦ "2" ¦ "3" ¦ "4" ¦ "5" ¦ "6" ¦
"7" ¦ "8" ¦ "9" ))
OddSingleDigit ::= IA5String (FROM ("1" ¦ " 3" ¦ "5" ¦ "7" ¦ "9" ))
```

Size constraint subtype.   This subtype is formed by applying a size constraint to the number of elements in a parent type. This subtype can be applied only to the type OCTET STRING, BIT STRING, SEQUENCE OF, SET OF, and character strings.

```
AreaCodeDigit ::= DigitString (SIZE (3))
SerialNumber ::= OctetString (SIZE (10))
```

In the above examples, AreaCodeDigit is a subtype of DigitString, which in turn is a subtype of IA5String, while SerialNumber is a subtype of OCTET STRING with the size restricted to 8 octets. The element that is constrained depends on the parent type. Table A.2 lists the element types that are constrained for all possible parent types.

Inner subtyping. The inner subtyping constrains the structured types by applying constraint conditions to the parent type, using the key words WITH COMPONENTS, OPTIONAL, PRESENT, and ABSENT. The constraint conditions test absence, presence, and/or values of the components of the parent type.

```
Statement ::= SEQUENCE OF GeneralString
CodeBlock ::= Statement (SIZE(1..10) ¦ WITH COMPONENT (SIZE(1..80)))
```

In the above example, the CodeBlock consists of up to 10 statements and each statement can be up to 80 characters long.

ASN.1 Macros

The ASN.1 macros extend ASN.1 syntax by allowing one to define a family of types and to assign values to the instances of the new types. In some aspects, it is similar to C++'s template classes out of which you can define specific classes. On the one hand, the ASN.1 macro provides flexibility, and on the other hand, it puts a burden on the ASN.1 compilers. A newly defined type can have a syntax that is different from the established one, and for two communicating entities to understand each other, the compilers at both ends must be extended as well to accommodate the new syntax. From this perspective, the ASN.1 macro should be used judiciously.

The ASN.1 macro has the general form as shown below that consists of TYPE NOTATION and VALUE NOTATION. The TYPE NOTATION is for writing ASN.1 syntax and the VALUE NOTATION is for assigning values to the created syntax.

TABLE A.2   Constrained Elements

Type	Constrained element
BIT STRING	Bit
OCTET STRING	Octet
character string	Character
SEQUENCE OF	Component value
SET OF	Component value

```
<macro-name> MACRO ::=
BEGIN
   TYPE NOTATION ::= <new-type-syntax>
   VALUE NOTATION ::= <new-value-syntax>
      <supporting-productions>
END
```

The ASN.1 macro example below is borrowed from RFC 1155, which defines both type syntax and value syntax for managed objects of the SNMP MIB. For instance, the syntax specifies that an SNMP managed object consists of six components, i.e., Access, Status, DecrPart, ReferPart, IndexPart, and DefValPart.

```
OBJECT-TYPE MACRO ::=
BEGIN
   TYPE NOTATION ::=
      "SYNTAX" type (TYPE ObjectSyntax)
         "ACCESS" Access
         "STATUS" Status
         DescrPart
         ReferPart
         IndexPart
         DefValPart
   VALUE NOTATION ::= value (VALUE ObjectName)
   Access ::= "read-only" | "read-write" | "write-only" |
                     "not-accessible"
   Status ::= "mandatory" | "optional" | "obsolete" | "deprecated"
   DescrPart ::=
"DESCRIPTION" value (reference DisplayString)
   ReferPart ::=
      "REFERENCE" value(reference DisplayString) | empty
IndexPart ::= "INDEX" "{" IndexTypes "}"
IndexTypes ::= IndexType | IndexTypes "" IndexType
IndexType ::= value (indexObject ObjectName) |
type(indexType)
DefValPart ::=
"DEFVAL" "{" value(defValue ObjectSyntax) "}" | empty
   DisplayString ::= OCTET STRING SIZE (0..255)
END
```

Basic Encoding Rules (BER)

The basic encoding rules (BER) provide a method for encoding values of each ASN.1 type into a string of octets. The structure of the BER has three parts: identifier, length, and contents (ILC). Table A.3 shows the BER simple encoding structure while Fig. A.3 depicts an encoding structure that is recursive, that is, the content part of the structure itself can be of the form identifier-length-content.

Encoding identifier

The general structure of the identifier field that is shown in Table A.4 consists of the ASN.1 tag class, an indication of whether the encoding is primitive or constructed (P/C), and the tag number. Bits 8 and 7 indicate one of the four

TABLE A.3   Simple Encoding Structure of the BER

Identifier	Length	Content

Identifier	Length	Identifier	Length	Content

Figure A.3   Complex encoding structure of the BER.

ASN.1 tag classes, as listed in Table A.5. Bit 6 indicates whether the encoding is primitive or constructed, as will be explained shortly. Bits 5 to 1 indicate the ASN.1 tag number. If the last five bits are all 1s, it means that the tag number is greater than 31 (the maximum number five bits can represent) and the next octet contains the tag number. Tables A.6 and A.7 show examples of tag numbers 26 and 101, respectively.

The P/C bit is for either the primitive or constructed encoding. The primitive encoding refers to simple atomic data (e.g., built-in types), while the constructed encoding refers to the data of a structured type.

Encoding length

The length field specifies the length of the content field in number of octets. The length field can be one of two types: definite or indefinite. The definite form, in turn, can be either short or long, as explained below.

- *Definite short form.*   The length of the content field is less than 128, and the length field consists of a single octet beginning with a 0.

- *Definite long form.*   The length of the content field is greater than 127, and the first octet contains a 7-bit integer indicating the number of the additional length octets. It is the additional octet that actually contains the length of the content field.

- *Indefinite form.*   The length of the content field is unknown, and bit 8 is set to 1, and the rest of the bits 7 to 1 are set to 0. The last octet of the data field has the hexadecimal value "00 00X to signal the end of the data content field.

Table A.8 shows an example of an identifier field (universal tag class, primitive tag=4) and a length field of the short form with length=11.

Encoding content

The rules for encoding data content are briefly summarized below, and for details, the reader is referred to ITU-T Recommendation X.219.

TABLE A.4   The Structure of the Identifier Field

8 7	6	5 4 3 2 1
Class	P/C	Tag number

TABLE A.5   Class Bits in the Identifier Field

Class	Bit 8	Bit 7
Universal	0	0
Application	0	1
Context-specific	1	0
Private	1	1

TABLE A.6   A Sample Identifier Field With the Tag Number 26

Class	P/C	1	1	0	1	0

TABLE A.7   A Sample Identifier Field with Tag Number 101

Class	P/C	1 1 1 1 1	0 1 1 1 0 0 1 0 1

TABLE A.8   An Example of an Identifier Field and a Length Field

0 0 0 0 0 1 0 0	0 0 0 0 1 0 1 1
Identifier field	Length field

- *BOOLEAN.*   The content consists of a single octet with a content of 0 for FALSE and a nonzero value for TRUE.

- *INTEGER.*   The content consists of one or more octets using a two's complement representation with the following restrictions:
  - For the first octet, all bits shall not be 1s
  - For the second octet, bit 8 shall not be 0.

- *ENUMERATED.*   Same as the encoding of the integer value with which it is associated.

- *REAL.*   If the value is 0, there are no content octets. Otherwise a real number is encoded as follows. The first octet indicates the type of encoding: binary, decimal, or special value encoding.

- When binary is used and if the mantissa is nonzero, the mantissa is determined by

$$M = S \times N \times 2^F \qquad \text{with } 0 \le F < 4 \text{ and } S = +1 \text{ or } -1$$

- where S is the sign, F is the binary scaling factor, and N is the nonnegative value.
- If the base is 10, the character-encoding schemes are used for the remaining content octets. If the base is 2, 8, or 16, the first content octet specifies the length of the exponent, sign of the mantissa, and the scaling factor to align the implied decimal point of the mantissa with the octet boundary. Following the first content octet is zero or one additional octet to specify the exponent length, followed by octets for the exponent and octets for the mantissa.

- *BIT STRING.*   For primitive encoding, the content consists of an initial octet followed by zero or more subsequent octets. The first octet indicates the number of used octets in the final octet, in the range of 0 to 7. The actual bit string starts with the second octet and continues through as many octets as needed. For constructed types, a BIT STRING is decomposed into substrings; each substring except the last must be an increment of 8 bits in length.

- *OCTET STRING.*   For primitive encoding, the content octets are encoded the same way as the value of the bit string. For constructed encoding, the octet string is decomposed into substrings; each substring may be encoded as primitive or constructed, but normally the primitive encoding is used. No significance is associated with the boundary between the data values encoded in the content octets.

- *NULL.*   There are no content octets.

- *SEQUENCE.*   The content octets are the concatenation of the encodings of the values of the elements of the sequence in their originally defined order. If the value of an element with the OPTIONAL or DEFAULT qualifier is not present from the sequence, no encoding is included for that element. If the value of an element with the DEFAULT qualifier is the value, then encoding of that element may or may not be included, depending on the implementation.

- *SEQUENCE OF.*   The content octets are the concatenation of the encodings of the values of the occurrences in the collection, and the order of the elements shall be the same as that of the data values in the SEQUENCE OF type. This encoding is constructed.

- *SET.*   The content octets are the concatenation of the encodings of the values of the elements of the sequence, in any order. If the value of an element with the OPTIONAL or DEFAULT qualifier is absent from the sequence, no encoding is included for that element. This encoding is constructed.

- *SET OF.*   The content octets are the concatenation of the encodings of the values of the occurrences in the collection, in the order chosen by the sender.

- *CHOICE.*   The encoding of the CHOICE value is the encoding of the selected alternative.

- *Tag.*   For the IMPLICIT tag, the encoding of the data value replaces the underlying tag value. The encoding of the length and content octets is the same as for the underlying value. For the EXPLICIT tag, the data octets are the same encoding as the underlying value.

- *ANY.*   The encoding of the ANY type shall be the complete encoding of the ASN.1 type that is substituted in.

- *OBJECT IDENTIFIER.*   This is a list of encodings of subidentifiers. The first two components are combined using the formula $(X*40) + Y$ to form the first subidentifier where X and Y are the values of the first and second object identifier components, respectively. Each subsequent component forms the next subidentifier. A nonnegative integer represented by the first seven bits of each octet is used to represent a subidentifier, and the last bit is for indicating whether this is the last identifier.

- *Character string.*   Each of the eight character string types is encoded as if it had been declared [UNIVERSAL x] IMPLICIT OCTET STRING where x is a universal tag for the character string type.

An Overview of the TCP/IP Protocol Suite

Internet Introduction

This appendix consists of four sections, i.e., an introduction and three other sections that provide an overview of protocols of the top three Internet layers. This introduction first gives a quick tour of Internet history and then an overview of the Internet architecture layers and the protocol suite. The remainder of this appendix will discuss the protocols at each of the layers.

Brief history

Transport Control Protocol (TCP) and Internet Protocol (IP) are two network protocols that are used together to refer to the whole Internet protocol suite and often the networks that use the protocol suite. The Internet protocol suite has its origin in a Department of Defence (DOD) research project called Arpanet in 1969. The goal of the project was to test computer-based communication for the purpose of joint research and collaboration between geographically dispersed locations, using computers of diverse platforms. The initial network included the computers in four university campuses (Stanford Research Institute, University of Utah, University of California at Santa Barbara, and University of California at Los Angeles). Twenty some years later, it covers virtually the whole world.

Compared to telecommunications networks, computers are a relatively new invention. It was not until the 1970s that computers were networked to talk to each other. Now computer networks, most prominently the Internet, have become ubiquitous and indispensable in daily life. The following list outlines the major events in Internet history.

- *1962.* Paul Barran at the Rand Corporation developed the idea of packet switching in which communications data are broken into small units called packets and then routed to their destinations individually, where they are assembled.

- *1969.*   Arpanet is formed, linking University of California at Los Angeles, Stanford Research Institute, the University of California at Santa Barbara, and the University of Utah. Under the sponsorship of the U.S. Defense Department's Advanced Research Project Agency, the network was originally intended to allow scientists to share research-related information over a network.

- *1974.*   The TCP/IP protocol suite was published by Vinton Cerf and Robert Kahn to allow packet networks to be linked to form a bigger network.

- *1979–1981.*   Various research and university networks began to form. Examples include Bitnet (for "Because It's Time") and CSNet.

- *1986.*   The National Science Foundation (NSF) launches NSFNet, linking five U.S. supercomputing centers.

- *1987.*   The NSF assumes management responsibility for an internet backbone.

- *1988.*   A computer virus called *worm* was introduced into the worldwide network and brought over 6000 host computers to a halt.

- *1989.*   Arpanet was formally shut down, giving rise to the Internet.

- *1991.*   The NSF lifted restrictions on commercial use of the Internet.

- *1992.*   The worldwide web was made public by researchers at CERN (European Laboratory for Particle Physics).

- *1993.*   The mosaic web browser was first released by the National Center for Supercomputing Applications at the University of Illinois.

- *1995.*   The Internet backbone goes commercial.

Internet components

The Internet is made up of network components of diverse types, and only the major ones are briefly introduced.

Backbone.   The part of the communications network that carries the heaviest traffic. The backbone is also that part of a network that joins local area networks (LANs) together. LANs are connected to the backbone via bridges or routers. A rough analogy is that the backbone is like an interstate highway and LANs are like local and state highways.

Host.   A computer with full two-way access to other computers on the Internet.

File server.   It typically is a combination of computer, data management software, and a large hard disk. A file server allows the user to store data (e.g., application data, e-mail messages) and maintain the application software shared among many users on a LAN.

Router.   A device that interfaces or interconnects two networks, usually two LANs. Routers have the capability to route the traffic and balance the load. It is often used in an enterprisewide network. Routers operate at the network layer of the OSI reference model and have more functionality than bridges.

Bridge.   A data communication device that interconnects two or more networks and forwards data packets between the networks. Bridges operate at the data-link layer of the OSI model.

Internet layered architecture

The Internet architecture, like that of the OSI network model, is layered. It has four layers: network access/local network, internet, host-to-host, and process/application. A comparison between the Internet layers and the OSI layers is shown in Fig. B.1.

The network access/local network layer provides the physical media and interface between the host computer and the LAN. The access protocols at this layer are between a communication node and an attached host. The TCP/IP suite does not provide any Internet-specific protocols at this layer. Instead, the protocols of the appropriate LAN such as Ethernet, token ring, and X.25 are used.

The internet layer provides the routing function that routes data packets through multiple networks between the originating and terminating hosts. The internet protocol, the protocol at the internet layer, is implemented within hosts and routers. A router connects two networks and relays data between them. The internet layer plays a role similar to that of the network layer of the OSI model.

The host-to-host layer is similar to the session layer of OSI architecture responsible for managing the host-to-host connections for assuring that data exchanged between hosts are reliably delivered. It is also responsible for delivering incoming data to the appropriate application.

Process/ application layer	Application
	Presentation
Host-to-host	Session
	Transport
Internet	Network
Network access or local network layer	Data link
	Physical link

Figure B.1   Internet layers in comparison with OSI layers.

At the top of the layered architecture is the process/application layer that contains protocols for user applications. A set of application level protocols has been defined for specific applications such as file transfer and network management. Three application level protocols are included in the TCP/IP protocol suite: SMTP, FTP, and TELNET.

Internet protocols

Figure B.2 lists the protocols of each Internet layer, and Fig. B.3 shows the dependencies between the protocols. Though some of the protocols are not part of the official TCP/IP protocol suite, they are introduced because of either their relationship to the network management protocol or their wide applications. The remainder of this appendix will focus on the protocols of the top three layers..

Application Layer Protocols

Only a brief description of several widely used process/application layer protocols is provided.

Simple Network Management Protocol (SNMP)

Initially designed for the management of TCP/IP-based network stack, SNMP is a widely used management protocol for the TCP/IP-based networks. Chapters 2 and 3 provide a detailed discussion of this protocol.

File Transfer Protocol (FTP)

This protocol enables a user at a terminal to access and interact with a remote file system. A client FTP at the file server can support multiple users concurrently and allows a user to specify the structure of the file involved and the type of the data in the file. A file can be structured or unstructured. An unstructured file can contain any type of data, binary or text, and it is transferred between the two FTP entities as a transparent bit stream. A structured file consists of a sequence of fixed-size records of a defined type.

Simple Mail Transfer Protocol (SMTP)

Also referred to as e-mail, SMTP manages the transfer of mail from one host computer mail system to another. It is not responsible for accepting mail from local users or for distributing received mail to its intended recipient. These are the responsibilities of the local mail system.

TELNET

TELNET enables a user at a terminal on one machine to communicate interactively with an application (e.g., text editor) running on a remote machine, as if the user terminal were directly connected to it. A client TELNET protocol is accessed through a local operating system by a user at a terminal. It enables

Process layer	FTP, SMPS, Telnet SNMP
Host-to-host	TCP/UDP
Internet	IP/ICMP
Network Access	Network access protocols (Token ring, Ethernet, etc.)

Figure B.2 The Internet layers and the corresponding protocols.

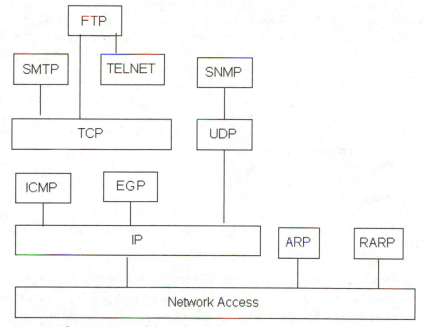

Figure B.3 Internet protocol dependencies.

a user to log on to the operating system of a remote machine and to initiate the execution of an application program on the machine.

Host-to-Host Layer Protocols: TCP and UDP

Two protocols of the host-to-host layer are the Transmission Control Protocol (TCP) and the User Datagram Protocol (UDP).

TCP

The TCP is a connection-oriented, reliable packet-switching protocol that provides support for communication between processes in host computers. First, the TCP is connection-oriented. The TCP provides for a set of application level addresses called ports within each host. Multiple processes can be associated with a port. The combination of the IP address and a port number identifies a socket, and a pair of sockets identifies a connection. A connection between two processes must be established first before they can communicate. Such a connection is also referred to as a *virtual circuit*. According to the TCP, a communication proceeds through three phases: connection establishment, data transfer, and connection take-down.

The TCP is reliable and guarantees the delivery of data. In a network, data segments may be damaged, lost, corrupted, or become out of the original order. The TCP provides a mechanism for acknowledgment and retransmission. The mechanism works as follows. A sequence number is generated for the first data octet of each data segment. Once the data segment is sent along with the sequence number, the sender starts a timer. The receiving end is expected to send an acknowledgment number that refers to the sequence number of the next expected data segment to be sent by the same sender. The sender retransmits the data segment if the expected acknowledgment does not arrive before the timer runs out. The TCP also has a provision for a checksum of the data to detect whether the received data has been corrupted.

The TCP provides for a control mechanism to regulate the traffic flow between two communicating parties. A variable window can be sent along with an acknowledgment by a receiver to indicate the pace at which it can receive data. If the receiver does not have much buffer space left, it can adjust the window to a smaller number.

The TCP is flexible. It does not require a particular protocol above it or the layer below it. It can be used on top of either a connectionless or connection-oriented protocol. The TCP segment format as shown in Fig. B.4 provides further details on the characteristics of TCP just described. A TCP segment has a TCP header part and a data part. The fields in the header are explained below.

- *Source port number.*   A 16-bit field for source port number.

- *Destination port number.*   A 16-bit field for destination port number.

- *Sequence number.*   A 32-bit number for tracking a data octet of a segment that is sent by the sender. In the case of synchronization or when the SYN bit is set, it is called the *initial sequence number.*

- *Acknowledge number.*   A 32-bit field that is sent by the receiver once a connection has been established. This is the number of the first data octet, which a receiver expects to receive next. For example, if a receiver has received data up to 1000 octets, the receiver assigns the acknowledge number to 1001 to indicate that the data of the next segment it expects to receive starts at octet 1001.

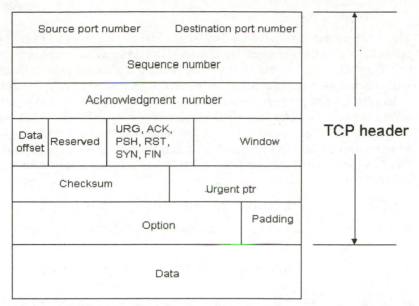

Figure B.4   TCP segment format.

- *Data offset*.   A 4-bit field to indicate the length of the TCP header measured in terms of multiples of 32 bits.

- *Reserved*.   Reserved for future extension and set to 0.

- *URG*.   A one-bit field for an urgent indicator. When this bit is set, the receiver should process this segment, interrupting the current activities.

- *ACK*.   A one-bit acknowledgment field. If this bit is set, an acknowledgment number is sent by the receiver.

- *PSH*.   A one-bit push function field to ask the sender to send whatever data are available without waiting for a full buffer.

- *SRT*.   A one-bit field used for resetting a connection.

- *SYN*.   A one-bit field used for synchronizing a sequence of numbers. This is used in the beginning to establish a connection by means of a three-way handshake.

- *FIN*.   A one-bit field used for closing a connection.

- *Window*.   A 16-bit field to indicate the number of octets a receiver is willing to receive.

- *Checksum*.   A 16-bit field to check whether data are corrupted.

- *Urgent pointer*.   A 16-bit pointer pointing to data offset following urgent data. This field is significant only if the URG bit is set.

- *Option*.   A variable-length field; one option currently defined is the maximum TCP segment size.

UDP

The UDP is a connectionless transport layer (i.e., host-to-host) protocol that provides best-effort service for transferring datagrams from one end to another. It provides an alternative to TCP that optimizes the overhead associated with each message transfer since no network connection is established.

In general the procedure to send and receive a packet works as follows. The data item is enveloped in a UPD header and sent to the IP layer. At the IP layer, an IP header is added. Then at the network access (NA) layer, an NA header is appended and the frame is sent across the network. At the receiving end, the NA header is removed first, then the IP layer header, and then the UDP header. See Table B.1.

- *Source port.*   A 16-bit port address of the source application protocol.

- *Destination port.*   A 16-bit port address of the intended recipient application protocol.

- *Length.*   The length of this frame in number of octets, including the UDP header.

- *Checksum.*   An optional field that has the one's complement of the checksum derived from a pseudo-header if this field is used and 0s otherwise. This checksum is for the entire UDP datagram.

Note the absence of the reliability-related fields such as the sequence and acknowledgment numbers in the UDP datagram header. Since the in-order, guaranteed delivery is not a requirement for UDP, the header is shorter and the overhead is smaller.

Internet Layer Protocol: Internet Protocol (IP)

This section introduces three Internet layer protocols, the Internet Protocol (IP), Address Resolution Protocol (ARP), and Reverse Address Resolution Protocol (RARP), with an emphasis on the IP. The IP addressing scheme is introduced first to provide the necessary background information.

IP addressing scheme

Each addressable network element within the Internet must be uniquely identified. The IP addressing scheme provides such capability. Since the addressing scheme lies at the heart of the Internet-based communication, a brief overview of the scheme is provided first.

TABLE B.1   UDP Datagram Format

Source port number	Destination port number
Length Checksum	
Data in octets	

An IP datagram contains both source and destination Internet addresses, uniquely identifying the source and destination hosts. Each IP address is 32 bits long and is usually expressed as four decimal integers, separated by a period. There are five classes (A, B, C, D, and E) of IP addresses, providing flexibility for various situations. The IP address of classes A, B, and C are made up of two portions: a network address, assigned by the Internet address administrative agency and host address, assigned by the local network administrator. Table B.2 shows the format of the five classes of IP addresses.

Class A addresses are used for very large networks and begin with a 0. The network identifier part takes up 7 bits, and the host address has 24 bits. Class B addresses are for medium-size networks such as a campuswide LAN whose first octet begins with 1 0 and whose network addresses range between 128 and 191 (decimal). As shown in Table B.2, the network identifier part has 14 bits and the host identifier part takes up 16 bits. The maximum number of hosts that can connect to a class B network is 2^{16}, and the maximum number of class B networks allowed is 2^{14}. Class C addresses are for small networks. The first octet begins with a 110 and has a very large network ID field and a small host field. It allows up to 2^8 hosts to connect to a class C network. The network addresses range from 192 to 254.

Class D addresses begin with a 1110, and the remaining 28 bits are known as a multicast address. When IP packets are sent to a group of stations, this is known as multicasting. Broadcasting is a variation of multicasting, in which all the stations in a subnet receive the broadcast IP packets. A class E address begins with 1111, and the rest of the bits are reserved for future use.

The main advantage of the Internet addressing scheme is flexibility in arranging networks. For example, if a network has a large number of workstations, the class A addresses may be more suitable than other address classes. On the other hand, if there is a relatively small number of workstations in a network, the class C addresses may be appropriate.

TABLE B.2   IP Address Classes

				Class A		
0	Network ID (7-bit)		Host ID (24-bit)			
				Class B		
1	0	Network ID (14-bit)		Host ID (16-bit)		
				Class C		
1	1	0	Network ID (24-bit)		Host ID (8-bit)	
				Class D		
1	1	1	0	Multicast address		
				Class E		
1	1	1	1	Reserved		

Internet Protocol

The IP is a packet-switched, connectionless, best-effort network layer protocol. It is a best-effort protocol in the sense that the IP does not guarantee end-to-end delivery and there is no acknowledge, flow control, retransmission, or error recovery. If the connection-oriented service is required, it must be provided by the upper-layer protocols such as TCP.

The IP provides a number of core functions and procedures necessary to the interworking between dissimilar networks. The main functions include

- *Fragmentation and reassembly.* This concerns the transfer of user messages across networks and subnets which support smaller packet sizes than the user data.

- *Routing.* To perform the routing function, the IP in each source host must know the location of the internet gateway or local router that is attached to the same network. In addition, the IP in each gateway or router must know the route to be used to reach other networks or subnets.

- *Error reporting.* The IP performs error checking in the process of routing and reassembling datagrams and may discard the datagram in error. It then proceeds to report such errors back to the IP in the source host.

The structure of the IP datagram is shown in Table B.3, which is made up of two parts: the header and the data portion. The fields in the header are briefly explained below.

- *Version.* IP version number, currently version 4.

- *Internet header length.* The length of the header in multiples of 32 bits. This field is 16 bits long and thus can accommodate up to 65,535 octets.

- *Type of service.* Instruction for quality-of-service parameters, such as delay, throughput, reliability, and cost.

- *Total length.* Length of the IP datagram in number of octets, including the length of the IP header.

- *Identification.* A unique identification for this datagram.

TABLE B.3   Structure of an IP Datagram

Version (4-bit)	Internet header length (4-bit)	Type of service (8-bit)	Total length (16-bit)
Identification (16-bit)		Flags (3-bit)	Fragment offset (13-bit)
Time to live (8-bit)		Protocol (8-bit)	Header of checksum (16-bit)
Source IP address (32-bit)			
Destination IP address (32-bit)			
Option+padding (32-bit)			
Data			

- *Flags.*   Option to indicate if fragmentation is allowed.

- *Fragment offset.*   The point in the datagram where this fragment starts measured in 64-bit units from the beginning of the datagram.

- *Time to live.*   The internet gateway hops and seconds this datagram has gone through.

- *Protocol.*   The protocol of the next layer up that follows the IP header, i.e., TCP or UDP.

- *Header checksum.*   A checksum on the IP header that may be recomputed at the receiving end to detect any header data corruption.

- *Source address.*   The internet address of the source host.

- *Destination address.*   The internet address of the destination host.

- *Option (variable).* Options from the sender, e.g., a route specification.

- *Padding.*   Some meaningless number (e.g., 0s) to make the IP header end on a 32-bit boundary.

- *Data.*   Actual payload of data, in multiples of octets, not exceeding 65,535 in total (header plus data).

Address resolution protocol (ARP)

Once an IP packet is received at a local host, the destination is expressed in terms of the IP address. The IP address needs to be mapped to an actual physical location for the intended workstation. The ARP provides such mapping through a lookup table that has a mapping between an IP address and its physical location. In general it works as follows. If the destination IP address in the received datagram is present in the mapping table, then the ARP simply passes the datagram address pointer with the corresponding physical address to the network access layer protocol (e.g., Ethernet, token ring). The network access layer protocol then either sends or broadcasts the datagram.

If the physical address is not present in the mapping table, the ARP will try to locate it by sending an ARP request message. The request message contains both its own IP-to-physical-address pair and the requested IP address. This can be done by broadcasting the request message to all the hosts in this network or sending it to the ARP in the gateway if the requested IP address belongs to another network.

Reverse address resolution protocol (RARP)

RARP is the reverse process of ARP, providing mapping from a physical address to the IP address. In general, it works as follows. The network server has a mapping between the IP address and the physical address for each of the hosts it serves. When a host first starts up, it broadcasts an RARP request message to the server containing its own physical address. Upon receipt of such message, the RARP in the server responds with a reply message containing both the IP address of the host and the IP-to-physical-address pair.

An Overview of the OSI Layered Network Reference Model

Introduction

As network communication systems became increasingly complicated, the International Standards Organization (ISO) realized the need to standardize the communication environment in the late 1970s and early 1980s. This led to the establishment of the Open System Interconnection (OSI) subcommittee to define such a standard communication environment. At the center of this standard communication environment is a layered network architecture model.

Among the principles of the layering architecture are, as specified in CCITT Recommendation X.200 (1988),

- Creating separate layers to handle functions that are apparently different in the process performed or technologies involved

- Collecting similar functions into the same layer

- Creating a layer of easily localized functions so that the layer could be totally redesigned and its protocol redesigned to take advantage of technological advances without affecting the neighboring layers

- Allowing changes of functions or protocols to be made within a layer without affecting other layers

- Creating for each layer boundaries with its upper and lower layers only

The result is a seven-layer logical network reference model as shown in Fig. C.1. This reference model provides a general framework for standardizing the communication protocols. Since the function of each layer is well defined, standard protocols for each layer can evolve independent of standards development of other layers.

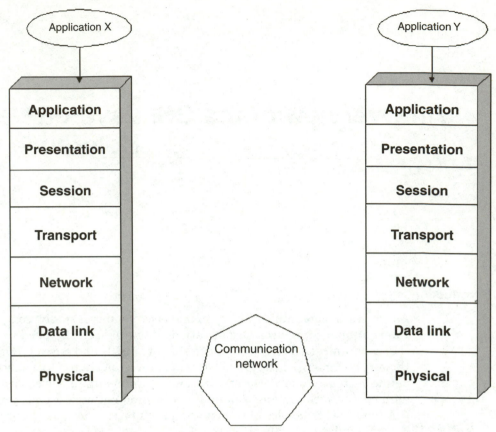

Figure C.1   Seven-layer OSI network reference model.

In this model each layer uses the service provided by the lower adjacent layer and in turn provides service to the upper layer. The relationships between the layers are shown in Fig. C.2. The service is passed from the layer N to the layer $N+1$ through the interface between them. The service is expressed in the form of the service primitives and parameters. The service primitives specify the functions to be performed; the actual forms of primitives are implementation-dependent. The service parameters pass data and control information between two adjacent layers. Four types of primitives are defined, i.e., *request, indication, response,* and *confirm.*

The sequence of steps for an information exchange from a source to a destination system using the service primitives is shown in Fig. C.3.

The steps for service primitive exchanges are

- The source layer N entity invokes its $(N-1)$ entity with a DATA.request primitive.
- The source layer $(N-1)$ entity prepares an $(N-1)$ PDU to be sent to its peer layer $(N-1)$ entity at the destination side.

- The destination layer $(N-1)$ entity delivers the data to the appropriate layer N entity via a DATA.indication primitive.
- If an acknowledge is called for, the destination layer N entity issues a DATA.response to the layer $(N-1)$ entity.
- The destination layer $(N-1)$ entity conveys the acknowledge in a layer $(N-1)$ PDU to the source layer $(N-1)$ entity.
- The acknowledge is delivered to the source layer N entity as a DATA.confirm.

Though the format of the protocol data units (PDUs) differs from protocol to protocol, several general characteristics of the PDUs of different protocols are identified as

- Having bounded size
- Containing control information that is used to coordinate the joint operations by both ends
- Containing user data from a higher layer

One layer may segment the PDU from the upper layer before sending it to the adjacent lower layer for various reasons such as size constraint, efficient error control with smaller PDUs, or buffer size constraint. The receiving end needs to reassemble the fragmented PDUs into the original one before sending it to the adjacent upper layer.

Description of OSI layers

Each layer performs a well-defined function in the context of the overall communication subsystem. The seven layers can be viewed as consisting of two cate-

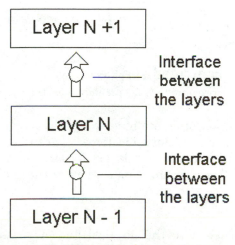

Figure C.2 Relationships between the layers.

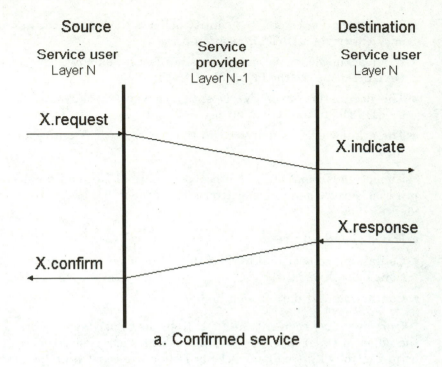

a. Confirmed service

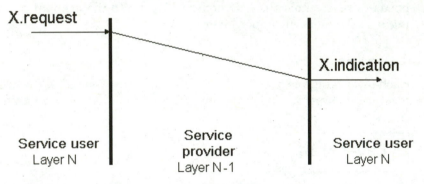

b. Non confirmed service

Figure C.3   Illustration of the use of service primitives for communication between two systems.

gories: network-oriented and application-oriented. While the bottom four layers are concerned with network aspects of communications between two systems, the top three layers, the session and application layers in particular, mainly deal with the communications details concerning applications at the end systems.

Physical layer

The physical layer is concerned with the physical and electrical interfaces between the user equipment and the network terminating equipment. It provides the link layer with the capabilities to transmit a stream of bits between two end network devices. It is the physical layer that specifies electrical representation of bits and the data transmission rate and defines the procedure for bit stream exchange across the network.

Data-link layer

The data-link layer builds on the physical layer and provides the network layer with reliable information transfer capabilities. It provides means to activate, maintain, and deactivate a physical link and is responsible for error detection and retransmission of messages in the event of transmission error.

The data-link layer provides two types of services: connectionless and connection-oriented. The connectionless service transmits information frames on the best-effort basis and discards a frame if a transmission error is found in the frame. The connection-oriented service provides an error-free information transfer facility. High level data link control (HDLC) is a widely used data-link layer protocol.

Network layer

The network layer provides for the transfer of information between end systems. It insulates the session layer from the details about the underlying data transmission and switching technologies used to connect end systems. At this layer, the computer system interacts with the underlying network to specify the destination address and to request certain network facilities such as priority. The services provided by the network layer include

- Perform routing functions using routing protocols based on the OSI Routing Framework as specified in ISO TR9575 for routing domains and interdomain and intradomain routing protocols
- Provide global addressing schemes
- Provide two modes of network layer services: connection-oriented and connectionless

Transport layer

The transport layer provides a connection-oriented transport service that ensures data are delivered error-free, in sequence, without loss or duplication. The transport layer acts as the interface between the application-oriented layers (i.e., session and application layers) and the underlying network-dependent protocol layers. It insulates the session layer from the network details and provides the message transfer capabilities that are free of the underlying network technology to the upper layers. The transport layer provides a number of classes of service to accommodate the different qualities of service provided by different types of underlying networks. There are five classes of services, ranging from class 0 to class 4, with increasing levels of quality of service. While the class 0 service is used for establishing basic connections and data transfer, the class 4 service provides full error control and flow control procedures.

- *TP 0.*   Simple class
- *TP 1.*   Basic error recovery class
- *TP 2.*   Multiplexing class
- *TP 3.*   Error recovery and multiplexing class
- *TP 4.*   Error detection and recovery class

Session layer

The session layer provides functions to control and coordinate dialog between applications in two end systems. It is thus responsible for setting up and taking down a communication channel between two applications. The session layer provides a number of optional services, including

- *Interaction management.*   The data exchange associated with a dialog may be two-way simultaneous (duplex) or two-way alternate (half-duplex). The session layer provides facilities for ensuring that data exchanges proceed in a synchronous way.
- *Synchronization.*   The session layer provides a mechanism for applications to periodically establish synchronization points during a lengthy network transaction. In case of a fault during the transaction, the dialog may start from the latest synchronization point.
- *Exception reporting.*   The session layer can report exceptions that took place during a transaction to the application layer.

Presentation layer

The presentation layer defines the format of the data to be exchanged and provides to application programs a set of data-transformation services. It defines the syntax used between application entities and provides for the selection of the representations used. ASN.1 is one such syntax used at this layer. Other services provided by the presentation layer include data compression and encryption.

Application layer

The application layer provides an interface for application programs to access the OSI environment and the required communication support to applications (user applications themselves aren't at this layer). At a conceptual level, a user application can be divided into communication and noncommunication components. The communication components are called application entities (AEs). An AE communicates with a peer AE using the protocols provided by this layer. Access to the application layer services is normally through a defined set of primitives, each with associated parameters that are supported by the local operating system.

In addition to information transfer and the user interface, the application layer also provides the following services:

- Identifies the intended communication entities by name or by address
- Determines the current availability of an intended communication entity
- Establishes authority to communicate
- Establishes an agreement on the privacy mechanism to be used in a session
- Authenticates the intended communication entities
- Selects communication policies with regard to initiation and release procedures
- Determines the data syntax for data exchange

Managed Object Class Dictionary

This appendix provides a dictionary of the commonly used managed object classes, attributes, and notifications. The primary criterion for including a managed object class in the dictionary is that the object class should be generic and designed for general management operations that are applicable to a variety of network technologies. Overwhelming majorities of the object classes are from the X.72x and X.73x series of ITU recommendations on system management functions.

The dictionary consists of three parts: classes, notifications, and attributes. A managed object, as is true of its counterpart in the general computing world, consists of a set of attributes that describe the internal properties of the object, and exhibit certain behaviors. The properties and behavior of the managed object class are represented by the attributes and notifications, respectively. Because the attributes and notifications are used in many different managed object classes as common components, their definitions are presented in sections separate from that of the managed object classes.

The primary goal of this dictionary is to provide a central repository for the references on the managed object class definitions that are scattered across many recommendations and are represented in GDMO format.

Managed Object Class Directory

A managed object class is presented in a table that consists of a set of fields describing an object class. The *superclass* field refers to the direct superclass only. The *source* field indicates the ITU recommendations in which the object class is defined. The *description* gives a brief description of the responsibility and purpose of the object class.

The *attributes* field lists all the attributes this managed object class may have and the management operations permitted on each attribute. Four types of management operations are possible, i.e., GET, REPLACE, ADD, and

REMOVE, which are equivalent to retrieve, modify, create, and delete operations, respectively. While the GET operation indicates that a manager can retrieve the attribute value, the REPLACE operation means the manager is allowed to modify, set, or reset the attribute value. While the ADD operation indicates that the manager can create the attribute when the object the attribute is in is being created, the REMOVE operation means that the manager is allowed to delete the attribute from the managed object. Note that no distinction is made between conditional and mandatory attributes as in GDMO, because there is no counterpart of such distinction in general object-oriented programming language.

Optionally a managed object class may have one or more notifications that the object can send to other management entities.

Alarm Record

Class name	alarmRecord	
Superclass	eventLogRecord	
Description	This object defines what information goes in the log as a result of receiving an alarm notification or alarm event report.	
Source	X.721, X.733	
Attributes	probableCause	GET
	perceivedSeverity	GET
	specificProblem	GET
	backedUpStatus	GET
	backUpObject	GET
	trendIndication	GET
	thresholdInfo	GET
	stateChangeDefinition	GET
	monitoredAttribute	GET
	proposedRepairAction	GET

Alarm Severity Assignment Profile

Class name	alarmSeverityAssignmentProfile	
Superclass	top	
Description	This managed object class specifies the alarm severity assignments for a set of managed objects. For each associated managed object, it lists the cause of an alarm and severity levels assigned to the alarm.	
Source	M.3100	
Attributes	alarmSeverityAssignmentProfileId	GET, SET
	alarmSeverityAssignmentList	GET-REPLACE, ADD-REMOVE
Notifications	objectCreation, objectDeletion, attributeValueChange	

Attribute Value Change Record

Class name	attributeValueChangeRecord	
Superclass	eventLogRecord	
Description	This object class defines what information goes into the log as a result of receiving an attribute value change notification.	
Source	X.721, X.730	
Attributes	attributeValueChangeDefinition	GET
	sourceIndicator	GET
	attributeIdentifierList	GET

Daily Scheduler

Class name	dailyScheduler
Superclass	scheduler
Description	This object class provides the capabilities for a manager to automatically control another object's activities for a period of 24 h. A sequence of daily schedulers can be run repeatedly. The bulk of its behavior is derived from its superclass scheduler.
Source	X.721, X.746
Attribute	sequenceOfDays

Daily Operation Scheduler

Class name	dailyOperationScheduler	
Superclass	dailyScheduler	
Description	This object identifies specific operations in the managed objects to be scheduled on a daily basis with the operationSpecifications attribute. Once the scheduled operation is completed, the operation result is returned to the managing system that scheduled the operation.	
Source	X.721, X.746	
Attribute	operationSpecifications	GET-REPLACE, ADD-REMOVE
Notification	operationResult	

Discriminator

Class name	discriminator	
Superclass	top	
Description	This managed object class allows an outside application to set the event forwarding criteria for event forwarding. In addition, it provides for the capability to control the event forwarding discriminator through setting the administrative state of the discriminator object.	
Source	X.721, X.734	
Attributes	discriminatorId	GET
	discriminatorConstruct	GET-REPLACE

Discriminator (Continued)

	administrativeState	GET-REPLACE
	operationalState	GET
	availabilityStatus	GET
	startTime	GET-REPLACE
	stopTime	GET-REPLACE
	intervalsOfDay	GET-REPLACE, ADD-REMOVE
	weekMask	GET-REPLACE, ADD-REMOVE
	schedulerName	GET
Notifications	stateChange, attributeValueChange, objectCreate, objectDeletion	

Event Forwarding Discriminator

Class name	eventForwardDiscriminator	
Superclass	discriminator	
Description	This object class is a subclass of the discriminator class and inherits all its attributes and notifications. In addition, it specifies the destination and alternative destinations for the event forwarding.	
Source	X.721, X.734	
Attributes	destination	GET-REPLACE
	backUpDestinationList	GET-REPLACE
	activeDestination	GET
	confirmMode	GET

Event Log Record

Class name	eventLogRecord	
Superclass	logRecord	
Description	This is a generic object class that defines what information goes into the log as a result of receiving notifications or event reports. Many specific types of log record object classes, such as alarmLogRecord, attributeValueChange-Record, and objectDeletionRecord, are all derived from this object class.	
Source	X.721	
Attributes	managedObjectClass	GET
	managedObjectInstance	GET
	eventType	GET
	eventTime	GET
	notificationIdentifier	GET
	correlatedNotification	GET
	additionalText	GET
	additionalInformation	GET

Log

Class name	log	
Superclass	top	
Description	The object class provides the capability for a managing system to control logging. Numerous attributes of the log managed object class can be put in three different categories: log state or status related, log control action, and log scheduling. The first category has three attributes: operational, administrativeState, and availabilityStatus. The log control actions include max log size, log threshold alarm, log full action. A third category of attributes allows a managing system to schedule logging activities on a daily, weekly, or more extensive period basis.	
Source	X.721, X.735	
Attributes	logId	GET
	discriminatorConstruct	GET-REPLACE
	operationalState	GET
	availabilityStatus	GET-REPLACE
	logFullAction	GET-REPLACE
	maxLogSize	GET-REPLACE
	currentLogSize	GET
	numberOfRecords	GET
	capacityAlarmThreshold	GET-REPLACE, ADD-REMOVE
	availabilityStatus	GET
	startTime	GET-REPLACE
	stopTime	GET-REPLACE
	intervalsOfDay	GET-REPLACE, ADD-REMOVE
	weekMask	GET-REPLACE, ADD-REMOVE
	schedulerName	GET
Notifications	objectCreation, objectDeletion, attributeValueChange, stateChange, processingErrorAlarm	

Log Record

Class name	logRecord	
Superclass	top	
Description	This is a generic managed object class that represents one record in the log.	
Source	X.735, X.721	
Attributes	logRecordId	GET
	loggingTime	GET

Monthly Scheduler

Class name	monthlyScheduler
Superclass	scheduler
Description	This object provides the capability for a managing system to control the operations of a managed object within the period of a month. The attribute sequenceOfMonths specifies a set of time intervals for each selected day within a month.
Source	X.721, X.746
Attribute	sequenceOfMonths GET-REPLACE, ADD-REMOVE

Monthly Operation Scheduler

Class name	monthlyOperationScheduler
Superclass	monthlyScheduler
Description	This object class identifies specific operations in managed objects to be scheduled on a monthly basis. Once the scheduled operations are completed, the operation results are returned to the managing system that scheduled the operations.
Source	X.721, X.746
Attribute	operationSpecifications GET-REPLACE, ADD-REMOVE
Notification	operationResult

Object Creation Record

Class name	objectCreationRecord
Superclass	eventLogRecord
Description	This object class defines what information goes into the log as a result of receiving an object creation notification or event report.
Source	X.721, X.730
Attributes	attributeList GET
	sourceIndicator GET

Object Deletion Record

Class	objectDeletionRecord
Superclass	eventLogRecord
Description	This object class defines what information goes into the log as a result of receiving an object deletion notification or event report.
Source	X.721, X.730
Attributes	attributeList GET
	sourceIndicator GET

Operation Result Record

Class name	operationResultRecord
Superclass	eventLogRecord
Description	This object is used to record the results of scheduled operations.
Source	X.721, X.746
Attribute	operationResult

Periodic Operation Scheduler

Class name	periodicOperationScheduler
Superclass	periodicScheduler
Description	This object identifies specific operations in managed objects to be scheduled on the basis of a specified period. Once the scheduled operations are completed, the operation results are returned to the managing system that scheduled the operations.
Source	X.721, X.746
Attribute	operationSpecifications GET-REPLACE, ADD-REMOVE
Notification	operationResult

Periodic Scheduler

Class name	periodicScheduler
Superclass	scheduler
Description	This object provides the capability for a managing system to schedule triggering of operations of a managed object based on a specified schedule. The scheduler can be suspended and resumed by the managing system through setting the administrative state (inherited from its superclass). When the scheduler is resumed, the triggering points can be synchronized through either period synchronization time or resynchronization mode.
Source	X.721, X.746
Attributes	timePeriod GET, ADD-REMOVE
	resynchronizeMode GET-REPLACE
	periodSynchronization GET-REPLACE

Scheduler

Class name	scheduler
Superclass	top
Description	This is a generic scheduler object class to be inherited by other more specific scheduler object classes (e.g., dailyScheduler, weeklyScheduler). It provides the specific schedulers with the capability that allows a managing system to suspend and resume the scheduler. It provides the naming attribute (i.e., the attribute that uniquely identifies a scheduler object instance) for a specific scheduler that is derived from it. In addition, it provides a set of notification capabilities to report a scheduler creation, deletion, attribute value change, or state change.

Scheduler (Continued)

Source	X.721, X.746	
Attributes	administrativeState	GET-REPLACE
	operationalState	GET
	schedulerId	GET
	scheduledManagedObject	GET
	startTime	GET
	stopTime	GET
Notifications	AttributeValueChange, stateChange, objectCreation, objectDeletion	

Security Alarm Report Record

Class name	securityAlarmReportRecord	
Superclass	eventLogRecord	
Description	This managed object class defines what information is stored in the log as a result of receiving a security alarm notification or security event report.	
Source	X.721, X.736	
Attributes	securityAlarmCause	GET
	securityAlarmSeverity	GET
	securityAlarmDetector	GET
	serviceUser	GET
	serviceProvider	GET

Relationship Change Record

Class name	relationshipChangeRecord	
Superclass	eventLogRecord	
Description	This managed object class defines what information is stored in the log as a result of receiving state change notifications or state change event reports.	
Source	X.732, X.721	
Attributes	relationshipChangeDefinition	GET
	sourceIndicator	GET
	attributeList	GET

State Change Record

Class name	stateChangeRecord	
Superclass	eventLogRecord	
Description	This managed object class defines what information is stored in the log as a result of receiving state change notifications or state change event reports.	
Source	X.731, X.721	
Attributes	stateChangeDefinition	GET
	sourceIndicator	GET
	attributeList	GET

System

Class name	system	
Superclass	top	
Description	This object class is used to represent a combination of hardware and software that forms an autonomous system capable of processing or transferring information. This object class is often used either as a superclass from which other object classes are derived or as a superior class that contains and names other objects.	
Source	X.720, X.721	
Attributes	systemId	GET
	systemTitle	GET
	operationalState	GET
	usageState	GET
	administrativeState	GET
	superFeatures	GET-REPLACE, ADD-REMOVE

Top

Class name	top	
Superclass	(none)	
Description	This is the top class that is at the very top of the inheritance hierarchy from which all other object classes are derived directly or indirectly	
Source	X.720, X.721	
Attributes	objectClass	GET
	nameBinding	GET
	allomorphs	GET

Weekly Operation Scheduler

Class name	weeklyOperationScheduler	
Superclass	weeklyScheduler	
Description	This object identifies specific operations in managed objects to be scheduled on a weekly basis. Once the scheduled operations are completed, the operation results are returned to the managing system that scheduled the operations.	
Source	X.721, X.746	
Attribute	operationSpecifications	GET-REPLACE, ADD-REMOVE
Notification	operationResult	

Weekly Scheduler

Class name	weeklyScheduler
Superclass	Scheduler
Descripton	This object provides the capability for a managing system to control the operations of a managed object within the period of a week. The attribute sequenceOfWeeks specifies a set of time intervals for each selected day within a week.
Source	X.721, X.746
Attribute	sequenceOfWeeks

Notification Dictionary

A managed object can emit a notification to report an event concerning the object that has taken place or the results of a management operation. The *description* field briefly describes the purpose and the management entities involved in a notification. The *notification attributes* field lists the attributes that are included as the notification parameters and that provide the information contents of the notification. A notification may be included in one or more managed object classes.

Attribute Value Change

Notification identifier	attributeValueChange
Description	This notification allows a managed object to report the changes in the value of managed object attributes. The selection of attributes that can trigger this notification may be application-dependent. The notification can be in either a confirmed or nonconfirmed mode.
Notification attributes	sourceIndicator, attributeIdentifierList, attributeValueChangeDefinition, notificationIdentifier, correlatedNotifications, additionalText, additionalInformation

Communications Alarm

Notification identifier	communicationAlarm
Description	This notification allows an application entity in the agent role to report an alarm that is primarily concerned with the faults associated with communication process or procedure. Examples of the causes for this type of alarm include loss of signal, loss of frame, and transmission error.
Notification attributes	probableCause, specificProblems, perceivedSeverity, backedUpStatus, backUpObject, trendIndication, thresholdInfo, notificationIdentifier, correlatedNotifications stateChangeDefinition, monitoredAttributes, proposedRepairAction, additionalText, additionalInformation

Environmental Alarm

Notification identifier	environmentAlarm
Description	This notification allows an application entity in the agent role to report an alarm that is primarily concerned with the faults associated with the environment or surrounding equipment. Examples of the causes for this type of alarm include high temperature caused by a conditioner failure, a fire, or toxic leak.
Notification attributes	probableCause, specificProblems, perceivedSeverity, backedUpStatus, backUpObject, trendIndication, thresholdInfo, notificationIdentifier, correlatedNotifications, stateChangeDefinition, monitoredAttributes, proposedRepairAction, additionalText, additionalInformation

Equipment Alarm

Notification identifier	equipmentAlarm
Description	this notification allows an application entity in the agent role to report an alarm that is primarily associated with faults of equipment. Examples of the causes for this type of alarm include power failure, transmission facility malfunction, and I/O device errors.
Notification attributes	specificProblems, perceivedSeverity, backedUpStatus, backUpObject, trendIndication, thresholdInfo, notificationIdentifier, correlatedNotifications, stateChangeDefinition, monitoredAttributes, proposedRepairAction, additionalText, additionalInformation

Integrity Violation

Notification identifier	integrityViolation
Description	This notification allows an application entity in the agent role to report an event that is primarily associated with violation of information integrity. Examples of this type of violation include illegal modification, insertion, or deletion of information.
Notification attributes	securityAlarmCause, securityAlarmSeverity, securityAlarmDetector, serviceUser, serviceProvider, notificationIdentifier, correlatedNotifications, additionalText, additionalInformation

Object Creation

Notification identifier	objectCreation
Description	This notification allows an application entity in the agent role to report the creation of an object instance that is of interest to the managing system.
Notification attributes	sourceIndicator, attributeList, notificationIdentifier, correlatedNotifications, additionalText, additionalInformation

Object Deletion

Notification identifier	objectDeletion
Description	This notification allows an application entity in the agent role to report the deletion of an object that is of interest to the managing system.
Notification attributes	sourceIndicator, attributeList, notificationIdentifier, correlatedNotifications, additionalText, additionalInformation

Operation Violation

Notification identifier	operationViolation
Description	This notification allows an application entity in the agent role to report an event that is primarily associated with malfunction, unavailability, or incorrect invocation of a service.
Notification attributes	securityAlarmCause, securityAlarmSeverity, securityAlarmDetector, serviceUser, serviceProvider, notificationIdentifier, correlatedNotifications, additionalText, additionalInformation

Physical Violation

Notification identifier	physicalViolation
Description	This notification allows an application entity in the agent role to report an event that is primarily associated with violation of physical resources such as forced and illegal entry into a system.
Notification attributes	securityAlarmCause, securityAlarmSeverity, securityAlarmDetector, serviceUser, serviceProvider, notificationIdentifier, correlatedNotifications, additionalText, additionalInformation

Processing Error Alarm

Notification identifier	processingErrorAlarm
Description	This notification allows an application entity in the agent role to report an alarm that is primarily associated with software or processing faults. Examples of the causes for this type of alarm include corrupt data, version mismatch, and file error.
Notification attributes	probableCause, specificProblems, perceivedSeverity, backedUpStatus, backUpObject, trendIndication, thresholdInfo, notificationIdentifier, correlatedNotifications, stateChangeDefinition, monitoredAttributes, proposedRepairAction, additionalText, additionalInformation

Quality-of-Service Alarm

Notification identifier	qualityOfServiceAlarm
Description	This notification allows an application entity in the agent role to report an alarm that is primarily associated with the degradation of the service quality. Examples of the causes for this type of alarm include excessive response time, bandwidth reduction, and congestion.
Notification attributes	probableCause, specificProblems, perceivedSeverity, backedUpStatus, backUpObject, trendIndication, thresholdInfo, notificationIdentifier, correlatedNotifications, stateChangeDefinition, monitoredAttributes, proposedRepairAction, additionalText, additionalInformation

Relationship Change

Notification identifier	relationshipChange
Description	This notification allows an application entity in the agent role to report the changes in the values of relationship attributes of the managed object. Examples of the relationship attributes include primary, secondary, peer, member, and owner.
Notification attributes	sourceIndicator, attributeIdentifierList, relationshipChangeDefinition, notificationIdentifier, correlatedNotifications, additionalText, additionalInformation

Security Service or Mechanisms Violation

Notification identifier	securityServiceOrMechanismViolation
Description	This notification allows an application entity in the agent role to report a potential security attack that has been detected by the security service or mechanism. Examples of the violations include authentication failure, nonrepudiation failure, and breach of confidentiality.
Notification attributes	securityAlarmCause, securityAlarmSeverity, securityAlarmDetector, serviceUser, serviceProvider, notificationIdentifier, correlatedNotifications, additionalText, additionalInformation

State Change

Notification identifier	stateChange
Description	This notification allows an application entity in the agent role to report changes in the values of managed object state attributes. It can be an operational, usage, or administrative state.
Notification attributes	sourceIndicator, attributeIdentifierList, relationshipChangeDefinition, notificationIdentifier, correlatedNotifications, additionalText, additionalInformation

Time Domain Violation

Notification identifier	timeDomainViolation
Description	This notification allows an application entity in the agent role to report that an event has occurred at an unexpected or prohibited time. Examples of such events include delayed information, expired key, and out-of-hour activities.
Notification attributes	securityAlarmCause, securityAlarmSeverity, securityAlarmDetector, serviceUser, serviceProvider, notificationIdentifier, correlatedNotifications, additionalText, additionalInformation

Attribute Dictionary

Attributes in the managed object classes or notifications represent the management information to be manipulated in management operations or exchanged between management entities. Attributes are classified into two categories: generic and specific. The generic attributes are meant for all applications, and they are like attribute types to be used to define specific attributes for particular applications. There are five generic attributes, i.e., counter, gauge, counter threshold, gauge threshold, and tide mark. A specific attribute can be included in one or more of the managed object classes or notifications.

activeDestination   This attribute identifies an application entity that is the destination for an event forwarding discriminator to forward the current event. This attribute is read-only for the managing system, and its value is assigned as a result of the system operation using the destination or backup destination list attributes.

additionalInformation   This attribute allows an event report to include the additional information in the form of three-item sets: an identifier, a significance indicator, and the problem information.

additionalText   A free-form text description providing additional information on the semantics of an object that contains this attribute.

administrativeState   This attribute represents the mechanism that allows a managing system to control the managed object independently of the operational and usage state of the object. The three possible enumerated values of the attribute are

- *Locked.*   The resource represented by the object is administratively prevented from providing services to its users.
- *Unlocked.*   The resource is administratively permitted to provide services to its users.
- *ShuttingDown.*   The use of the resource is allowed for the in-progress users, but new service requests will be rejected.

alarmStatus   This set-valued attribute indicates the status of a current alarm:

- *Under repair.*   The resource is currently under repair.
- *Critical.*   One or more critical alarms have not been cleared.
- *Major.*   One or more major alarms have not been cleared.
- *Minor.*   One or more minor alarms have not been cleared.
- *Alarm outstanding.*   One or more alarms have been detected, and the resource may or may not have been disabling.

allomorphs   This attribute identifies a set of object classes that are allomorphs of this managed object.

attributeIdentifierList   This attribute identifies a set of attributes that are the target of a management operation. Examples of such management operations include monitoring object attributes for their value changes.

attributeList   This attribute specifies a set of initial attributes to be included in an object when the object is first created. This provides the option of selectively instantiating the attributes of an object.

attributeValueChangeDefinition   This attribute consists of a sequence of three-attribute sets: attribute identifier, old attribute value, and new attribute value. Each set describes a single attribute value change.

availabilityStatus   This attribute is a sequence of integers identifying the condition that affects the resource.

- *In test*.   The resource is undergoing a test procedure.
- *Failed*.   An internal fault prevents the resource from operating.
- *powerOff*.   The resource is not powered on.
- *off-line*.   The resource is taken off-line and needs to be taken back on-line to be operational.
- *offDuty*.   The resource is made inactive based on a predefined schedule.
- *dependency*.   The resource is not operational because another resource it depends on is not available.
- *degraded*.   The service provided by the resource is degraded.
- *NotInstalled*.   The resource represented by the managed object is not installed or the installation is not complete.
- *logFull*.   The resource is in a state where the log is full.

backedUpStatus   This attribute of the Boolean type indicates whether or not the object emitting the alarm has been backed up and thus whether the system as a whole is capable of continuing to provide services in the event of a failure.

backUpDestinationList   This attribute consists of an ordered list of application entity titles that are alternative event forwarding destinations if the destination specified by the destination attribute fails.

backUpObject   This attribute is present only if the backedUpStatus attribute is set to TRUE. It specifies the managed object instance that is providing backup services for the managed object that is emitting the alarm or event notifications.

capacityAlarmThreshold   This attribute consists of a set of integers that specifies capacity levels at which an alarm notification will be generated.

confirmedMode   This attribute indicates whether an event is forwarded by the event forwarding discriminator object in a confirmed or nonconfirmed mode.

controlStatus   This attribute is a sequence of integers identifying the control status of a resource due to an administrative action. This attribute can take on one or more of the following values:

- *Subject to test*.   The resource is available for normal use but tests may be carried out on the resource simultaneously, and consequently the resource may exhibit certain unusual behaviors.

- *Part of service locked.* The part of the service provided by the resource is administratively locked. An example is incoming or outgoing calls barred.
- *Reserved for test.* The resource has been administratively rendered unavailable for normal use because of a test procedure.
- *Suspended.* The service provided by the resource has been administratively suspended.

correlatedNotifications   This attribute is often included in a notification to specify a set of notification identifiers and optionally the associated managed object instance names to which this notification is correlated. A notification is considered correlated to other notifications if it needs information from other notifications to provide a comprehensive view of an event.

counter   This is a generic attribute that is used to define other application-specific attributes for managed object classes. A counter is an abstraction of an underlying counting process. An attribute of the counter type is a single-valued, nonnegative integer that can only be incremented one at a time, with an initial value of 0. A counter attribute has a maximum value, and the attribute wraps around when the maximum is reached. There are two kinds of counters to meet different needs, i.e., settable and nonsettable counters. The difference between the two is that a settable counter attribute can be set by a management operation while a nonsettable attribute can only be read by a managing system.

counterThreshold   This is a generic attribute that provides a general mechanism for generating notifications for the changes in the attribute values of the counter type. The counterThreshold is a set-valued attribute that models different comparison levels, with an integer offset value. A counterThreshold is associated with a defined notification and it can be switched on or off by the management operation. For example, once a comparison level is reached by the corresponding counter attribute, a predefined notification may be generated if the counterThreshold attribute is switched on.

currentLogSize   This attribute specifies the current size of the log in octets.

daysOfWeek   This attribute defines the days of the week during which logging operations are scheduled to occur. This attribute, if not present when the object is created, will default to all seven days of the week.

destination   This attribute identifies one or more destinations to which the discriminator object forwards event reports.

discriminatorConstructor   This attribute specifies tests for determining whether or not to forward an event by the discriminator object.

discriminatorId   This attribute specifies a unique identifier that can be used to name an instance of the discriminator managed object class.

eventTime   This attribute contains the time when the event was generated. The time is expressed in the ASN.1 generalizedTime type.

gauge   Like the counter attribute, the gauge is a generic attribute that is used to define other application-specific attributes for managed object classes. A gauge is an abstraction of the value of a dynamic variable. The value of an attribute of the gauge type can change in either direction, up or down. There is no constraint on how much the value can change each time as long as the change is within the boundary of a maximum and minimum. When the change takes the gauge attribute value beyond the maximum or minimum, the value will be left at the maximum or minimum, instead of wrapping around.

gaugeThreshold   This is a generic attribute that provides a general mechanism for generating notifications for the changes in the attribute values of the gauge type. A gaugeThreshold is a set-valued attribute that models different comparison levels, with an integer offset value. To avoid the repeated triggering of event notifications when the gauge makes small oscillations around a threshold value, threshold values are specified in pairs with one being a high-threshold and the other being a low-threshold value. The difference between the two values is the hysteresis interval.

intervalsOfDay   This attribute defines the list of time intervals (interval-start and interval-end times of day) for which the log will exhibit the logging-on condition, if the current day is one of the days that is selected within the corresponding daysOfWeek. During excluded intervals the log exhibits the logging-off condition. If not specified in the create request, the value of this attribute defaults to a single interval encompassing the entire 24-h period of a day.

logFullAction   This attribute specifies the action to be taken when the maximum size of the log has been reached. Two possible actions are

- *Wrap.*   The oldest log records, as identified by the log record identifier, will be replaced by the new log records in a round-robin fashion.
- *Halt.*   No more new records will be logged.

logId   A unique identifier for the log object.

loggingTime   This attribute identifies the time when the record is put into the log.

logRecordId   This attribute uniquely identifies each record in the log.

managedObjectClass   This attribute specifies the managed object classes as part of the parameters of the discriminator construct.

managedObjectInstance   This attribute specifies the managed object class instance as part of the parameters of the discriminator construct.

maxLogSize   This attribute specifies the maximum size of the log measured in octets.

member   This attribute identifies an object that has the role of a member object in a group relationship. A group relationship is a relationship between two managed objects where one, the member object, belongs to a group represented by the other object, the owner object.

monitoredAttributes   This attribute specifies one or more attributes of the managed object that is being monitored for alarm or a specified event.

notificationIdentifier   This attribute provides an identifier for a notification that can be carried in the correlated notifications attribute. This identifier must be unique across all notifications of a particular managed object.

numberOfRecords   This attribute specifies the current number of records contained in the log.

operationalState   This enumerated attribute describes the operability of a resource with two possible values:

- *Disabled.*   The resource is totally inoperable and unable to provide the normal service to its users because of the occurrence of certain events (e.g., alarms).
- *Enabled.*   The resource is partially or fully operable.

owner   This attribute identifies an object that has the role of an owner object in a group relationship. A group relationship is a relationship between two managed objects where one, the member object, belongs to a group represented by the other object, the owner object.

peer   This attribute identifies a managed object that has the role of a peer in a peer relationship. A peer relationship is a symmetric relationship in which pairs of similar managed objects or peer objects communicate with each other.

perceivedSeverity   This attribute takes on one of the six defined severity levels of an alarm:

- *Cleared.*   One or more previously reported alarms have been cleared.
- *Indeterminate.*   The severity level cannot be determined.
- *Critical.*   A service-affecting condition has occurred and an immediate corrective action is required.
- *Major.*   A service-affecting condition has developed and an urgent corrective action is required.
- *Minor.*   A non-service-affecting fault condition has occurred and corrective action should be taken in order to prevent a more serious (for example, service-affecting) fault.
- *Warning.*   A potential or impending service-affecting fault has been detected, and an action should be taken to further diagnose (if necessary) and correct the problem in order to prevent it from becoming a more serious service-affecting fault.

primary   This attribute identifies a managed object that has the role of the primary object in a fallback relationship. A fallback relationship is an asymmetric relationship in which the second of a pair of managed objects is designated as a fallback or alternative to the first managed object or primary object.

probableCause   This attribute identifies a probable cause of the alarm. The identified probable causes are listed in App. 10A.

proceduralStatus   This attribute indicates a phase of a process the resource is going through in order to become operable. The following generalized phases are identified:

- *Initialization required.*   The resource requires an initialization before becoming operable.
- *Not initialized.*   The resource requires an initialization in order to become operable but the initialization has not been performed yet.
- *Initializing.*   The resource requires an initialization process and the process is under way.
- *Reporting.*   The resource has completed certain operations and is in the process of reporting the operation results.
- *Terminating.*   The resource is in a termination phase.

proposedRepairActions   This attribute consists of a set of integers or object identifiers that specifies a set of repair actions such as switch to standby unit, retry, or replace the unit.

providerObject   This attribute identifies one or more managed objects that act in the service provider role in a service relationship. A service relationship is an asymmetric relationship in which the first of a pair of managed objects is a provider object to the second object, or a user object.

relationshipChangeDefinition   This attribute consists of a sequence of three-item sets: attribute identifiers, old attribute value, and new attribute value. Each individual set describes a single relationship attribute value change.

relationships   This attribute contains all the relationships of a managed object. The relationships include service relationship, fallback relationship, and peer relationship.

schedulerName   This attribute specifies the name of the scheduler managed object associated with the discriminator object.

secondary   This attribute identifies a managed object that has the role of the secondary object in a fallback relationship. A fallback relationship is an asymmetric relationship in which the second of a pair of managed objects, or secondary object, is designated as a fallback or alternative to the first managed object, or primary object.

securityAlarmCause   This attribute identifies probable causes of a security alarm. The causes are classified into five categories:

- Integrity violation
- Operational violation
- Physical violation
- Security service or mechanisms violation
- Time domain violation

securityAlarmDetector   This attribute identifies the entity that first detected the security alarm.

securityAlarmSeverity   This attribute indicates the level of significance of the security alarm as viewed by the managed object. The following five levels of severity are defined:

- *Indeterminate.*   A security attack has been detected, but the level of the adverse impact on the system integrity is unknown.
- *Critical.*   The system security has been severely compromised, and the system may no longer be assumed to be operating correctly in support of the security policy.
- *Major.*   A security attack has been detected, and the important system security mechanisms or information has been compromised.
- *Minor.*   A security attack has been detected, and not-so-important security mechanisms or information has been compromised.
- *Warning.*   A security breach attempt has been detected, but the security mechanism is not believed to have been compromised at the moment.

sequenceOfDays   It specifies a sequence of time intervals for a day. Each interval is specified with an interval start time and interval end time.

serviceProvider   This attribute identifies the entity that provides the security service responsible for generating the security alarm.

serviceUser   This attribute identifies the entity whose request for security service leads to the generation of the security alarm.

sourceIndicator   This attribute, often contained in a notification, identifies the type of the management operation that led to the generation of the notification.

specificProblems   This attribute, often contained in an alarm, provides further information on the probable cause of the alarm.

standbyStatus   This attribute is single-valued and used to indicate one of the following standby status values when the backup role exists for this resource:

- *Hot standby.*   The resource is not providing services but is operating in lock step with another resource that this resource is to back up. The resource in hot standby status can immediately take the service-providing role when the need arises.
- *Cold standby.*   The resource is intended to back up another resource but is not synchronized with that resource. The resource in the cold standby status must go through an initialization process before it can take over the role of the resource it backs up.
- *Providing service.*   The backup resource is providing service and is backing up another resource in the meantime.

startTime   This attribute defines the date and time at which an unlocked and enabled managed object starts functioning. If the value of the startTime attribute is not specified in the create request, its value defaults to the time of creation of the managed object, thus causing it to function immediately.

stateChangeDefinition   This attribute consists of a sequence of three-parameter sets: attribute identifier, old attribute value, and new attribute value. Each set describes a single state attribute value change.

stopTime   This attribute defines the date and time at which a managed object stops functioning. If the value of the stopTime attribute is not specified in the create request, its value defaults to continuous operation. Continuous operation is represented by a null value for the stop time. A change in the stopTime attribute results in an attribute change notification.

supportedFeatures   This attribute identifies features within the system that can be managed.

systemId   This attribute specifies a unique identifier for the system managed object instance.

systemTitle   This attribute specifies a unique title for the system managed object instance.

thresholdInfo   This attribute is often contained in an alarm to describe the threshold that caused the generation of the alarm. The attribute consists of four parameters:

- *Triggered threshold.*   The identifier of the threshold that caused the alarm.
- *Threshold level.*   In the case that the threshold is of the gauge type, the threshold level consists of a pair of values: the first is the value of the crossed threshold and the second is the corresponding hypothesis.
- *Observed value.*   The value of the gauge or counter that crossed the threshold.
- *Arm time.*   The time when the threshold is reset after the previous threshold crossing, thus allowing generation of an alarm if the threshold is crossed.

tideMark   The tideMark provides a mechanism to record the maximum or minimum value reached by a gauge during a measurement period. An attribute of the tideMark type is defined as either maximum tideMark or minimum tideMark and it must be associated with a gauge. A maximum tideMark attribute value changes only when the gauge value increases beyond the current tideMark value, and a minimum tideMark attribute value changes only when the gauge value decreases beyond the current tideMark value.

trendIndication   This attribute indicates the trend of alarms associated with a managed object: more severe, no change, or less severe.

unknownStatus   This attribute is used to indicate that the status of the resource is unknown.

usageState   This single-valued attribute indicates whether the resource is busy or idle with the following three values:

- *Idle.*   The resource is not in use at the moment.
- *Active.*   The resource is in use and has sufficient capacities to provide services to additional users.
- *Busy.*   The resource is in use and does not have any spare capacities to provide for additional users.

userObject   This attribute identifies one or more managed objects that act in a service user role in a service relationship. A service relationship is an asymmetric relationship in which the first of a pair of managed objects is a provider object to the second object, or a user object.

weekMask   This structured attribute defines a set of mask components, each specifying a set of time intervals on a 24-h time-of-day clock, pertaining to selected days of the week. The weekMask attribute defaults to the scheduling criteria of always being on at the log object creation time..

Abbreviations

AC	Authentication center
ACSE	Association control service element
AIN	Advanced Intelligent Network
APDU	Application protocol data unit
ASE	Application service element
ASN.1	Abstract Syntax Notation One
BML	Business management layer
CCITT	Consultative Committee for International Telephony and Telegraphy
CM	Computing module
CMIP	Common Management Information Protocol
CMIS	Common Management Information Service
CMISE	Common Management Information Service element
CORBA	Common Object Request Broker Architecture
DCE	Distributed computing environment
DCF	Data communication function
DCN	Data communication network
DMS	Digital multiplex switch
DPE	Distributed processing environment
EFD	Event forwarding discriminator
EIR	Equipment identification register
EIU	Ethernet interface unit
EML	Element management layer
ENET	Enhanced network
ETSI	European Telecommunication Standard Institute
FP	File processor
FTAM	File transfer, access, and management
IAB	Internet Architecture Board
IDL	Interface Definition Language

IN	Intelligent Network
INA	Integrated network architecture
INM	Integrated network management
ISO	International Standards Organization
LAN	Local area network
LAP-B	Link access procedure balanced
LPP	Link peripheral processor
LVM	Logical volume manager
MAF	Management application function
MAPDU	Management application protocol data unit
MD	Mediation device
MF	Mediation function
MIB	Management information base
MOI	Managed object instance
MTP	Message transfer part
NE	Network element
NEF	Network element function
NM	Network management
NMC	Network management center
NMF	Network Management Forum
NML	Network management layer
NMS	Network management system
NSS	Network and switching subsystem
OAM&P	Operation administration, maintenance, and provisioning
ODP	Open distributed processing
OMC	Operations management center
OMG	Object Management Group
ORB	Object request broker
OS	Operations system
OSF	Operations system function
OSI	Open system interconnection
OSS	Operations support system
PDU	Protocol data unit
PSTN	Public switched telecommunications network

ROSE	Remote operations service element
RPC	Remote procedure call
SCCP	Signaling connection control part
SMAE	System management application entity
SMAP	System management application process
SMASE	System management application service element
SMF	System management function
SMFA	System management functional area
SML	Service management layer
SS7	Signaling System No. 7
TMN	Telecommunications management network
WS	Workstation
WSF	Workstation function
QA	Q adapter
QAF	Q-adapter function
QoS	Quality of service
WAN	Wide area network

References

Adams, E., and K. Willetts (1996). *The lean communications provider*. New York: McGraw-Hill.

Beckman, P. (1968). *Introduction to Queuing theory and telephone traffic*. New York: Golen Press.

Black, U. (1995). *Network management standards*. New York: McGraw-Hill.

Booch, G. (1994). *Object-oriented design with application,* 2d ed. CA:Addison-Wesley.

Byrne, C. (1994). "Fault management," in *Telecommunications network management into the 21st century*,"Aidarous, S., and Plevyak, T. (eds.), IEEE Press.

Case, J., F. Fedor, M. Schoffstall, and J. Davin (1990). *RFC 1157: The simple network management protocol*.

Flood, J. (1995). *Telecommunications switching, traffic and networks*. New York: Prentice Hall.

Gardner, R., and D. Harle (1996). "Methods and systems for alarm correlation." *Proceedings of IEEE Computer Communication Conference*. pp. 459–464.

Gaynberg, I., R. Lawrence, M. Gabuzda, P. Kalan, and N. Shah (1997). *The intelligent network standards: their application to services*. New York: McGraw-Hill.

Glitho, R., and S. Hayes (1996). "Approaches for introducing TMN in legacy networks: a critical look," *IEEE Communication Magazine*. March: 55–60.

Glotho, R., and S. Hayes (1995). "Telecommunications management network: vision vs. reality," *IEEE Communication Magazine,* March: 47–52.

Green, J. (1996). *The Irwin handbook of telecommunications,* 2d ed., Irwin Professional Publishing, 1996.

Hoorne, H., and A. Kuiper (1998). "Charging and billing within AMT network," An NMF white paper.

Jacobson, I., M. Christerson, and G. Overgaard (1992). *Object-oriented software engineering*. MA: Addison-Wesley.

JIDM (Joint Inter Domain Management task force) (1998). "Interaction translation—Initial submission to OMG's CORBA/TMN interworking RFP."

JIDM (1994). "A JIDM report: translation of GDMO/ASN.1 specification into CORBA-IDL."

Kershenbaum, A. (1993). *Telecommunications network design algorithms*. New York:McGraw-Hill.

Kliger, S., and S. Yemini (1994). "A coding approach to event correlation." In *Integrated network management IV*, Chapman & Hall, pp. 266–277.

Kumar, B. (1994). *Broadband communications: a professional's guide to ATM, frame relay, SMDS, SONET and BISDN*. New York:McGraw-Hill.

Lee, K., Y.L. Choi, and S. Lee (1998). "Loop expanding fast restoration," *Proc. of NOMS,* pp. 513–522.

Litcher, J. (1997). "Number portability as an enabler for competitive local service," *XVI World Telecom Congress Proceedings*, 21–26 Sept., Toronto.

McCloghrie, K., and M. Rose (1990). *RFC 1156: Management information base for network management of TCP/IP-based Internets*.

McCloghrie, K., and M. Rose (1991). *RFC 1213: Management information base for network management of TCP/IP-based Internets: MIB-II*.

Meyer, B. (1997). *Object-oriented software construction*. NJ: Prentice-Hall.

Mills, D. (1984). *RFC 904: exterior gateway protocol formal specification*.

Nakamura, R., H. Ono, and K. Nishikawara (1994). "Reliable switching system recovery." *Proc. of IEEE GlobeCom 94*. pp. 1596–1600.

OMG (The Object Management Group) (1998). *The common object request broker: architecture and specification*.

OMG Telecom Domain Task Force (1996). *CORBA-based telecommunications network management system. A white paper*.

OMG Telecom Domain Task Force (1997). *Interworking between CORBA and TMN Systems—Request for proposal*.

Petermueller, W. (1996). "Q3 object models for the management of exchanges." *IEEE Communications Magazine,* March: 48–60.

Postel, J. (1981). *RFC 792: Internet control message protocol.*

Rambaugh, R., M. Blaha, W. Premerlani, F. Eddy, and W. Lorensen (1991). *Object-oriented modeling and design.* NJ: Prentice Hall.

Rose, M. and K. McCloghrie (1990). *RFC1155: Structure and Identification of Management Information for TCP/IP-based Internets.*

Rose, M. and K. McCloghrie (1991). *RFC1212: concise MIB definitions.*

Sidor, D. (1995). "Managing telecommunication networks using TMN interface standards." *IEEE Communications Magazine,* March: 54–60.

Siegel, J. (1996). *CORBA fundamentals and programming.* New York: Wiley.

Stallings, W. (1999). *Cryptography and network security: principles and practices.* 2d ed. NJ: Prentice Hall.

Stallings, W. (1992). *SNMP, SNMPv2, and CMIP—The practical guide to network management standards.* MA: Addison-Wesley.

Tang, A. (1994). *Open networking with OSI.* NJ: Prentice Hall.

Tracey, L. (1997). "30 years: A brief history of the communications industry." *Telecommunications Magazine.* June: 24–36.

Udupa, D. (1996). *Network management systems essentials.* New York: McGraw-Hill.

Weese, S., T. Barrett, A. Hazra, and R. Eisemann (1996). "TMN and the implementation of electronic bonding." *IEEE Communications Magazine.* Sept.: 76–81.

Wirfs-Brock, R. (1990). *Designing object-oriented software.* NJ: Prentice-Hall.

Referenced Standards and Recommendations

ATIS (Alliance for Telecommunications Industry Solutions) (1997). *Initial report to the North American Council on number pooling*, version 2.

ATM Forum AF-NM-0020-001 (1998). *M4 interface requirements and logical MIB: ATM Network Element View*.

ATM Forum AF-NM-0073-000 (1997). *M4 network view CMIP MIB specification*, version 1.0.

ATM Forum AF-NM-0080-000 (1998). *ATM remote monitoring SNPM MIB*.

Bellcore GR-1093 (1994). *Generic state requirements for network elements*.

Bellcore, GR-471-Core (1995). *Operation technology generic requirements (OTGR), section 1: Introduction*.

Bellcore GR-1286-CORE (1993). *Advanced intelligent network (AIN) operation system (OS) service control point (SCP) interface generic requirements*.

Bellcore GR-1343-Core (1996). Core generic requirements for the automatic message accounting data networking system (AMADNS).

Bellcore GR-2802-CORE (1993). *Advanced intelligent network (AIN) 0.X generic requirements*.

Bellcore GR-2869 (1991). Generic requirements for operations based on the TMN architecture.

Bellcore Special Report, SR-NPL-001623 (1990). Advanced intelligent network: release 1, network and operations plan.

CCITT Recommendation E.164 (1991). *Numbering plan for the ISDN era*.

ITU-T Recommendation E.175 (1993). *Models for international network planning*.

ITU-T Recommendation G.803 (1993). *Architectures of transport network-based synchronous digital hierarchy (SDH)*.

ITU-T Recommendation I-210 (1993). *Principles of telecommunication services supported by an ISDN and the means to describe them*.

CCITT Recommendation M.3010 (1996). *Principles for a telecommunication management network (TMN)*.

ITU-T Recommendation M.3020 (1995). *TMN interface specification methodology*.

ITU-T Recommendation M.3100 (1995). *Generic network information model*.

CCITT Recommendation M.3180 (1992). *Catalogue of TMN management function*.

CCITT Recommendation M.3200 (1992). *TMN management services: overview*.

CCITT Recommendation M.3300 (1992). *TMN management capabilities presented at the F interface*.

CCITT Recommendation M.3400 (1992). TMN management functions.

CCITT Recommendation M.3600 (1992). *Principles for the management of ISDNs*.

CCITT Recommendation M.3602 (1992). *Application of maintenance principles to ISDN subscriber installations*.

ITU-T Recommendation Q.751.1 (1995). *Specifications of signaling system No. 7: network element management information model for message transport part (MTP)*.

ITU-T Recommendation Q.751.2 (1995). *Specifications of signaling system No. 7: network element management information model for signaling connection control point (SCCP)*.

ITU-T Recommendation Q.752 (1997). *Specifications of signaling system No. 7: modelling and measurements for system signaling No. 7*.

ITU-T Recommendation Q.754 (1995). *Specifications of signaling system No. 7: SS7 management application service elements definitions*.

ITU-T Recommendation Q.811 (1997). *Specifications of signaling system No. 7: lower layer protocol profiles for the Q3 and X interfaces*.

ITU-T Recommendation Q.812 (1993). *Specifications of signaling system No. 7: upper layer protocol profiles for the Q3 interface*.

ITU-T Recommendation Q.821 (1993). *Specifications of signaling system No. 7: stage 2 and stage 3 description for the Q3 interface—alarm surveillance.*

ITU-T Recommendation Q.822 (1994). *Specifications of signaling system No. 7: stage 2 and stage 3 description for the Q3 interface—performance management.*

ITU-T Recommendation Q.823 (1996). *Specifications of signaling system No. 7: stage 2 and stage 3 functional specifications for traffic management.*

ITU-T Recommendation Q.824.0 (1995). *Specifications of signaling system No. 7: stage 2 and stage 3 description for the Q3 interface—customer administration—common information.*

ITU-T Recommendation Q.824.1 (1995). *Specifications of signaling system No. 7: stage 2 and stage 3 description for the Q3 interface—customer administration—integrated service digital network (ISDN) basic and primary rate access.*

ITU-T Recommendation Q.824.2 (1995). *Specifications of signaling system No. 7: stage 2 and stage 3 description for the Q3 interface—customer administration—integrated service digital network (ISDN) supplementary services.*

ITU-T Recommendation Q.824.3 (1995). *Specifications of signaling system No. 7: stage 2 and stage 3 description for the Q3 interface—customer administration—integrated service digital network (ISDN), optional facilities.*

ITU-T Recommendation Q.824.4 (1995). *Specifications of signaling system No. 7: stage 2 and stage 3 description for the Q3 interface—customer administration—integrated service digital network (ISDN)—teleservices.*

ITU-T Recommendation Q.1211 (1993). *Introduction to Intelligent Network Capability Set 1.*

ITU-T Recommendation Q.1221 (1995). *Introduction to Intelligent Network Capability Set 2 (Draft).*

ITU-T Recommendation T.60 (1993). *Terminal equipment for use in the telex service.*

ITU-T Recommendation T.100 (1993). *International information exchange for interactive videotex.*

CCITT Recommendation X.21 (1992). *Interface between data terminal equipment and data circuit-terminating equipment for synchronous operation on public data networks.*

ITU-T Recommendation X.25 (1993). *Interface between data terminal equipment (DTE) and data circuit-terminating equipment (DCE) for terminals operating in the packet mode and connected to public data networks by dedicated circuit.*

ITU-T Recommendation X.121 (1996). *International numbering plan for public data networks.*

CCITT Recommendation X.200 (1988). *Reference model of open systems interconnection for CCITT applications.*

CCITT Recommendation X.208 (1988). *Specification of abstract syntax notation one (ASN.1).*

CCITT Recommendation X.209 (1988). *Specification of basic encoding rules for abstract syntax notation one (ASN.1).*

CCITT Recommendation X.210 (1988). *Open systems interconnection layer service definition conventions.*

CCITT Recommendation X.213 (Annex A) (1992). *Open systems interconnection—network service definition.*

CCITT Recommendation X.214 (1995). *Open systems interconnection—transport service definition.*

CCITT Recommendation X.219 (1993). *Remote operations: model, notation, and service definition.*

CCITT Recommendation X.225 (1994). *Connection-model protocol specification.*

CCITT Recommendation X.226 (1994). *Connection-oriented presentation protocol: protocol specification.*

CCITT Recommendation X.290 (1992). *OSI conformance testing methodology and framework for protocol recommendations for CCITT applications—general concepts.*

ITU-T Recommendation X.500 (1993). *The directory: overview of concepts, models and services.*

CCITT Recommendation X.501 (1988). *The directory—models.*

ITU-T Recommendation X.509 (1993). *The directory—authentication framework.*

ITU-T Recommendation X.511 (1993). *The directory—abstract service definition.*

CCITT Recommendation X.700 (1988). *Management framework definition for open systems interconnection (OSI) for CCITT applications.*

CCITT Recommendation X.701 (1991). *Open systems interconnection—systems management overview.*

CCITT Recommendation X.710 (1991). *Common management information service definition for CCITT applications.*

CCITT Recommendation X.711 (1991). *Common management information protocol specification for CCITT applications.*

CCITT Recommendation X.720 (1992). *Open systems interconnection—structure of management information: management information model.*

CCITT Recommendation X.721 (1992). *Open systems interconnection—structure of management information—part 2: definition of management information.*

CCITT Recommendation X.722 (1992). *Structure of management information: guidelines for the definition of management objects.*

CCITT Recommendation X.730 (1992). *Open systems interconnection—systems management—part 1: object management function.*

CCITT Recommendation X.731 (1992). *Open systems interconnection—systems management—part 2: state management function.*

CCITT Recommendation X.733 (1992). *Open systems interconnection—systems management: alarm reporting function.*

CCITT Recommendation X.734 (1992). *Open systems interconnection—systems management: event report management function.*

CCITT Recommendation X.735 (1992). *Open systems interconnection—systems management: event report management function.*

CCITT Recommendation X.736 (1992). *Open systems interconnection—systems management: security alarm reporting function.*

ITU-T Recommendation X.737 (1995). *Open systems interconnection—systems management: confidence and diagnostic test categories.*

ITU-T Recommendation X.738 (1993). *Open systems interconnection—systems management: summarization function.*

ITU-T Recommendation X.739 (1993). *Open systems interconnection—systems management: metric objects and attributes.*

CCITT Recommendation X.740 (1992). *Open systems interconnection—systems management: security audit trail function.*

ITU-T Recommendation X.741 (1995). *Open systems interconnection—systems management: objects and attributes for access control.*

ITU-T Recommendation X.742 (1995). *Open systems interconnection—systems management: usage metering function for accounting purposes.*

ITU-T Recommendation X.744 (1996). *Open systems interconnection—systems management: software management function.*

ITU-T Recommendation X.745 (1993). *Open systems interconnection—systems management: test management function.*

ITU-T Recommendation X.746 (1995). *Open systems interconnection—systems management: scheduling function.*

ITU-T Recommendation X.790 (1995). *trouble management function for ITU applications.*

CCITT Recommendation X.800 (1991). *Security architecture for open systems interconnection for CCITT applications.*

ITU-T Recommendation X.810 (1995). *Security frameworks for open systems: overview.*

ITU-T Recommendation X.811 (1995). *Security frameworks for open systems: authentication framework.*

ITU-T Recommendation X.812 (1995). *Security frameworks for open systems: access control framework.*

ITU-T Recommendation X.813 (1996). *Security frameworks for open systems: non-repudiation framework.*

ITU-T Recommendation X.814 (1995). *Security frameworks for open systems: confidentiality framework.*

ITU-T Recommendation X.815 (1995). *Security frameworks for open systems: integrity framework.*

ITU-T Recommendation X.816 (1995). *Security frameworks for open systems: security audit and alarm framework.*

ITU-T Recommendation X.902 (1995). *Open distributed processing—reference model: foundations.*

ITU-T Recommendation X.903 (1995). *Open distributed processing—reference model: architecture.*

ITU-T Recommendation X.904 (1997). *Open distributed processing—reference model: architectural semantics.*

CCITT Recommendation Z.300-Series (1988). *Man-machine language.*

NMF 040-1 (1997). *ASN.1/C++* application programming interface part 1: specific interface, Issue 1.0.

NMF 040-2 (1997). *ASN.1/C++* application programming interface part 2: generic interface, Issue 1.0.

NMF 041 (1997). *CMIS/C++* application programming interface, Issue 1.0.

NMF 042-1 (1997). *GDMO/C++* application programming interface part 1: architecture, Issue 1.0.

NMF 042-2 (1997). *GDMO/C++* application programming interface part 2: C++ class reference, Issue 1.0.

NMF 042-3 (1997). *GDMO/C++* application programming interface part 3: specific aspects, Issue 1.0.

NMF 043 (1997). *TMN/C++* application programming interface, Issue 1.0.

NMF 035 (1995). *Switch interconnection management configuration management ensemble.*

NMF 504 (1998). *SMART Ordering—SP to SP interface business agreement.*

NMF 701 (1998). *Performance reporting definitions.*

NMF GB903 (1995). *Implementing OMNIPoint: A safer, faster path to improved service management.*

NMF TR 114 (1994). OMNI*Point integration architecture.*

Index

ABOUT THE AUTHOR

Henry Haojin Wang obtained his MS and Ph.D. degrees in computer science from the University of Massachusetts at Amherst and Texas A&M University at College Station, respectively. Currently he is a network management system architect at Santera Systems Inc. Previously he had been with Samsung Telecom America, Nortel, and IBM and worked in areas such as telecommunications network management, next generation telecommunications networks, circuit-switched telephony systems, and network simulation tools. In addition to a wide range of telecommunications industry experience, Dr. Wang also taught at the University of Texas at Dallas and Nortel on subjects ranging from network management, call processing, ISDN, SS7 signaling network, to object-oriented design and analysis.